Bird Trapping and Bird Banding

In order to band a heron Christine Charlotte Prinzessin von Hessen-Kassel (1725–1782) takes a brass plate, dated 1764, from the hat of the Oberfalken- und Forstmeister Freiherrn Ludolph von und zum Canstein. The bird is held by the Falkenjunker L. H. von Osterhausen. Looking down on the ceremony is Ernst Ludwig, Erbprinz von Sachsen-Gotha. Other onlookers are Constantin Landgraf von Hessen-Rotenburg and (in green forestry uniform) the Hofjägermeister Friedrich W. G. von Oynhausen.

Bird Trapping and Bird Banding

A Handbook for Trapping Methods All Over the World

Hans Bub, *Institut für Vogelforschung*
Vogelwarte Helgoland Wilhelmshaven

Translated by
Frances Hamerstrom *and*
Karin Wuertz-Schaefer

Illustrations by Winfried Noll
and Eitel Raddatz

Forewords by George Jonkel
and Chris Mead

Cornell University Press

Ithaca, New York

English translation first published 1991 by Cornell University Press.

Library of Congress Cataloging-in-Publication Data

Bub, Hans.
[Vogelfang und Vogelberingung. English]
Bird trapping and bird banding : a hand book for trapping methods all over the world / by Hans Bub ; translated by Frances Hamerstrom and Karin Wuertz-Schaefer ; most drawings by Winfried Noll and Eitel Raddatz. 328 p. 16,7 × 24,0 cm.
Translation of: Vogelfang und Vogelberingung.
Includes bibliographical references.
ISBN 0-8014-2525-5 (alk. paper)
1. Bird trapping—Handbooks, manuals, etc. 2. Bird banding—Handbooks, manuals, etc. I. Title.
QL677.5.B77413 1991
598′.07232-dc20 90-34188

Printed in Hong Kong by Everbest Printing Co. Ltd. through Four Colour Imports, Ltd.

Contents

Foreword

Human beings have been capturing birds with special techniques for a thousand years or more throughout many cultures. During this time we have learned much about bird populations, habitats, diets, life histories and behaviours just by catching—or trying to catch—birds. But until now, no one has written such a complete and detailed single book on bird capture.

Originally published in four volumes in German, HANS BUB's **Bird Trapping and Bird Banding** has been integrated into one volume and translated into English by FRANCES HAMERSTROM and KARIN WUERTZ-SCHAEFER. Many additional bird banders and researchers will now be able to consult and study this fine, comprehensive reference book.

BUB covers in detail most taxonomic groups of land and sea birds and almost all known bird capture techniques, ancient and modern. In doing so, he makes it clear that bird capture is no armchair pursuit. Researchers looking for new data may need to endure the rigors and dangers described herein—landslides, treacherous trees and cliffs, leeches and snakes, and the piercing talons and beaks of birds.

A word of caution: the techniques described need to be adapted to comply with national laws and international treaties. In North America, the Migratory Bird Treaty Act, which implements treaties between the United States and Canada, Mexico, the Soviet Union and Japan, restricts how, by whom and when migratory birds can be captured. The act designates most bird species as migratory—nonmigratory birds being the Rock Dove, House Sparrow, European Starling and the various resident game, or gallinaceous, birds. In North America the taking of migratory birds requires a permit, and most national bird banding programs stipulate that bird trapping must relate to research or management. Consequently, the use of live lure birds, described extensively in this book, is illegal in North America. And stuffed migratory birds can be used as decoys only with a special permit issued by the U.S. Fish and Wildlife Service or the Canadian Wildlife Service.

Though this book is primarily about the capture of live birds, it also includes other information useful and of interest to professional and amateur ornithologists.

GEORGE M. JONKEL,
Chief, Bird Banding Laboratory
U. S. Fish and Wildlife Service
Laurel, Maryland

Foreword

A comprehensive book such as this represents the distillation of knowledge culled from generations of bird catchers all over the world. Many of the methods were pioneered in more barbarous times, when man's wits were pitted against nature and the ultimate fate of the birds being caught was to be eaten. Happily times have changed. This book has been written to bring together the many catching methods that ringers can use to get their hands, and then their bands, on wild birds.

Of course, for the scientific bird ringer, the safety of the bird is paramount. Ringers should have deeply felt humanitarian scruples about causing any damage or suffering to the birds they catch. They should also be certain that the birds they mark are normal representatives of the natural, wild bird populations: unless they are, and remain so after being marked, the results from ringing would be false. Each national ringing scheme throughout the world has its own rules and regulations to ensure that these goals are achieved. Some of the catching methods described and illustrated in this book may be risky to the birds, but I am sure that the author has served the ringers well by including them. As ringers are an ingenious breed, the ideas conveyed are bound to be refined and modified in many different ways.

Clearly ringing has developed since the days of the pioneers, now almost a century ago. Then the hopeful ringer placed rings on the legs of as many birds as possible and sat back to wait for recoveries to be reported. Now many ringers, though grateful for a recovery from a member of the public, are providing their own data through rehandling their own birds or resighting those marked with colour-rings, wing-tags, plumage dyes and the like. Some are even keeping surveillance on their birds with radio transmitters. For such detailed studies there is a pressing need for special catching techniques which allow the ringers to handle particular birds of crucial interest to their studies. This is one of the great assets of this book—specialist ringers desperate for ideas on how to catch their most elusive birds are likely to find something useful within its covers.

Happily, over the past two decades, more and more countries throughout the world have adopted detailed and comprehensive laws promoting the conservation of the natural environment and birds in particular. In most countries the use of many, often all, of the techniques described here to catch wild birds is illegal except with an official licence. Such licences are generally available only to scientific bird ringers who have been trained through and are operating under the National Bird Ringing Scheme. Readers of this book who are not authorized bird ringers are warned that it is very likely that any attempt they might make to catch birds using the techniques described here would be illegal. This would certainly be the case in Britain and most European countries. However, readers who are fascinated by this documentation of man's ingenuity in catching birds and who have thereby had a real interest in ringing birds kindled within them should contact their National Ringing Scheme. In most countries new recruits are welcomed and, after a period of training, can take up what has been, for thousands of amateur ornithologists throughout the world, a particularly absorbing and rewarding hobby.

CHRIS MEAD,
British Trust for Ornithology,
Tring, England,
General Secretary EURING
and Head of British Ringing Scheme.

Introduction

A few thoughts should be set down as a preface to this book. Bird trapping and banding have become an indispensable feature of ornithology. Birds were once trapped by ornithologists purely for the purposes of collection, but today this aspect has become almost entirely unnecessary. For the past 100 years trapping has been in the service of bird banding, without which whole areas of ornithology today would be impossible. We only have to consider its great importance, along with other marking techniques, in migration research, and a great many questions remain to be answered in that field alone.

The trapping of birds has always been a part of the ancient hunting instinct common to mankind in every corner of the world. For many thousands of years birds were caught purely for food. This has largely become redundant since almost all human communities turned to agriculture and the domestication of other animals for their nourishment. I will touch on the question of hunting in this introduction only to say that the hunting of birds, with a few exceptions, should be brought to an end. Even most duck and geese populations can be regarded as under threat, especially when we consider how their habitats are being continuously eroded.

Before and during the Middle Ages trapping was also employed for the purposes of falconry and obtaining cage-birds. How does the situation look today? In many countries the majority of species are protected. However, even in the so-called industrial countries, many millions of birds are captured or shot every year, and thus lose their lives. It seems not to have got through to the mass of people in those countries that all birds need our protection because of the multitude of threats to the natural world in all parts of the globe. In Europe and North Africa this is a problem primarily in the Mediterranean countries and their adjoining islands, where migrants particularly are ruthlessly persecuted. Where is the reverence for creation which is at the heart of all religion, and of which we have been made so aware by such figures as the doctor and theologian ALBERT SCHWEITZER, or the veterinary surgeon and zoologist BERNHARD GRZIMEK? The same question is posed when we look at the wholesale destruction of the rain forests in South America, Africa and Asia. No matter how much we do as individuals, the scale of the task is such that only the international organizations, like the U.N. or the World Wildlife Fund, can undertake the really vital work.

There was a great variety of bird trapping techniques even in early history, and they too will be dealt with in this book. The reader will gain a picture of the importance of trapping in the past by the number of references in the bibliography at the end of the book.

With the introduction of banding at the close of the 19th and beginning of the 20th Century, the capturing of birds took on an entirely new dimension. Only a living bird, and a bird with a band, is of any use in helping to solve the problems we face, and so some trapping methods had to be altered and improved. Some pieces of equipment, for example the Italian trammel nets, were adopted unchanged since the birds are taken alive and uninjured.

Bird banding has become so important for scientific research and bird protection that it no longer requires any justification on our part. In every country where banding is practiced it is done so within a legal framework, and banders work according to strict rules. There is no need today for us to ask ourselves whether or not banding is a responsible activity; we have known for many years that it is. How to hold a bird in the hand is a matter of experience, and the banding itself is governed by precise regulations. The important thing is that beginners learn by supervised practice and that the knowledge of experienced banders is always at their disposal. The bander who has had tens of thousands of birds in the hand knows best how a captive bird is to be held. The following publications should be mentioned here: Wildlife Management Techniques, edited by SANFORD D. SCHEMNITZ of the Wildlife Society (USA); The Ringer's Manual, compiled by ROBERT SPENCER (BTO) and the BTO Guide No. 16, Bird Ringing, by CHRIS MEAD.

It is always possible to change the banding procedure for individual species should new facts come to light through experience. For instance the question has arisen in Europe whether a banded kinglet *Regulus* sp. could damage its eggs while on the nest. It is also worth mentioning that many birds do not seem to object to being trapped, otherwise they would not turn up again and again in the trap simply searching for the food they once found there. Obviously this behaviour varies among individuals and species.

Technical procedures in banding stations were simpler in the past, but due to the tremendous increase in banding operations (worldwide 70 million birds banded to date) the organizational processes have become much more complex. This has led to the introduction of electronic data processing. Every modern technology we employ must ensure that the information from band returns is maximised, and this means that alongside expert data processors there will be equally experienced ornithologists.

Some recent developments, and some not so recent, from the United Kingdom and the USA deserve mention here. It is generally accepted that banding in these two countries is a model of excellent, well-supported organization and administration. I would like at this point to let an American ornithologist speak, namely Frederick C. Lincoln, who, with S. Prentiss Baldwin, must be regarded as an outstanding figure in the first three decades of North American banding. I am thinking particularly of the Manual for Bird Banders published by the two men in 1924. A banding journal has existed since that same year, when the first issue of the Bulletin of the Eastern Bird Banding Association appeared. In his article Bird-Banding; Its First Decade under the Biological Survey (Bird-Banding, Vol. 2, 1931: 27-32) Lincoln, even then, expressed ideas which still serve us as arguments against the tiny number of opponents banding still has. Beginners can still read his words with profit. It is important for European banders, or those in other parts of the World, to realise that banding had already reached a high level of sophistication in the 1920's in North America. Baldwin had already begun his work on the House Wren *Troglodytes aedon* around 1920, and Margaret Morse Nice had an entire issue (No. 2, 1925) of the Bulletin of the Eastern Bird Banding Association devoted to her study of the Song Sparrow *Melospiza melodia*, a study that was to become world famous. The research at that time is well-documented in the following three publications:

1. Lincoln, F. C. 1928. A Bibliography of Bird Banding in America. The Auk, XLV: Suppl. p. 75.
2. Baldwin, S.P. 1931. Bird Banding by Systematic Trapping. Sc. Publ. of the Cleveland Museum of Natural History, Vol. 1, No. 5: 125-168.
3. Baldwin, S. P. Bird Banding in America. Baldwin Bird Research Laboratory, Cleveland, Ohio.

The last-named book is a splendid volume containing 41 papers by 15 authors, collected and donated by S. P. Baldwin, and is a mine of information for those seeking a picture of banding in the US before 1930.

It is interesting to note how bird trapping by North American ornithologists began to develop in new directions even in its early stages, and US banders are still well-known today for their readiness to experiment. This individualistic development cannot be ignored when taking a historical overview of banding in North America. It is difficult now to separate the influences exerted by the various bird trapping traditions brought to the US by immigrants from different parts of the World. The directions taken in the USA were doubtless not created in isolation, as Lincoln showed in 1925 (Some Results of Bird Banding in Europe. The Auk, XLII: 358-388.)

On the occasion of the 1925 conference of the A. O. U. in New York, Lincoln presented a paper with the title Bird Banding – in Progress and Prospect, which was published in The Auk of April, 1926. This paper dealt with the first years of banding under federal direction. In 1931 the same author gave a resumé of the progress achieved up until that year:

This retrospect is viewed with pardonable pride by those charged with the direction of the work, and, it is believed, by every one who has actively participated. As was stated in the earlier report, "Bird Banding as a method in ornithology has come to stay," and while no one seriously thinks that it will supplant any other method, we do consider it the most important tool that has been placed in the hands of the ornithological craftsman since the art of taxidermy enabled him to preserve specimens for future study. Banding data may be little used by the majority of the professional ornitholo-

Fig. 1. The "prey" of an Italian bird hunter. Millions of songbirds are still being killed every year. Photograph taken by an Italian bird protector.

gists of the present day, as most of them already have more work outlined than they may be reasonably expected to complete in the normal span of years allotted to productivity. We would not urge that these specialists add to their other work by "taking on" further investigations to be conducted by this method, but their advice is invaluable, and it is with a feeling of sincere gratitude that I record the many valued suggestions that have come from ornithologists who are not themselves enrolled under the banner of Bird Banding.

There is, however, a growing generation of ornithological students constituting what may be termed the "new school". These, fearing that the science may reach a stage of stagnation under the methods of the "old school", are casting about for new lines of research. They find much that is still unknown, unknown because the methods formerly in vogue have not permitted precise investigation in all possible fields. To these, systematic banding is hailed as a means of solution for interesting and important problems.

As in all other innovations, difficulties both real and fancied had to be overcome, and for fear of burdensome repetition, I hesitate to recapitulate the trials and tribulations of the first two or three years. Usually, however, they had interest or humor, and so in a measure lightened their own weight.

Paramount among them was the struggle to obtain a source of supply for the necessary bands. Practically every manufacturer of aluminum smallware in the country was approached, either by letter or by personal visit, and finally in desperation the first Biological Survey (now U. S. Fish and Wildlife Service) bands were ordered from England. This resulted in the curious error in which, through a transposition of two letters, the address, "Biol. Surv., Wash., D. C." was transformed into apparent cooking instructions, the finder of a banded bird being admonished to Wash, Boil, and Surv.

At this point a word of commendation is due to Mr. Theo. A. Gey, of Norristown, Pennsylvania, whose company for many years made practically all of the bands issued by the Service. Not only did he invent and build the ingenious machine that produces the bands, but he has consistently shown an interest in the work that is far above mere commercialism. In addition to producing bird bands his interest has extended to active support of one of the regional associations, of which he is a sustaining member.

Opposition to banding work upon humanitarian grounds by well-meaning but uninformed persons caused a little anxiety, as data were then lacking for refutation of all charges made that banding would result in wholesale destruction of bird-life. Any investigations of wildlife that involve actual handling by man result in some mortality among the subjects, but with the years of experience in back of us, and the carefully accumulated data bearing upon causes of avian mortality, it is a cause for much satisfaction to record that bird deaths directly traceable to the operator, his traps, or the bands, are so few as to represent a negligible quantity. Firearms excepted, the so-called "domestic cat" is a greater menace to our avifauna; however, the destruction of the birds' habitats by humans all over the world is a worse threat than all other causes put together.

In some quarters this opposition continues to smoulder, showing a tendency to flare up when some new means for the capture of birds is advocated. To these opponents I may say that there is no organization more concerned for the welfare of the wildlife of North America, including its birds, than the U.S. Fish and Wildlife Service, and personal contact with a great many banders at their trapping stations has demonstrated that the Service's cooperators are almost without exception worthy champions of the highest conservation principles. Accordingly, the bird-students of America may be assured that no trap or other device will be recommended for general use until it has been carefully tested to learn that it will not have an objectionable effect upon our birds. Some methods that are efficient means for the capture of birds by the poacher may when properly used prove of great value to the bird bander.

At one time, during the development period, it was contended that the rapid increase in feeding and trapping stations would destroy the economic value of insectivorous birds. I should like to be so sanguine that we shall ever have such a vast percentage of the bird population of America under our control, but study of the facts thus far made available leads me to believe that, no matter how greatly our banding stations are increased in number, or how efficient and energetic our operators may become, the sum total of the birds banded in any single year will be but "a drop in the bucket" of the total avian population. Occasionally individual birds act almost human and develop insatiable appetites for certain types of bait. Ordinarily, however, station operators have learned to their cost that, when natural food is abundant, it is very difficult to entice birds to the traps. Nevertheless, we are improving in our ability to furnish "appetizers", so this utopian idea may yet come to pass.

This book deals only with those trapping methods which, when properly carried out, do not endanger the lives of the birds caught. Any additional information required would be supplied with the equipment. Apart from the trapping techniques in general use, such as mist nets, there are many described in this book hardly in use today. However, under special circumstances they could still be profitably employed, and it should be our aim to ensure that these methods are not lost.

Many kinds of trap cannot be purchased commercially, but have to be constructed by the banders themselves, or at least be built according to their plans. Small traps can be made following simple illustrations, but larger ones should be observed in operation at a banding station before undertaking construction, so that expensive and time-consuming mistakes can be avoided. Experienced operators will be glad to be of assistance.

I should like to thank the publishers, Urania-Verlag of Leipzig and Ziemsen-Verlag of Wittenberg-Lutherstadt, both of the G.D.R., for enabling this American/English edition to be published. Frances Hamerstrom, of Plainfield, Wisconsin, on whose advice I first went to the US in 1962 to learn more about banding in North America, was the first to undertake the translation of Vogelfang und Vogelberingung. Frederick S. Schaeffer reviewed the German edition at length in The North American Bird Bander, 1981, Vol. 6: 116–117. For various reasons Mrs. Hamerstrom had to cease her translation work, and the greater part was then carried out by Karin Wuertz-Schaefer, of Mt. Airy, Maryland. George Jonkel gave the project his great support and helped to establish contact with many banders. JOHN WESKE gave valuable advice and, along with ROBERT SCHAEFER, read the manuscript. KATHY KLIMKIEWICZ, GLADYS COLE, BILL CLARK, DOTTIE MENDINHALL and many others assisted greatly by showing us the different trapping methods used in the United States. This introduction was kindly translated by the Scottish ornithologist BRIAN HILLCOAT.

Bird names were taken by KARIN WUERTZ-SCHAEFER from CLEMENTS, J. 1981. Birds of the World: A Checklist, 3 ed., and place names from the National Geographic Atlas of the World, 3 ed. revised, 1970.

It should be the wish of us all that the tradition of bird trapping be continued, to the benefit of ornithology but also for our own enjoyment. Let us hope that this book can play its part.

H. B.

1. Basics of bird catching and banding

1.1 Baits and devices

Live lure birds have always been important to bird catchers. In medieval Europe birds were generally caught with live lures, and the roots of this catching method reach back in time to the ancient Egyptians (Fig. 2) and the Chinese as reported in surviving documents (see LIPS, 1927; HOFFMANN, 1960). It may even be a more ancient practice, as the use of lure birds is known to almost all native peoples (Fig. 3 and 173).

Modern day bird banders often use lure birds to increase their catches or to trap specific birds.

1.1.1 Live lure birds

Bird catching with certain traps can often be enhanced by using live lure birds. It is impressive to see a migrating flock peel down out of the sky

Fig. 2. Illustration of Egyptian bird hunting with boomerang and probably a lure bird during the XVIII dynasty. Photo: R. ROSER.

Fig. 3. A native of New Zealand snaring birds. In the middle a lure bird, a captured bird on each side. After ANELL, 1960.

after having spotted the lures on the trapping area. Gregarious species such as corvids, finches, buntings, thrushes, shorebirds, ducks and geese decoy especially but other birds are also often attracted.

Some birds make good lures and adapt easily to life in cages. Others are all but impossible to keep. Conditions and requirements vary according to the seasons. To make the best use of his lures, the bird catcher must understand their behavior and calls (including warning cries that many birds possess) in relation to age, sex, season, weather, food and region.

Lure birds can be useful year round. During migration they attract their conspecifics as well as raptors. Well-adjusted healthy lure birds facilitate bird catching as they act and call naturally, making strange surroundings attractive to migrants looking for food and resting places. Often the wild birds lose their fear to such an extent that they try time and again to eat the food of the lure birds through the bars of the cages.

Fig. 4. Lure bird cage of the Palau islands in the South Pacific. The cage is made of tree bark. After ANELL, 1960.

This observation may have given rise to the cage traps in ancient times.

In the breeding season caged singing males and females are set visibly between two net walls. The owners of the territory will try to chase the supposed rival and thus wind up in the nets. This is particularly true of Chaffinches, Icterine Warblers, Black-cap Warblers, European Robins, larks and European Quail. The last can also be caught with females. Basically all birds will defend their territories against intruders. Some finches often are still migrating in flocks when others of their conspecifics have already set up territories. For these it is important to set out single males which will provoke aggression. W. STÜRMER noted that more than one lure bird is ignored by the territory's owner.

In winter lure birds show others where food is present and entice them into funnels or traps.

What specifically should be kept in mind? The lure birds should be plainly visible on the trapping area. Migrating flocks do not react to calls alone, but also to the sight and behavior of their friends below. Exceptions will be mentioned later as well as many other things concerning lure birds. Some camouflage may be necessary in some instances such as when catching with a double clap net.

Nets and traps are set up before the lure birds are brought out. The birds remain in the dark until then so that they do not tire themselves calling beforehand. Besides, birds attracted then fly away again while the bander is still busy setting up his nets. Keeping lure birds in the dark prior to putting them out only makes sense if catching does not extend beyond a few hours.

Lure birds must be on hand in time. It is best to keep one or two members of several species at all times. If no aviary is available, a large cage will do. Make sure though that the cage is not overcrowded and that birds kept together get along well.

If one is mostly trapping for a certain species, it is best to obtain conspecific lure birds. Both sexes are usually suitable. SUNKEL uses both a male and a female when catching Common Bullfinches, Eurasian Siskins and similar species. The pairs kept in a flight cage are separated into single cages when catching. For the best effect they are then placed on the trapping area so that they can hear but not see each other. The number of lure birds depends on the bander, his trapping area and banding projects.

The lure bird cages are of different sizes. For a few hours cages 20×10×12 cm are large enough for finches, but usually the measurements

Fig. 5. Lure bird cages 7 m high beside a mist net 8 m high and 25 m long (near Hilden, Rheinland, W. Germany). Photo: K. STORSBERG.

Fig. 6. Lure bird cages near food-bearing shrubs with mist nets placed at right angles by H. MÜLLER in the Fichtelgebirge, W. Germany. Photo: H. BUB.

should be at least 22×16×17 cm for the smaller birds.

Each lure bird that is used for more than two to three hours is supplied with food and water. During inclement weather or when it is very hot, leave them out only for shorter periods. Snow can be substituted for drinking water in winter.

GIESE and his co-workers elevate their entire trapping station up to 7.5 m into trees. The lure bird cages are raised on metal arms with pulleys (Fig. 5). Sometimes the cages remain close to the ground next to the traps (Fig. 6).

High sets are most successful near fruiting alders and birches for catching migrating redpolls and siskins. Flocks tend to come to the ground only to watering places or when the alder seeds have fallen.

Common Bullfinches and thrushes are attracted by the calls of their own kind regardless of height, but since thrushes are repelled by the sight of others hanging and struggling in high nets, their lure runway cages (Fig. 7) should be on the ground near the nets. Thrushes are best attracted by songsters early in the morning. Thus bird catchers in the Harz Mountains (Germany) distinguish between mere 'lurers' and singers.

Not all vocalizations attract: warning cries may 'spook' uncaught birds, and too soft calling is often ineffective. Whether or not the lure birds actually sing is often the decisive factor. The trapper must select his singing males (Chaffinches, Icterine Warblers, etc.) in spring but he is also interested in having singing lure birds in fall. As even good singers tend to be silent in fall, bird catchers in the early days tried to deceive the birds by keeping them in dark quarters

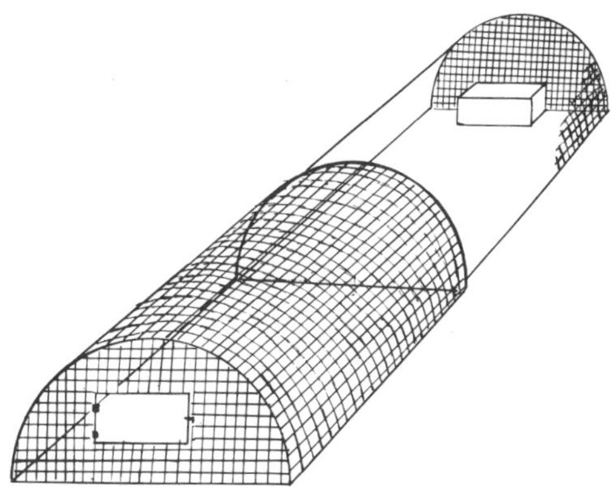

Fig. 7. Runway cage 80–100 cm long for thrushes. One end can be pushed over the other.

during spring and summer; they were introduced to sunlight and a false spring at the time of the autumn migration. Modern banders consider this practice inhumane and have desisted. Instead we are exploring dietary changes in a number of species to induce autumnal song. Hormones, mixed in with the feed, or testosterone with oil rubbed on the breast and belly skin can also induce autumnal singing.

Lure birds need to be conditioned to sing wherever they are put. VON PLEYEL (1901) writes of Viennese bird lovers who carried their good singers out each Sunday so they would learn to sing anywhere, for without such training birds are silent in unfamiliar terrain. HOFFMANN (1960) describes how the Chinese also practiced these walkabouts with their birds.

Examples of the power of a good singer to attract follow: with a male European Goldfinch, taken from the flight cage and kept in a cage hung under the eaves of a roof for ten days before trapping started, SADLIK caught 65 European Goldfinches in April and May 1968 and over 60 in the same months in 1969. Most of them were paired females. Fewer males were caught as the songster was on the ground and males tended to fly over the 2.8 m high net whereas females were caught in direct flight toward the lure.

In the autumn SADLIK conditions European Goldfinch and Brambling lure birds in the same manner. He catches mostly young birds, as they are far more abundant—in 1969, 276 goldfinches were caught without help of feed-bearing bushes. The lure bird is close to a funnel trap near the apex of the right angle. Common Serins and Bramblings are more apt to enter the funnel trap than to fly into the net.

Some species actually interfere with catching. For example, GIESE never has Bramblings nearby when setting for crossbills. Strongly territorial birds can also disrupt catching at winter feeding places by driving away their own and other species.

SUNKEL (1927) caught a European Reed Warbler that was attracted to the set by the calls of another already in the bag. There are other similar occurrences, for example, with tits and Common Nightingales. Twites, Eurasion Siskins, and crossbills will even sit on the bander's foot if it is near the lurebird.

Crossbills are among the few species that need no conditioning, as fresh-caught individuals do not flutter about in a cage but start calling others straightway. Birds already caught in cage traps or in funnel traps—for example, sparrow traps—often assume a lure bird role and attract more. FRÜNTE in Westphalia (Germany) made a little-known but interesting observation when catching Ruffs. As soon as one Ruff was caught in a clap net, others crowded around it so tightly that it was possible to make a big catch with a large pull net.

In the course of special studies a brood sometimes fledges before the young are banded. Soon after fledging the young can often be induced into a trap or net by utilizing any adult of the same species as a lure. Lure birds present many opportunities. They are not always caged, but can be tethered in the trapping area by means of light harnesses of Moroccan leather. The banding station near Wassenaar (Holland) uses harnessed lures for finches, Common Starlings, and thrushes. The birds quickly get used to the harnesses (Fig. 402), which can be worn for months at a time. Harnessed birds sing little and are used in conjunction with caged lure birds. The harness is slipped first over the wings, and then the legs are gently pushed through 'leg holes' (harnesses are removed wing first). Once the harness is in place, the bander does the initial 'preening' of the feathers over the harness. The birds are kept on short tethers and are well acclimated. Thus they are quiet and behave normally. This also applies to Eurasian Golden Plovers and Northern (Common) Lapwings. The use of lure birds in this manner is ancient, going at least as far back as the Middle Ages in Europe. AITINGER (1653) devotes a whole chapter to them.

The Sparka, according to BÖHME (1952), is a device used in Russia by which tame lure birds can be activated (Fig. 8). A flexible stick 1.5 to 2 cm thick is bent into a U-shape and both ends are stuck into the ground. In the middle is a revolving crossarm to which a thin rod is attached at right angles. This rod serves as a perch for the

Fig. 8. Lure bird at a sparka. After BÖHME, 1952.

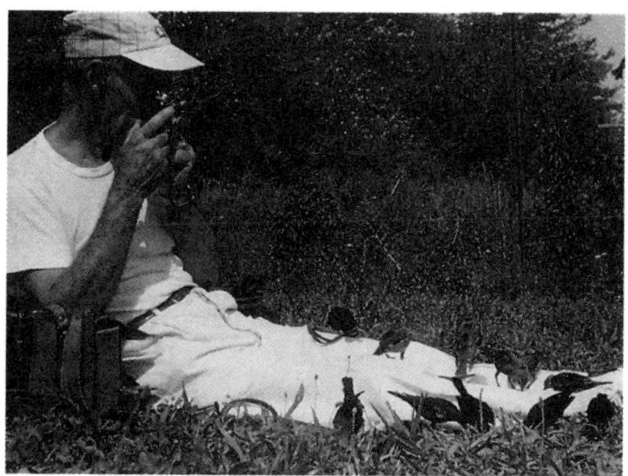

Fig. 9. Hungry cowbirds feed in a funnel trap oblivious of the bander. They even run about on his legs. These serve as lure birds too. Photo: H. E. Burtt, 1965.

bird and is caused to seesaw by means of a pull string. The catcher can move the perch from his blind, making the bird fly and attracting others by its motion, and also its calls. As soon as a flock has been lured into the trapping area, the catcher stops pulling the string.

A similar device was used in America to catch Passenger Pigeons. A lure pigeon was tethered to a shingle placed over a thick stick so that it could be caused to flip up and down by means of a pull string. The device was known as a stool and the pigeon was the stool pigeon.

Italian bird catchers place thrushes in runway cages (Fig. 7) approximately 80 cm long when extended to full length. Those used in the Flemish part of Belgium are larger (1 m wide and 2.25 to 3.50 m long) and are called 'Loopnetjes'. Sometimes rectangular runway cages (1.9 × 0.7 m) are also used. The mesh should be fine enough so that the birds cannot stick their heads through it.

Lure birds purchased from pet shops tend to be soft, as they have been kept in heated quarters for a long time. One must guard against putting such birds outdoors without a gradual period of acclimatization. The life of a cage bird is unlike that of a wild one: small quarters have allowed less freedom of movement and thereby altered his metabolism so that he is more apt to suffer from dampness or wind.

Pet shops often feed birds soft diets, and care must be taken to introduce them gradually to a diet of hard seeds. The birds should not be crowded, and their drinking water should be uncontaminated. Pet shops often treat birds with sulpha and antibiotics, so if a lure bird becomes ill, such cures are less efficacious. When purchasing a bird, note that it does not sleep by day, that it is sleek and its plumage undamaged, and that droppings do not adhere near the vent. Take the bird in hand as though to band it and blow the breast and back feathers apart. A healthy bird's breast bone is not a sharp keel, nor does it have a swollen red belly. Fat, in autumn, is no sign of poor health.

We can stress Neunzig's (1927) advice when purchasing: before the bird lover obtains a bird, he should learn its natural history, its feed in captivity, the suitable cage size and its other prerequisites. Everything should be ready before obtaining the bird.

One should buy from a reputable dealer, rather than shopping for the cheapest birds. In general, the beginner would do well to buy where lively birds in good plumage are in large, clean cages, rather than where a sickly bird humps about with ruffled, dirty feathers.

If one purchases lure birds, one does not have to go to the trouble of getting them adapted to cages, for they are already used to them. One's first lure birds have to be bought unless one can obtain them from another bander. A group of banders can share lure birds and undertake their care jointly.

The birds must be protected from all predators. For raptors, a raptor trap is recommended. Raptors so caught are transported, except during their breeding season, to another locality for release. The usual traps are set for rats, weasels, polecats, etc. Predator-proofing should be planned when first setting up a permanent trapping station. One way of protecting lure birds from raptors, shrikes, corvids, rats, cats, weasels, etc. is to place them in a small cage within a larger predator-proof cage. We sometimes place a wire hood over a lure-bird cage. Lure-bird compartments in some traps are covered with especially fine meshed wire. Small predators are not to be underestimated. Weasels, owls, and shrikes sometimes manage to kill or injure birds through fine mesh.

We usually do not trap during heavy rains, but sometimes we need to protect the lure birds from light rains: a pane of glass over the cage usually suffices. Complete plexiglass hoods are not practical: song does not penetrate well. We put them over cages only in heavy showers. They must be removed promptly afterwards, as they quickly get too hot for the birds in sunlight.

Fig. 10. A little owl decoy, from AITINGER, 1653.

Lure birds are sometimes used for weeks at a time, as in Twite research. Whether or not to replace them or to keep using the same individual depends on the birds' response and their luring ability. Any birds that are replaced are put back into the aviary: having become accustomed to free meals, they are not put back in the wild to fend for themselves—besides a bander becomes attached to his good lurers. The decision as to whether or not to release most of the lure birds at the end of the trapping season depends on circumstances.

Lure birds are thus an important factor in bird catching. In summary, they accomplish the following:

1. During migration and in winter they attract not only their own kind, but also raptors, shrikes and other species to the catching area.
2. Their presence makes shy birds less apt to fly away and makes them more trusting and more at home in unfamiliar surroundings.
3. During the breeding season, lure birds used singly awaken territorial aggression and lure aggressors into the nets.

Much emphasis must be given in this chapter to attracting birds with owls. It has long been known that many species mob owls and that their alarm calls attract others to the scene, all gathering around the owl. Owls were thus used even in early times for bird catching (Fig. 10). Mobbing consists of alarm cries, scolding and mock or real attacks on the owl's head so that the latter has to move to avoid getting hit. Especially during their breeding season, songbirds gather to attack the owl.

In some regions of Central Europe, Common Scops Owl, Eurasian Pygmy Owl and Tengmalm's (Boreal) Owl were once, and still may be, used. For example, in the Ticino (Switzerland) forests and in Italy Common Scops Owls were once tamed and often used as decoys (TSCHUDI, 1860). These species are now scarce and are seldom used anymore, but some are still offered in Italian markets for bird catching. The Little Owl has proven to be a fine decoy—be it a live or a wooden one.

CURIO (1963) gives clues that may be adapted by the practical trapper. The alarm call of nesting European Willow Warblers when approached by a Common Cuckoo is different from the call given when they see a perched European Sparrow Hawk. A stuffed Common Cuckoo elicits alarm, threat and attacks by many host species, but European Reed Warblers, Dunnocks, Northern (Winter) Wrens, Common Redstarts and Common Blackbirds do not seem to react. A Great Spotted Woodpecker—robber of broods of both open and hole nesting species—is promptly attacked near the nest by both adults. Birds also mob a Red-backed Shrike, but always keep a distance of 3-4 m. The woodpecker is soon driven off by the attacks, but the shrike tends to keep his perch placidly. Neither enemy needs to move in order to be effective, though the woodpecker tends to peck and the shrike just sits. Birds attack a perched woodpecker decoy (in hunting position) rather than one placed in climbing position, but both positions elecit alarm calls. Female shrikes are about half as effective as the males and draw about the same reaction as a white plaster model with a black eye stripe. The brightly colored male Eurasian Roller is about half as effective as a female shrike and the reaction appears to be due to shrike-like head markings, rather than to the vivid coloring. Many European songbirds mob this species—perhaps as many as mob owls.

European Sparrow Hawks and European Hobbies, in flight or when perched, draw attacks from Barn Swallows and Common House Martins. A stronger reaction is drawn by a perched European Sparrow Hawk than by the larger Northern Goshawk or Honey-Buzzard. According to CURIO (1963), yellow eyes, a hooked beak not too large in proportion to the head, position of the decoy and plumage draw mobbing. The decoy's position and stance must be somewhat natural. A Pied Flycatcher does not react at all to a stuffed Red-backed Shrike hanging upside down. Tame Common Jackdaws and Carrion Crows, however, fear just the head of a Tawny Owl or Northern Goshawk, be it real or made of plaster; their reaction to the whole bird is only slightly stronger. Here again the eyes may play an important role as the back of the head elicits weaker reactions than the front view and a Northern Goshawk elicits weaker reactions than a Tawny Owl. It is astonishing that many finches simply ignore raptors, whereas titmice, *Sylvia* warblers, old world flycatchers, wrens, nuthatches, woodpeckers, hummingbirds and others have developed highly specific mobbing and attack reactions toward Tawny Owls or Northern Goshawks.

Songbirds react toward enemies more strongly during the breeding season than the rest of the year. W. STÜRMER suggests attracting migrants to an owl decoy by playing taped mobbing cries; otherwise it may take the birds quite some time to discover the enemy.

CURIO (1963) explains the innate recognition of an enemy. Yellowhammers, Song Sparrows, Chaffinches, Common Bullfinches, Woodchat and Red-backed Shrikes and Common Jackdaws as well as others recognize owls naturally, without any previous bad experiences. Curve-billed Thrashers do not mob owls smaller than Great Horned Owls, but react to them in the same manner as they do to snakes, and they avoid sitting buteos. Sitting raptors all have the same effect on young ignorant Song Sparrows, Common Jackdaws, crows, shrikes and Common Turkeys. Behavior towards enemies seems to mature slowly. Fledgling Chaffinches first look curiously at a Tawny Owl, fly from it when they are 2-4 weeks out of the nest and then mob it; the older they get the more they mob. The behavior of young Yellowhammers matures earlier, but Song Sparrows do not develop owl mobbing behavior until they are several months old.

How does a bird react to an enemy not present in its own range? This relates to the question of how quickly innate behavior changes. CURIO (1963) tested two subspecies of the Pied Flycatcher (*Ficedula h. hypoleuca* and *F. h. iberiae*) with decoys of the Red-backed Shrike and the Tawny Owl. *Hypoleuca* mobbed both, *iberiae* only the owl. This corresponds to the geographic distribution of the birds: the shrike as well as the owl inhabit much of the same area as *hypoleuca*, while the Red-backed Shrike is missing in Spain (the two shrikes [*Lanius senator* and *L. exubitor meridionalis*] occurring there live in different habitats).

When the Pied Flycatcher lost its mobbing behavior towards shrikes could only be determined if the time of its colonization of Spain were known. The Red-backed Shrike never touches Spain in distribution or migration, and the first 'settlers' of *iberiae* in Spain were never again bothered by *Lanius collurio*.

Recognition of the Common Cuckoo as an enemy is similar. Many European species will be alarmed by *Cuculus canorus* at their nests, but North American birds do not mob a Common Cuckoo decoy—no parasitic cuckoo occurs in America. CURIO, through observation and experiments, has gained much insight into bird behavior, some of which might have been known by tradition to early bird catchers.

SUNKEL (1955) pointed out that behaviorists can learn much from bird catchers. Vocalizations can be defined and are used in relation to age, sex, season, weather, food, and habitat. Practical behavior studies used to be handed down by word of mouth by many classes of earlier societies.

Modern workers can gather much data about the reaction toward a particular species. They can thereby vary their catching methods to suit the time of day, age and sex of the mobbers, etc. M. RIEGEL caught 2,539 birds of 25 species with a single live Little Owl decoy!

HARTLEY (1950) and HINDE (1954) published on enemy recognition. Barn Owls produce weak mobbing—perhaps because their plumage is not coarsely enough marked. NICE and TER PELKWYK (1941) produced alarm in hand-reared Song Sparrows by shading the edges of flat owl models. They also increased the reaction toward stuffed owls by adding moving wings. Early bird catchers used movable wings on stuffed owls.

CURIO (1963) found owl eyes the decisive releaser for attacks by Red-backed and Woodchat Shrikes and that two eyes produced a stronger reaction than one. W. SUNKEL trapped successfully using a one-eyed Little Owl decoy and F. HAMERSTROM caught Northern Harriers (Marsh Hawks) with a stuffed Great Horned Owl having no eyes at all.

The size of the owl decoy need not be precise, but in general small birds do not mob Eagle

Owls, whereas crows and diurnal raptors hardly mob small owls; the latter are more apt to make a meal for them. According to Curio, the optimum decoy size for Pied Flycatchers lies between the Eurasian Pygmy Owl and the Tawny Owl. Shrike-sized birds, such as thrushes, seldom mob shrikes except at twilight, and may even seek out their presence. This has been demonstrated by the nesting associations of the Barred Warbler and the Red-backed Shrike and the Fieldfare and the Northern Shrike.

Small birds can detect whether or not an enemy is in a hunting mood. Songbirds know whether a Eurasian Pygmy Owl is hunting mice or birds (Schnurre, 1942). Their reaction to a Little Owl decoy in this respect is not known. These owls are most active at dusk, but it is our impression that these decoys—alive or stuffed—are promptly attacked at any time during the day. Of course, it is best not to feed the owl too much when it is on the job.

Sound plays a part in successful catching too. It seems that owl calls are learned, which explains the fact that songbirds frequently mob owl calls without seeing an owl—or even an imitation of an owl's call with no owl present. In the Swiss Jura mountains songbirds react to the Eurasian Pygmy Owl only where it normally occurs. This diurnal owl often can be seen from afar as it calls from the tip of a spruce.

It is advisable to bear in mind that once mobbing is underway, more individuals and more species are apt to join in. In North America, where it is an introduced species, the House Sparrow mobs only when native species have already started the chorus. According to Hamerstrom (1957) American Robins mobbed her trained Red-tailed Hawk when she whistled it into the glove. Later the robins started mobbing on hearing the whistle alone. More needs to be discovered about the learning processes of predator recognition and mobbing (Curio, 1963).

Owls to be used for decoys are best taken from the nest as flightless, downy young. Wiedermann (1914) recommended taking larger young, but owls taken older remain man-shy and wild. Young Little Owls are easy to rear. They eat June beetles and dung beetles except for the wings; they eat partially plucked headless and eviscerated House Sparrows chopped up for them; chopped mice; whole young mice and tidbits of raw heart. The owlets gape for food as soon as their tactile bristles are touched. Little Owls require water only during the molt. Barred Owls want water rather often, and most owls like to bathe. Hamerstrom (1970) describes the hand-rearing of a Great Horned Owl from day one.

Owls, unlike diurnal raptors, are not easily tamed or trained when taken after they have fledged, nor do wild-caught owls take to sitting on a block perch—instead they seek cover at least as long as the bird catcher is nearby. Italian bird catchers prefer tame, hand-reared Little Owls and whenever they have to resort to a wild-caught owl they use the short tether and furthermore release the owl at the end of the migration season. In order to insure plentiful reproduction of owls, Italians place dark nesting nooks here and there in their houses; some of the young are selected for use as decoys (Detmers, 1905).

Vallon (1883) described the then prevalent method of training decoys: the owls were taken from the nest just before fledging and were kept in a cage for a week. Then they were put on a bolster atop a thick, meter high post. The bird was fitted on both legs to leather jesses from which a cord ran. This was fastened to the bottom of the post while training; during bird catching the trapper, hidden in a blind, held the end of the cord. It took a few hours a day to train the owl in time for decoying in October. The post was fitted with an iron point so it could be stuck into the earth. The trapper carried the owl away a few paces and tossed it toward the perch; he kept doubling the distance so that the owl had to fly to the perch. It took at least a month before the training from hand to bolster was completed. The next step was to give the cord a strong pull causing the owl to flutter about a bit and then fly to the perch. This step took a lot of time too and undoubtedly not all owls succeeded in doing what was expected of them. During catching the trapper pulled the cord from his blind as soon as he decided his quarry was near enough.

Kruis believes that this ritual is unnecessary: he points out that practically every owl takes to a perch. Those few that act scared and don't might as well be released.

The only training that is needed is to get the owl tame, accustomed to the jesses and leash and to being carried. Owls should not be leashed until they are almost fully grown. Get them tame, give them rides in the car, expose them to children, dogs and noise. Then they will

Fig. 11. Creel to carry a little owl decoy in and in which it can take cover from enemies. Photo: H. Bub.

make placid, well adjusted decoys. Owls like to perch, especially in open country where they will take to any perch offered.

Sunkel carries his decoy owl to the trapping site in a wooden box; when he arrives he simply opens the door of the box. The leash is fastened inside the box but the owl easily takes its perch on top for a better view of the surroundings. Leaving the box on the ground is not advisable unless there is little or no vegetation. The box also serves as an escape cubby from predators. Little Owls are easy prey for European Sparrow Hawks. Stürmer uses a fisherman's creel instead of a box and so do Italian bird catchers. The owl can take cover in it quickly (Fig. 11).

The creel is weighted with a stone so that the owl cannot tip it over and earth or stones are placed around it but in such a way that the owl cannot hide behind them. When catching is over, the owl usually jumps right into the container and is ready to be carried home.

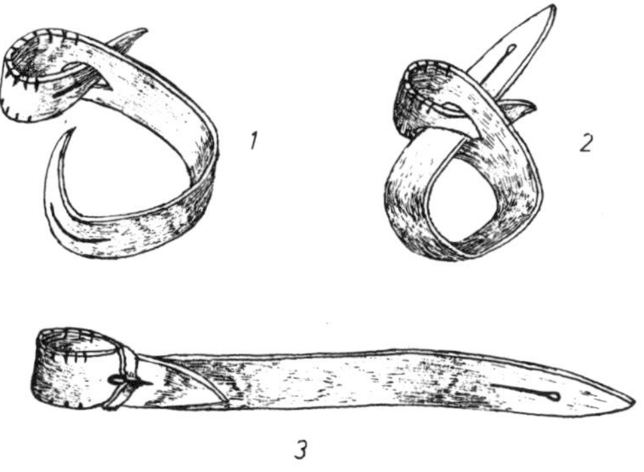

Fig. 12. Jess knot. After Mebs, 1964.

The Little Owl wears soft leather jesses on both legs identical to those used by falconers. The jesses are not fastened with slip knots as they could tighten too much. They are fastened with jess knots; for one version see Fig. 12. To fit a Little Owl each jess is about 16 cm long, the strap is 6-8 mm (10 mm at the foot) wide and 1.2 mm thick; the closed knot measures 10 mm in diameter. The distance between the two front slits is about 22 mm. Fitting a jess is rather like applying a bird band: it should never constrict, nor should it be too loose.

Although European falconers ordinarily remove the jesses from their birds during the months of molt, the Little Owl decoy can wear its jesses throughout the year. It is merely necessary to keep checking from time to time to make sure that the leather has not hardened; if it has, it should be oiled or the jesses should be replaced.

When the owl is to be tethered, the ends of the jesses are expertly affixed to a swivel (Fig. 13). A swivel consists of two metal rings

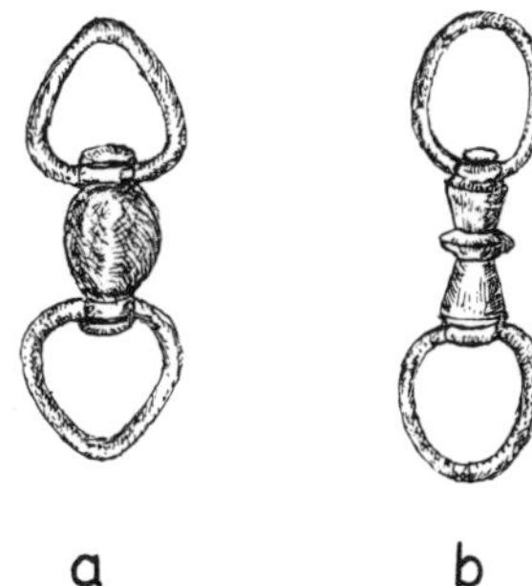

Fig. 13. (a) European ball bearing swivel; (b) Oriental swivel. After Waller, 1962;

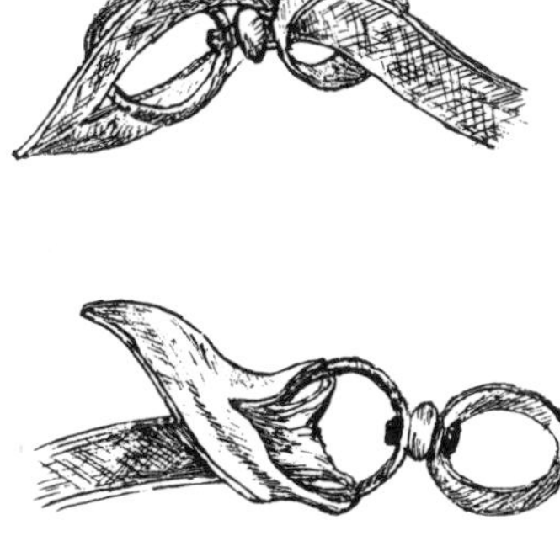

Fig. 14. The end of the jesses are fastened expertly in one end of the swivel. After Mebs, 1964.

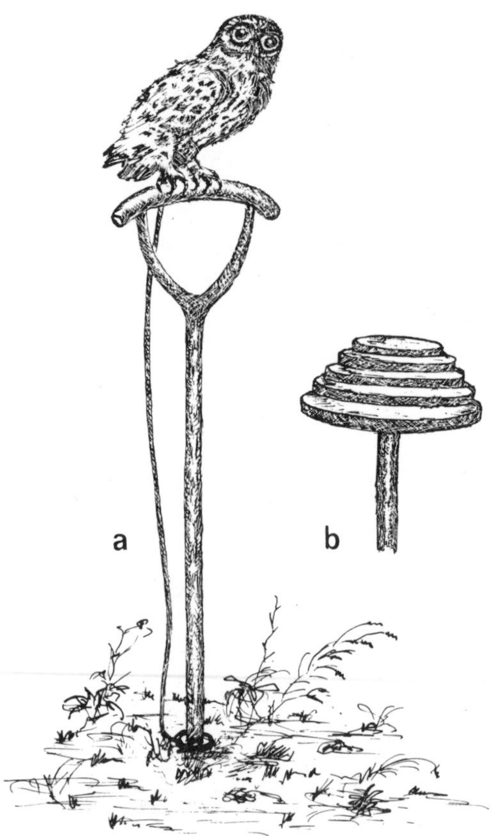

Fig. 15. (a) Little owl decoy on a T-perch; (b) the top of a 5 m telescoping perch used in Italy. The decoy is fastened with a short tether.

fastened together so that they will rotate around each other. It connects both jesses which are pulled through one ring by their slit ends, one after the other. Then the swivel is pulled through (Fig. 14) the slits. The other swivel ring holds the leash, an approximately 1.5 to 2 m long, 5 mm wide and 3 mm thick leather thong (a thick knot prevents the leash from sliding through the ring). The swivel is intended to prevent the twisting of the jesses when the bird is on the perch. Nonetheless, it can occasionally happen even with the swivel. Some catchers with Little Owls use regular rawhide shoe laces for the leash. The leash is fastened to a ring at the base of a block perch. Block perches 1 to 1.5 m high or T-perches (Fig. 15) are used, but block perches are better. DETMERS (1905) suggests keeping the owl in a cage. This method should not be used except in extraordinary circumstances. Untamed birds should not be kept, in cages or otherwise.

A good tame decoy owl bobs on its perch in a lively manner and flies down and back up to it; it pops its beak and calls as soon as the songbirds start to mob. It does not try to fly away nor hide, nor does it do so when people approach, but a man-shy Little Owl has the knack of becoming inconspicuous very quickly, promptly reducing its efficiency as a decoy.

Mobbing songbirds are always prepared to flee from on owl. A person can approach more closely while mobbing is going on, but the songbirds are more aware of the nets and even perch on shelf strings or on the tops of the poles.

The Little Owl should be fed a varied diet. We suggest 2 sparrows one day, 2-3 sparrows the next, and heart meat of beef, calf, horse or pig on the third. Furthermore the birds should have a piece of fresh heart the size of a small apple at least twice a week. A steady diet of heart for up to two weeks does not impair the owl's health, and the meat may even be somewhat smelly. Little Owls refuse thawed heart form the freezer even when hungry. Mice are as good as sparrows in the diet, and mouse-sized chunks of lean beef, veal or horse meat will do. Such pieces do not have to be rolled in feathers.

Food requirements vary. On warm days owls will eat less than on cold ones. They also tend to eat more during the molt. According to KUNZE (1957), well-nourished decoy owls can get along for several days without food, but then they need water.

Little Owls will not feed from anything much bigger than a mouse or sparrow. Larger pieces need to be cut up. Head, intestines and wings are always removed from sparrows or birds killed by cars. Head and intestines present dangers of infection, especially in sparrows, and the wings are mostly bones. Dead birds should be plucked somewhat as the owl otherwise eats too many feathers and not enough meat to satisfy calorie requirements.

It is best to feed natural foods. KÖNIG (1969) found that Little Owls feed primarily on insects in summer, especially on beetles and grasshoppers, but also on caterpillars and earthworms. They also take small mammals, especially mice and shrews, and to some degree birds up to thrush size. Lizards and frogs are preyed upon occasionally.

Care must be taken not to feed owls poisoned

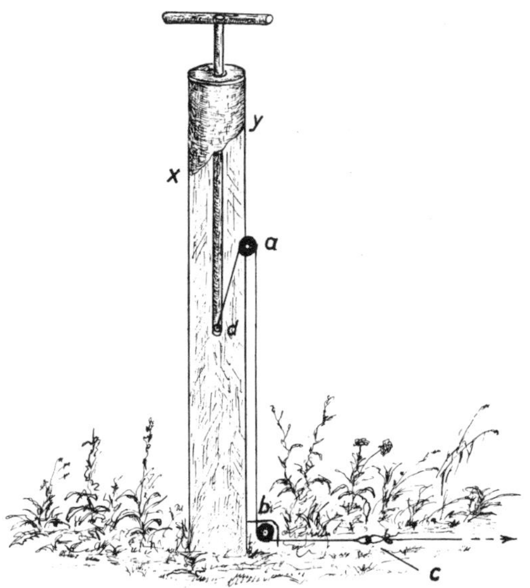

Fig. 16. Agitation mechanism for an owl decoy block perch. Below x and y it is shown in cross section to reveal attachment of cord (d), pulleys (a) and (b), line leading to blind (c). After RAESFELD, 1942.

mice—especially those that have been poisoned with anti-coagulants and which often live for days after ingesting them. If such mice are fed to an owl, it, too, will suffer from internal bleeding and die suddenly. Poisoned sparrows present the same problem: they may not die promptly and it is best not to collect owl food in an area where a neighbour is poisoning.

SUNKEL noticed that his Little Owls got close to the edge of the pen to get wet whenever it started to rain. HEINROTH's handreared Little Owls never bathed, but liked to have a hose played over them and spread their wings and tails as though rain-bathing. They behaved like Long-eared and Short-eared Owls, whereas Tawny, Barred and Great Horned Owls like a good bath in a shallow pan. It behooves the owl keeper to supply suitable bathing facilities and to remember to put his bird out to rain bathe as the falconer does his.

Decoys should be checked for ectoparasites from time to time. Pet stores often sell suitable remedies.

Suitable quarters for owls vary. SUNKEL keeps a pair of Tawny Owls in a large cage in which they also breed. The cage has several places for hiding and nesting. The Tawny Owl can be kept in a large 100×50×50 cm box that resembles its natural day roost. W. STÜRMER closes the front of his box with broomstick-sized vertical bars instead of wire. The box is lined with paper and contains two broomstick-sized perches. Unless one is breeding owls, it is best to keep them separately, for otherwise the stronger may kill the weaker.

During pleasant weather the owl is put on a block perch in partial shade. A shelter nearby keeps it from panting and needing water. Decoy owls, like all birds restricted in movement, are in danger from accipiters.

For decoying the owl is placed in woodland openings, near edges of woods or shrubbery, near stream banks or in parks and gardens. If push nets are used the terrain should be open so that the excited and scolding birds can only sit on the nets themselves. Mist nets or trammel nets can be placed under trees or near shrubs. If mist nets are placed in a triangle it pays to set up some perches inside so that the mobbing songbirds will sit on these instead of on the top shelf strings. Another net erected close to any nearby shrubbery is apt to catch too. The owl is placed inside the triangle after all the nets are in place and the vicinity of its perch should be free of vegetation and hummocks. It takes to its perch right away to protect itself from enemies on the ground.

SADLIK set up various trapping stations 100 to 200 m apart in a wooded lot near Kötzschau (near Halle, Germany). He trapped regularly every two to three weeks. In this way he netted the entire Icterine Warbler population and had many recaptures.

Each trapping clearing should be at least 8 to 10 m across. Woodland paths are often ideal, especially if the bushes along the edges are not higher than the top of the net, and preferably are lower. Paths for trammel or mist nets are cut between the lower vegetation and the woods.

Rather bare ground with scattered bushes also makes a good trapping site. A stout 1 m long stick to which the owl's leash is tied is put slantwise in the ground. Nearby cover is removed so that the owl cannot hide itself. As soon as enough songbirds are sitting near the owl the catcher drives them into the nets.

SADLIK uses three types of sets (Fig. 17). Some small birds such as *Sylvia* and *Phylloscopus* warblers fail to fly directly over a Tawny Owl decoy, but often fly from bush to bush and end up in the nets set between them. It is well to use a

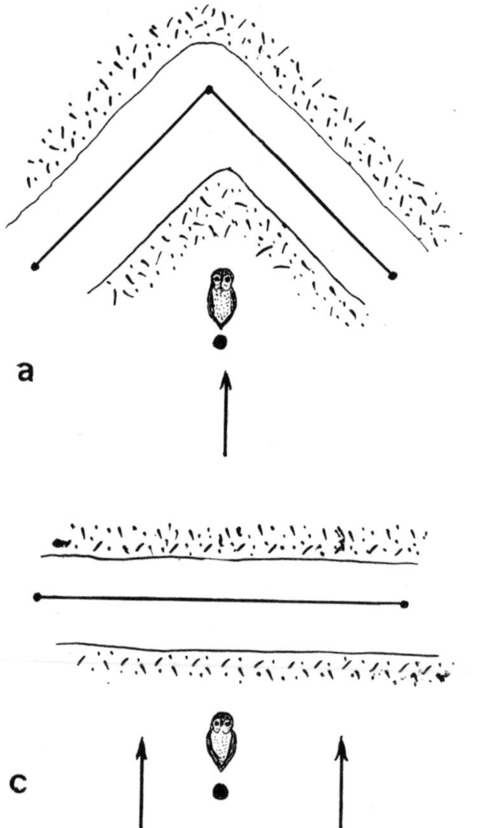

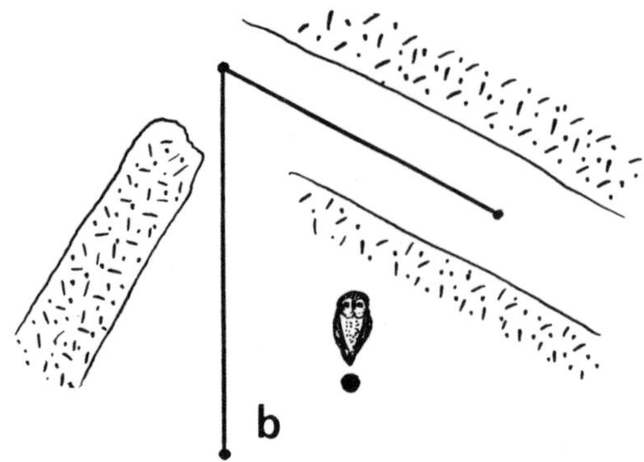

Fig. 17. Various sets for mist netting with a decoy. Arrows indicate the direction of the drive. After J. SADLIK.

12 m net, high enough above the decoy to catch those species (Eurasian Golden Oriole, Hawfinch, Spotted Flycatcher, trushes, European Goldfinch, etc.) that fly over it. SADLIK preferred to trap near breeding Common Starlings as these excellent mobbers quickly attract other species to mob too.

It is best not to interrupt the mobbing by trying to take birds out of the nets. Wait until your catch is complete and you are ready to stop catching birds. Otherwise the birds will stop mobbing and fly away.

Guidlines help, but each bander must develop his feel for where to place his set.

Early morning up to about 10 a.m. is the best time. Some banders consider cool, clear days most favorable. One must adapt to the daily activity pattern of the birds. In summer no one would try to decoy songbirds late in the morning or at noon. Successful catching follows rules, but often the why's and wherefore's are not known.

When placing a decoy one should realize that songbirds may remember a place where they have encountered an enemy. CURIO (1963) found that ravens and Song Sparrows, with some individual exceptions, shun places where they have had a fright. An European Robin was seen to dive at a spot where it had seen a stuffed Common Cuckoo before. Carrion Crows watch the woods into which a Northern Goshawk has disappeared, ready to mob it again when it reappears. On the other hand, tits, Common Blackbirds and sparrows go their own way as soon as a predator is out of sight. W. STÜRMER warns against placing an owl decoy too close to a songbird nest if the young have not yet hatched or the nest is being built, lest the adults desert. A stuffed Common Cuckoo reputedly does not cause desertion.

After a while mobbing will wane and the birds inexplicably will leave the owl, sometimes after a few minutes. CURIO (1963) and HINDE (1954) have suggested that the departure occurs when an untethered owl would fly away. A skilled trapper will not leave his decoy in one place too long: moving it 20 to 30 m often incites the small birds anew and brings in more as well.

According to SUNKEL (1956) a lively decoy is mobbed by almost all small birds and also by some of the larger ones like corvids and woodpeckers. It gives the ornithologist unusual opportunities, not only for catching, but also for observing seldom seen warblers, etc. just as in earlier times the double clap net and the snaring of thrushes turned up uncommon migrants.

In general, the Tawny Owl is suitable for decoying most small birds up to thrush size. The finches are an exception: many of these react little or not at all. Other species of finches scold but keep their distance, as KRUIS noted with Hawfinches in Czechoslovakia. Crossbills usually take no notice. The Tawny Owl is fine for catching Tree Pipits and *Regulus* warblers. Eurasian Golden Orioles can be caught effectively with two 6 m nets set at an angle or with one 12 m net, but they must be the first to arrive. If thrushes set up a commotion first, the orioles just scold, but do not fly into the net. KRUIS nets many Black-cap Warblers with his owl in late summer. One Czechoslovakian bander uses his decoy successfully for shrikes, Common Hoopoes, and Eurasian Rollers. Surprisingly, MICHELS caught more White (Pied) Wagtails with a stuffed decoy than with a live owl.

SADLIK near Kötzschau (Halle, Germany), caught Eurasian Golden Orioles and found that they reacted strongly to a Tawny Owl decoy around the end of June or the beginning of July in the vicinity of their nests and up to 100 m from the nest. It is just at this time that the young orioles fledge. The parents dive screaming at the owl and are caught in a net placed behind it. Whenever SADLIK knows exactly where the nest is he puts up a single 6 m mist net. The orioles usually dive in from a high perch, so the net is placed in a opening 5 to 10 m from the forest edge or from a clump of trees. The Tawny Owl sits on a slanting perch 1 m behind the net. Because other adults may come to mob too, up to four adults may be caught in a short time. Occasionally young orioles leave the nest prematurely. If SADLIK finds one he places it in a cage near the owl and usually catches both adults within a half hour.

Creepers and Great Spotted Woodpeckers decoy well to a Tawny Owl. Creepers, especially, fly between tree trunks near the owl and land in the nets, or they may be driven into them.

The Great Spotted Woodpecker lowers itself slowly down a tree trunk when it has spied the decoy. When it is low enough we scare him into the net. This must be done promptly as Great Spotted Woodpeckers do not like to stay low on a trunk long. The Green Woodpecker scarcely reacts to a Tawny Owl decoy and tends to remain almost motionless in the tree tops.

Shrikes, according to SADLIK, react much as Eurasian Golden Orioles do: they can be caught soon after their arrival in May. The net should be set parallel to a hedgerow but only 3 to 5 m away. The owl is tethered so that the net is between it and the hedge. Catching is also good when the net is set at right angles to the hedge with a third of the net set within the hedge itself. Then the owl is placed about 1 m from the exposed end of the net. The parallel net set was more successful. Shrike sets can also be placed between two clumps of bushes.

Ortolan Buntings, like other buntings, are first-rate mobbers: they circle the decoy by running and sometimes by flying (low and from the back). They show little fear of man and can easily be driven into a net both in the breeding season and at other times of the year. To catch them, place two nets at right angles to each other in open country. SADLIK had an Ortolan Bunting follow him 200 m as he was carrying an owl decoy to a new set (where the bunting was caught immediately). Ortolan Buntings seldom perch on branches, bushes, etc., unlike Yellowhammers, so one does not have to be choosy about selecting the set. The best sets are perpendicular to the edge of a woods, but one has to adapt to the terrain.

Ordinarily, birds react to the decoy without any driving. SUNKEL (1927) described a drive in which a bander lies flat on the ground a few meters from the owl ready to jump up when mobbers alighted between him and the net. In this way he caught Crested Tits, for example. These birds are so excited when mobbing that they do not readily take off.

Under no circumstances should Little Owls or other owl decoys be used in connection with double clap nets or pull nets. Trapping with an owl decoy and attracting to bait, water, or conspecifics are entirely different principles. Lure birds cannot do their job well with an owl decoy nearby: they become quiet and timid.

Substitutes for live owl decoys are sometimes satisfactory. These include stuffed owls, and wooden, papier machè or even foam rubber replicas. Some banders agitate the wings with a pull string, quite rightly assuming that although a motionless owl attracts small birds, a moving owl has a stronger drawing power. Other banders put their decoys on a see-saw (Fig. 18). The lower end of a slightly bent 4 mm wire is weighted. The decoy base rides in a flat groove

Fig. 18. Stuffed little owl on a perch. Photo: A. Präkelt.
Fig. 19. A mechanized wooden decoy. After H. P. Arentsen.

so that it can't fall off the perch. When rocked, the decoy's motion seems quite lifelike and resembles the bowing of a live owl.

Italian bird catchers make very good decoys of stuffed owls with movable heads and wings set in motion with a pull string (Fig. 20). A stuffed owl first requires the availability of a dead owl, and the Little Owl is no longer common in Central Europe. Therefore we are pleased to note that an Italian company in Brescia makes good foam rubber owls with movable wings. Such a decoy owl could be useful for banders. Many banders in cities cannot keep a Little Owl properly, and an artificial decoy is often adequate.

Bings puts the stuffed Little Owl on a tension spring which is fastened on a small pole. He moves the bird with a thin string. A light wind can also move it.

H. F. Arentsen in the Netherlands catches birds with a wooden Little Owl (Fig. 19) and has had good success. This bird is approximately 20 cm high and has light yellow eyes. Important are three little mirrors (the number can be increased) on each wing. They are intended to incite the birds' curiosity. Each mirror has several facets like a diamond to reflect the highest possible percentage of light. In Italy even multicolored mirrors are available. The wings are fastened by hinges to the body so that they lift when the body is turned. They should hang down vertically when the body is in a rest position, but they should stand out from it somewhat. To accomplish this, a rubber plug of 1 to 2 cm length is fastened to the lower part of the body. An old record player motor can be used to activate the wooden owl. In France such mechanisms can be bought commercially.

The call of the Little Owl can be imitated with an owl call or by mouth when decoys are used. Little Owl calls can be bought. They are not always necessary.

Other owls are also suitable for catching small birds (especially passerines). Of the bigger owls Kruis and Stürmer consider only the

Fig. 20. Stuffed little owl with movable parts. Photo: H. Bub.

Tawny Owl suitable. STÜRMER believes that birds smaller than trushes tend to ignore it and mob only after the actual prey birds such as thrushes and jays have started the action. The general excitement then carries them away. All corvids and many raptors react to the Tawny Owl and especially to the Northern Eagle Owl. Few small birds get close to a Long-eared Owl —it is probably too large for them—and they react even less to Barn Owls.

1.1.2 Artificial decoys and lures

The use of live lure birds can be traced far back into the Middle Ages and antiquity. But bird catchers in former times also used decoys or stuffed lure birds. According to BALSS (1947), ALBERTUS MAGNUS (1193–1280) used owls other than the Northern Eagle Owl.

Although live lure birds are usually necessary for efficient songbird catching, decoys are indeed a great help in catching bigger birds such as ducks, geese, doves, crows, and shorebirds. This will be described later. Commercially manufactured ducks, crows, Wood Pigeons (sitting upright, with spread wings, or with movable wings), can be bought in some European countries. So can Northern (Common) Lapwing decoys, and even Eagle Owl decoys—some with movable wings. Banders can also make their own decoys. The shorebirds (Fig. 21) (Eurasian Curlew, Common Redshank and Ruddy Turnstones) are of wood with coarsely painted natural plumage. They were used traditionally in southern and western France, and are still in use in other European countries as well as in other continents. The Danish ornithologist PETERSEN (in SCHILDMACHER, 1965) saws out wooden duck decoy silhouettes, adds cork to the sides so that they will float upright, and attaches a metal keel. Then he paints them in natural plumage. The ducks ride at anchor. The anchor cord is fastened toward the front of the duck so that it always turns into the wind just like a real bird. Wooden goose decoys are also used—PETERSEN mentions Brant Geese in particular. During spring migration they are easy to attract on their breeding grounds. Australian banders (CAMPION, 1964) caught flocks of non-breeding Little (Least) Terns with decoys and a rocket net.

Decoys may also be cut out of sheet metal and set up with a stick in shallow water near the shore. They are probably most effectively used

Fig. 21. Wooden decoys. Photo: A. PRÄKELT.

at dawn or dusk. Before shorebird hunting was prohibited, they were successfully used by gunners in the prairie states of North America.

According to GHIGI (1933), Italians use cork decoys as well as stuffed birds to catch Northern (Common) Lapwings, Eurasian Golden Plovers, doves and others. They are set up facing the wind. Asians (BÜTTIKER, 1959) fashion clay duck bodies and add head and necks of cow dung. Recently foam rubber, rubber and plastic decoys have appeared on the market.

Furthermore, stuffed decoys can be used in conjunction with live lure birds and a double clap net or with a funnel trap. It is best to dis-

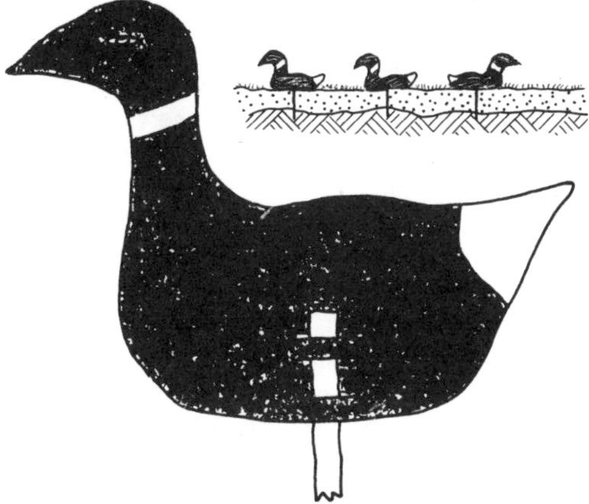

Fig. 22. Decoy for Brant Geese. After NØRREVANG & MEYER, 1960.

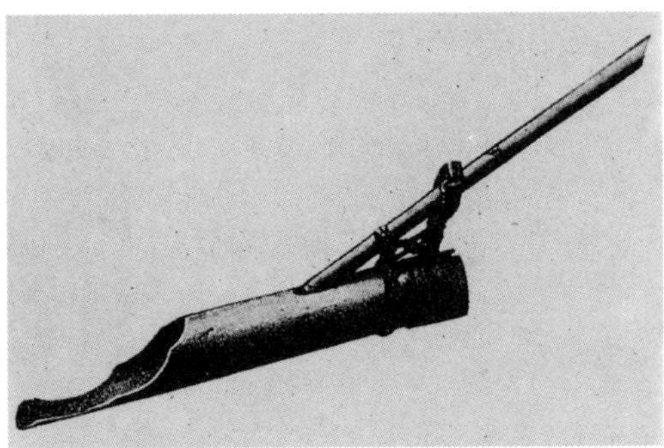

Fig. 23. Malayan bird call (Northwest Borneo). Length of the blow pipe 238 cm, length of the sound horn 47 cm; the material is bamboo, the string rattan. After LIPS, 1927.

play decoys in various positions—flying as well as sitting and standing. Bad weather is hard on stuffed birds. They are not necessarily better than decoys.

For trapping waterfowl, some of the decoys should be placed so that they can be agitated from a blind.

Artificial decoys may be 'animated' by calls, but some species (especially marsh and water birds) perceive their own kind optically and are not much influenced by calls.

Calls were widely used in former times. Australian aborigines used emu calls, and other primitive people had lure calls (Fig. 23). CHR. L. BREHM (1855) talks about catching tits:

"The tit call is made from the ulna of a goose and is approximately 3 inches (= about 8 cm) long. Not far from the upper end a cut is made, and the bone is filled with wax from this cut to the upper end. With a penknife a low, rather wide opening is made in the wax on the upper surface of the bone. This opening, large or small, tunes the call and lures the tits. This is rather easy as the lure call of the tits (in Thuringia, Germany, the Great Tit) and others consists of only two main notes. The lower note is produced by covering the lower small opening with the index finger. If the note becomes too low, another small opening is cut and the note becomes higher because of the air outlet. With practice, it is easy to imitate tits. Tits are not very particular and are easily attracted. This may be due to their curiosity. Obviously a perfect imitation of tit calls works best."

MODERSOHN (1870) mentions the Little Owl call. It consists of two pieces of wood between which a call-producing piece of inner bark from a cherry tree is strung. To make such a call, take a 14 cm piece of wood, such as fir or beech, of 20 cm diameter. Cut out a center piece of approximately 45 mm down to the heartwood. Fit another piece exactly into this cut. Scrape the red layer of an 8 mm wide faultless piece of cherry bark perfectly smooth and insert it as tightly as possible under the small wooden piece. If the bark is not tight or if either of the wooden pieces are not smooth enough a jarring sound will be present.

MODERSOHN talks further about catching which, in his time, was done with bird lime on the so-called sprig tree. We would accomplish the same with stationary and push nets. "After hiding in his hut everything that may seem suspicious to the birds, the catcher begins to imitate the Little Owl with the call. As soon as the birds answer—jays always will—he crawls into the blind. To produce the call of the owl which sounds like 'kiwitt', the notes are produced simply by blowing. For the 'whoowhoo' call to sound low enough, the call has to be enclosed in the hollow of both hands. The pitch of the call can be changed by enlarging or narrowing the opening between the wood pieces. After the birds have responded to the call from a distance for a short time, they will fly to the tree, scolding and highly agitated."

Many species will scold owl voices as well as the owl itself.

Today specialty stores carry metal or plastic calls for snipes, crows, Eurasian Jays, accipiters, ducks, gallinaceous birds and plovers as well as for songbirds such as larks, Common Bullfinches, thrushes, Common Starlings and others. There we can also obtain hare and deer calls and other lures for mammals. The Belgian firm of Priem-Verlinde in Brugge carries a large selection. Specialty shops for hunting supplies usually only carry lure calls for gallinaceous birds, ducks, doves, Eurasian Jays, crows, and Eurasian Woodcock.

The calls must be cared for properly. No dirt, hair, feathers, or other foreign matter should get inside. Always carry them in a small box or plastic bag except when trapping. Most types can then be carried on a string.

USINGER (1963) writes about their use in his book which is highly recommended. Some banders make their own calls according to old traditions. The use of these instruments is probably

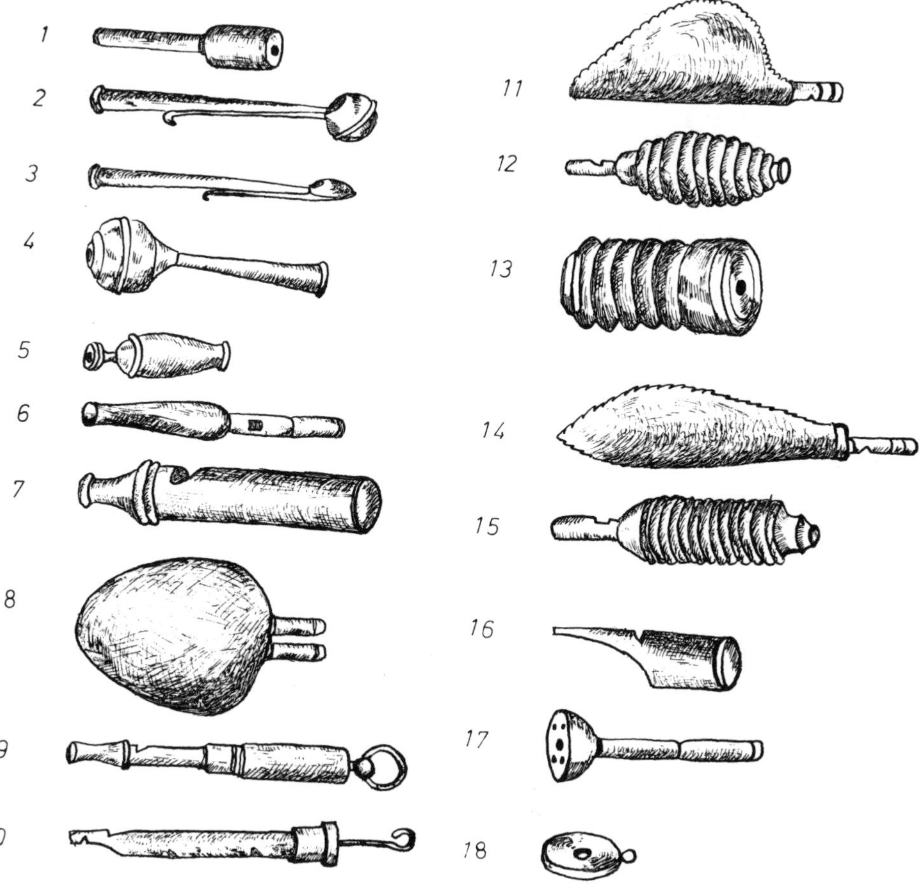

Fig. 24. Calls for lure birds and animals: (1) and (2) Skylark; (3) Golden Oriole; (4) Red-legged Partridge; (5) rabbit; (6) Hazelhen; (7) Eurasian Curlew; (8) ducks; (9), (11), (12) thrushes, especially Fieldfare; (10) European Starling; (13) European Blackbird; (14) and (15) Quail; (16) (Greater) Golden Plover; (17) Northern Lapwing; (18) various sizes for songbirds.

more common in those areas where the trapping of birds has been pursued traditionally. Few of today's bird banders use a call, but a return to old ways seems to be under way. P. BECKER uses a widely available plastic mouse call on which he produces a great variety of high notes with his tongue and lips by regulating the intensity of his breath: he imitates the calls of Dunnocks, creepers, tits, Eurasian Nuthatches, Common (River) Kingfishers, and other birds. Using the calls near set nets, a great variety of birds is caught. Penduline tits come from great distances. The little mouse call proved useful even as a lure for owls, and it can be a great boon on banding expeditions (H. RINGLEBEN).

In the United States the National Audubon Society sells a call with a great variety of high notes that appeal to many birds.

The Belgian ornithologist SPAEPEN (1952) proved how productive netting with calls can be. They were his most important aid. In eight years he caught about 2 800 Tree Pipits, and he emphasizes that lure birds are not necessary when catching Yellow Wagtails, and Tree and Tawny Pipits. SPAEPEN considers calls very important, but warns that proper use requires practice, time and experience. He believes many bird catchers never learn to use them well and that a good ear together with some musical training is valuable.

In 1926 a cartoon in a hunting magazine showed a hunter usuing a "radio call" to deceive and shoot a buck. NIETHAMMER (1955) talks

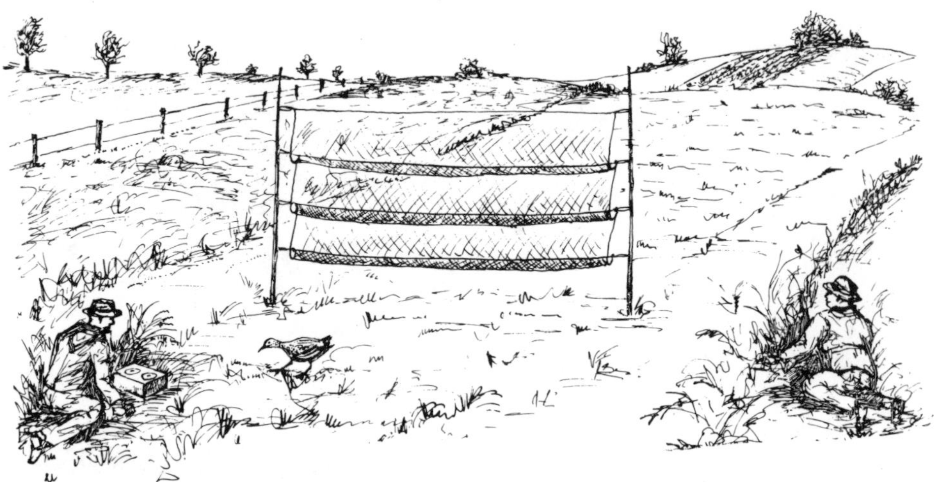

Fig. 25. Night hunt with tape recorder for the Corncrake. The bander at right stays close to the net pole in order to notice the captured bird immediately. After a description by H. PRÜNTE.

about a similar experience with the Striped Bunting and the Black-headed Gonolek during a trip through northwestern Africa in 1953. After their own recorded songs were played back to them, both came right up to the tape recorder, the male Striped Bunting even following it into the kitchen. More about this later. In the meantime, the tape recorder has become a very useful addition to ornithology (RINGLEBEN, 1957) and bird banding, during the breeding season as well as during migration. AUSOBSKY (1964) even wrote an entire book on "hunting for animal voices with a tape recorder". The quality of the recorder is not so important according to SCHUPHAN, who gained his experience working on the Rock Bunting.

A portable tape recorder with high speaker volume is best as bird voices do not carry far in the open. The tape must be played back as loudly as possible. SCHUPHAN uses a parabolic reflector to make his recordings, but often songs from commercially available records can be used after having been recorded on a tape. "Generally I use mist nets, more rarely push nets. For mist nets one needs an attractive background. Often a small bush is sufficient. Before setting up the net I check with the tape whether a male Rock Bunting is in residence. Sometimes I have to try different locations before the male will notice the taped song. Once the bird has been found I quickly set up the net at a suitable location, put the recorder with the running tape about 1 m in front of it, and retire about 10 m, using all available cover. Usually the excited bird immediately aproaches the tape recorder and can easily be shooed into the net. It often gets caught by itself. Going on such expeditions in March or April I take along the male caught in the neighboring territory. A lure bird—in a collapsible cage—plus the recorded song facilitate the catching even more. This way I have had success until June. The early morning hours are best. It is rather difficult to get the males to stay close to the net for longer periods during noon hours. Sometimes the female gets caught along with the male." SCHUPHAN also catches Cirl Buntings, Common Nightingales, European Nightjars and Eurasian Golden Orioles in this manner. Common Hoopoes and Common Scops Owls could always be located and relocated with tape recordings. PRÜNTE caught Corncrakes with a portable tape recorder in Westphalia (Germany) (Fig. 25). To catch these birds—the best times are the end of May through the middle of June—it is necessary to start after 22.30 h (10:30 p.m.). Two persons and a car are needed for the operation. As soon as a calling bird has been located, both catchers approach it silently with the 12 m long mist net. So that no time will be lost, the net is kept on the poles even in the car. About 5 m from the bird both catchers push their poles into the ground. While one hunches close to one of the poles in order to take the captured bird out quickly, the other retires about 8 m from the net with the tape recorder and takes up his place toward the middle of the net. Prünte does not play the calls constantly but switches the recorder off in regular intervals. Usually both catchers will notice—especially during the call intervals—the bird approaching the tape recorder. The Corncrake will walk under the net and, as soon as it is noticed at the recorder, is shooed into the net. The excitement of the birds varies, and sometimes they can be caught by hand. So far 3 to 5 birds were captured in one night.

In this way we can catch many birds. Many males (all?) will react more to their own song than to the species'. This is very important if catching a specific male is contemplated in which case his own song should be recorded beforehand. Kneutgen (1969) cites the White-rumped (Common) Shama as an example. Singing the same theme is not only a prerequisite for pair formation, it also insures the stability of the pair bond. A bird's own themes coming from someone else's bill are the greatest fight-inducing provocation for a Shama, as Kneutgen has shown repeatedly. If one plays his own song back to a Shama male while he is fighting with another male, he will abandon the fight immediately, dash to the tape recorder and start pecking at the speaker. "For the pair bond this means that neither partner will sing the other's themes in that bird's presence—except during pair formation—as this would provoke an immediate attack. But as soon as the male has been away for half an hour or so, the female will call the male with his own song. The male promptly returns with loud territorial song. If the male comes back ready for combat, the female has two options depending on his distance: if he returns after a few seconds and from a short distance, the female can assume the mating position. Instead of attacking, the male then begins preening or eating.—If the male is distant and his answering song sounds faint, territorial singing ensues on both sides. When he returns, she assumes the mating position singing beligerently. This usually leads to copulation.—Lorenz observed repeatedly the female calling back the absent male by singing his song."

In this context it is interesting to note how birds are able to distinguish very similar sounding songs. Thielcke (1969) reported on the reaction of Coal and Great Tits to the song of nearly related birds from Afghanistan:

The song of *Parus ater* and *Parus melanolophus* cannot be distinguished by the human ear or by audio-spectograms (sonagrams). Consequently Coal Tits *(Parus ater)* in Bavaria (Germany) react similarly or very nearly so to the songs of Spotwinged Black Tits *(Parus melanolophus)* as to their own species song. The difference between the songs of *Parus melanolophus/ater* and *Parus major decolorans* are minimal to us. Nonetheless, Coal Tits are able to distinguish clearly their own species song from that of *decolorans*. European Great Tits *(Parus major)* react to songs of Afghan Great Tits as if they were a different species.

Stürmer uses tapes to stimulate listless lure birds or to release other birds from raptor-induced rigidity. When this goal has been achieved, the tape recorder should be switched off immediately.

Recordings and lure birds should never be used at the same time. With a tape recorder we will obtain good results for many species.

Of course, there are many opportunities to catch birds without live or artificial lure birds, particularly during migration when birds look for resting places at the border of woods, in bushes, and at lake shores. Feeding and watering stations are important and necessary means to attract birds.

1.1.3 Food and water

Food is at least as important as lure birds in many cases. Sets for shorebirds and others can often be located in suitable habitat with a good food supply. For other birds here are some general hints.

A variety of seeds are suitable for seed eaters. Often commercial wild bird food consisting mainly of sunflower and red and white millet seeds can be used. In the case of traps glueing some seeds (with non water-soluble glue) to the trigger is advisable. Seeds sprout easily in hot humid weather and trap areas have to be "weeded" periodically.

Birds can also be attracted with various plantings. Annuals, perennials and larger fruit-bearing bushes and trees quickly attract many different birds. Such plantings are valuable for all bird catching activities, be they nets or large funnel traps such as at Helgoland (Germany).

Seeds and some berries can also be spread on the ground. The feeding area should be kept largely snow-free during the winter months.

Seeds already on the ground and covered during snowfall should be stirred up with a rake or shovel. Dark patches in the snow tend to attract birds. Pull nets and bow nets are ideal sets for such conditions. If a lure bird is also used, its cage must be within the range of the nets.

Some berries may be dried for later use. It is best to air dry them as they may get too hard in the oven.

Birds do not always accept the food we put out for them. During the winter of 1926 Sunkel (1928) waited in vain for Common Bullfinches to come out of the trees to his carefully set up feeding and trapping station below. They instead were eating the seeds and leaf buds of the trees.

Having dealt with plant food, we now have to discuss animal food. The first place still belongs to the mealworm, the larva of the meal beetle *(Tenebrio molitor)*. The worms can be bought at most pet shops. If you need mealworms only occasionally (for example for bow nets), it is easiest to buy them. A box of 50 or 100 worms will go a long way. Only if you need mealworms year round in sizable quantities will it be worth your while to raise them yourself.

Mealworms should only be raised in wooden boxes, preferably at least two so that the entire culture may not be wiped out by disease and parasites. The corners of the boxes should be covered with tin or plaster to prevent the worms from chewing at the wood. The boxes must be covered to prevent escapes, but also must admit air—gauze works well. The two boxes should be set up three months apart.

First line the box with unprinted paper preferably corrugated paper. Then fill it three quarters full with corn meal and add some rolled oats. It might be advisable to heat the corn meal and oats in the oven to destroy possible parasites. To keep the mixture from compacting add a few pieces of corrugated paper to the meal. On top of all put a clean white piece of cotton (old sheet). Now put your mealworms in the box—preferably all approximately the same age to prevent "cannibalism".

The worms and beetles also need moisture. Feed them with an apple or a potato which has a number of holes bored into it with a fork or knife (make sure the holes are large enough for the worms to have access to the interior of the potato). Put the food directly on the meal mixture.

Keeping the worms warm or cool (in a heated room or in a cold garage for example) will speed up or delay their development.

Parasites of mealworms can be destroyed by putting a moist cloth over the container and warming the colony thoroughly, preferably overnight. In the morning the cloth with the parasites can be removed. Over two or three days this will result in the destruction of the entire brood of parasites. Another possibility is to expose the mealworm container to strong sunlight. Adult parasites are captured, and their eggs and larvae destroyed by the heat. The mealworm colony is not affected by this procedure.

Difficulties exist in fastening a mealworm to a trap, as the worm should not be injured.

Lötzsch (1937) tied a mealworm to a thread and squeezed the thread into the split end of a small stick.

Faust (1937) tied a mealworm to each end of a green thread around the 4th or 5th section. He attached the thread to the split end of a small stick so that the worms were barely dangling. They will live long and move vigorously.

The brothers A. and E. Fricke fastened the mealworm around its 3rd last section in the loop of a thin wire bent to fit the bait. They also use this for earthworms, feathers, etc. They sometimes use two mealworms at once. The worms move for five to six hours. The wire can easily be attached to the trigger. Sunkel (1937) used two thin wires such as a cello string and twisted them together. Untwisting them a bit results in an opening for holding the mealworm. The wires are then carefully twisted tight to hold the worm without injuring it. Sticking the wire in a bit of cork attached to the trigger is easier than trying to force it into wood.

A "contemporary" method of baiting with mealworms is to put a few individuals in a small box of clear plastic.

J. Winkler, bander and maker of traps for many years, uses a 50 mm long and 5 mm wide "mealworm holder" (Fig. 26). He uses it mainly for bow nets but it is also suitable for other traps.

Some birds take to mealworms right away, while others refuse them altogether. The worms should always move so that birds will recognize them as food. Some birds may prefer small earthworms to the mealworms.

Other bait includes moths and butterflies, grashoppers and crickets, large and small beet-

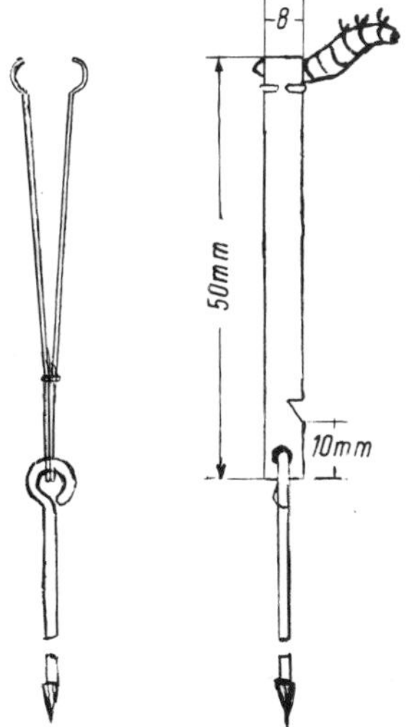

Fig. 26 Mealworm holder. The mealworm is always fastened lying on its back.

les. Buse (1927) trapped Northern Wheatears with a bow net (60 cm×60 cm) and baited with ants. Sunkel (1927) reports swarming ants getting into a funnel trap on the island Mellum (Germany). Many songbirds in turn were caught.

Bait which attracts insects (such as honey) should be placed so that birds cannot reach it. This prevents dirty and sticky feathers. Flies should be kept in clear plastic containers (with a few holes) and sheltered against sunlight.

Bait for raptors includes Common Starlings, House Sparrows, Rock Doves, mice, and hamsters. Frogs can be used as bait for herons, etc.

An unusual but nonetheless useful bait for seed eaters is a salt lick. Doves prefer the taste of anise. Crossbills seem to like this particularly. To get at certain minerals they will also eat old mortar or plaster, ash and urine-soaked snow (Hayder 1954, 1960).

Soviet ornithologist Gavrilov (1968) noticed many Common Rosefinches and European Goldfinches flew to those spots in a village on a small lake in the Kazakh SSR where they could find salty spots on the ground. They especially liked to visit the area where villagers salted down their fish catches.

The ornithologists mixed two buckets of earth with 1 kg salt on the ground. They then poured water on it and stamped it solid. They had located this "salt lick" in a wood of young trees and the Common Rosefinches began using it the same day. Between 27 June and 25 July 1966 Gavrilov and his colleagues trapped 863 birds (392 males, 260 adult females and 211 grey birds, probably second year birds) with mist nets. They did not trap every day. On 18 July they trapped from 7 a. m. to 7:30 p. m. (19.30 h), but generally they only trapped one to five hours mostly during mornings. The largest daily total was 254 birds on 18 July. Recaptures were rare.

The behavior of the birds was interesting. Their enthusiasm for the salt lick was so great that they were not to be deterred. Often a bird would bounce out of the net, sit on the ground before it and try again after a while. Some birds tried four and five times to get to the lick before getting caught. Others sat before the net for some time, then walked underneath it to the lick. Greatest attendance at the lick was between 5 and 10 a. m.

1.1.4 Camouflage of traps

In many instances additional camouflage is not necessary if the equipment already has a color that blends into the environment. This is usually the case for nets and funnels; traps and clap nets often need only minimal disguise. Corvids and gulls are very suspicious.

We will give additional details on camouflage of catcher and equipment later.

1.1.5 Projectiles

Projectiles aim at scaring the bird which will then not fly up and away, but dart, in straight flight, into set nets and funnels. Catching birds in a roccolo, the Italians throw a round woven bast mat on a stick or a stylized accipiter (Fig. 27). At the banding station on Helgoland, Germany, throwing disks have been used for years. They are made of galvanized steel, measure 20 cm in diameter and are 0.5 to 1.0 mm

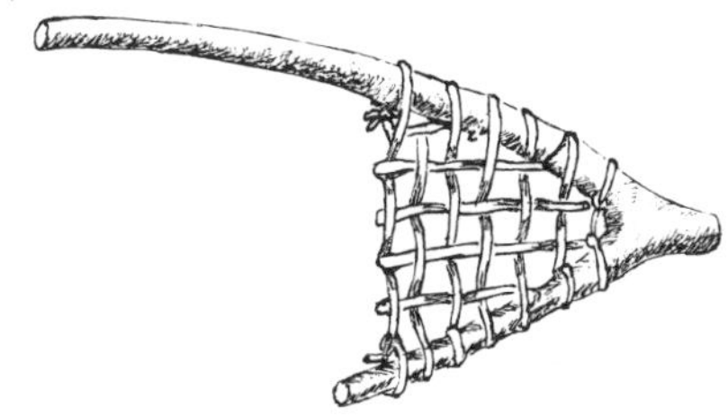

Fig. 27. Stylized sparrow hawk as projectile.

thick. According to Drost (1932) they have the advantage of being easy to store in large numbers and they are also easy to throw. In addition, they are relatively easy to retrieve as they rarely get stuck in trees and stand out in color from the surroundings.

Other catchers throw a stone wrapped in white cloth into the air or carry a long stick with the cloth fastened to the top.

The artificial accipiter is well known. It is made of cardboard and glides on a wire from a high perch to a lower one, thereby frightening birds. It is worthwhile to carry such a device while driving birds through bushes or along hedgerows.

1.2. Clothing

It is not advisable to wear bright clothes while catching birds. Sturdy shoes or rubber boots are a must. Anglers' waders are necessary for some catching procedures. These waders permit entering waist-high streams. Protruding buttons on clothing can be rather annoying when working with stationary nets.

1.3 Binoculars

Binoculars are indispensable for the bird catcher. Often it is necessary to check out the catching location from a distance. Binoculars should have a magnification of at least 6. The well-known publication "British Birds" occasionally gives its readers further information: vol. 71, 1978, p. 429-439; vol. 76, 1983, p. 155-161; vol. 78, 1985, p. 167-175.

1.4 Dangers to the bird bander

Two dangers to the bander may arise from his activities: the catching equipment itself and the environment. Large bow nets as well as pull nets and double clap nets with springs require caution. The force of their bows can be great enough to break bones.

Other dangers abound in the environment. Wading in marshes and water requires caution. Banders should watch for broken glass, tin and wires in places of public access to water. Wear rubber boots or at least sturdy sneakers when wading. Slimy and moss-covered stones are another hazard in water. Before working under the earth walls in which Bank Swallows (Sand Martins), Common (River) Kingfishers and European Bee-eaters build their nests, make sure there is no danger of a sudden slide. A horse and wagon team was once buried at a Bank Swallow colony.

Banders can suddenly sink in the water. Roping up may be necessary. Leeches live in some marshes and ponds. Do not pull them off by force thus causing an infection. They will fall off as soon as they are full. Kuhk and Sonnabend though did pull them off without ill effects, but heavy bleeding occurred. Others frighten them with burning cigarettes or a few drops of alcohol. Poisonous snakes and rabid animals create further dangers.

Climbing trees requires care, experience and a good physical condition. One should be alert to sudden attacks by parent birds. William and Coan (1973) told of the repeated attacks of a female Common Buzzard which refused to accept a blind. The attacks on the hatching day of the eggs were launched noiselessly on the observer leading the way and caused three deep bleeding wounds in the neck and a deep hole behind the ear. Once Coan was attacked while sitting under a tree and changing films. In order to get close enough, the buzzard flew lying on her side, the lower wing tip almost touching the ground. She missed Coan's ear by only a few centimeters.

Sudden fright can result in falls from ladders during checks of nesting holes and boxes. Old birds, fledgling young, hornets, wasps, bees or bumblebees may fly out when one approaches or opens a nest box. The initial shock is usually more dangerous than the actual danger itself, but bees, wasps and hornets do attack—the last almost always and directly. They even pursue their victim a short distance. Immediate flight is best and leads quickly out of danger: do not try to slap or drive them away! Return later for things left behind.

Ladders used in forests for the checking of nests or nest holes should always have metal spikes at the lower end to prevent slipping on the forest floor.

Rock climbing in mountains and on coastal cliffs is dangerous. Banders catching birds on extensive mudflats during low tide should carry a compass day and night. Sea fogs rise quickly, and whoever loses his orientation there is in gravest peril.

Captured birds may also injure banders. The talons of raptors can cause toxemia—always carry a disinfectant. Herons, bitterns, cormorants, grebes and similar birds will thrust after face and eyes as fast a lightning. When banding owlets, the parent may attack and cause head and especially eye injuries. The Tawny Owl is notorious for such behavior, particularly just after the owlets have left the nest hole (and may still be easily caught by hand). During the day chances of an attack are much smaller than at dusk.

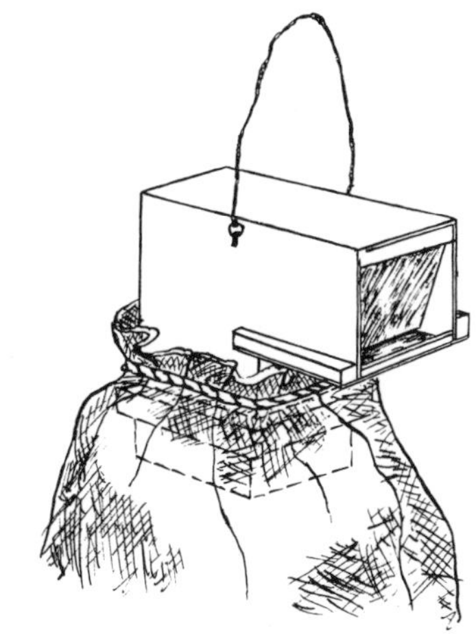

Fig. 28. Special box funneling captured birds into bags. After Drost, 1930.

2. Holding birds after catching until banding and release

2.1 Holding birds

Each bird caught for banding must be released immediately after it has received its band and after the necessary entries have been made in the banding records. In those instances when many birds are caught simultaneously, they must be kept in suitable places. Bags made of thin linen or curtain material are best. The weave must be porous so that the birds will not suffocate. Bags 20 × 30 cm in size hold 5 to 7 small birds easily and, lying on the ground, up to 8 or 10. The bags must be cleaned regularly and washed from time to time. They should be made of white or brightly coloured material which will prevent the bander from stepping on them accidentally. While banding Bank Swallows (Sand Martins) Bub often had 30 to 40 bags full of birds. Bags of different colors could be useful in population studies. Individuals to be examined more closely are easier to keep track of. Or one could number the bags instead of having different colours.

Large catches need special adaptations. Drost (1930) first used bags at the banding station on Helgoland (Germany). To avoid the time consuming tying and untying, the empty bag has a wooden 'door' (Fig. 28). The birds enter on one side through a hinged drop door. The vertical section of the box opens upon a bag which is tied around this opening. A wooden strip prevents the bag from slipping off even when full. Bags containing a number of birds are put down flat on the ground so that they have enough room.

Birds caught in cool or cold weather should not be kept in warm rooms for long periods as the sudden temperature change may be unhealthy for them. A short stay in a warm room is harmless though. It is extremely important that birds potentially dangerous to each other are not kept in the same bag. Each shrike, for example, always gets a bag to itself. European Greenfinches, Bramblings, crossbills, and Hawfinches must be separated from small birds and each other as they may inflict considerable injury and even death on others. Even tits, especially Coal and Blue Tits, may be dangerous. It is also advisable not to keep small birds confined together with larger ones such as thrushes.

Birds with long wings and tails (cuckoos, falcons, etc.) should never be put in bags as the tips of the primaries and/or rectrices may be bent or broken.

While bags of varying sizes are useful, wooden cages are preferable for regular banding activities at permanent catching places. Such boxes have been used successfully for years on Helgoland. The holding cages as described by Drost are 50 cm long and 15 cm high and wide (Fig. 29). The partitions are made of wood, the bottom of fine wire mesh and the ceiling of oilcloth. At one end is a hinged drop door that opens to the inside and falls closed by itself. At the other end, near the cage floor, is a thick clear plastic strip letting light into the cage to

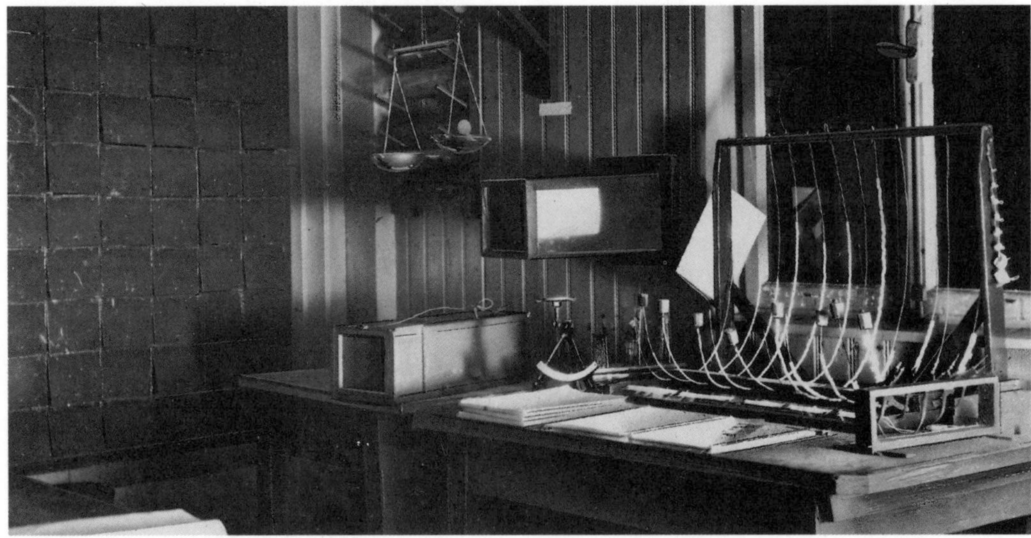

Fig. 29. Banding room of the Ornithological Research Station on Helgoland until 1945. The features shown here: the large holding cage with the many individual compartments for birds on the left. On the left hand table is a transporting cage, above that the "freedom tunnel" for banded birds. This particular set-up for keeping bands proved very convenient. The bands can quickly be taken off the wires during banding. Corks on the wire ends prevent bands from slipping off during breaks. Photo: R. Drost.

lure birds away from the entrance. Otherwise the cage is dark so that the birds will be quiet. Taking birds out of these boxes is much quicker than removing them from bags. Holding cages made of wood are preferable to those made entirely of wire or gauze, as the birds sit in the dark and see neither the catcher nor their surroundings.

The box size mentioned earlier is intended mainly for small songbirds. For thrushes the box may be twice as wide and may have two entrance doors. In this case caution should be taken that no bird escapes when adding or taking out others. In addition, watch that the wind does not push the doors inward. At the banding station on Helgoland these cages are used not only for holding but also to drive birds into from a funnel. Of course, care has to be taken that thrushes and warblers or other delicate little birds do not occupy the same cage. The larger birds might trample the little ones.

Another possibility for keeping large numbers of birds between catching and banding is a large array of holding cages such as those attached to the wall at the banding hut on Helgoland (Fig. 29). Each of the 80 compartments here was 20 cm wide, 15 cm high and 20 cm deep. These openings are rather large and care must be taken that there are no escapes. A variety of compartments with smaller openings may prevent this problem.

When catching Herring Gulls during nesting time, Bub uses boxes with two compartments

Fig. 30. Shown below is a holding cage for larger birds (2 compartments), on top is a transporting cage for larger songbirds. The flap door is on the right, a sliding door—left rear—for releasing birds after banding. Dimensions for the lower cage are 65×60×30 cm (for Herring Gulls), the upper cage measures 48 × 48 × 18 cm. Maximum width should not exceed 30 cm. Photo: A. Präkelt.

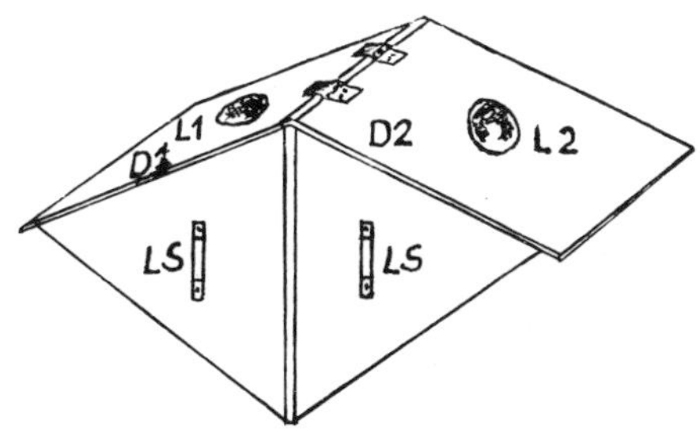

Fig. 31. Two compartmented holding cage; D 1 and D 2—hinged covers; L 1 and L 2—holes in cover for taking birds out; LS—leather loops. After MOGALL, 1956.

(Fig. 30) so that the first bird caught can be kept isolated until its mate has also been captured. It is important that the first bird cannot hurt itself in such a cage during the waiting period. Therefore all cages should have a soft ceiling (burlap or oilcloth). Birds the size of a Northern Lapwing or larger (sandpipers, gulls, etc.) should never be kept in bags as the feathers may be damaged and the birds themselves appear very distressed by them.

The two-compartment container (Fig. 31) developed by MOGALL (1956) to hold nestling birds for banding seems very practical. It was originally created for banding the young of hole-nesting birds, but also proved useful in banding other nestlings. The box consists of two equally large compartments separated by a vertical wall. For each compartment a rectangular wooden board (D1, D2) fastened to the dividing wall with hinges serves as roof and lid. Each lid has a hole (L1, L2) the size of a fist in the middle. A cylindrical sleeve is fitted to the inside of each opening. A belt can be pulled through two leather loops on one side wall. When taking nestlings out of their nest boxes, the bander wears the box on his back and puts the young birds into the right compartment. On the ground, the young birds are taken from the one side of the box, banded, and put into the second compartment. Now the box is worn in front and the nestlings can conveniently be returned to

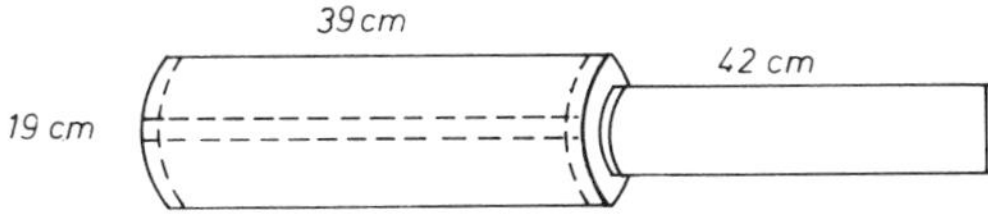

Fig. 32. Australian holding cage. After LANE, 1965.

their nest holes high up by taking them out of the right compartment with the right hand. The diagonal outside walls are covered with plastic sheeting on the inside thus preventing the birds from climbing up. A height of 20 cm and a width of 25 cm is large enough for small songbirds.

LANE (1963) describes a simple container (Fig. 32) used by Australian banders to hold small birds. The cage consists entirely of wire flyscreen. The screen should be sewn by hand with a sturdy thread and short knotted stitches. LANE gives the following instructions: Cut 2 pieces of wire mesh 26×26 cm each and 1 piece 39×55 cm. Form the larger piece into a cylindrical shape over a can with a diameter of about 19 cm. Sew the material along the overlapping side. Now sew one 26×26 cm piece onto the left side of the cylinder after having inserted a stainless steel ring. This will help the container keep its shape. Remove the can before finishing the right side in the same manner as the left. Cut a round opening about 12 cm in diameter into one side and fit a sleeve to it (Fig. 32, on the right). The inside of this sleeve must be smooth so that neither birds' feathers nor feet will get caught. The measurements may be increased for larger birds.

This container is light, well ventilated, can hold many small birds, takes up little space, and is fairly sturdy. LANE keeps several of these containers on hand to separate birds according to size and other criteria. He uses a dark cloth as a cover. Cleaning these cages, though, seem rather difficult to us as this type of screening does not permit excreta or feathers to fall through. They will become very dirty quickly. For this reason the floors of holding and gathering cages on Helgoland are made of hardware cloth only. In addition, wire screen is dangerous to small birds as claws may break or even be pulled out.

Holding gallinaceous birds requires special techniques. Before putting them in boxes or holding cages—if that seems necessary—the American ornithologists F. and F. HAMERSTROM pull the lower half of a child's stocking over the birds' heads. The upper end of the stocking has an elastic band. The birds remain perfectly quiet in this manner. In rare cases, if the top of the cage is not soft and yielding enough, these birds will crack the skin of their heads on it. The skin can be sewn back together with a sterilized

Fig. 33. A device for handling shearwaters. After SHALLENBERGER, 1971.

needle and 2 or 3 stitches. The wound will heal quickly.

During the colder part of the year, wet birds should be dried in a warm room. They should stay there only as long as absolutely necessary.

The method used for holding raptors such as hawks and falcons at Cedar Grove on Lake Michigan (U.S.A.) is surprisingly simple. Small accipiters fit nicely into long sausage cans. For larger hawks like buteos, two empty coffee cans were taped together after the bottom of one had been removed. Sometimes 20 or more raptors were kept, lying in semidarkness and without starting to struggle. A very nice arrangement with the added benefit of not disturbing the plumage of the birds in the least.

2.2 Handling during banding

The many and often rather different groups of birds must be handled according to their specific requirements. Basically, each bird must be held so that it does not feel unduly constrained.

Small birds should be held in the 'hollow' hand. They should be held lightly so that they don't get too hot. Of course, the bird should not escape either. Wrens on the other hand must feel, according to RUSCHKE (1963), that they are being held. A slightly relaxed hold will immediately induce the bird to struggle wildly for freedom. Which bander has never had a wren escape? In a systematic banding program escaped birds are a great loss, particularly since they often avoid recapture. L. v. HAARTMANN insured against such instances during his study of Pied Flycatchers in Finland by banding inside a net cage.

All raptors, snipes, gulls, etc. should be grasped around the abdomen while stretching the legs toward the back. Be careful of the talons of raptors which may cause toxemia or other severe infections. Iodine should therefore always be around. Raptors usually do not bite. If necessary, after catching raptors, give them a stick to grasp and then quickly and firmly take hold of them. Do not hesitate. Be very careful with cormorants, herons, grebes and related species as they may stab toward the face and eyes with lightning speed. Large gulls should also be handled carefully as they sometimes snap quickly at nose, lips or the arteries on arms. Sturdy leather gloves are useful for some species.

Ducks and geese are best grasped with both hands. Special care should be taken that the feathers on their breasts and bellies are not ruffled. Their water repellancy may suffer otherwise. This applies also to other similar water birds. Grouse and gallinules (moorhens) should be grasped by their feet.

Having a bird lying on its back while banding is the best method, although basically there is nothing wrong with holding the bird in a different position (Fig. 34). Exceptions are large birds such as herons, storks, etc. if two people are banding the bird. Lying on its back, the bird is usually quieter. Point the cloaca away from you to avoid getting dirty. It is advisable to wear old clothing and possibly a rubber suit when handling large birds.

Some species peck with their bills and bite into fingers. The pinching and pecking of tits is bearable, but bites by larger finches such as Hawfinches, cardinals, grosbeaks and the like as well as shrikes can prove to be very uncomfortable. Take their heads between forefinger and thumb immediately so that they cannot move. Put larger birds such as crows and gulls upon their backs on your lap and stick their heads under your arm. There they can work on your jacket. One can protect oneself even better by carefully putting a rubber band around their

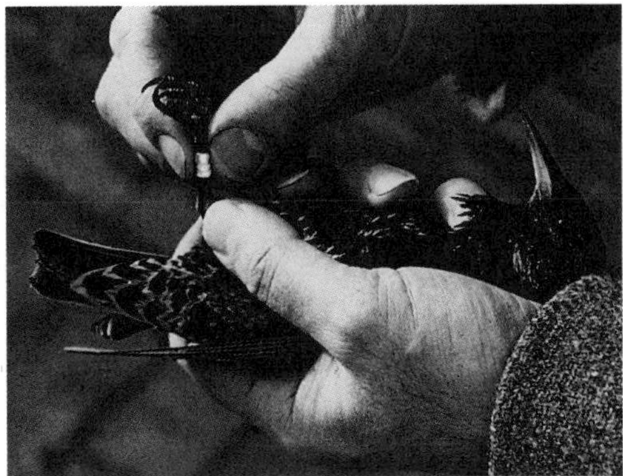

Fig. 34. Banding a starling. Photo: E. SCHONART.

bills. But do not forget to remove the rubber band before releasing the bird! According to LOCKLEY and RUSSELL (1953) hoods turned out to be very effective in quieting down geese, gulls, and gallinaceous birds (see also HAMERSTROMS).

When banding indoors, be careful to avoid letting birds escape from their cages or bags so that they do not fly against windows. Close bags tightly again after each bird until the last has been taken out.

2.3 Release after banding

Trapped birds should be banded immediately or as soon as possible and released. Usually, and especially with individual catches, they can be released without any precautions. The released birds must be able to find an appropriate hiding place, though, and should not be exposed to any dangers by sudden freedom, such as passing cars, waiting raptors or even members of their own species, which sometimes attack a liberated bird.

In a number of cases, however, certain precautions should be taken. Birds caught at or near their nests should not be released in the immediate proximity. Those Herring Gulls we caught with nest traps were immediately wrapped in burlap or put into a holding cage to prevent other Herring Gulls from seeing them. After banding they were released in a way that went largely unnoticed by their cohorts.

We advise against releasing banded birds at the trapping place so that they will not come to dislike it.

The report of K. MEUNIER of Schleswig-Holstein (Germany) illustrates how closely some species observe the fate of their fellows. In the spring a Carrion Crow ate from an egg injected with a narcotic and remained there unconscious. It was found soon, and a helper covered the bird immediately although no other crow seemed to be around, and took it to the house. About one to two hours later approximately 15 Carrion Crows had assembled—something that had never happened before—clamoring noisily. They had noticed everything.

During peak migration days the birds caught at the Helgoland banding station (Germany) are not released into the garden but to the opposite side. Thus the banded birds do not interfere with new catches.

If lure birds are used at the banding site, it is better not to release birds close by. Many small birds—especially the seed eaters—may be recaught immediately, although most banded birds continue to migrate right away. According to H. WEBER the ones that stay can interrupt banding by flying towards the new arrivals and enticing them to continue their journey. On the other hand, according to our own experiences, birds that stay at the banding site for a few days or weeks may make a welcome addition to the caged lure birds. Our colleague O. MARTEN noticed this especially with Twites. Releasing birds 100-300 m away from the banding site is sufficient. Of course, whether a bird stays at the banding site—provided food is available—also depends on the weather conditions. If snow covers the ground birds are reluctant to leave a feeding station.

If a flock of birds of the same species has been caught, it is important to release them again together. Small birds are not kept in cages where they can flutter about wildly and possibly injure themselves, but are returned to a holding

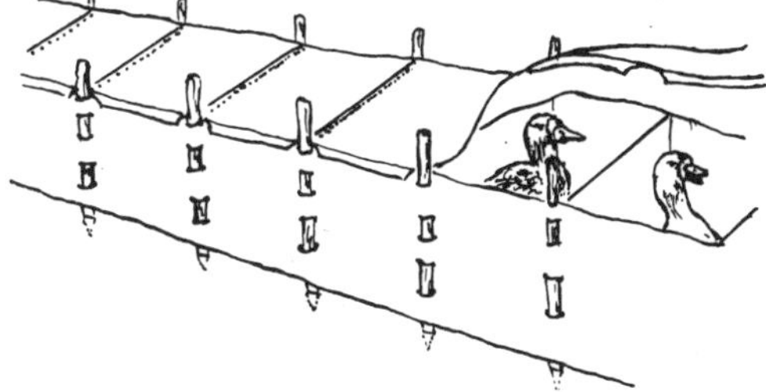

Fig. 35. Holding cage for geese. After LOCKLEY & RUSSEL, 1953.

cage or a carton with air holes and a small door.

Geese caught with rocket nets in England were processed according to these guidelines. Each banded goose was put into a 'box'. This collapsible holding cage is made of fabric and bamboo and has 25 single compartments. Each is 45×30×35 cm (Fig. 35). By unrolling the ceiling cloth the compartments can be filled one by one. The cage does not have a floor. The last thing after banding is to lift the entire holding cage, and the geese fly off together.

3. Catching methods

The human penchant for order has not passed by the many traps used for catching animals, be they mammals, birds or fish. Historians of hunting methods have classified them in different ways. The Swede S. Lagercrantz (1937) has published one version which is by no means the only valid one. Lindner (1940) added his own comments and corrections to that system. The form which any such listing takes depends mainly on the point of view of the observer.

If we consider only the methods of catching birds, we recognize two branches:

1. Methods for covering birds: First there were the deadfall traps (Fig. 36) which kill and the first primitive fall traps (Fig. 36) which captured individual birds either alive or dead. At the end of this development today we have bow nets, pull nets and double clap nets. Even further advanced are rocket and cannon nets which quickly cover large aggregations of birds.
2. Methods for holding birds: First came snares and lines of snares which were set on the ground and above it. Today we have developed a great variety of stationary nets (trammel nets, mist nets, etc.) in various sizes. Here too we have the snare catching individual birds, while in long net walls large numbers of birds get caught.

There are numerous branches from these two main stems. Cage traps for example are a branch of the fall traps as shown in Fig. 37. Even funnel traps with their fixed walls may be counted among these.

If indeed we are looking for the one root of bird catching we may take that to be 'grabbing by hand'. The bird incapable of flight—the unfledged young, the injured or the molting bird—is captured with bare hands, either by the individual hunter or his group.

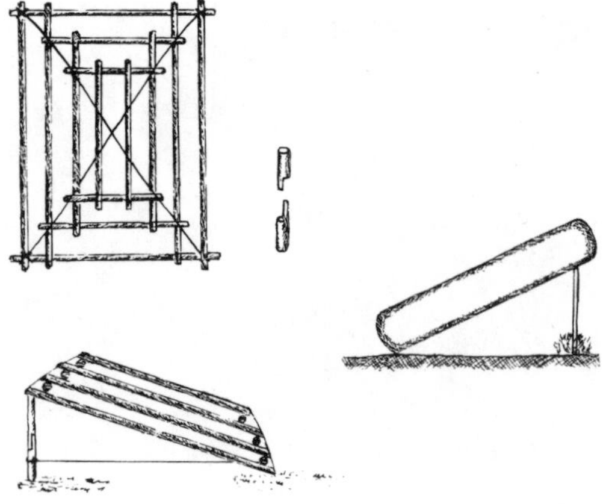

Fig. 36. Fall trap of the Guiana Indians on the right, after Roth, 1897; on the left, the deadfall trap of the Wahehe, after Nigmann, 1908.

Some additional historical information will be given for most of the traps described in this book.

3.1 Fall traps

3.1.1 Forerunners

What did the first fall trap look like? Nobody can answer this question today, but the traps of various primitive peoples show us the direction evolution took. Fig. 37 on the right shows a deadfall trap of the Wahehe (East Africa) and a simple trap made of small pieces of wood by the Guiana Indians. Lips (1927) claims that this design is probably of foreign origin, but in this context his statement is of little consequence. The construction material is no surprise as only roots, twigs, and branches of trees or other plant material were used for many years.

Anell (1960) wrote about the corapa net (Fig. 38) of the Maori in New Zealand, which may also be regarded as a forerunner of regular fall traps and that vast array of traps which 'cover'. The corapa net, 40-50 cm high, has a frame made of a U-shaped willow twig. The ends of the twig are fastened to a strong stick. The net itself is made of flax. The concealed hunter holds the net almost vertically with the help of a string. As soon as a bird feeds on the

Fig. 37. Wangoni bird trap. After LIPS, 1927.

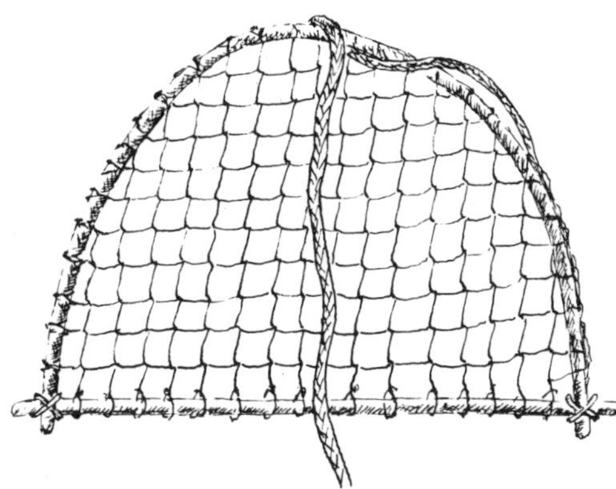

Fig. 38. The corapa net of the Maori. After ANELL, 1960.

bait in front of the net, the hunter releases the string and the net covers the bird.

AITINGER (1653) shows us an improved corapa net (Fig. 39), an intermediate stage between it and the double clap net or the pull net. We have to note that, surprisingly, similar methods were developed independently of each other in different parts of the world.

Apparently not all native peoples were familiar with fall traps. HAVESTADT (1929) reports from Abyssinia that European fall traps were quickly adopted in order to catch weavers, waxbills, whydahs, Red-cheeked Cordon-bleus and Cut-throats as well as many starlings. The wire mesh frames were operated from a hiding place. The lure bird was kept in a small cage or else was tethered.

Not much is known of the development of fall traps in Europe. It seems certain that similar types were used since medieval times. MÈRITÈ (1942) talks about a thrush trap from Normandy (France) (Fig. 40) but nobody knows whether it originated there or was brought back from faraway countries. This is a particularly relevant question in coastal regions.

In his painting 'Winter landscape with ice

Fig. 39. A "pull net" of the 17th century. After AITINGER, 1653.

Fig. 40. Thrush trap from Normandy. After MÉRITE, 1942.

skaters and a bird trap' PETER BREUGHEL the Elder (1525?–1569) shows a 'fall trap' made of reeds, which undoubtedly killed the birds immediately. It resembles more the deadfall trap in Fig. 37.

It would seem that catching birds with fall traps was only practiced occasionally in the Middle Ages and the following centuries and was perhaps not more than an amusement for children and adolescents. AITINGER (1626, 1653) has handed down many illustrations of bird catching with double clap net, stationary net, funnel, cage, lime twig and noose, but no fall trap. Apparently it was not a very productive hunting method. BECHSTEIN writes in 1806 about a drop net for Yellowhammers and a pheasant trap. Both are fall traps and are triggered accordingly. The pheasant trap should be covered with linen so that the birds will not lose feathers when getting caught. After this historical digression, let us return to modern fall trap designs which are so very necessary for serious banding.

3.1.2 Simple fall traps

The fall traps in Fig. 41 and 42 are variable in size, although the minimum dimensions should not be below 60×60 cm. For normal use we recommend a size of 1×1 m. Remember that many birds visit feeding stations in winter. The success can be greatly increased at favorable locations if you have two or more traps at your disposal.

The Helgoland Ornithological Research Station (Germany) uses traps 2 m long and 1 m wide. In one such trap baited with food and a lure bird BUB caught 120 Twites within a week, half of that number in one evening. The trap was set near a roost. Such late evening catches are released the next morning as banding activities should ordinarily cease one hour before sundown.

The height of the traps is between 15 and 25 cm. Small traps should have a door either on top or on one side. Larger traps have an opening on one side for the gathering cage. Such a cage is described in detail under water lures.

Normally hardware cloth works fine as cover for a trap. Both frame and wire must receive a coat of paint for camouflage. The frames are made either of steel wire which is soldered together at the ends or of thin wood strips. The cover of the top may also be made of wire glass which has the added advantage that the food in winter is not so easily covered by snow.

The main attraction is food, especially if the ground is covered with snow. When trapping birds, the food should be put in the center of the trap. If birds need to get used to it, a bit of food may also be scattered around the outside of the trap at first. Lure birds in cages nearby very effectively show other birds the food source, especially on snow-free days.

Which species are easily caught this way? All birds coming to winter feeding stations or otherwise finding food on the ground, from European Robins to Common Partridges. Even Eurasian Kestrels have been caught when trying to strike lure birds during periods of high snow, and the undesirable Rock Doves in large cities can be captured with large traps (see also BRUNS, 1959).

Trapping can be done from various distances

Fig. 41. Set fall trap. The drawcord is shown in front.

Fig. 42. Fall trap after CROSBY, 1924. Photo: S. P. BALDWIN.

as the situation requires. The prop should be about 20-30 cm long. The trigger string must be sturdy and inconspicuous. When pulling the string, all birds should be well under the trap so that they don't get hurt.

HOESCH (1958) calls drop traps with pull-strings the surest method to catch birds in Southwest Africa. The bait is water, a scarcity. Hardly a bird will come for seeds as those are plentiful everywhere. The selected site has to be prepared for a week so that the birds will grow accustomed to finding water here. The catcher sits in his car about 30 m away, thus having a shady place. With this method one can select which of the drinking birds to trap. Of course, the trapping possibilities are limited because some birds, such as those drinking dew, are independent of watering places. Only seed and

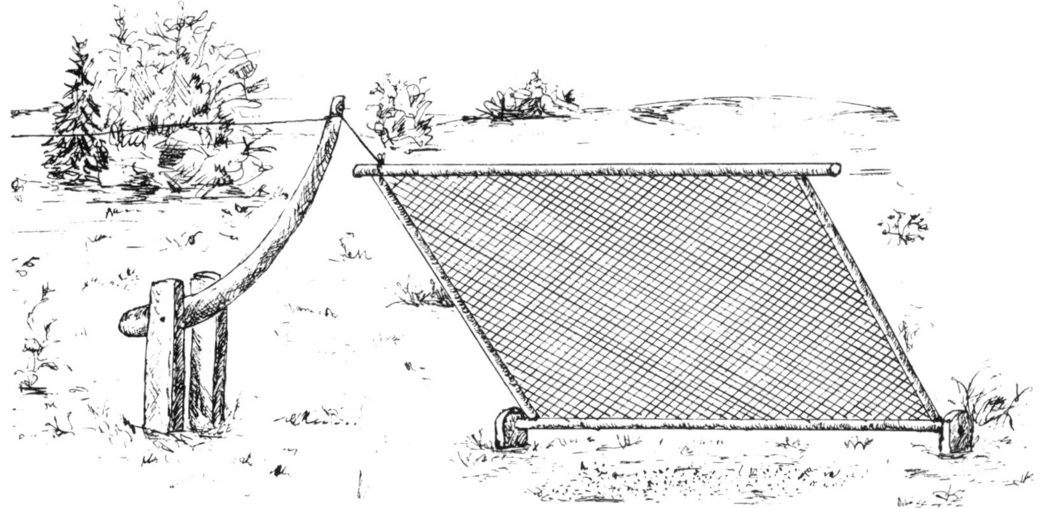

Fig. 43. Fall trap, size 1-1.5 m square. After NEUNZIG, 1927.

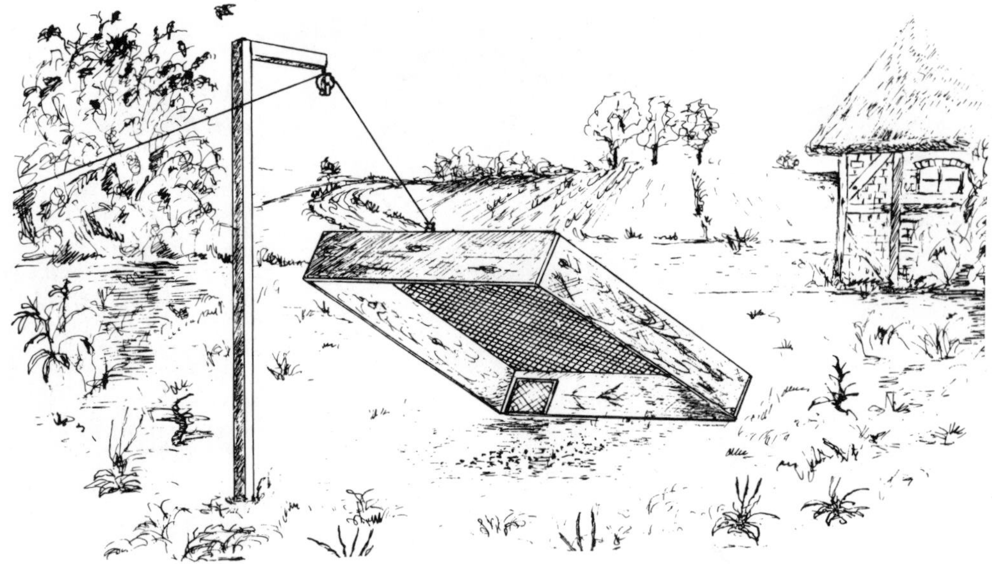

Fig. 44. Fall trap lowered over a pulley.

fruit eaters will come regularly to drink.

The traps mentioned so far are sprung with a prop and pullstring. But traps can also be sprung slowly without any sudden movements. In this case the string runs through an opening of an upright stick planted in the ground (Fig. 43) or over a pulley (Fig. 44). The latter is often the better method. Often the bird will not even notice that it has been caught and will continue to eat, drink or bathe as Beals (1939) tells us. He operated several traps from a little hut. From 1929-1933 Beals caught 2 200 birds with wooden trigger traps and had only 9 injuries. He caught 6 000 birds from 1934-1938 without mishap.

3.1.3 Automatic fall traps

These traps (Fig. 45) are triggered by the bird itself as it runs against the string underneath the trap and thereby causes the 2-piece wooden prop to collapse. The trap is set up near the edge of water along which shorebirds like to run. It is also useful for the inland bander for catching those species. This trap is easily made from a piece of wire mesh (80×100 cm). On each of the four corners a small piece (20×20 cm) must be cut out. The sides are bent down and wired together at the corners. Thus several traps can easily be made in a very short period of time.

The size does not have to be uniform. Traps 60×50×30 cm should be sufficient for most requirements. For the smallest shorebirds even smaller trap sizes may be big enough. The traps can be made with or without frames. Without frames they are lighter and are more easily released. The two trigger pieces must fit together perfectly and must be somewhat taller than the birds to be caught. The trigger string, fastened to one side of the trap, must be sturdy and inconspicuous.

Some banders use nylon strings for this purpose. Hanging some reeds or grasses over the string provides good camouflage. The trap should be fastened to the ground on one side so that larger birds cannot move it. On the island of Mellum (Germany) H. Bohlken (1934)

Fig. 45. Fall trap triggered by the bird itself. After Bohlken, 1934.

Fig. 46. Fall trap for rails. Photo: P. BECKER.

caught 154 birds within 7 weeks and with very few traps (4, later 8), among them 62 Common Sandpipers, 21 Common Redshanks, 19 Ringed Plovers and 12 Common Oystercatchers. In the region of Hildesheim (Germany) P. BECKER used this type of trap (Fig. 46) to catch small rails. The 'baskets' are made of 10 mm wire mesh and fastened onto a 4 mm wire frame. The 2-piece wooden prop is about 25 cm long (upper piece 10 cm, lower piece 15 cm) for a trap the size of 50×35×12 cm. Using a 4 cm long and 0.5 cm thick wire, one can fasten a mealworm to the front end of the string. The worm should be held by its tail and should move vigorously. BECKER makes his traps of varying sizes so that at least six nest into each other, and then carries them along in a rucksack. The largest trap is 50×35 cm, the smallest 38×26 cm. The length of the wooden props must, of course, be modified accordingly (see above).

If necessary, leads must be set up to guide the birds towards the traps. From August 13 to October 6, 1963, BECKER caught 41 Water Rails and 62 Spotted Crakes. In addition he caught Bluethroats, *Acrocephalus* warblers, White and Yellow Wagtails and Common Snipes. It is advisable to move the trap 6 to 8 m after catching a rail.

In spring and fall the edges of ponds and flooded areas are suitable. The traps should be set up under bushes whenever possible. In summer the birds are trapped in their nesting territories. The traps are most effective if they are placed directly over the paths of the rails. If there is more than one path, the others should be blocked by leads. The traps should be checked every half hour.

HOUWEN in Belgium also uses such traps, but feeds Common Gallinules (Moorhens) bread and corn daily at the same time to accustom them to the trapping site.

For a pyramid trap (Fig. 47) HOLLOM (1950) suggests the following measurements: sides 80 cm, the slats leading to the top about 70 cm each. The perch inside the trap is made of a flexible rod. The hooks, made of branches, are

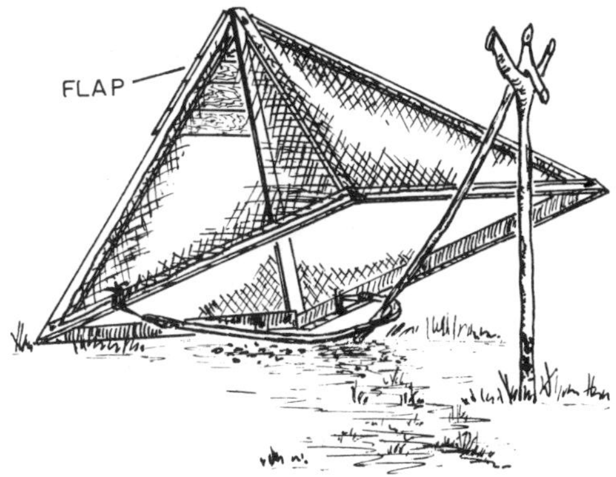

Fig. 47. Pyramid trap, triggered by the bird. After HOLLOM, 1950.

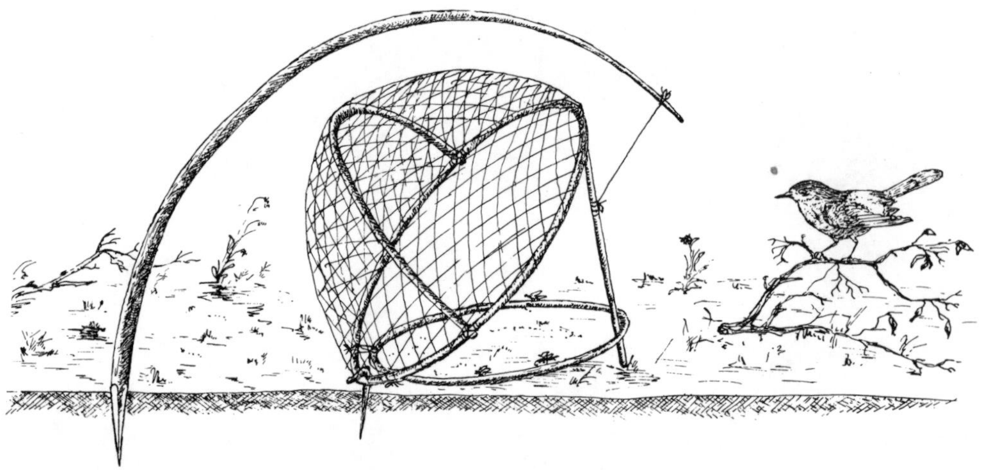

Fig. 48. An automatic trap for nightingales and other small birds. After OSTAJOW, 1960.

45 and 50 cm long. As soon as a bird alights on the perch, the movable branch is released and the trap falls shut. The trap, held by a small notch in the movable branch, can be set so delicately that even a Blue Tit can trigger it. In general, this kind of trap is suitable for crows, Common Jackdaws, jays and other corvids. Eggs are very good bait for them.

The Soviet ornithologist OSTAJOV (1960) built a small automatic trap for Common Nightingales and related species (Fig. 48). Fasten a standing bow over a wooden frame of 40–50 cm

Fig. 49. Raptor trap; a the trigger mechanism. After J. F. NAUMANN, 1826.

diameter. Tighten a narrow mesh net over this half globe. Bend a thin rod into a smaller circle and fasten it loosely at two places to the frame. Stake the net globe to the ground at one point. Drive an elastic switch into the ground in back of the net globe, bend it over the globe and tie its upper end to the support rod with a string. The lower raised circle carries the bait such as mealworms, ant eggs, beetles or cockroaches.

The raptor trap of J. F. NAUMANN (Fig. 49) is an example of an efficient trap made of simple materials (J. F. NAUMANN 1826; C. L. BREHM, 1855). The net is attached to a post about 2.4 m in length. The vertical net poles each are 2.4 m long, the upper horizontal bar is 1.5 m. The two vertical poles are fastened at the bottom to a 1.8-2 m long 'roller', a somewhat thicker wooden bar. The lure birds (a dove or several sparrows) are kept in a very fine wire cage out of reach of raptor talons. A nice example to show that ornithologists were already concerned with the safety of their lure birds 200 years ago. As soon as a raptor dives down to the cage, which hangs just above the ground, he pulls the tip off the trigger mechanism (see Fig. 81 and 81 a) and the net, pulled in addition by a large stone, falls immediately, covering the raptor.

3.1.4 Other types of fall traps

HOLLOM (1950) and HOLLOM and BROWNLOW (1955) describe another fall trap. (Fig. 50) It consists of a rectangular net fastened to bamboo poles on three sides. There are no sides. 1.20×1 m is a practical size. The longer bamboo pole connects with the two poles at the sides. The rear of the net is weighted down with stones or fastened to the ground with hooks. Birds will hop under the net quickly as it is open on three sides. A disadvantage is that the net moves with the wind. Taking out birds can be done with a

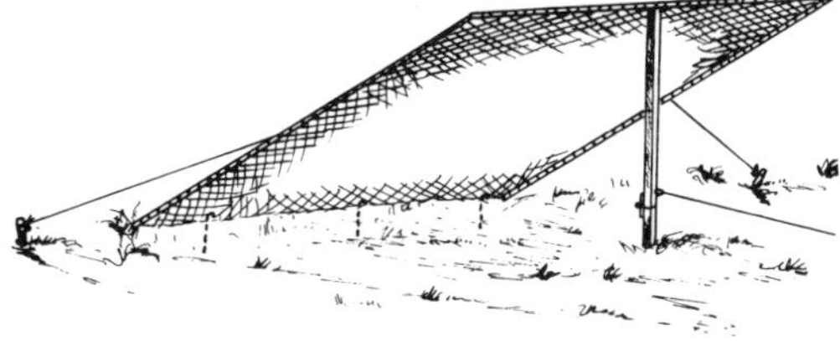

Fig. 50. Trap without sides made out of netting. The rear of the net is weighed down with rocks or fastened with hooks. The size is approx. 1.8 m square. After BROWNLOW, 1955.

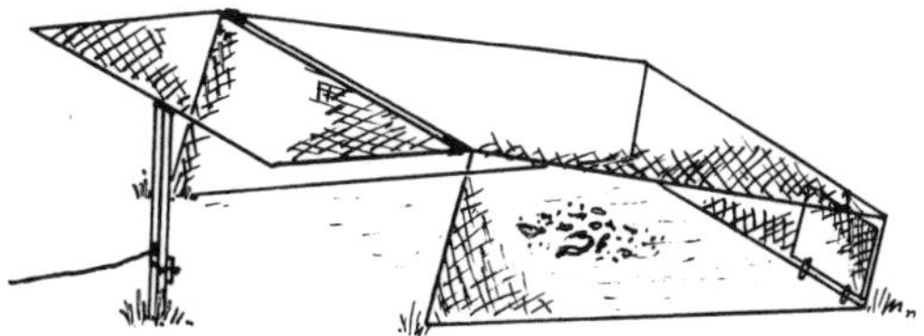

Fig. 51. Trap with front flap. After HOLLOM, 1950.

Fig. 52. Combination trap. After HOLLOM, 1950.

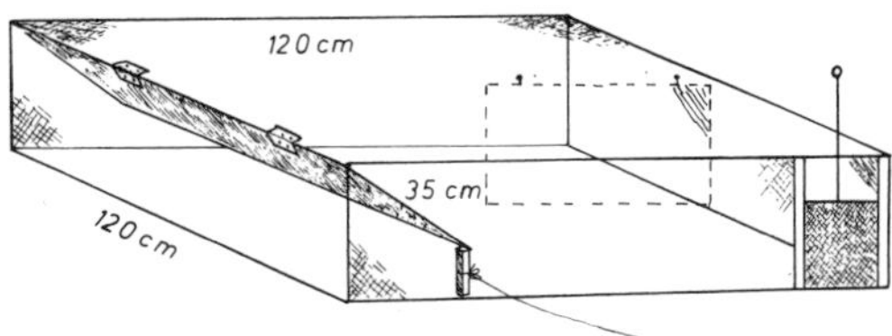

Fig. 53. Trap with drop door closing from inside. Ebba News, 1963.

small landing net or simply by hand. This particular trap is easily transported by car as the side poles fold against the longer top pole and the net can then be wrapped around all three. Such a folding net can also be fitted with a complete frame.

If only one side of a fall trap opens (Fig. 51) access for birds is limited but they cannot escape sideways. With larger traps, it might be better to increase access by having openings on two sides.

HOLLOM (1950) describes a similar but more elaborate trap (Fig. 52). The upper flight entrance may appeal to some birds. The funnel entrances on the ground permit birds to enter from the sides. Usually, however, only one of these two possibilities can be used at any one time. The holding cage should be stationed in the rear near the door. If the doors are open the presence of the catcher is necessary. This trap is 3×1.3×1.5 m and can be used in the same way as other related traps.

The door of the trap in Fig. 53 closes from the

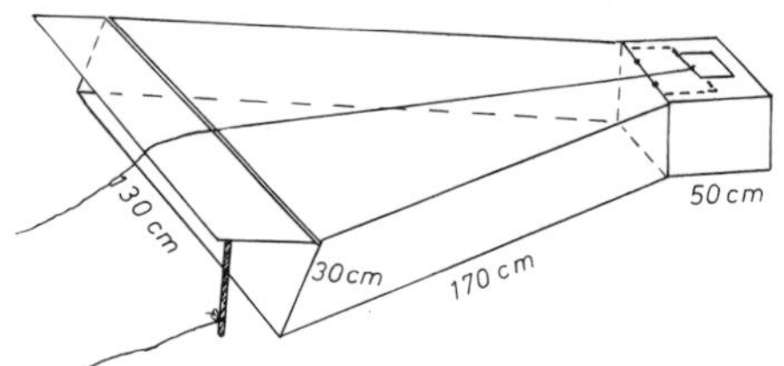

Fig. 54. Funnel trap. Ebba News, 1963.

inside thereby making escape more difficult for birds. We would hesitate to recommend this method for all occasions. The funnel-shaped trap in Fig. 54 closes from the outside. Both trap types were developed by North American banders (EBBA News, 1963).

The trap by Hollom and Brownlow (1955) shown in Fig. 55 is also funnel-shaped. This trap is 2-3 m long and can be transported. The length can, of course, be varied to fit needs.

In this context I would like to mention a trap which is not a true fall trap, but its principle recalls the trap in Fig. 51. Collenette (1929/30) set up a wire cage 1.6 m long and 0.7 m wide and high in his garden. In one winter Collenette caught the following species: Common Song Thrush, Mistle Thrush, Redwing, Common Blackbird, Common Starling, European Robin, Dunnock, Chaffinch, Great, Blue and Coal Tits, House Sparrow and Great Spotted Woodpecker. The birds were fed regularly at the banding site and trapped from time to time.

P. Houwen builds similar traps 10×1×1 m. By feeding Common Gallinules (Moorhens) several days prior to trapping he once caught 33 at the same time. The trap is sprung from a hiding place by pulling the prop away from the open door.

3.1.5 Bell-shaped nets

These are not true fall traps, but they have their logical place here. The purpose of this net is to cover feeding birds on the ground quickly and effectively. The set-up of a catching place is illustrated in Fig. 56. The 'bell' is a folding metal frame. The net is fastened to this frame. At the corners are rings for the four poles supporting the net. As soon as the guy line on top is released, the net falls quickly to the ground.

All birds seeking their food on the ground or accustomed to feeding and watering stations can be caught with this net trap. Before setting up the net the birds should already be used to the locality. Lure birds are sometimes useful.

The method is not new. In 1797 Bechstein

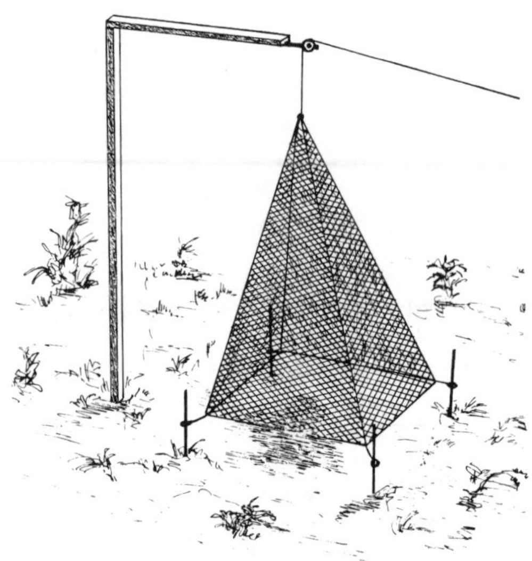

Fig. 56. Bell-shaped net.

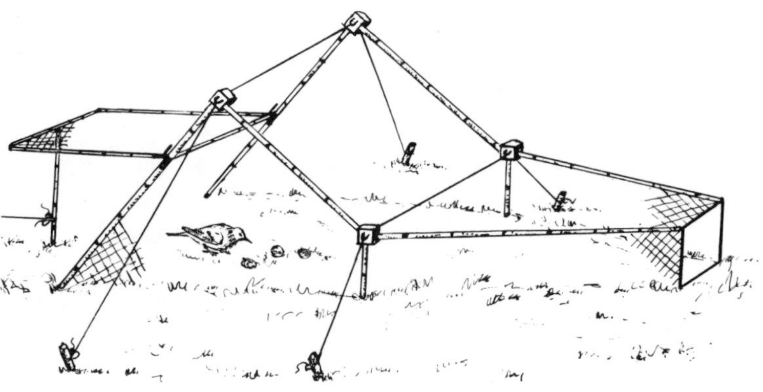

Fig. 55. Collapsible funnel type trap. After Hollom & Brownlow, 1955.

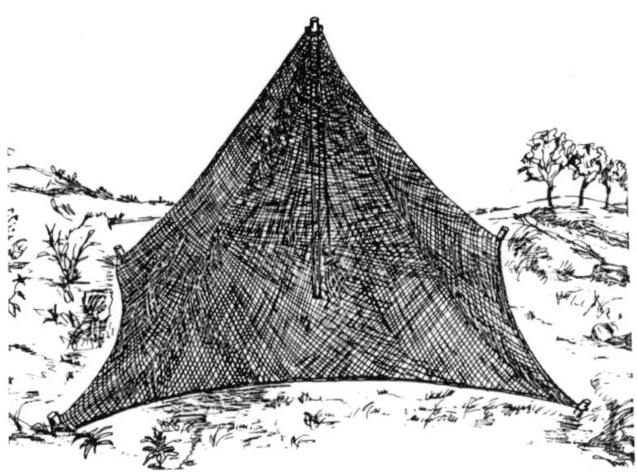

Fig. 57. Bell-shaped net. After Bechstein, 1797.

described a bell-shaped net, although of a somewhat different kind (Fig. 57). The four corners of this older net are fastened to stakes. The pole in the middle lifts the net just high enough to permit Common Partridges to pass through unhindered. The partridges, for which this net was designed, trigger it by eating or pulling on the grain bundle on the pole. The ring hangs only loosely on top after a feeding routine has been established. The net must be wide-meshed, and it seems that only trammel nets were used.

3.1.6 Window traps

Shaub (1948) developed a combination window trap and feeding station. Measurements depend upon the particular window size. Floor and sides are made of wood, the parts needing wire mesh (top and the middle partition) are indicated in Fig. 58. The birds can be taken out via the two rear doors. The trigger mechanism is shown in the drawing. The wire drop door is held by a pin guided through brass screw eyes. These are screwed into a slat. The slat is fastened to the partition with 2 ½ cm long wood screws. The pin is a brass wire 3 mm thick and 20 cm long. The door holder (1) is made of sheet brass and must be attached to the door in such a way that the pin fits through the hole when the door is open. The pin receives a brass stop (2) so that it cannot slip out of the screw eyes. (This stop can only be soldered onto the pin after it has been threaded through the screw eyes.) The 'door latch' (3), a very important part, is made of hardwood, as is the slat which is fitted with the screw eyes. The back end of the door latch is attached to the slat with a round-headed screw so that it turns easily downward with the falling door. The back end of the door latch will slide through a mesh of the wire. This prevents birds from pushing the door open from the inside. The front end of the 'door latch' should be slightly slanted and should be fitted with a small nail. Pinch off the head of the nail. The nail should then still protrude about 2.5 mm.

Drill all holes for screws in the hardwood parts first, as the wood will split otherwise. Fasten small rubber pieces at the lower ends of the bevels of middle partition and side walls. They soften the fall of the door. All wooden parts must be glued and nailed together. Attach the trap to the window sill with two 2 in. angle irons (5) and one brace (6), supporting the projecting part of the trap. The brace is supported by small wood blocks nailed to the (wooden) house wall below the window and to the bottom of the trap. To close the trap, pull at the string (4) which

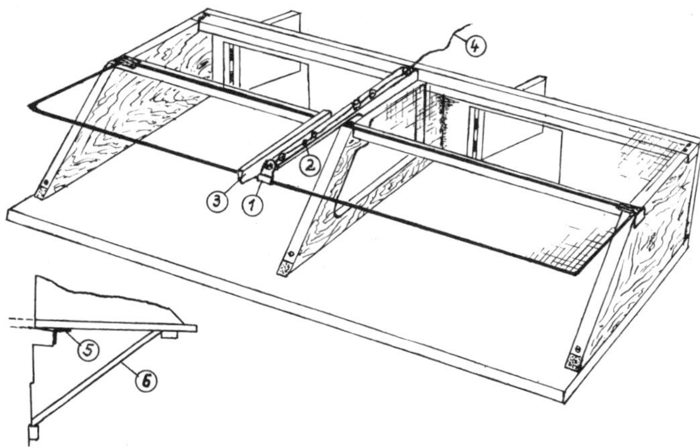

Fig. 58. Window trap. After Shaub, 1948.

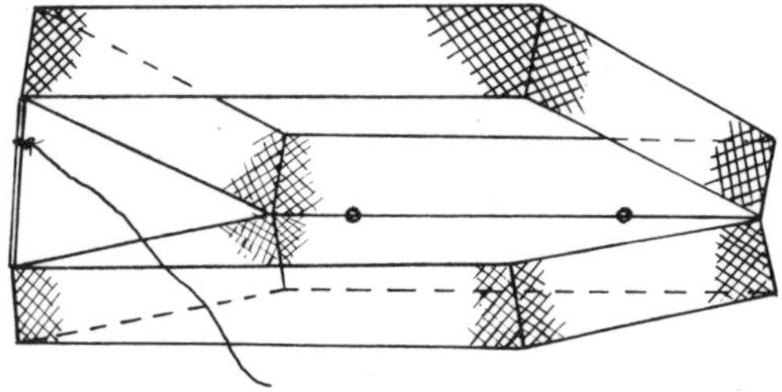

Fig. 59. Window trap. This trap can also be used on the ground. After H. RITTER.

leads under the window into the room. Glass curtains on this window hide the person in the room.

During the winter of 1969/70 H. RITTER of Braunschweig (Germany) caught, among other species, 150 Twites at the window of his apartment on the 3rd floor. He developed the trap (Fig. 59) himself. He took two wire traps 45×30 cm, secured them to each other at one side and kept them opened at the other with a stick. After birds are caught, the trap can be taken into the room. Under certain circumstances, this trap can also be used on the ground.

4. Small and medium sized funnel traps

Many funnel traps for bird catching were developed over centuries. The various ornithological research stations, especially those at European seacoasts which catch and band thousands of migrants each year, have set up large stationary funnels. Besides those there are a number of small and medium sized movable funnel traps. We will talk about the latter first.

4.1 Funnel traps for song and other land birds

4.1.1 Small funnel traps

The small funnel trap is a very successful trap for small birds. It was developed by DROST (1933) on Helgoland (Germany). The island has only very little fresh water and the trap, with drinking water as bait, therefore attracts birds strongly. Of course, water cannot be used everywhere as a lure. Often scattered food will entice birds into the funnel, especially in winter. If food is used, be careful not to scatter any outside the trap. The attraction of this set-up can be increased greatly for finches if a lure bird is present. At time it may even be necessary.

Fig. 60. Small funnel trap set up with a lure Twite. The evening primrose branches on trop are intended as an additional lure. In front of the funnel is a protective cover for the lure bird cage. Photo: H. BUB.

At first funnel traps were rectangular. Later they were made only in triangular form (Fig. 60), a design that has served very well. Unfortunately, birds can also escape again from the trap especially if they have learned from previous experience. Others get caught time and again. If you notice birds that slip out of the funnel at your approach, we recommend setting up a screen of reed, canvas or other such materials a few meters in front of the trap. One can approach unnoticed under cover of this wall and then suddenly jump out.

Many species are attracted to the small funnel trap: finches, buntings, larks, *Phylloscopus* and *Sylvia* warblers, Northern Wheatears, chats, Common and Black Redstarts, European Robins, Dunnocks, wrens, old world flycatchers, shrikes, etc. J. Sadlik used a singing Ortolan Bunting as a lure bird and caught a number of conspecific individuals.

The greatest test of these traps has been Bub's systematic banding of Twites. In the course of several years, more than 100,000 Twites were banded at 130 different sites with the help of them. Some stations banded over 1,000 birds in one winter using two traps. Ornithologists and banders Dr. D. Missbach, A. Hilprecht, K. Lechner, H. Stein of Magdeburg (Germany) and others had spectacular success during the winter of 1965/66. They trapped and banded mainly on the flat roof of a high building, the sides of which were used by Twites as roosting places. Over 5,000 individuals were caught. Once they caught 650 Twites within three days.

At banding stations where shrikes are caught the traps must be very well attended so that these birds do not kill others caught at the same time. R. Schnorr baits a bow net with a dead sparrow 2 m away from the funnel trap. Very quickly he caught three Northern Shrikes and 2 Black-billed Magpies. H. Wüllner opens the sliding door to the wooden holding cage enough for a Twite to slip through in an emergency, but so that no shrike can enter.

Even small raptors try to enter the funnel trap. The entrance holes for small songbirds should not be larger than 5×5 cm. In addition, we recommend a protective cover of fine-meshed wire for the lure bird cage.

In Wilhelmshaven, Germany, we built funnel traps 180×180 cm for Common Gallinules (Moorhens). We attached four connected hanging wires to the funnel entrance, preventing

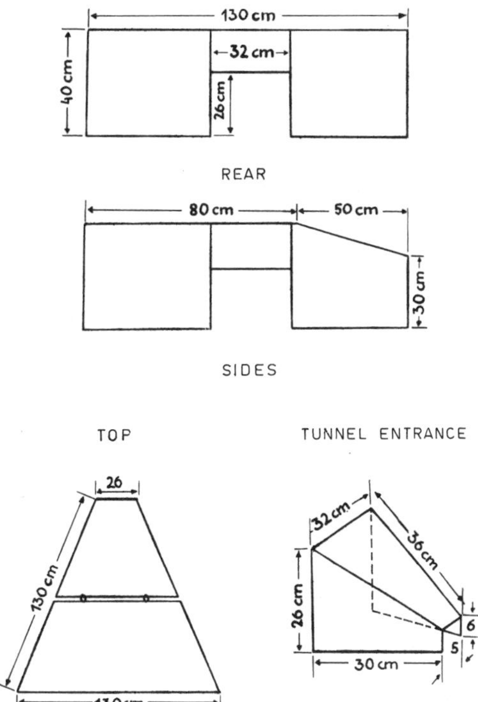

Fig. 61. Parts of the small funnel trap.

birds from escaping. One can also use similar funnel traps to catch woodpeckers at an ant hill by placing the trap over it.

To facilitate the transport, our small funnel trap can be taken apart: 1 back, 2 sides, the top in 2 parts, and 3 entrance funnels. The frame for these parts (Fig. 61) consists of 6 mm galvanized wire and should be soldered together. It can also be made of thin wood strips. Wire mesh no larger than 12 mm works fine. All parts of the funnel should be painted to protect and camouflage them. Besides the entrances at ground level, one can also add a funnel to the top as some birds such as *Phylloscopus* warblers like to enter that way.

To make leaving the trap more difficult for birds, you can fasten a circle of horizontal wires, 6-8 cm long, to the inside of the entrance. The distance between the wires is 7-10 mm.

Normally each small funnel trap should have a gathering cage into which the birds can be driven and from which they can be taken one by one. The cage shown in Fig. 60 is 50×29×28 cm. If birds are to be driven into

Fig. 62. Small collapsible funnel trap. Photo: K. GREVE.

this cage, open the sliding door (use a strong piece of string to pull it up). The rear of the cage is closed off with a thick piece of glass or, better, with some wire mesh. The birds see a possibility for escape toward the lighter side and usually fly quickly towards it. Close the sliding door again and the birds can be taken out via the opening on top. If no gathering cage is available, install a wire door on the top of the funnel trap. This opening mut be wired closed when catching birds. Taking out birds this way is more complicated. On the other hand, a small funnel trap without a cage has the advantage of being more inconspicuous. Stationary funnel traps, however, should never be without a gathering cage.

K. GREVE constructed a collapsible small funnel trap (Fig. 62). For birds up to finch size the dimensions for the three sides are 1 m long and 25 to 27 cm high. The entrances are 22-25 cm at the beginning and are placed in the center of each side. Wood strips 20×20 mm thick are best for the frame. The finished frame can be covered with wire mesh or netting. Fasten small hooks at 10 cm intervals to the top wood strip on the sides. Now the 'roof netting' can be 'hooked up' and will not need a separate frame. The three walls can also be connected with hooks. Attach the gathering cage to the open end of the trap. The entrances are worked in with the netting and thin wire. You may want to omit the gathering cage as it may be awkward to carry should you walk or ride a bicycle to the banding site. The other pieces can be fitted into a sturdy canvas bag and carried on one's back.

Finally a word about getting rid of 'trouble-makers'. If rats or weasels enter the small funnel trap (particularly if a lure bird is present), the usual traps for those species will work fine. Eurasian Kestrels, too, like to hang around the traps. Outside the breeding season they can be caught with a raptor trap and 'deported'. But such intrusions should not discourage banders from using this trap. It is easily made, cheap and one of the most successful traps. It can be set up any time of the year. During the breeding season, frequent checks are necessary so that nesting birds are not kept away for long from eggs or young. If this is impossible it is best to cease trapping for the time being.

E. RADDATZ also built a collapsible small funnel trap (Fig. 63) with 12×25 mm wood strips as frame. Make sure the funnel entrances are long enough. Otherwise the birds will easily find their way out again.

W. DE JONG, BUB's Dutch cooperator on Twites, began by using the rectangular small funnel trap, but soon constructed a round model (Fig. 64). The round funnel trap has the advantage that birds hop all the way around it

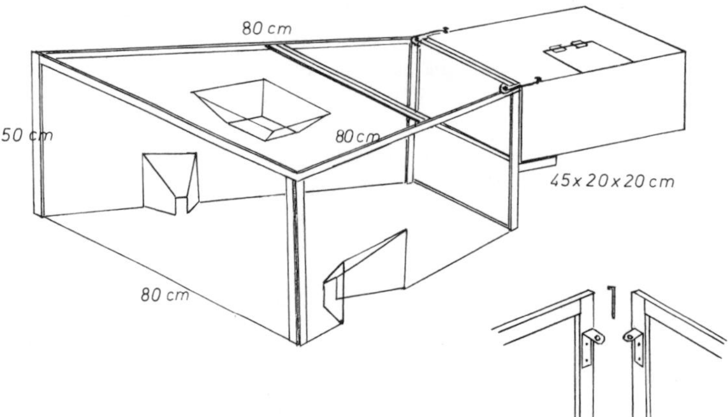

Fig. 63. Collapsible small funnel trap. After E. RADDATZ.

Fig. 64. Round small funnel trap. Photo: H. BUB.

which is not always the case with rectangular ones. The bottom frame should be interrupted at the entrances. The frame can be made of lighter material. A. KOOY (1967) built a collapsible round funnel trap. It is a variation of the time-tested rectangular version.

4.1.2 Small kidney-shaped funnel trap

This small funnel trap (Fig. 65), built by the Biological Station Steckby, is 30 cm high, 75-78 cm long and, at its narrowest section 45-55 cm wide. At this point a 12 cm long funnel entrance to the inside was installed. On the outside it is 10×15 cm, on the inside 5 cm wide by 6-8 cm high. Opposite the entrance is a door through which birds are removed. At first this funnel—with a lure bird inside—was intended to catch shrikes. The design is appropriate for this purpose. Later mostly Corn Buntings in their breeding territories were captured with another Corn Bunting as lure bird. The lure bird cage had a cover of nylon gauze.

4.1.3 North Carolina dove trap

Fig. 66 shows a funnel trap built to catch Mourning Doves. The piece of wire mesh from which the funnel is made has these dimensions: lower width 31 cm, upper width 29 cm, sides 25 cm.

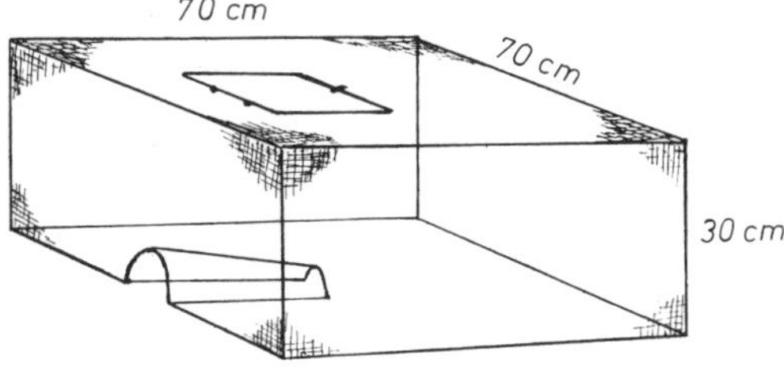

Fig. 66. North Carolina dove trap. Ebba News, 1964.

4.1.4 Sparrow traps

This trap (Fig. 67) consists of two sections. The birds enter via a funnel into the first section. Their search for an exit usually leads them through the second funnel into the second section of the trap. The second entrance ends

Fig. 65. Small kidney-shaped funnel trap. Photo: M. DORNBUSCH.

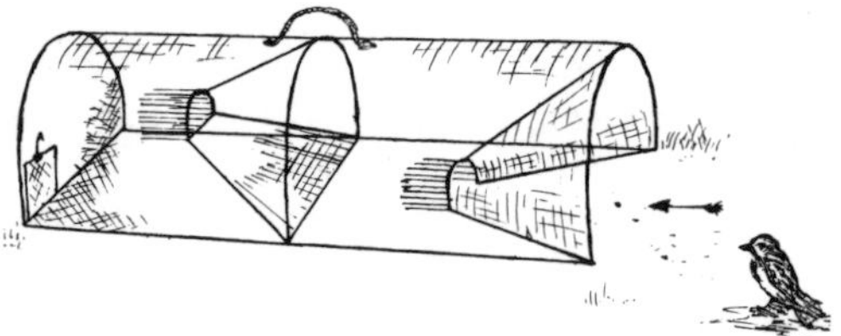

Fig. 67. Sparrow trap. After HOLLOM, 1950.

about 10 cm above ground so that only rarely will birds find their way out again. This second compartment has a wire bottom, the first one does not.

According to HOLLOM (1950) the most useful measurements are a length of 120 cm, a width of 50 cm and a height of 40 cm. The funnel entrances reach 30 cm into the trap. Beside the regular garden birds, Black and Common Redstarts, Northern Wheatears, Common Skylarks and Redwings were caught in this funnel trap. With the appropriate bait it is very reliable.

LANE and LIDDY (1965) altered this design. They replaced the one large funnel entrance with three lower ones, one on each side.

4.1.5 Waxwing funnel trap

We cannot mention every small bird funnel trap design, but would like to show a few others besides the very important small funnel trap. The Modesto trap (Fig. 68) designed by FELTES (1936) in North America, is about 70×70×25 cm. In California, from February until May 1935, banders caught about 6 000 waxwings in several of these funnel traps (the exact number of traps set is unkown). The highest daily total was 462 birds. Berries and raisins were used as bait.

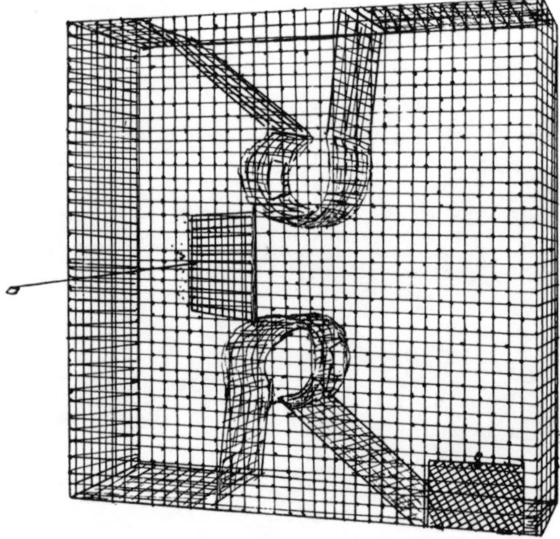

Fig. 68. Waxwing funnel trap made of wire netting, view from angle to the right. After FELTERS, 1936.

4.1.6 Small feeding station funnel trap

MALEK (1966) used this very simple trap over a feeding station. He caught buntings, nuthatches, Common Cardinals, and even a Blue Jay squeezed through the entrance (Fig. 69). The height of the trap is 30-40 cm, the diameter 40 cm. Birds are taken out through the flap door which closes with a spring.

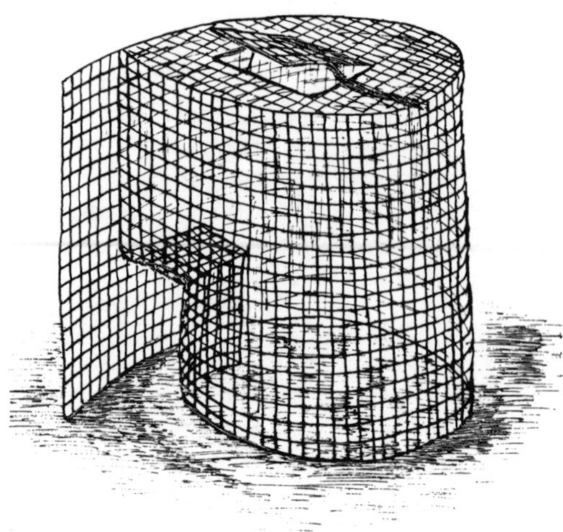

Fig. 69. Small feeding station funnel trap. After MALEK, 1966.

4.1.7 Tunnel or Mason trap

This trap (Fig. 70) named by HOLLOM and BROWNLOW (1955) after its builder Mason, is one of the best known and used traps in North America. It is easy to make and it has no bent or movable wire parts.

The important part of the trap is the tunnel leading from one side of the trap to the other. In the middle it is closed off by one vertical piece of wire mesh or two slanted pieces (see Fig. 70). The tunnel has no roof except at each end (7-8 cm). The wire mesh from which the tunnel is to be fashioned must be twice the height of the entrance plus its width. Fig. 71 shows how to make the tunnel.

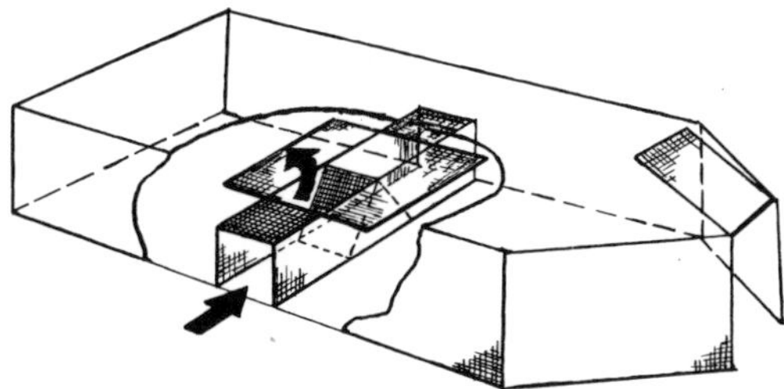

Fig. 70. Tunnel trap. After HOLLOM & BROWNLOW, 1955.

The strips (b) of the middle (open) part of the tunnel should be folded to the outside forming a shelf. Birds trying to fly from one end of the trap to the other will usually rather fly over the tunnel than sit on these shelves. The width of the tunnel can vary between 5×5 cm and 10×10 cm. Birds above a certain size cannot enter the smaller tunnel. On the other hand, a larger tunnel has the disadvantage that small birds can easily turn around and hop out. It is best to make the tunnel somewhat higher than wide. The tunnel may be in the middle of the trap as shown. Or one can install it towards the pointed end (leading to the gathering cage) thereby gaining more space for bait on the other side.—The trap should be at least 1.5×1 m in size and 40-50 cm high.

Within a short period of time R. MOHR caught 50 Bramblings while it was pouring rain. He thought that the trap worked generally well during rain. We have not heard of similar instances so far.

North American banders came up with the novel Fabian trap. Water seems to be the primary bait. Birds, once caught, cannot escape again through the single tunnel entrance (Fig. 72).

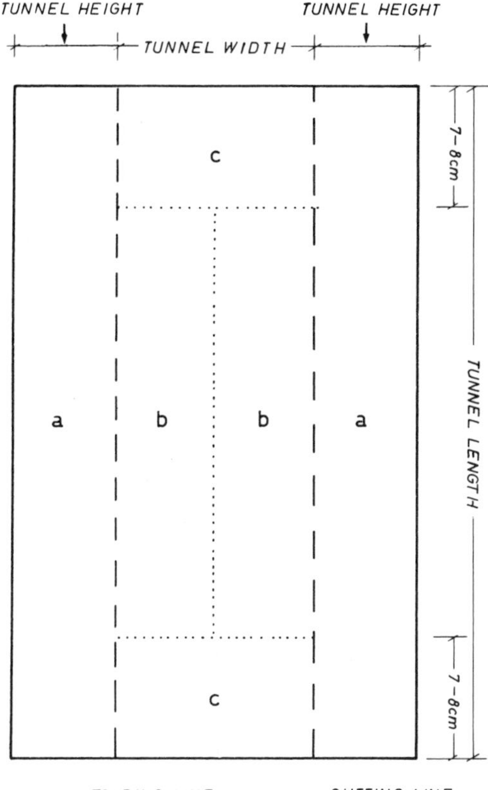

Fig. 71. Pattern for the Mason trap tunnel. (a) fold 90° down'; (b) fold 180° up; (c) remains as tunnel roof. After W. SCHLOSS.

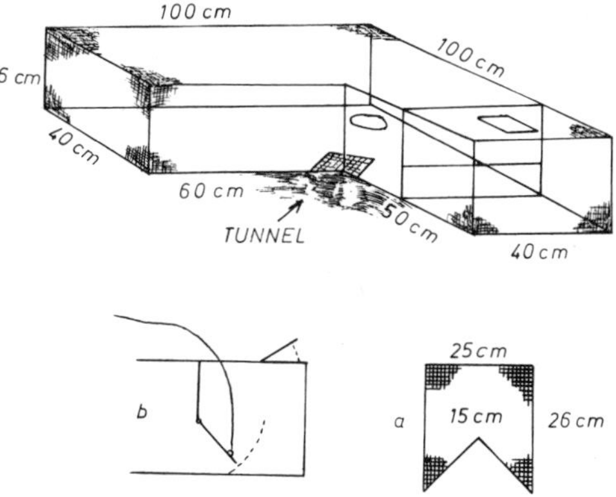

Fig. 72. Fabian trap. Ebba News, 1964.

4.1.8 Cohasset trap

This is a good trap for small birds such as warblers, etc., which are attracted to the water located under the funnel entrance. The diameter of the trap is 70 cm, the height 50 cm. A small platform for the gathering cage is useful. The cage, of course, facilitates removal of the birds from the trap. The birds can also be caught in the trap with a small landing net though this may prove difficult for some species. Except for the wooden platform for the gathering cage, this trap is made entirely of hardware cloth (16 mm mesh) and wire (see Fig. 73).

A much larger trap (diameter 1.80 m, height

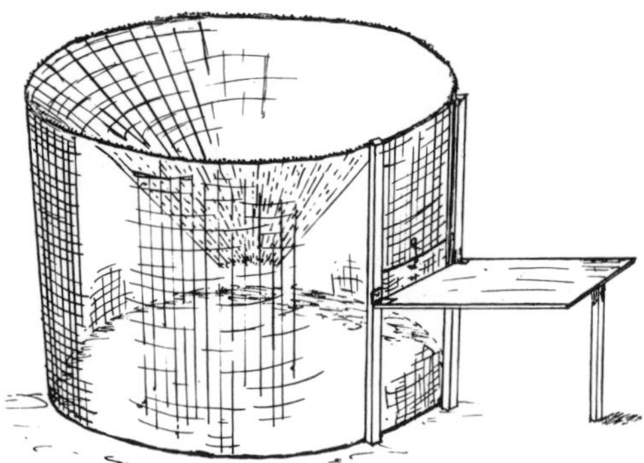

Fig. 73. Cohasset trap. After LOCKLEY & RUSSELL, 1953.

1.80 m), built on the same principle and with the same funnel-type entrances in the center of the roof, was used successfully to catch corvids. Often crows and Common Jackdaws will hesitate to enter the trap until a live crow has been place in the trap to lure them. Meat scraps make good bait. For traps this size the funnel entrance should have a diameter of 60 cm tapering down to 40 cm. The length should be 50 cm. We will deal with other ways to catch corvids in the section on large funnel traps.

4.1.9 Funnel trap for buzzards

Whitehead (year unknown) of North America describes a funnel trap for vultures which will probably work for kites as well. The trap should be set up in an open area near some trees or woods. The trap is 4.5 to 6.0 m long, 3.0 to 4.5 m wide and 1.7 m high. Cover the frame with wide-meshed wire or netting. The frame itself is made of 5×5 cm wood strips. Carrion is good bait. The birds reach the meat through a funnel which leads into the trap from the front end. The funnel entrance is 1.3 m long, and 1 m wide and 80 cm high at the beginning. At the entrance into the trap it narrows down to 40–50 cm width and 30 cm height. Up to 100 or 150 birds in a week were caught with this kind of trap.

4.1.10 Funnel traps with roof entrances

The necessity of keeping a hunting Sparrow Hawk well supplied with sparrows spurred K.-H. PETER of Cologne, Germany, to construct a very useful trap (Fig. 74). It is 50×50×20 cm. The entrance is made of chicken wire and is 17 cm in diameter. The frame is 3 mm galvanized wire. The trap has a chicken wire bottom. Using a lure bird in a special compartment or small cage will attract other birds.

For the capture of small birds on the coast of Flevoland, Holland, W. DE JONG built a trap

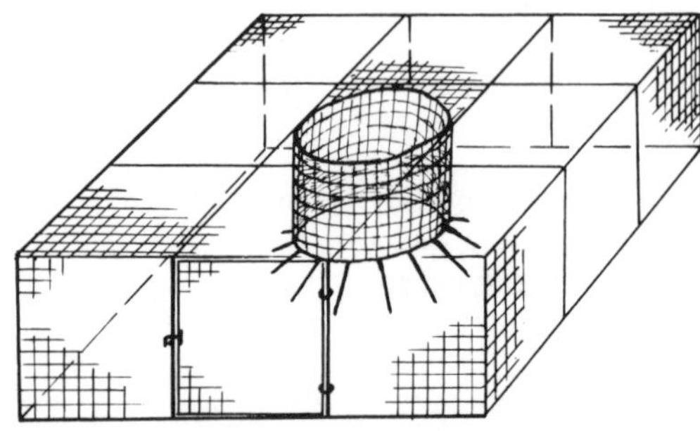

Fig. 74. Sparrow trap. After K.-H. PETER.

Fig. 75. Funnel trap with top entrance for small birds. Photo: H. BUB.

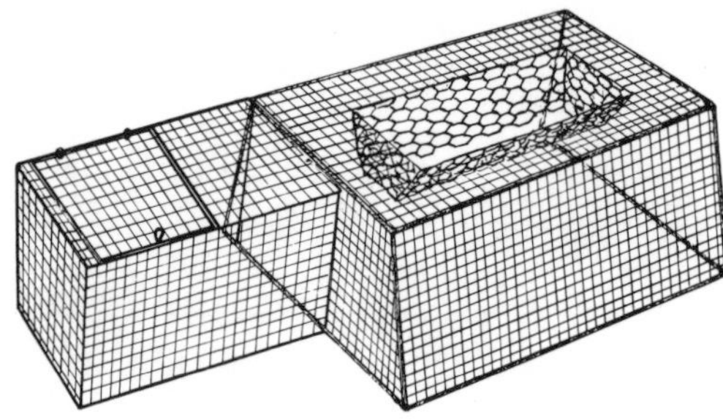

Fig. 76. Funnel trap for raptors with entrance from above.

80×30×25 cm (Fig. 75). The entrance is on top, 40 cm long and 10 cm wide tapering to a width of 5 cm at the opening into the trap.

Using the same principle, DE JONG constructed a funnel trap for Marsh Harriers and other raptors (Fig. 76). The dimensions were: length on top 70 cm, on the bottom 90 cm; width on top 50 cm, on the bottom 70 cm; height 40 cm. On top the entrance is 40 cm long tapering to 30 cm; the width is 17 cm tapering down to 8 cm at the opening into the cage. The funnel opening spreads for the entering bird. The lure bird cage is directly under the entrance and is ¾ buried in the ground. The trap must be secured to the ground with stakes.

4.1.11 Hegereiter trap

According to LOCKLEY and RUSSELL (1953) this trap was designed for Ring-necked Pheasants and Common Partridges, but can also be used for finches, doves, Common Jackdaws, crows and even rabbits (and rats, if the bottom is made of wire mesh). Fasten a grating of strong wire (for large birds the distance between bars is 3.5 cm, for small ones 1.8 cm) with hinges so that it swings to both sides. Bait birds or other small animals into the trap with raised door until they are used to picking up their food inside. Then lower the door leaving it free to swing to either side until the birds are freely walking in and out of the trap. When trapping, close the 'door' with a wedge so that it will only swing to the inside. The lower edge of the door consists of pieces of wire without crossbars. This arrangement lets the door glide over birds and animals. For large birds the trap size is 90×60 cm. (Fig. 77).

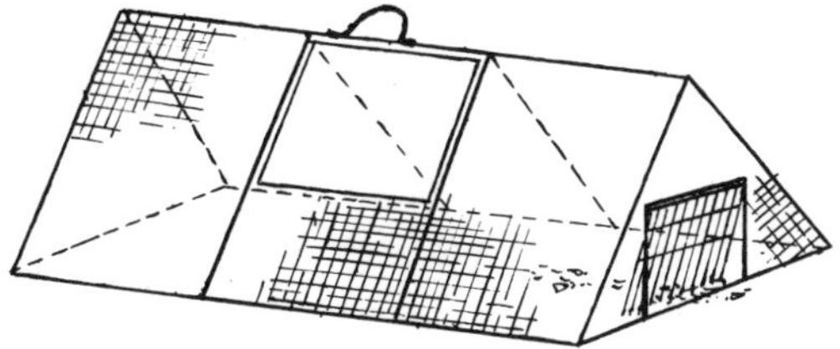

Fig. 77. Hegereiter trap. After LOCKLEY & RUSSELL, 1953.

4.1.12 Swinging wire trap for small birds

C. H. CHANNING (1964) built a similar trap. If one is trapping larger birds, increase the dimensions (Fig. 78).

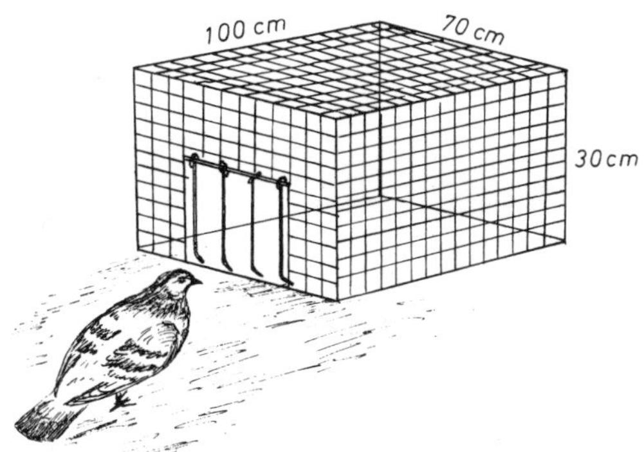

Fig. 78. Swinging wire trap for small birds. After CHANNING, 1964.

4.1.13 Weir baskets or fish traps

Amazingly, song birds and rails living in reeds are attracted to drum nets (Fig. 79) or otherwise do not fear them. AITINGER (1653) mentions catching Common Starlings in Westphalia and Bavaria, Germany, in such funnel traps. PRESCHER (1933) had many experiences with this trapping method. The fishermen of Nowe Warpno, Poland, at the Oder Haff (Baltic Sea) had up to 300 weir baskets set up to dry in which they caught 750 birds of 35 species in 1932. In all, 75 different species were captured in these traps, mostly warblers. Other birds caught (without respect to numbers) included: Common Starlings, Yellowhammers, Common

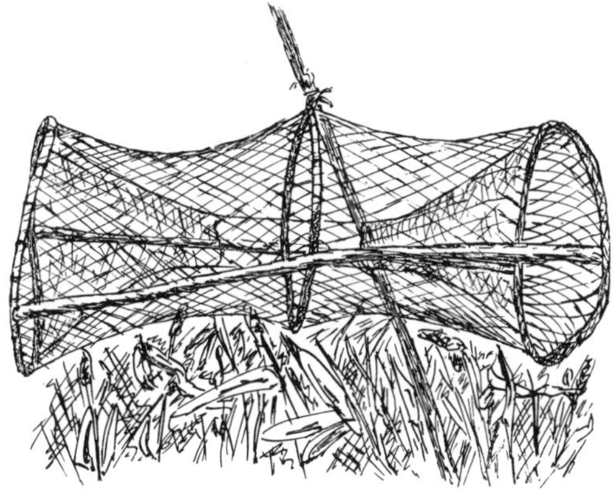

Fig. 79. Drum net, useful for trapping small birds living in reeds. After v. SANDEN, 1936.

Reed Buntings, wagtails, pipits, tits, *Phylloscopus* warblers, European Robins, Bluethroats, Northern (Winter) Wrens, European Greenfinches, Pale Grasshopper Warblers, Eurasian Tree Sparrows, House Sparrows (only a few of the latter), Pied Flycatchers, Goldcrests, one Pallas' Warbler, Water Rails, even a few Carrion Crows and one Eurasian Jay.

W. von Sanden (1936) set up several of these funnel traps (Fig. 79) in marshes. He caught up to 20 birds a day. For small birds he set the traps 1 to 2 m above the ground along natural lanes and corners. The traps were set at ground level for rails. Lanes leading to them were made by beating down the reeds. The effectiveness of these traps is due to two reasons: (1) the birds are guided to the traps via the paths made through the reed, and (2) many insects stay around the fine-meshed netting. Marinkelle (1957) made the same observation on Borneo (Kalimantan) and Sumatra (Sumatera) where natives use large frame nets—drying in the sun and swarming with insects because of the strong smell—to catch birds. Within a few hours Marinkelle himself captured dozens of birds in just such a net. He is probably right in assuming that only the numerous insects can have attracted the birds.

The Dutch ornithologist Koridon (1958) achieved some interesting and remarkable results. In 1957, in 10 funnel traps—always left in the same place—he caught over 1 000 birds, mostly warblers and Bearded Tits. The birds were not in the least afraid of the traps. Especially on windless days the warblers and tits stayed near the nets, which were literally crawling with insects. In contrast to those on Borneo and Sumatra, his nets were clean and did not smell. It seems flies like the warmth of the nets whether or not they were used for fishing. As soon as a bird noticed the entrance to the trap, it would fly in. Bearded Tits liked to sit on top of the funnel pulling at the threads, which had to be replaced from time to time. Short strings in front of the funnel entrance were a good lure for Bearded Tits, enticing them into the trap. Once a bird was caught, it would call unceasingly luring all Bearded Tits in the area to the net. For rails, Koridon set the traps at ground level and added leads (Fig. 80). Reed beds as well as other terrain with marshy character are suitable. Rats and other predators had to be dealt with occasionally.

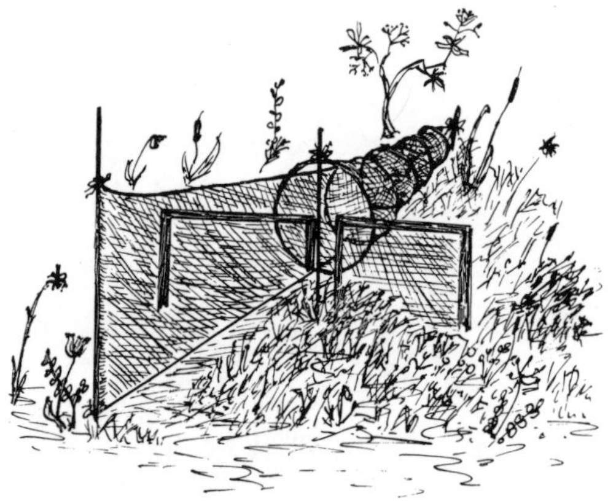

Fig. 80. Weir basket or fish trap, set up for bird catching. After Koridon, 1958.

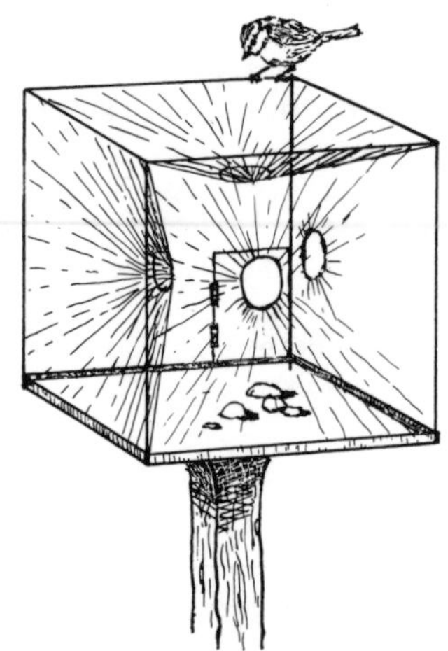

Fig. 81. Pole trap for tits. After Hollom & Brownlow, 1955.

4.1.14 Pole trap for tits (Fig. 81)

This is a small trap in the form of a cube: the sides are each 40-50 cm long. One or several of the sides are bent towards the inside and have an entrance in the middle. One straight wall contains the door through which birds can be removed.

4.1.15 Simple wire-mesh funnel trap

The Sempach Ornithological Research Station (1932) in Switzerland suggests to their collaborators that they be prepared for bad weather which forces thousands of birds to interrupt

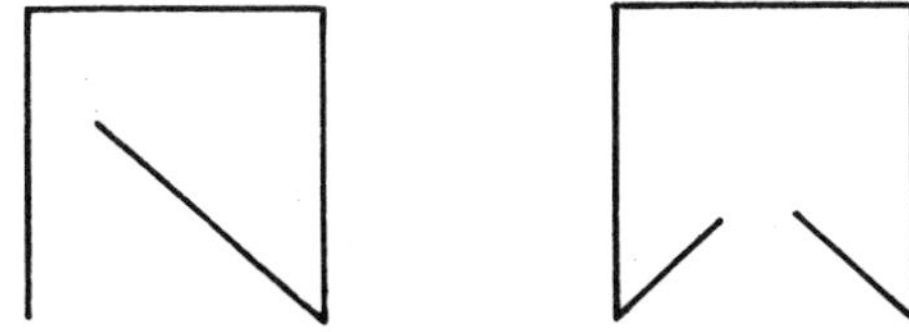

Fig. 82. Simple wire mesh funnel trap. After Ornithological Research Station Sempach, 1932.

their migration. They recommend a very simple trap.

Shape some wire mesh 8 m long and 1.5 m high as shown in Fig. 82. Two additional pieces of wire mesh (2×1 m) make the top for the trap. Wire them in place. Of course, these measurements can be increased. Scatter the bait in front of the entrance and in it and especially inside the trap. With such a trap, during deep snows, more than 400 birds were caught in March 1931 in Sempach: Common Skylarks, Common Reed Buntings, Corn Buntings, Northern Shrikes, Rooks and Carrion Crows. Birds in the entrance are scared into the trap itself when the bander appears suddenly. The bander enters the trap, immediately pulls a wire screen over the entrance, and then proceeds to catch the birds.

4.1.16 Siskin trap

Drost (1940) built a large cage (Fig. 83) at the banding station on Helgoland (Germany). In this trap he caught, among others, many Eurasian Siskins and Hoary (Arctic) Redpolls. Drost says specifically: "Constant checking and resetting of treadle traps is not always possible. Therefore I decided to build a cage with funnels permitting entrance but not exit. At the outside the funnels are 25 cm wide and 15 cm high. The inside opening, 10 cm away, measures 5 cm square. These measurements make an escape of the birds all but impossible. To make perfectly sure I added small glass boxes, open at the bottom—see the sketch. To help the birds find the entrance, I attached perches at the outside. A lure bird must be present and food visible. It is advisable to attach perches and a food tray inside the trap at the height of the entrances. The overall size of the siskin trap is 100×150 cm wide and deep, it is 200 cm high and has 3 wire mesh sides with 6 entrances."

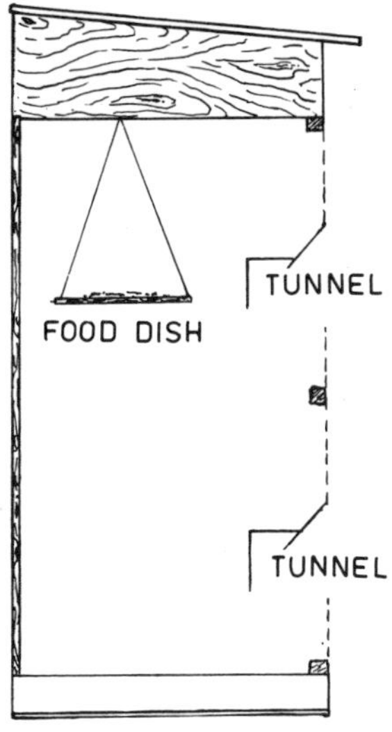

Fig. 83. Siskin trap. After Drost, 1940.

4.2 Traps for gallinaceous birds

Hamerstrom and Truax (1938) introduced several traps designed particularly for the capture of Greater Prairie Chickens. These traps will probably also be useful for pheasants, partridges and other gallinaceous birds.

Gallinaceous birds flutter wildly as soon as they realize they are caught. All traps therefore should be covered loosely with netting. One can use burlap for the roof but not for the sides; otherwise the birds cannot see into the trap. The birds can still hurt themselves on the wire funnels. This gave Hamerstrom and Truax the idea to experiment with swinging wire entrances which turned out to be safe and effective. Such swinging wire entrances have long been used in pigeon lofts.

As yet no method has been developed for keeping 'troublemakers' away from funnel traps. Raptors, rabbits (which rip the netting), and small dogs can enter the trap without any difficulty. Regular checking of the traps is therefore important.

4.2.1 Stationary funnel trap

This trap is shown in Fig. 84. Drive the number of posts necessary into the ground before it freezes, and stretch old fish nets over these

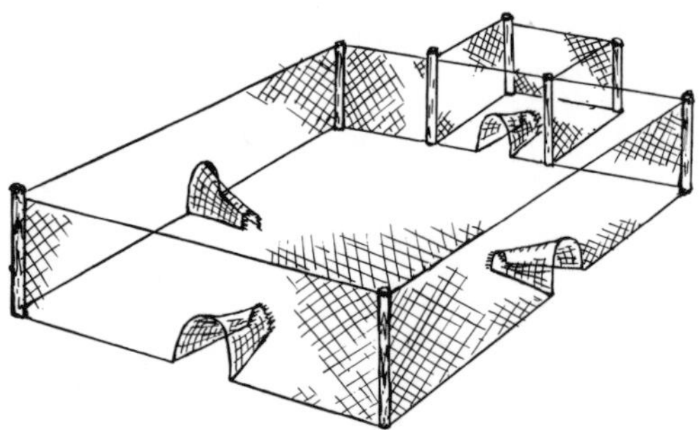

Fig. 84. Stationary funnel for small gallinaceous birds. After HAMERSTROM & TRUAX, 1938.

posts. F. SCHMIDT (op. cit.; HAMERSTROM and TRUAX, 1938) who first constructed this trap caught about 900 Greater Prairie Chickens in four winters.

Length and width of the larger trap are between 2.40 m and 3.60 m, the height is 0.6 to 1.0 m. The gathering cage for captured birds is about 1.2×1.2 m. The funnel entrances are 30 cm high at the beginning tapering down to 13 cm. These measurements are for small gallinaceous birds. SCHMIDT also used a smaller movable version. Here the posts were driven into deep snow. The birds do not hurt themselves on funnels in sufficiently large stationary traps.

4.2.2 Swinging wire trap (Fig. 85)

This trap consists of a row of swinging wires along the sides of a narrow rectangle. The trap is 3 m long, 80 cm wide and 55 cm high. The top and the gathering cage are covered with burlap. You can use a sliding door instead of the drop door to close the gathering cage. Several of the wires at each end plus every fifth one are immobile—otherwise the birds can lift them with their wings and thus escape. This trap was very successful catching prairie chickens. In winter it could be used to catch Common Coots and Common Gallinules (Moorhens) as well as hungry ducks. Only very few birds hurt themselves. One can also use smaller dimensions.

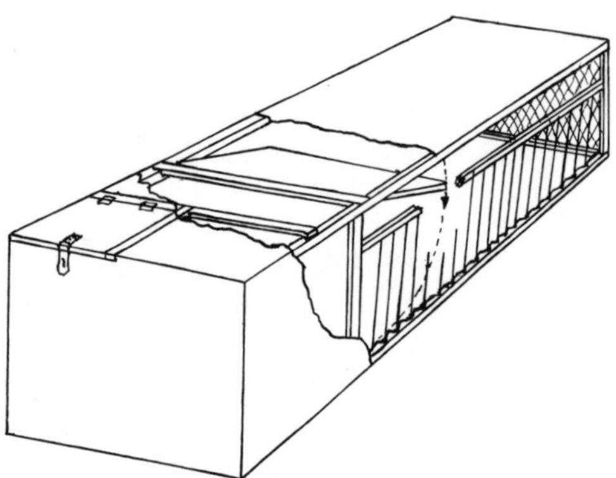

Fig. 85. Swinging wire trap. After HAMERSTROM & TRUAX, 1938.

4.2.3 Wing funnel trap (Fig. 86)

According to HAMERSTROM and TRUAX (1938) this trap is particularly effective when gallinaceous birds visit a heavily frequented feeding station. The previous traps are intended more for smaller feeding places in the field.

The wing funnel trap is collapsible and therefore always easily transported. As all sides are of non-rigid material (the top is loosely stretched burlap, the sides netting) injuries are kept to a minimum. The entrance of swinging wires turned out to be extremely successful. If you set this trap up in winter, protect it with a screen

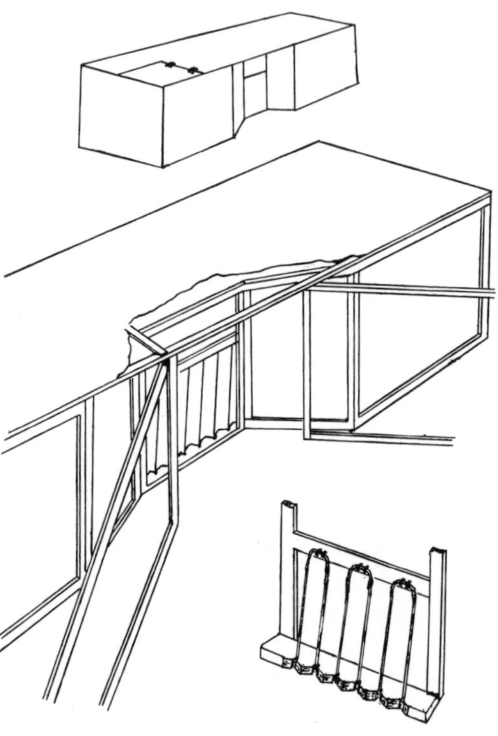

Fig. 86. Wing funnel trap. After HAMERSTROM & TRUAX, 1938.

against an accumulation of snow on top. The trap was used only for prairie chickens but is probably useful in capturing any birds that look for protection from the weather when feeding. As seen in Fig. 86 the trap has 3 'gateways' but only one entrance. Birds hopping about between the two wings or between a wing and a side wall are guided towards the entrance. By eliminating the leads, one has a simple trap with only one entrance (see Fig. 86, top). The slanted edges leading to the entrance replace the wings. Birds are taken out via the door on top of the trap. The left part of this trap acts as gathering cage through a sliding door inserted just beyond the entrance. Birds are driven into the left section by approaching the trap from the right.

The measurements of the trap are: 2.8×0.6×0.5 m. Increase the dimensions if trapping at a well-frequented feeding station for pheasants. Depending on the kinds of birds to be caught, all dimensions must be adjusted accordingly, including those for the entrance. The entrance for the trap shown is 30 cm wide and high. The distance between wires is 5 cm.

Good bait for gallinaceous birds is buckwheat or other grains, and other seeds for smaller species. The seeds should be distributed in the cage so that they cannot be reached from the outside. A thin trail of seeds leading to the entrance helps the birds find their way into the trap. It is a good idea to leave the trap open for a few days so that the birds will get used to feeding beside or in it. Raise the entrance wires on swinging door traps and lift the netting on the side or end of funnel-type traps so that birds will have no trouble finding their way out. When actual trapping starts as few seeds as possible should be outside the trap.

F. and F. N. HAMERSTROM have begun using, almost exclusively, a trap 190×110×35 cm for prairie chickens (Fig. 87). Sharp-tailed Grouse require somewhat larger traps. The frame consists of wooden strips 7 cm wide. Cover this frame with netting (mesh size 4 cm) loosely on top, tighter on the sides. The wire mesh entrance is an untapered tunnel almost 60 cm wide and 26 cm high, leading 35 cm into the trap. Make sure the entrance is not too high. This arrangement has the advantage that the birds will trample the wire down while walking about and thus close off their only escape route.

This trap is baited with corn cobs and set up in the colder seasons. Captured birds can be removed with a landing net. One can transport up to five birds in one large bag.

Fig. 87. Prairie chicken trap for flocks. Photo: H. BUB.

4.2.4 Quail traps

TARSHIS (1956) has written about several quail traps which he used mainly to trap California Quail. These traps resemble the small funnel trap but vary in several details due to their specific application.

The frame of the trap shown in Fig. 88 is made of 6 to 8 mm thick galvanized wire. It is 1.5×1.2×0.35 m. The frame is covered with 20 mm wire poultry mesh. Quails are suspicious of smaller mesh. If the mesh is too large, young birds can escape and adults can stick their heads through. This is dangerous to the quails if a raptor arrives at the scene.

The entrance funnel is 25×10×20 cm and tapers down to the size of a fist (which is approximately the size of a quail body). In front of the left triangle on the inside is a drop door

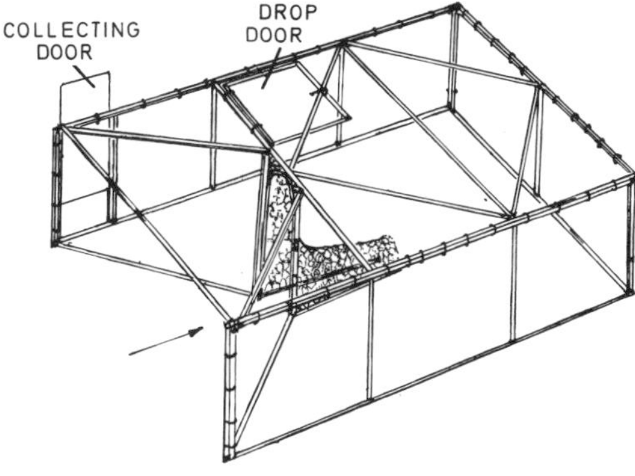

Fig. 88. Collapsible quail trap. After TARSHIS, 1956.

made of wire frame and covered with wire mesh. When catching birds, this door is fastened to the roof of the trap. To remove the quails, drive them into this one corner and close the drop door. This makes it easier to herd the birds through the collecting door and into the gathering cage. The side walls as well as the drop door are fastened to the roof with metal rings. Thus the trap can be folded flat and transported easily. The funnel entrance must be carried separately. Although the trap is light and therefore easily carried, the disadvantage is that it catches few birds at one time. It was used successfully in the open field and near water holes.

Another quail trap is larger and has a wooden frame. Length and width are 3 m each; the height is 35 cm. In the front left corner is again a gathering door, but the drop door is missing. The trap should be set up in a furrow. Watch out that the birds cannot run underneath it. The trap is set up in a good quail location for longer periods of time as its size makes it hard to transport.

Here are a few hints for quail trapping: study closely where the birds stay, where they feed and drink. Once you have found a good location for trapping, start feeding them there—chicken scratch, wheat, barley, rice, onion and lettuce seeds. The bait should be similar to the natural food in that particular environment. Set up the trap after the quails have started feeding on the bait, usually after two or three days, but ascertain this first through observation. Do not install the funnel and leave all doors open. Continue baiting inside the trap. You can also set the trap on stones or wooden blocks for the first few days so that the birds can walk underneath as they please. You may camouflage the trap somewhat with local plants, but too little is better than too much. Baiting must continue throughout the entire trapping period. The funnel entrances should be free from obstructions and, if possible, be on a quail crossing. The traps must be checked twice daily, in the morning between 8 and 10 and at dusk. Bait left overnight is usually eaten by other animals. Do not leave birds caught in the morning in the trap all day, or birds captured in the evening overnight. Use a large mesh on top of the trap to permit songbirds to escape.

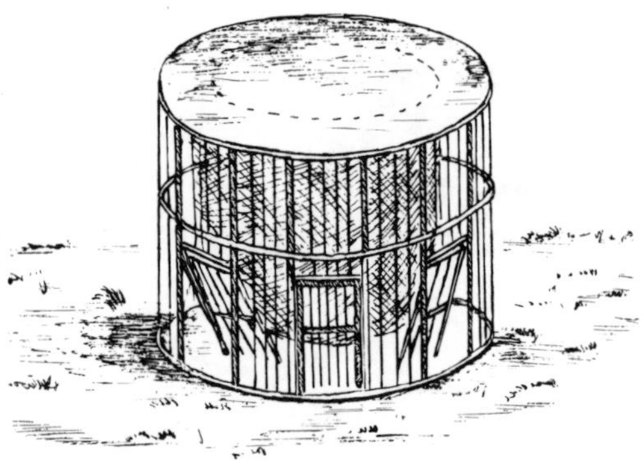

Fig. 89. Old quail trap. After BECHSTEIN, 1797.

We will talk about catching quail with stationary nets later. Another old trapping method used for European Quail should be mentioned here (BECHSTEIN, 1797) which can still be useful today. Inside a cage (Fig. 89) is a female quail covered with a cloth. Around the outside of the cage are a number of doors permitting the males to enter.

4.3 Shorebird funnel traps

The many species of shorebirds, which deserve special attention because of their migration routes and worldwide distribution, have long been banded only in small numbers because of a lack of suitable and successful traps. In Europe the Swedish, Danish and Norwegian banding stations have taken up shorebird-catching and during the last 20 years have developed very successfull traps. We will look at a number of ways to catch shorebirds in funnel traps here. Banding this group of birds is not restricted to sea coasts, as inland locations can also yield good results.

4.3.1 The Ottenby funnel trap

The Ottenby Ornithological Station on the Swedish island Öland developed a funnel trap (Fig. 90), patterned after Low (1935), which caught considerable numbers of shorebirds there. B. DANIELSON (1958) reports banding the following birds up to 1957 (1957 figures in brackets): Ringed Plovers, 1,983 (447); Wood Sandpipers, 4,641 (264); Common Redshanks, 2,040 (138); Little Stints, 652 (167); Dunlins, 22,580 (2,143); Ruffs, 1,795 (333).

The funnel trap consists of two walls (mesh 16 mm), a roof (mesh 50 mm to let songbirds escape if they are not specifically being trapped

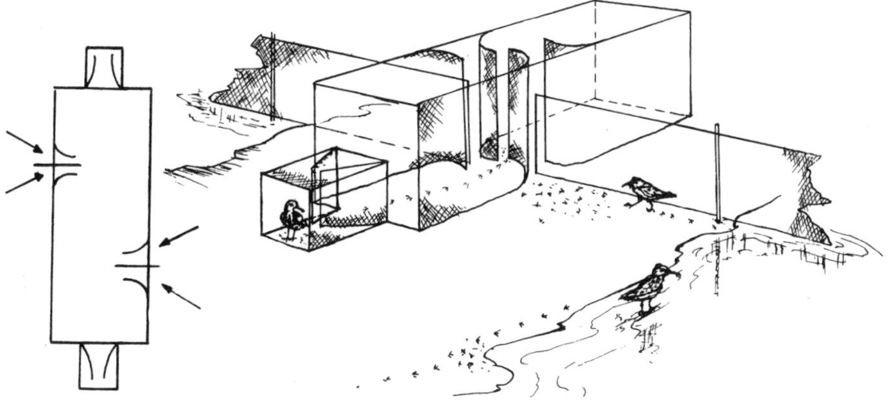

Fig. 90. Ottenby funnel trap. After HOLLOM & BROWNLOW, 1955.

for), 2 leads (mesh 25 mm) and 2 gathering cages from which the birds are removed. The mesh of the gathering cages may not be larger than 16 mm so that the birds cannot stick their heads through. Each funnel wall needs 4 iron rods to fasten it to the ground. The size of the funnel trap varies. Suggested measurements are given in Fig. 91. Experiments by the Helgoland Ornithological Research Station at the Jadebusen (near Wilhelmshaven, Germany) showed that large shorebirds such as Eurasian Curlews and Pied Avocets get caught only accidentally. Funnel traps of the size mentioned should therefore be sufficient for catching a number of

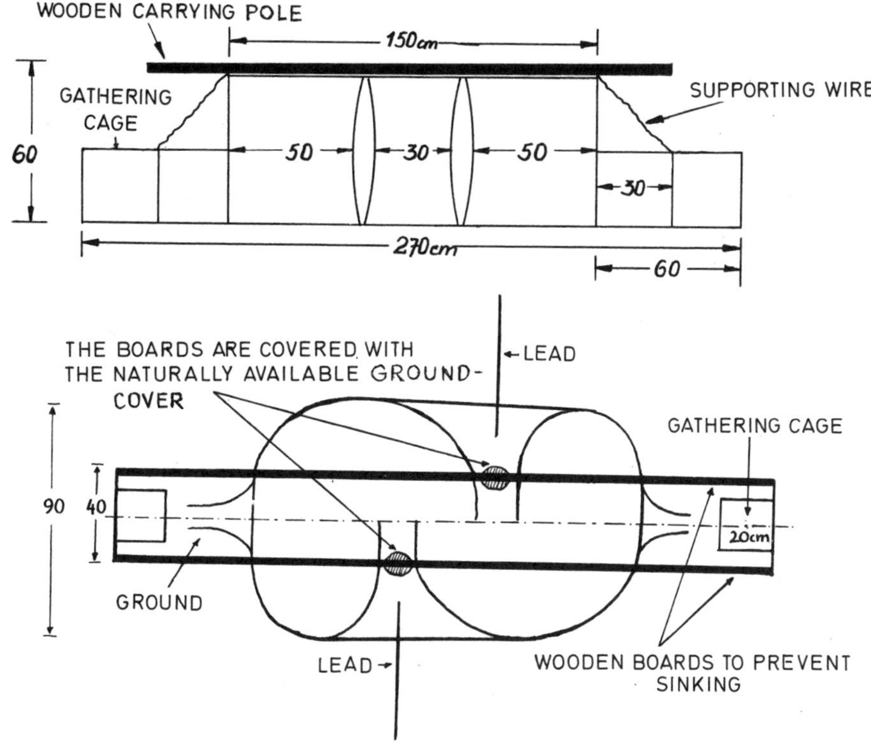

Fig. 91. Model of the Ottenby funnel trap with wooden boards. After W. JENNING.

smaller shorebirds (sandpipers, plovers, etc.) at one time.

J. Jenning has the following to tell about bird catching at Ottenby on the Baltic Sea:

"We set up our traps on rotting seaweed, especially floating seaweed islands that rise and fall with the tide and thus leave us independent of the water level. (The Helgoland Ornithological Research Station preferably traps on rotting brown seaweed on the dune of Helgoland, Germany, as the larvae of the seaweed fly—a delicacy for all shorebirds—live there.) To safeguard against sinking—a constant danger here—two wooden boards are attached under the trap (Fig. 91). They facilitate handling of the trap by just one person. A height of 20 cm is enough for the leads. The distance of the leads from the funnel walls measures the same as the funnel itself. 3 mm thick wires 70 cm long are used to prevent the lead from toppling over even if the ground shifts (waves, undercurrents, etc.). Bend the wire into a hook on one end and then guide it through to the top and edge of the lead. Shorter wires without the hook do topple more easily. The length of the lead is 3 to 4 m. Two arrangements of the funnel traps are very effective: either in a row or in double rows (Fig. 92). If there are many birds around, one can combine both arrangements. Additional leads in combination with natural obstacles can largely prevent birds from bypassing the funnel traps. In this manner one can completely surround an especially attractive location so that the birds which under such circumstances would only fly in an emergency must walk through the traps to go elsewhere.

If the wind is at their back, shorebirds require more space to land. Identify favorable places through observation: the more attractive the location, the closer the traps may be grouped together. Sometimes the funnel traps must be set up to suit specific species. For example, the Red Knot likes fresh seaweed. In that case the funnel trap must be set up so that one gathering cage with a wire mesh bottom is suspended above the edge of the seaweed. This is particularly easy to arrange during rising tides. But the bander must stay close by so that the birds will not drown. The Spotted Redshank favors quiet, vile-smelling lagoons in the seaweed bank where it finds many *Eristalis* larvae. The redshanks walk—they almost swim here—on submerged seaweed. The funnel traps must be set up on poles in this case. Therefore the bander will often catch Common (Green-winged) Teals and Spotted Redshanks together."

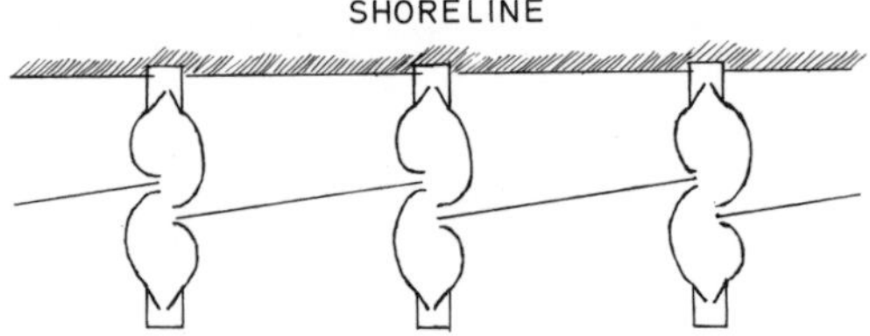

Fig. 92. Row of Ottenby funnel traps. After W. Jenning.

These funnel traps can be set up wherever shorebirds are present—with or without seaweed. The arrangement depends upon the catcher's preference—for example, the traps may also be grouped in a circle around a standing pond.

The major difference between the North and Baltic Seas is the tide, which is either entirely missing or of very minor consequence in the latter. Therefore, experiences from the North Sea coastline are of interest as well. In the fall of 1957 the Helgoland Ornithological Research Station set up 22 funnel traps on the Southeast coast of the Jadebusen (Germany). From the middle of August until the middle of October about 1,500 shorebirds were caught, among others 1,000 Dunlins, 200 Ringed Plovers, and 100 Curlew Sandpipers. The funnel traps were set up in the *Salicornia* zone as well as in the pure mud flats. People sink 30 to 40 cm into the very muddy ground there. This made trapping rather difficult especially as the funnel traps had to be checked regularly day and night because of the tides.

The immediate surroundings of the funnel traps were completely trampled down during set up, but no negative effects on catching were noted. The danger to captured birds from the mud was great, however. The possibility of mired birds was largely eliminated by our coworker W. Schurig who 'trampled' the mud down inside the cages along the wires. These 'channels' always contained water thereafter which saves the birds from the mire.—The leads to these traps were 8 to 10 m long. The length was appropriate for the requirements.

According to Schurig, the number of birds caught depends mainly on the tides, whether rising or falling, and the weather, and rather less on the number of birds present. The best catch-

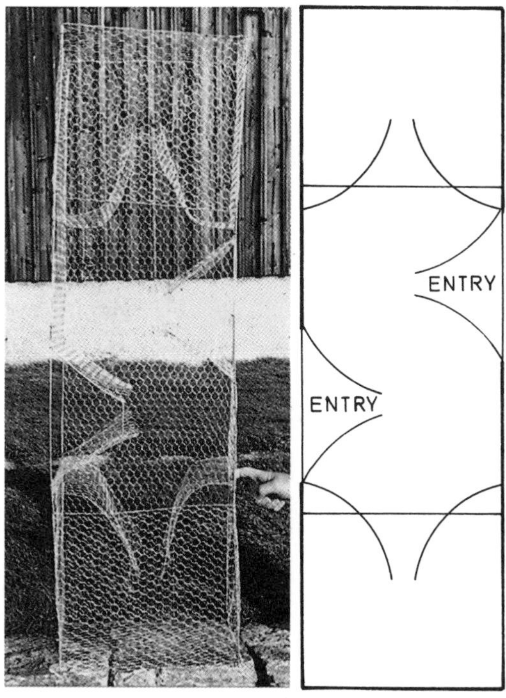

Fig. 93. New Ottenby shorebird funnel trap. Photo: R. de Vries.

ing results were expected when the tide came in in the morning and the evening. It is then that the birds are especially active looking for food. If the water continues to rise or it is raining hard and the mudflats do not drain on time, the search for food concentrates mainly in the *Salicornia* zone, i. e. between the funnel traps. The same thing happens if strong onshore winds delay the drainage of the water. The number of birds caught was usually higher after a high tide than before. Sometimes, especially during rain, many birds were caught between high and low tides.

Lately the station at Ottenby has further simplified their shorebird funnel trap (Fig. 93) according to R. de Vries. Now it is easier yet to set up or move. Both gathering cages for the captured birds now have a wire floor while the entrances are made of wire screen only (usual entrance width 6 cm) without rigid framing or reinforcement. Even small ducks get caught because the funnel yields to their body pressure. The two funnel edges are connected to each other only at the bottom. The funnel trap measures 135×40×30 cm. The gathering cages are 35×40 cm. The door to remove the birds is 10×20 cm. 12 mm chicken wire and 3-4 mm steel wire are suitable building materials. Leads are very useful for these funnel traps unless groups of them are set up together.

On seacoasts Ottenby funnel traps must be used in greater numbers if sizeable bandings are to be achieved. Distances between traps depend upon localities. In tidal areas—such as the North Sea—the funnel traps should be arranged in such a way that they can be easily checked during rising tides. Birds should never be endangered by the water. It is therefore advisable to find out about normal tidal conditions beforehand. These precautions are usually not necessary at inland locations although the water level may also change in sewage treatment pools and retaining ponds.

4.3.2 Revtangen shorebird funnel trap

The Stavanger banding center catches many shorebirds each year on their Revtangen field station (Fig. 94) on the southwest coast of Norway. The funnel traps are very similar to the small funnel trap (Fig. 95). The funnels are enlarged to admit shorebirds (6-7 cm for small ones and 10-15 cm for larger ones). The funnel trap consists of a wire-covered frame. The inside openings of the funnels have no frame so that they can easily be adjusted to the size of birds to be caught. A gathering cage of wire is attached to the trap only at the time when the birds are to be removed. Until then the opening is closed with a flap door. During times of large catches, all traps are first checked and the funnels are closed with clumps of grass to prevent escapes.

Bernhoft-Osa (1955) was for many years in charge of catching and banding near Revtangen on the island Jaeren. He writes:

"During the 12 years of work about 25,000 birds have been ringed at Revtangen most of them being waders. The best result for one single year was obtained in 1951: 4700 birds being ringed that year. The main reason why the waders prefer the rest here is the abundance of algae driven ashore. Amidst all the stones and rocks there is a small bay or lagune, and here, all the year round, a bridge of algae, about 100 meters in length is continuously kept fresh by new supplies drifting ashore. Even by rough sea this sediment of algae will seldom be

Fig. 94. Catching area near Revtangen on the island of Jaeren on the Norwegian southwest coast. In the foreground a group of shorebird funnel traps. Photo: H. Holgersen.

washed away and if it is, new quantities of algae will usually be driven ashore within a few days. After the fresh algae has turned foul in a few days, larvae of flies (*Coelopa eximia* and *C. frigida*) will soon be crowding in huge quantities, offering an abundance of food for resting waders. Not only the waders, however, appreciate the larvae. When the tide is coming in, or when the waves disturb the algae, lots of larvae drift out to sea, and then the various species of gulls and ducks, especially flocks of Eider Ducks partake of the larvae floating on the surface.

"A plentiful supply of larvae makes the waders stay at Revtangen; in the autumn on some days, you may find from about 3 000 up to some 10 000 waders resting along the beach. Numbers like that, however, are rather exceptional. The trapping of birds is almost entirely dependent on the present supply of larvae. In periods when

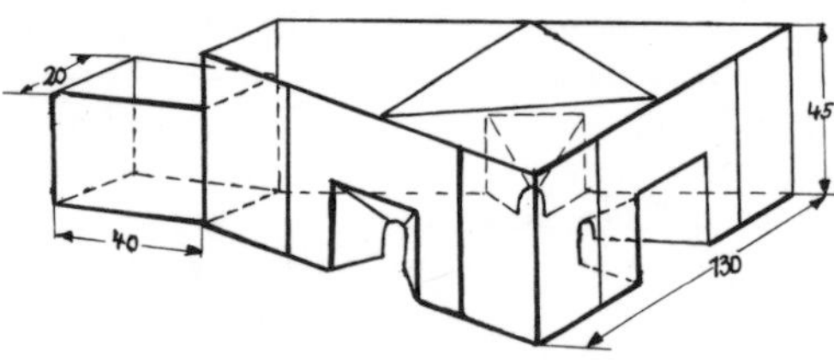

Fig. 95. Revtangen shorebird funnel trap. After Bernhoft-Osa, 1955.

Fig. 96. An arrangement of Revtangen shorebird funnel traps. After Hollom & Brownlow, 1955.

they are not available, even if pupae and flies are found, it is very difficult, as a rule, to get waders in the traps, although large flocks of them are to be seen. Even if larvae are plentiful, trapping may turn out very difficult or be entirely spoiled if the birds have settled down on places where the algae is so wet that it is impossible to get out with traps. Best results are obtained when larvae are limited to odd places in the algae. Then the birds seek for food in quite small areas extending to but a few square meters as a rule. When sunshine and dry weather

Fig. 97. Shorebird funnel traps used on the Australian west coast. After SERVENTRY, 1962.

lasts for some days, the top layer of the algae becomes so hard, that the birds are unable to get hold of the larvae. Then one can make a good taking by digging aside the dry surface of algae on just a small area. After the birds have discovered the place, the traps should be immediately introduced. It has been proved that Sanderling, Ringed Plover, Bar-tailed Godwit and Golden Plover prefer to take their food on places where the algae is covered by flying sand. They like to thrust their bills into the sand and pull out the food. Therefore, I have frequently made, so to say, artificial trapping places on the beach. I dig up some square meters in the surface of sand, and proceed to spread a layer of algae containing a rich supply of larvae out on the sand. The algae is then covered with moderatly thick layer of sand as soon as the birds have discovered the place, traps are hastily introduced. It was mostly owing to these 'artificial' catching places, that in the autumn of 1948 I was fortunate enough to catch no less than 138 Bar-tailed Godwits.

"The traps proved very efficient especially if used in groups (Fig. 96). It has occured, that the result of a single day's trapping, although I was alone at the work, has amounted to between 300-400 birds. In favorable conditions I have sometimes succeeded in trapping practically all waders resting near the station. It might be

Fig. 98. Shorebird funnel trap, temporary assembly. Photo: I. SCHUPHAN.

wiser to place the traps near each other in semi-circles, forming, as it were, a system of labyrinth. The traps will be best placed with the open side of the semi-circle facing the direction from where the birds make their approach, as a rule straight against the wind. The point is, here as elsewhere, and always when catching waders, to work as fast as possible, and not leave the birds time to find another preferable place in which to settle down. It is an advantage also to place the traps in several places at such a distance between them that the birds will not be disturbed whilst one is engaged with the trap on one spot."

SERVENTY and others (1962) reported on shorebird-catching near Pelican Point at the Australian west coast. They set up the funnel traps in rows. In 1958/59 they caught 439 shorebirds in 12 traps; in 1959/60 900 birds in 21 traps (Fig. 97).

I. SCHUPHAN constructed collapsible shorebird funnel traps (Fig. 98 and 99). He used 4 to 5 mm thick galvanized or ungalvanized wire and soldered the ends together to make the frame, with approximately 28 m of wire being required for one trap. SCHUPHAN used 10 to 12 mm chicken wire over the frame. The trap is 50 cm high and the measurements for the gathering cage are 40×30×30 cm. Each piece gets a coat of paint for protection and camouflage.

Eyelets to connect the frames are made of 3 mm galvanized wire. The top of the funnels are kept apart by 12 cm long wires. Four to six hooks 30-50 cm long hold the trap in place on the mudflat. Chicken wire leads 5 m long and 25 cm high stretch from the center of the funnels perpendicular to the land or the water. Wires 40 to 50 cm long keep them standing vertically on the mudflats.

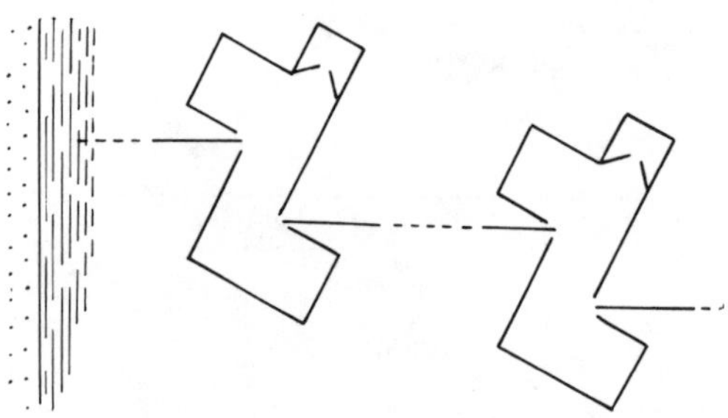

Fig. 99. Arrangement of funnel traps set up on mud flats. After I. SCHUPHAN.

Fig. 100. Skokholm shorebird funnel trap. After HOLLOM & BROWNLOW, 1955.

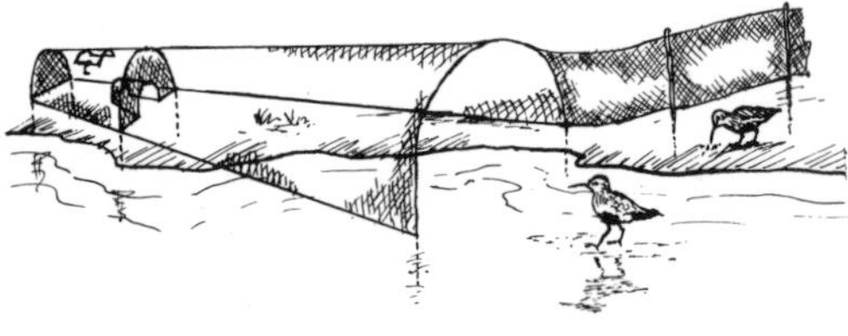

Fig. 101. Skokholm shorebird funnel trap. After HOLLOM, 1950.

4.3.3 The Skokholm shorebird funnel trap

Both funnel traps shown here (Fig. 100 and 101) can be used to good advantage at river and pond banks. It is best to drive the birds slowly into the trap or watch closely their going in because they can easily find the exit again. It may be advisable to add a second compartment in the trap in Fig. 100.

4.3.4 Skokholm pond funnel trap

This funnel trap is 3.5 m long, 1 to 1.5 m wide and 0.5 m above the water level. The last quarter

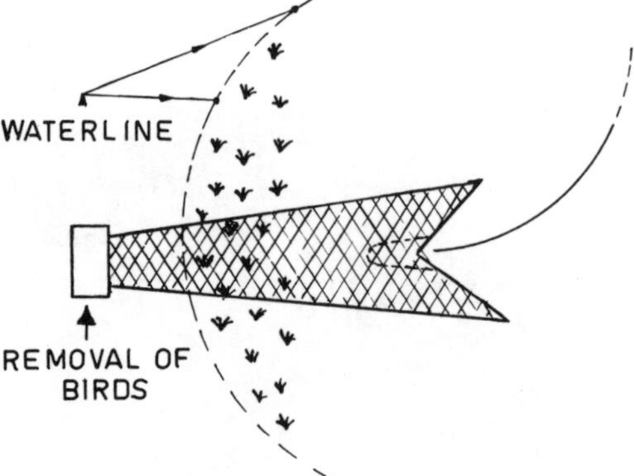

Fig. 102. Skokholm pond funnel trap. After LOCKLEY & RUSSELL, 1953.

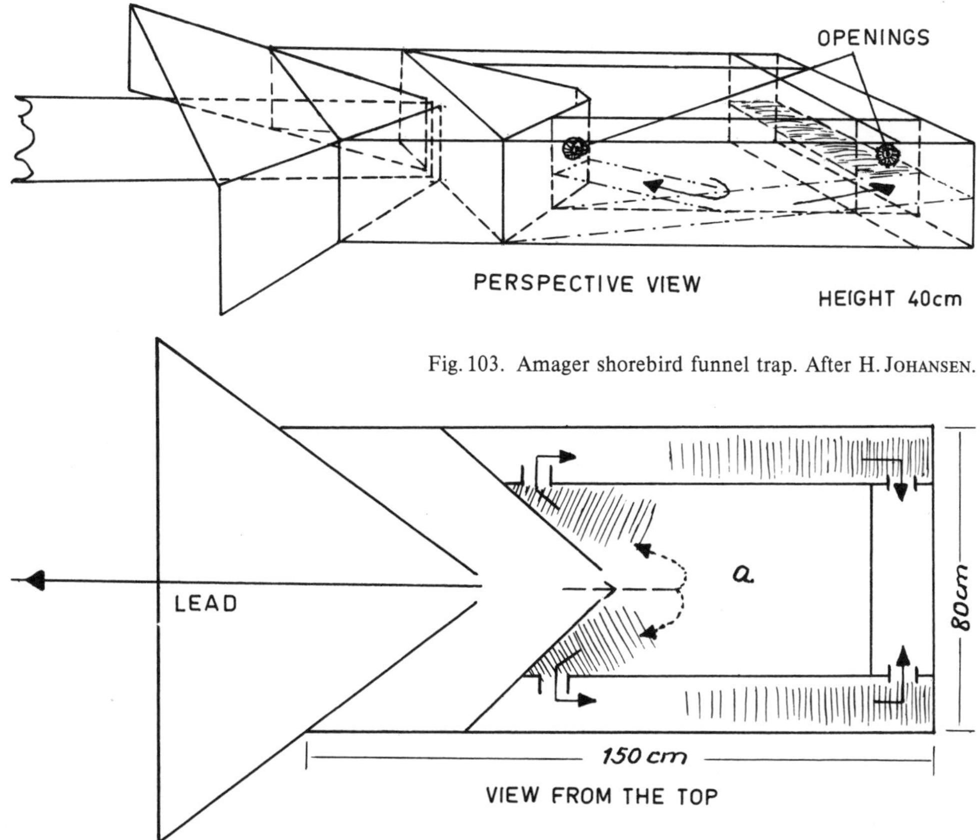

Fig. 103. Amager shorebird funnel trap. After H. JOHANSEN.

with the gathering cage is usually set on dry land. A 5 m lead, attached to poles, guides shorebirds and ducks to the funnel. The wire mesh should not be larger than 16 to 20 mm (Fig. 102).

4.3.5 Amager shorebird funnel trap

Another important northern shorebird banding station is on Amager Island not far from Copenhagen (Denmark). Along the shoreline Ottenby funnel traps are used, while funnel traps of a new and different type are used near the pools and lakes farther inland (Fig. 103 and 104). The innovation is a kind of stairway which the birds use without objection. Escape is now impossible and more importantly, the birds can no longer mire in the mud. It is necessary, though, to cover the gathering cage (Fig. 105) with netting so that nervous birds do not hurt their heads.

The Amager shorebird funnel trap (Fig. 103 and 104) shows that sandpipers, plovers and others 'climb' readily if they can get farther that way. Again, leads are important.

Catching results can be considerable. In 1957 25 funnel traps caught about 4,000 shorebirds,

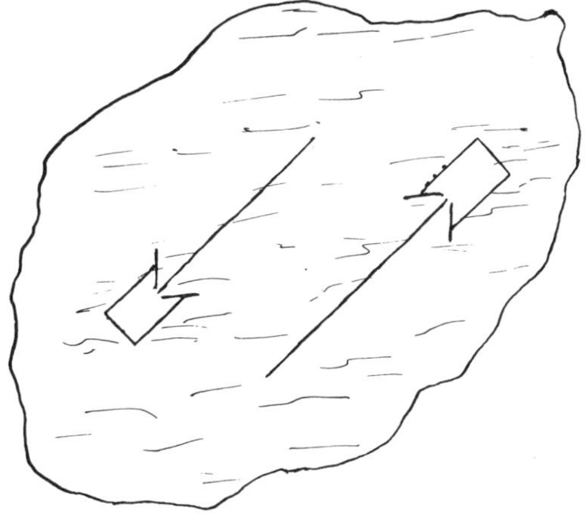

Fig. 104. Set Amager shorebird funnel trap with leads. After H. JOHANSEN.

Fig. 105. Gathering cage of an Amager shorebird funnel trap. Both entrances to the elevated section can be seen. Photo: A. NØRREVANG.

among them 1,860 Dunlins, 600 Common Redshanks, 360 Ringed Plovers, 170 Common Sandpipers, 160 Northern (Common) Lapwings, 140 Wood Sandpipers, 100 Ruffs, 75 Black-bellied (Gray) Plovers, 70 Common Snipes, 60 Common Greenshanks, 50 Eurasian Golden Plovers, and so on. The larger shorebirds were caught in bigger traps, similar to the Ottenby funnel trap, however.

In this context we should emphasize the very practical gathering cages which Low (1935)

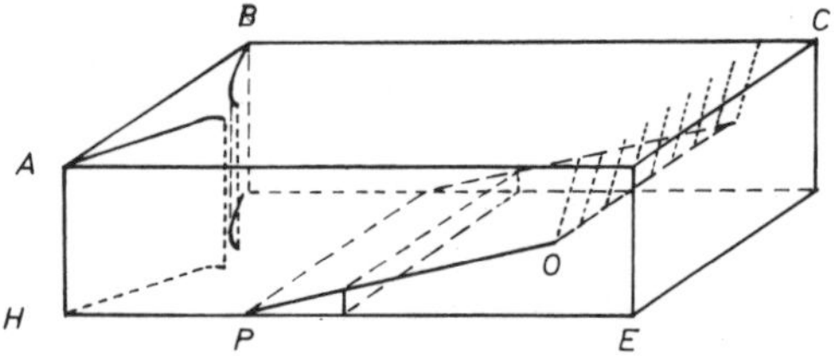

Fig. 106. Funnel trap gathering cage. After Low, 1936.

used for shorebird funnel traps (Fig. 106). After the birds have entered the cage through the funnel, they walk up the slanting wire floor (P-O) and slip through the wires hanging from the ceiling. Thus they are caught irrevocably. Such gathering cages, which also have a wire bottom, are highly recommended even if they take a bit longer to make. The cage also has a door to remove the birds. Here are the dimensions of Low's original: A-B 41 cm, A-H 30 cm, B-C 91 cm, H-P 25 cm, P-O 51 cm.

4.3.6 Ledskär shorebird funnel trap

Swedish ornithologists FREDGA and FRYCKLUND (1965) successfully caught small shorebirds with yet another trap near Ledskär (Fig. 107). Except for Common Redshanks, Wood Sandpipers and Ruffs, the larger birds were only rarely caught in these funnel traps. Fig. 141 shows the dimensions and construction of the trap. The wire

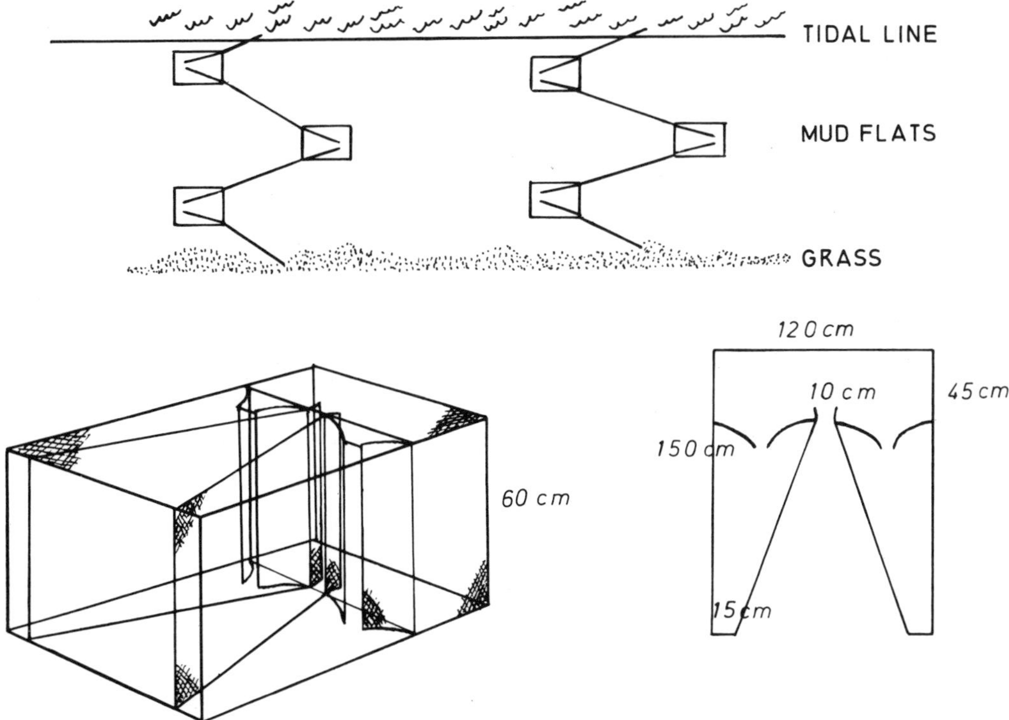

Fig. 107. Ledskär shorebird funnel trap. After FREDGA & FRYCKLUND, 1965.

mesh is 12 to 15 mm. Leads between the funnel traps are 10 to 12 m long and 30 cm high. FREDGA and FRYCKLUND used mist nets and bow nets at the same time.

5. Large funnel traps and sets with long leads

The principle of the funnel trap, or fyke net, is simple. Birds easily find their way into the funnel through its broad entrance, but frequently find their way out only through its smaller end, which leads to a gathering cage. Some funnels are built with the end narrowed to a push-through type hole so that escape is impossible.

As far as we know the earliest funnels were constructed of stones and stakes reinforced with woven vegetation; the quarry was driven into the funnel. After passing the narrow aperture it encountered pits, snares or other devices to keep it from escaping. Large mammals and fish as well as large running birds were caught in this way (Fig. 108). Modern methods still include some primitive materials for funnels, in particular reeds and willow withes such as fishermen use. Many bird catching devices have been adapted from fishing techniques. Indeed,

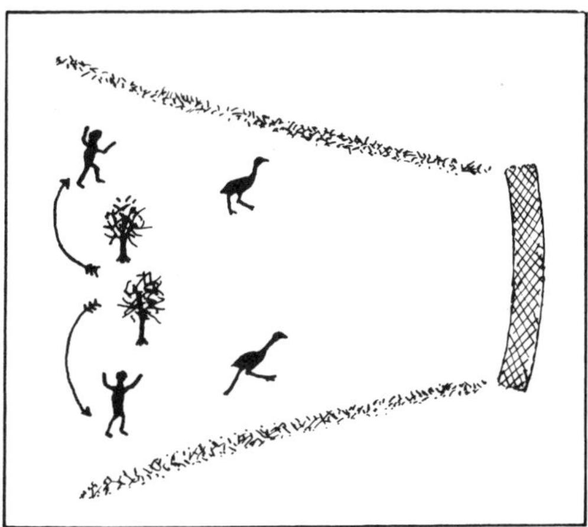

Fig. 108. Hunting emus in Queensland, Australia. After W. E. ROTH, 1897.

Fig. 109. An ancient fyke net (Hamen) for catching partridges from the Hunting Book by W. Birkner 1639. After Lindner, 1940.

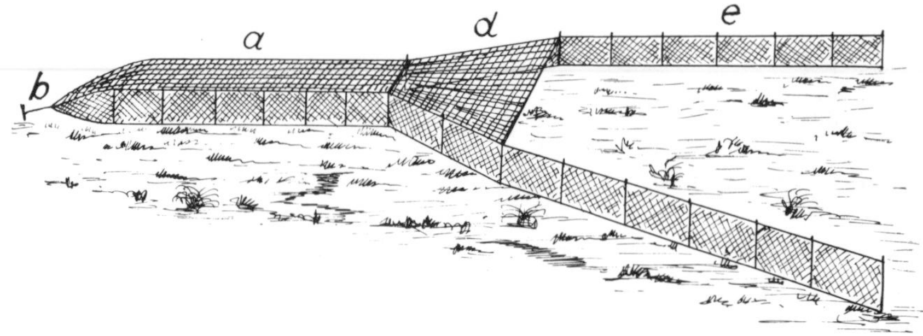

Fig. 110. A fyke type set, after Winckell called "drive net". (a) "Hamen" or throat 15-17 m long, (b) "Heftel" or tether, (d) funnel (entrance to the throat), (e) leads about 15 m long. After Winckell, 1878.

we can assume that fish catching not only preceeded bird catching, but started the catching of live birds. The first live birds ever caught probably—and by pure chance—were captured in a contraption intended for fish.

Fyke net sets had special categories such as the stake net from fish catching (Fig. 109) which was a set for partridges. Winckell (1874) called such a set a 'drive net' (Fig. 110).

Lindner (1940) shows a 'drive net'—a simple funnel used in 1682 for catching otters. The funnel principle was also used by the North American Indians. Bison herds were driven into a huge funnel. At intervals Indians hid behind rocks or vegetation, ready to jump out and frighten the bison onward to the mouth of the funnel. Sometimes they were driven into a corral and sometimes to a cliff—toward which they stampeded—at last falling to their deaths.

As in the case of the simple funnel, the great fyke and angle nets of today evolved from ancient hunting methods, used by our ancestors and by primitive peoples (Fig. 1 and 2). Some evolved from coverts designed for driving game, and others (the angle nets) from fences for herding big game in the Middle Ages.

5.1 The Helgoland trap

In 1911 Weigold built a prototype of the simple fyke net—a giant funnel—to catch and band at least a fraction of the birds migrating across the island of Helgoland (Germany) (Fig. 111 left) In

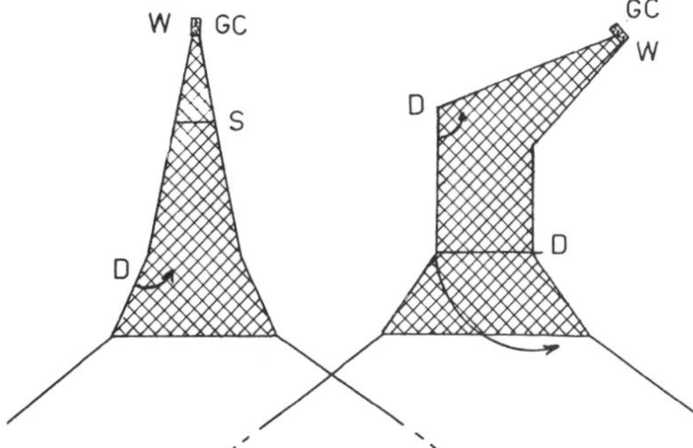

Fig. 111. Left, funnel trap: right, angle trap. (D) door, (S) shutter, (W) window, (GC) gathering cage. After Drost, 1926.

the twenties Drost (1926) developed the angle trap (Fig. 111 right and Fig. 112), which proved ideal for catching great multitudes of birds and which has surpassed all other traps in usefulness. Almost 300,000 birds were caught on Helgoland from 1911 to 1970. Many ornithological stations in Europe and elsewhere have installed replicas of these now famous 'Helgoland traps'.

Drost recognized that one often has trouble catching birds in straight funnels, for unlike fish, birds notice the ever narrowing sides. Therefore he placed the throat of the funnel off at an angle giving the birds the impression that they were only going around a corner. Furthermore he placed the catching compartment about one meter above ground—a position more suited to the flight patterns of most species. Having to reckon with individuals that sometimes flew back out of the funnels, especially when there was a strong influx, he constructed an extra door to prevent escapes (Fig. 111 and 112).

It is only worthwhile to build a Helgoland trap where countless birds migrate through or put in to rest. Such places occur along sea coasts, on islands, and on the shores of larger lakes and rivers. On Helgoland, the catching garden is on the banding station's property (Fig. 113 and 114 Ten to fifteen thousand birds are caught annually—for the most part small birds, but also European Sparrow Hawks, owls, Common Cuckoos, pigeons, Eurasian Woodcock, etc. Sparrow Hawks and other raptors undoubtedly seek cover in the garden's trees and are attracted by the abundant quarry as well. Until 1945 there was a small artificial pool in the garden that attracted the odd duck or shorebird. Now the water surface is far larger and bordered with marshes and swamps. These changes have clearly increased the catch of swamp and aquatic birds.

For migrants the garden serves as an oasis as it is the only luxuriant cover on the island (Fig. 114) for birds of brush and woodland. The plantings are deliberately manipulated to pres-

Fig. 112. Helgoland trap in which thousands of birds were caught before 1945. Photo: R. Drost.

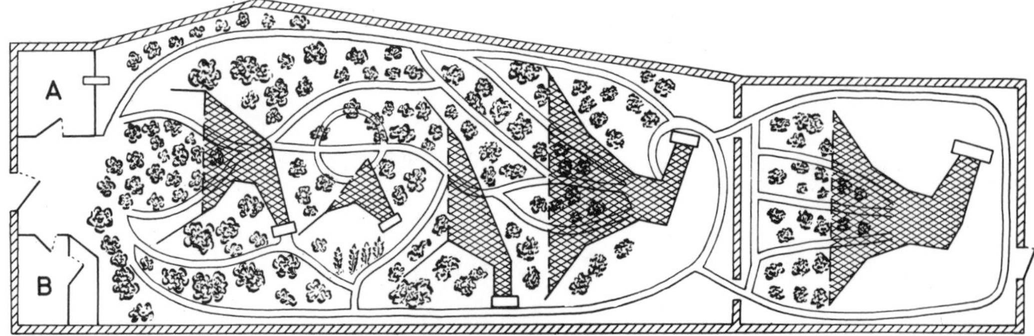

Fig. 113. Catching garden of the Helgoland Ornithological Station showing five traps as they were placed in 1944. (A) and (B) banding hut. Adapted from LOCKLEY 1953.

ent a variety of types. Several hundred birds are captured on good migration days. BUB was present on the day of the biggest catch: October 12, 1940. One thousand five hundred and ten birds of 40 species were taken, among them 760 Common Song Thrushes, 550 European Robins, 66 Redwings, 57 Chaffinches and 22 Ring Ouzels. The greatest surprise on this overcast day with its continuous drizzling mist was a Radde's Warbler. Previously there had been only two records for Europe. On such days the island shimmers with resting and moving birds.

On a big flight day such as this we never stopped catching and banding from 8:00 until 24:00 and several people were needed to keep up. As soon as most of the birds have dropped in, a crew of three or four drivers herds the birds towards the funnels. If the drivers move too fast, birds escape into the shrubbery; if too slowly, the birds double back from the funnel entrance.

Disks are thrown above the birds to discourage them from flying up and away.

The angle near the throat of the funnel also makes it more difficult for the birds to double back. In Figure 112 the trappers are ready at the pull strings to release first the door at the throat and next the door of the gathering cage. Once the throat door pull has been released, the birds are easily driven into the gathering cage. A long pole with a flag or small board fastened to its

Fig. 114. View of the catching garden on Helgoland as it appeared in 1935. Photo: R. DROST.

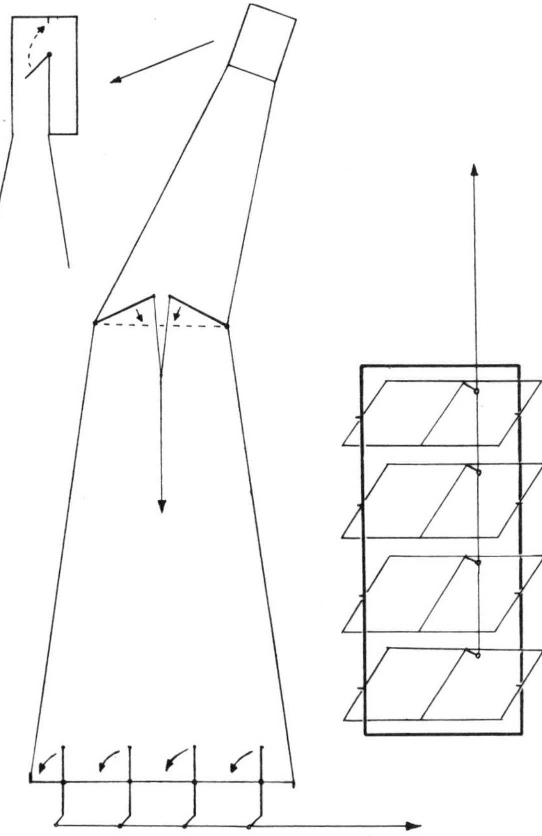

Fig. 115. Helgoland trap with four part door for closing the entrance. After K. SCHWAMMBERGER.

end is kept right by the last door. This 'persuader' encourages recalcitrant individuals to enter the gathering cage.

A North American fyke trap (EBBA News 30, 1967, pg. 103) narrows to one meter so that one person can fill the opening in front of the gathering cage with his body. But this technique serves only for special purposes.

K. SCHWAMMBERGER devised a 4-part door (Fig. 115) that closes the entire 2×4 m opening of the funnel with a pull string up to 50 m long. The closure is so perfect that the trap can also serve temporarily as an aviary. This method of closing off a funnel is practical, but it is of dubious value for the 15-20×5.6 m Helgoland entrances. We recommend it for smaller traps, for example to prevent escapes of particularly desired species.

Another method of cutting down escapes is to bend the funnel wall diagonally inward for one meter at the entrance. Such projections have little merit for broad entrances like those at Helgoland.

It is important that the catching compartment, or gathering cage, is well designed. The compartment shown in Fig. 112 has proven sound over many years. Twigs are painted on the pane of glass at the back of the compartment to keep certain species from hitting the pane too hard. The birds are deflected to the right and fall into a chamber below. In the meantime the exit flap is shut. Next the birds fly toward a second pane of glass. The sides of the catching compartment are of wood so that the birds fly only toward the light. In Fig. 112 a third, still lower chamber may finally receive the birds. They can, however, be removed directly from the middle chamber. The open door of the middle chamber can be seen in Fig. 112. The compartments are separated from each other by slides. The middle chamber has an extra slide which is channeled so that it can be pulled to and fro to concentrate the catch in a small compartment, from which it is easier to remove the birds.

The trapper on the right in Fig. 112 is holding up a rod divider grate which is used to separate the larger birds such as thrushes from the smaller ones which might otherwise get trampled. The rods of the divider must be spaced so

Fig. 116. Gathering cages of a Helgoland trap at the Swedish Ornithological Station at Ottenby. Photo: B. DANIELSSON.

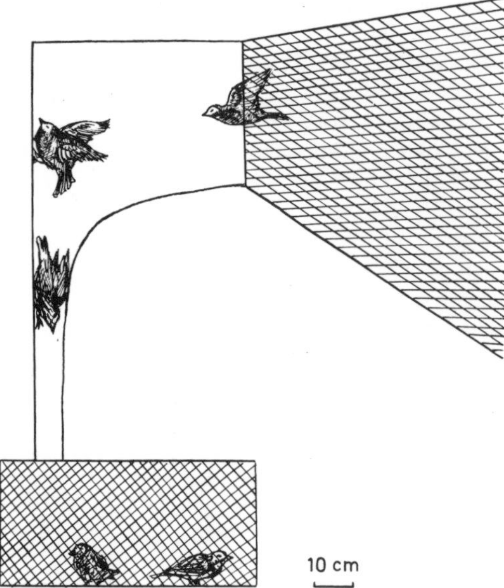

Fig. 117. Gathering cage of a fyke net. After GAVRILOV, 1968.

that birds the size of Common Song Thrushes cannot slip between them. The middle compartment is about 1 m long and 0.5 m wide. It is closed by means of a wire, running over a pulley with a counter weight at the other end.

Obviously the gathering cages do not always have to be of the same size, but certain specifications are essential. At Ottenby (Sweden) curved plexiglass is used instead of plate glass at the funnel's end to keep the incoming birds from striking too hard (Fig. 116).

This curved plexiglass deflector is a useful innovation. It is important that the glass be clear so that the birds can be driven into the cage easily. The Russian ornithologist GAVRILOV (1968) describes a wholly different gathering cage (Fig. 117). At the end of the funnel the birds fly against a clear, flexible plastic sheet and slide down a tunnel into a gathering cage, from which they cannot escape. He uses a variety of gathering cages, any of which fit the base of the tunnel. This tunnel has been highly successful. Finches, buntings, starlings, larks, bee-eaters and others, flying in quickly and directly, have not injured themselves. The sides of the upper cage are of plywood.

The mesh size of the wire netting should not exceed 16 mm in diameter. The last quarter of the funnel has 12 mm wire mesh so that even Northern (Winter) Wrens are caught. The wire is painted a greenish brown for camouflage and to keep it from rusting.

The funnel has no fixed measurements, but is adapted to the terrain. A 20 m wide mouth is considered large. The smallest mouth recommended is 4-8 m. The height of the funnel should be 2-2.5 m toward the end and 3-4 m at its mouth.

Russian ornithologists have erected a fyke net of sizable proportions (length 100 m, width

Fig. 118. Big funnel trap of the Soviet station Rybatschiy on the Kurskaya Kosa. Photo: V. PAYEVSKY.

Fig. 119. A gully trap (angle type) on Fair Isle. Photo: R. H. DENNIS.

30 m, height 12 m) at Rybatschiy (formerly Rossitten) on the Baltic Sea. (Fig. 118) This may well be the largest fyke net built so far. According to BELOPOLSKI and ERIK (1961), 45,851 birds of 107 species were caught in a few years. On their best day, they caught 2,764 birds.

Fyke sets need not be permanent installations. Transportable versions for expeditions and outlying stations are practical. The frames of the fyke nets used at the outlying stations of the Helgoland Ornithological Station are 1 m wide and 2 m high. Those who work far from settlements and supplies build frames of local materials and cover them with netting rather than poultry wire. Netting is light and easily spread, but it has the disadvantage of frightening the birds by blowing about in the wind.

The catching garden on Helgoland has trees giving the impression of a low wood interspersed with shrubbery. The last part of the funnel, however, has little cover—only a few single bushes. The cover must not be so dense that the birds sit tight when the driver appears. His noise may not be enough to get them moving if their feeling of well-being is too great. As the illustrations have shown, the sides of the funnel should be bordered by shrubbery. There should, however, be no cover in front of the gathering cage to hinder the birds on their way.

Bait can help. It pays to plant berry-bearing bushes in front of and in the funnel entrance and sometimes in other locations as well. Other feed attracts when snow is on the ground. Lure birds, preferably of many species, increase efficiency greatly. One of the successful traps at Wilhelmshaven (Germany) had an aviary of lure birds on one side. Some of the big fyke nets, such as in Helgoland's catching garden, have cages of lure birds inside. We use a cage within a cage with a space of 10–20 cm between them. The bottom of the aviaries must be protected from rats and weasels.

Lure bird cages, even those within the funnel, are camouflaged and located according to the size and positions of the funnel. Cages with lures should not be hung on the walls of the funnel or birds will just flutter about outside trying to reach them. If possible, the lure birds should be so placed that they are not disturbed by the trapping, although many become accustomed to this commotion.

British ornithologists on Fair Isle built a trap in the mouth of a gully adapting its contours to the gully's edges. This gully trap (Fig. 119) is not really an angle trap, but we include it here. With this gully trap they caught the following species in 1949: 95 Northern Wheatears, 95 Common Blackbirds, 76 Twites, 70 Meadow Pipits, 50 Redwings, 22 Goldcrests (WILLIAMSON, 1951).

Fig. 120. African quail net. Photo: R. A. ROSER.

5.2 Shapes of simple funnel nets

There are many possibilities for using simple funnel nets, the forerunners of the angle traps just described. They are useful as stationary sets or for expeditions. A primitive form is the African quail net in which the funnel idea is only partially developed (Fig. 120). The Bedouins in Northern Egypt built quail traps with a small net along the coastal sand dunes between Alexandria and Mersa Matrouth (R. A. MOSER, 1960). Solitary clumps of grass serve as catching spots as quail use these for cover. The Bedouins watch from a promontary to see where the quail are putting in after their long flight across the sea. When the trapper grabs at the bird from behind the clump, the quail struggles toward the light and gets caught in the net.

A variant of this method is the 'thrush bush' of Helgoland. Although fyke, mist and funnel nets have replaced thrush bushes for banding, in earlier times thrush bushes (Fig. 121) were of great importance to the islanders, especially for catching the many migrating thrushes. Thrush bush catching was still practiced for banding on the Island of Mellum (Germany) one of the substations of Helgoland until a few years ago. The first Orange-flanked Bush-Robin we ever caught was trapped in a thrush bush. Such a simple device can prove useful if one doesn't wish to build a fyke net or has no mist or trammel nets available, or it can be used as an auxiliary trap. It works not only on barren coasts and islands, but also on heaths, steppes, tundras or on big plowed fields. A full description of this technique is given by GÄTKE (1891).

A piece of ground about 6 m long and 2.5 to 3 m wide is rather thickly studded with standing branches in a semicircle. BUSE (1915) noticed that the Helgolanders used old Christmas trees to build their thrush bushes. This was not always the case as they sometimes built their semi-circular hedge of elder (MERKEL, 1927). The thrushes must be able to move about freely on the ground under the 'bush'. The shrubbery should stand straight at one long side, while that on the other should slant toward it. A strong net is pulled over the slanting side so that it extends from the tip of the bush down to

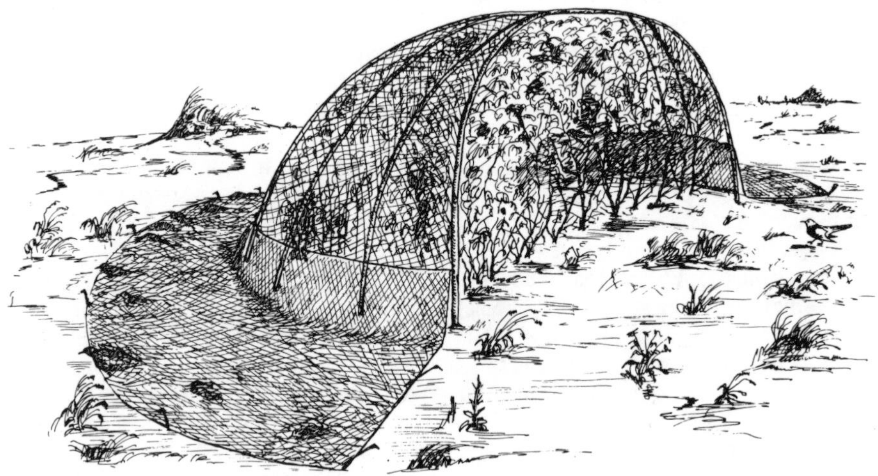

Fig. 121. Helgoland thrush bush—from sketch by H. WEIGOLD.

Fig. 122. Davison prairie chicken net. Photo: F. COPELIN.

0.6 m from the ground in a long semicircle. Next another net, loosely gathered on a rope, is placed to overlap the bottom of the first net slightly at its top; its bottom is spread on the ground and extends about 2 m from the base of the bush where it is stapled to the ground. In earlier times, it was held down with rocks.

The set must be placed so that thrushes can see it from a distance and tend to enter the open side. Bushes growing naturally are even more attractive. They may be available in somewhat protected gardens, but not on barren coasts where the raging storms of autumn and winter demolish them.

GÄTKE describes the technique. "There may be about 20 such thrush bushes on the island. Catching with them is rewarding: here where the soil above the rock is as bare as the surrounding sea after the potatoes and few cabbages have been harvested, the thrushes—accustomed to shady woodlands—are strongly attracted by dry branches and shrubs stuck into the ground. They willingly hasten toward them. Once they are within, they can easily be urged, by means of a long, light stick, to go under the part of the net that lies on the ground. There they usually stick their heads through the meshes and can't pull them back out. When the migration is strong, a couple of hundred thrushes may be captured in such a bush in one morning; on some days, when the weather is not so favorable, one must be satisfied with 30 to 50. In addition to thrushes, other birds get under the net by chance: often a Eurasian Woodcock, a Wood Pigeon, the Corncrake and its relatives, all species of shrikes, the Long-eared Owl and not infrequently a European Sparrow Hawk in pursuit of its quarry."

A thrush catcher once had exemplary good fortune: he caught 73 Ring Ouzels at one 'run in'. Ordinarily 10 to 15 of this species are considered an especially good catch. The Ring Ouzel responds particularly well to the calls of their own kind.

Fig. 123. Italian funnel net for quail catching. Photo: A. TOSCHI.

Fig. 124. Simple funnel trap on Helgoland in 1911. Photo: H. WEIGOLD.

A further development toward the modern fyke net is the DAVISON net (Fig. 122) used for catching Lesser Prairie Chickens in the scrub-oak thickets of Oklahoma (U. S. A.). He covered clumps of oaks with fish netting from which a funnel projected for two or three meters. The birds sought the shade under the oaks. For catching other species it would be better to raise the funnel 1 to 1.5 m above the ground. The fyke net idea shows clearly in the quail trap illustrated by TOSCHI (1959) (Fig. 123).

WEIGOLD built regular fyke nets on Helgoland in 1911 (Fig. 124), but his trap still had the gathering cage on the ground. Now it is elevated

Fig. 125. Simple funnel trap–but covered with netting – on the North Sea island of Scharhörn in 1947. All the poles were of driftwood picked up nearby. Photo: H. BUB.

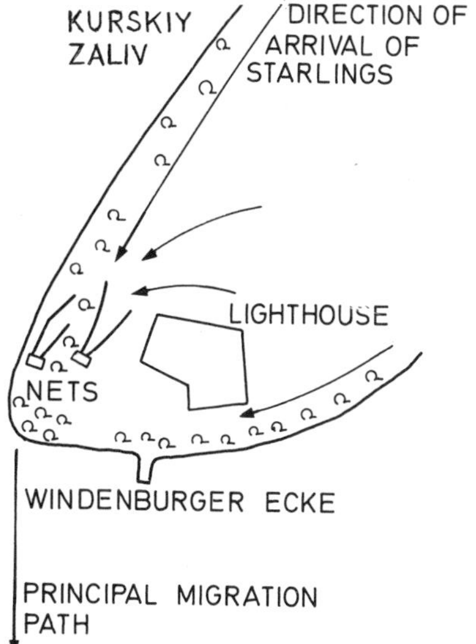

Fig. 126. Arrangement of starling nets. After H. KRÄTZIG.

Fig. 127. Fyke net for catching moor hens. After J. TAAPKEN.

about 1 m on posts (Fig. 112). Materials were scarce in 1947, but nonetheless the trap (Fig. 125) was built in two days. One of the merits of the fyke trap is that it can be built in a short time with few materials.

KRÄTZIG (1936) built a larger, somewhat similar trap in the early summer of 1934 at the Windenburger Ecke, Baltic Sea, to catch as many Common Starlings as possible (Fig. 126). Each of the two 20 m long net funnels had an entrance 10 × 3 m. They narrowed sharply toward the gathering cage opening. Interchangeable gathering cages were used so cages could be replaced without delay when the starlings came in flocks. The flight usually started after sunup. A day's catch of 500 starlings was not uncommon. The best daily catch was 1,640 starlings during the early summer migration.

Strategic use of natural opportunities is also shown by J. TAAPKEN who watched Common Gallinule (Moorhen) behavior near a body of water in Holland. When alarmed, the gallinules fled from fields into a shrub thicket and were easily driven into his trap (Fig. 127). Lead fences kept them from getting to the water.

Simple funnel traps with long leads have proven productive. British ornithologists have used this principle in a double funnel trap (Fig. 128) and caught many Northern Wheatears and other species. Birds moving along the stone wall from either direction were captured. The trap can also be adapted to hedges. Then its size depends on the length of the hedge row. For long hedges, opposing funnels are set with gathering cages between them (Fig. 129–130).

Fig. 128. Double funnel trap for catching wheatears and other species on Fair Isle. Photo: R. H. DENNIS.

Fig. 129. Side view of a double funnel in a hedge row. After A. PRÄKELT.

Under certain circumstances simple funnel traps can be set up adjacent to each other whereby one side is saved.

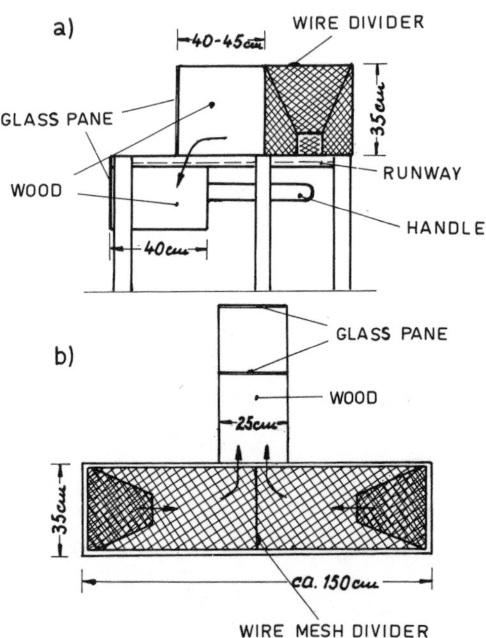

Fig. 130. Gathering cage of a double funnel trap (side view). The lower compartment is removed to unset the trap. (b) Gathering cage from above. After A. PRÄKELT.

5.3 Portable funnel traps

In rushes, shrubbery and such, portable sets can be essential. The funnel trap, made of poultry wire, used at Sempach (Switzerland) can be transported in sections (Fig. 131). The required number of sections (S) of 1×2 m wire mesh supported by iron rod frames, the funnel (F), and the gathering cage (GC) can be tied or wired together.

At Lake Sempach a portable funnel trap (Fig. 132) covered with netting was successful. The trap entrance consisted of a chamber 2 m high, 4 m wide, and 4 m long made of latticework frames fastened to each other. The sides and roof were covered with tightly stretched netting. The end of an attached 3 m long net bag is kept open by a wooden hoop 1.2 m in diameter through which the birds must pass to reach a fish net funnel 2 m long, which is stretched slanting upward and is held in position with 3 or 4 posts. Captured Common Starlings tend to fly upward. After the birds are in, the entrance is closed by pulling net 'curtains' shut. The 'curtains' are of fish net strung on metal rings and run on two strands of wire. As no part of the trap has a bottom it can be moved and set down anywhere in the rushes. The vegetation inside the trap provides good perches for the starlings.

The bottom of the trap is the surface of the water. The trappers must work quietly for the

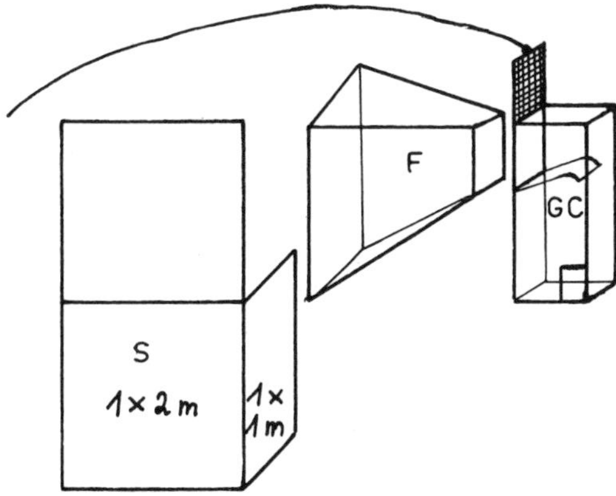

Fig. 131. Portable wire covered funnel trap. After Orn. Stations Helgoland, Rossitten and Sempach, 1935.

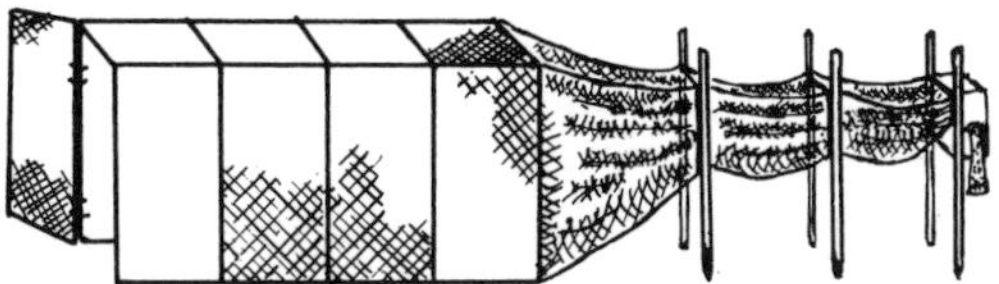

Fig. 132. Portable fish net trap. After Orn. Stations Helgoland, Rossitten and Sempach, 1935.

first starlings are already near the entrance. The driver moves slowly, the catcher crouches hidden near the curtains ready to pull them shut as soon as a few hundred starlings have settled on the rushes inside the chamber. The driver is stopped by a prearranged whistle. The starlings try to escape and most of them work their way up into the funnel. Then the catcher opens the curtains again and signals the driver to continue. The biggest catch comes at the end when the driver is near the trap. The last few starlings flying around in the funnel after the curtains have been closed for the last time are caught with landing nets or scared into the fish net funnel by this activity. The fish net funnel with its load of starlings is then detached and carried ashore. Sometimes it is worthwhile to make another drive from the opposite direction promptly in which case the trap is reverse and a spare fish net funnel attached. At the Sempach Ornithological Station the best catch was 450 Common Starlings in a short time.

A. Hilprecht (1937) used a large funnel trap for catching Common Starlings in rushes in the Magdeburg (Germany) region, "over 1,000 Common Starlings could be captured in one evening. The trap is a simple funnel: entrance 14 m wide, 6.5 m high, length also 14 m (Fig. 133). The funnel tapers to a 1 m square plexiglass chamber, the top of which is 4.5 m above the ground. An automobile headlight powered by a storage battery is hung behind the chamber and its beam entice the masses of starlings that are driven toward the funnel into the chamber. They slide down the chamber's smooth sides and through a chute, walled with oil cloth, and end up in a gathering cage 5 m long, 1 m wide and 3 m high, where they are finally caught, for they cannot find their way back out.

"The whole trap is portable and can be set up or taken down in a short time. The net-covered portions of the funnel are carried by telescoped and threaded seamless drawn steel rods. The funnel is drawn into position by tightly pulled cords. The steel rods have a diameter of 4 cm and a wall thickness of 2 mm."

"In order to simplify the use and transportation, the translucent chamber could also be disassembled. The sides are hinged and can be folded. Be it noted, however, that the top consists of a dark cloth as a translucent pane would be superfluous here."

"Experience has shown that a white light frightens the birds and they shy away, whereas a mellow yellowish-red light has the desired effect."

"For catching songbirds the plexiglass chamber was important and will remain important to the bird banders of Magdeburg (Germany), but for catching Common Starlings the chamber is not big enough to hold the masses for starlings—the birds turn back and stream toward the entrance. The catch can be speeded up by using a translucent wall 3-4 m high and a correspondingly broader oil cloth chute."

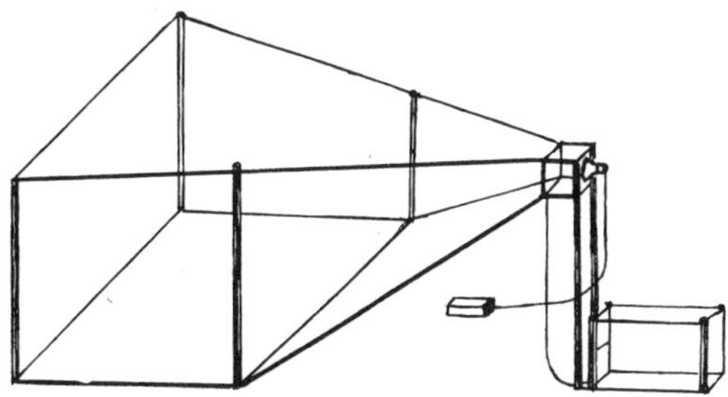

Fig. 133. Large funnel trap for catching starlings in the rushes. After A. Hilprecht, 1937.

5.4 Slit funnel traps

Funnels ending in slits often produce noteworthy catches. They can be placed either vertically or horizontally in thickets or rushes—especially for small birds. A gathering cage is placed at one corner of the trap, as taking the birds out with a dip net is too difficult.

A horizontal funnel 1 m high, 4 m wide and 4 m long caught about 200 Common Starlings on several occasions (Fig. 134). The trap is carried to a starling roost at night and laid on the rushes that the birds have bent down. Then the birds are slowly driven toward the funnel. At its entrance they sit on blades of rush or on perches (P) leading toward the 10 cm slit and then press on into the trap. The bigger the trap, the less apt they are to find their way out. As soon as the driver reaches the trap he ties the wire-bordered slit shut and carries the trap to shore.

Another slit funnel trap is described by BAER (1941-42) in Switzerland (Fig. 135). The trap sits about 50 cm above ground on posts. It does not have to be any special size, but should be several meters in breadth and length. The entrance slit must not be more than 20-25 cm. To unset the trap, the glass pane at the end of the gathering cage is removed. Rails are also caught in this trap. BUYSSE (1968) devised a practical trap for catching rails in lanes cut in the reeds in Belgium (Fig. 136). The trap is placed in the lane, lightly covered with reeds and visited four times a day to remove the birds. The frame is of 7 mm iron rod and the covering of 16 mm galvanized poultry wire.

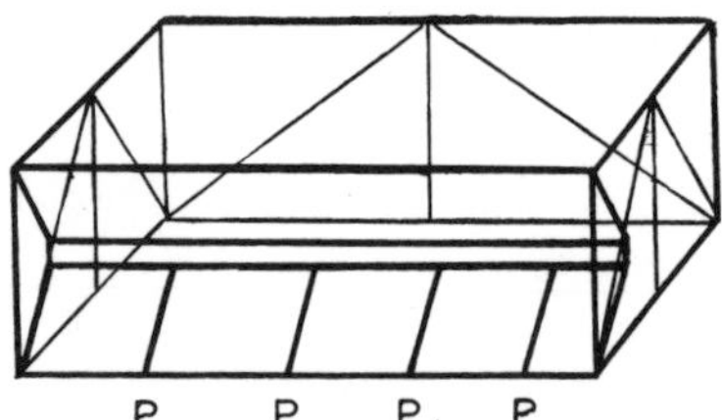

Fig. 134. Horizontal slit funnel trap. After Helgoland, Rossitten and Sempach Ornithological Stations, 1935.

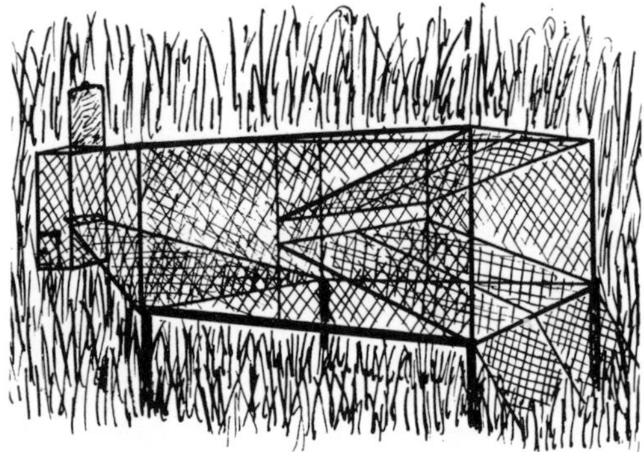

Fig. 135. Slit funnel trap on posts. After BAER, 1941/42.

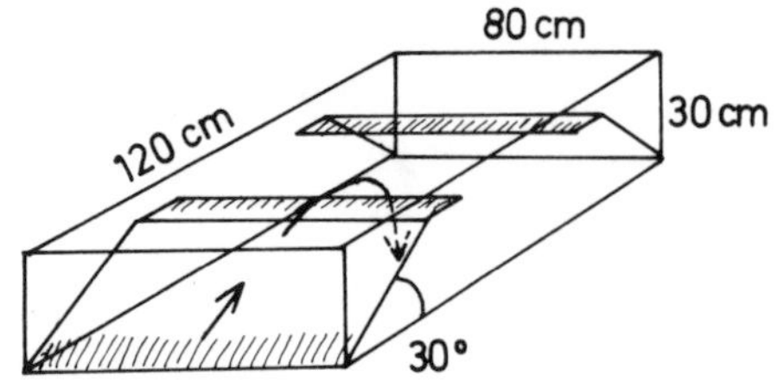

Fig. 136. Slit funnel trap for catching rails in reed lanes. After BUYSSE, 1968.

5.5 The orchard trap

This is a funnel trap first used by G. BOARDMAN in England (HOLLOM, 1950) to protect fruit (Fig. 137). A large cage (7×7 m) is erected using posts 3.5 m high and it is covered with wire mesh painted green. A funnel inside the trap leads to the gathering cage. When the catcher steps into the trap the birds usually fly to the left; if not they can be driven into the funnel. The gathering cage has a thick glass wall at the rear. Both the entrance to the funnel and the open door can be shut with pull cords. Many species, such as Common Starlings, tend to enter a trap from above and many were thus caught.

The doors are all left open when the trap is not in use, in part so that the birds can consume noxious insects and also can feed on the bait. In addition to garbage, peanuts and grasshoppers are put on feeding trays to attract other birds; bird baths present still another attraction. The first year this trap was set up it caught more than 200 birds.

5.6 Drop curtain trap for geese

JACK MINER of Kingsville, Ontario (Canada), caught hundreds of Canada Geese more than 30

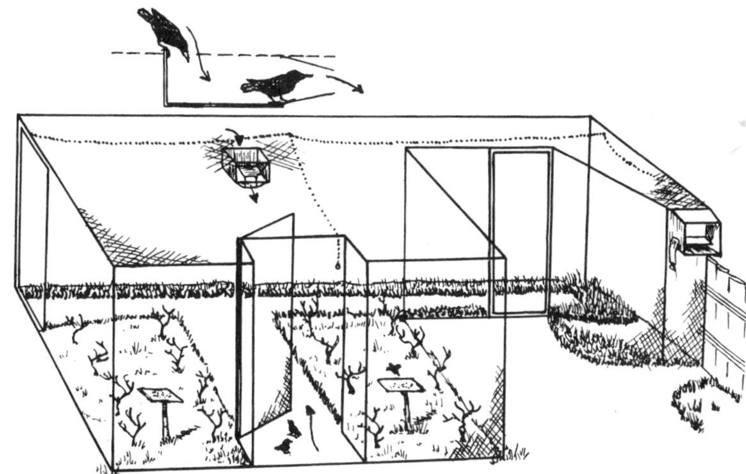

Fig. 137. Orchard trap. After HOLLOM, 1950.

years ago in shallow ditches on which nets were spread (LOCKLEY and RUSSELL, 1953). The nets were supported by horizontal rods resting on 2 m poles driven into the ground at the water's edge. The ditches were baited with corn. When the geese got under the net, the side walls (12, 24 and even 36 m long) were dropped by pulling a cord and the geese were caught. Next they were driven into a smaller compartment and banded.

The dropping of the long sidewalls made an enormous racket, and this giant trap is clumsy. The method is suitable only for a large permanent stations.

H. C. HANSON describes a variation of this technique (Fig. 138) after LOCKLEY and RUSSELL (1953). For the frame, posts of wood or steel are driven into the ground 2.4-3 m apart. The side walls are of 40 mm mesh poultry wire. The roof is of netting supported on strong brace wires which also lead to the ground at a 45° angle and serve to hold the posts in position as well (Fig. 138). The geese are baited into the trap with feed. The falling net curtain entrances are dropped by a pull cord running over pulleys. HANSON used four (about 10×12.5 m), two at each end of the trap. Our sketch shows only one curtain. HANSON discovered that a smaller trap with only one curtain was more successful. The drop curtain door can be used on other traps.

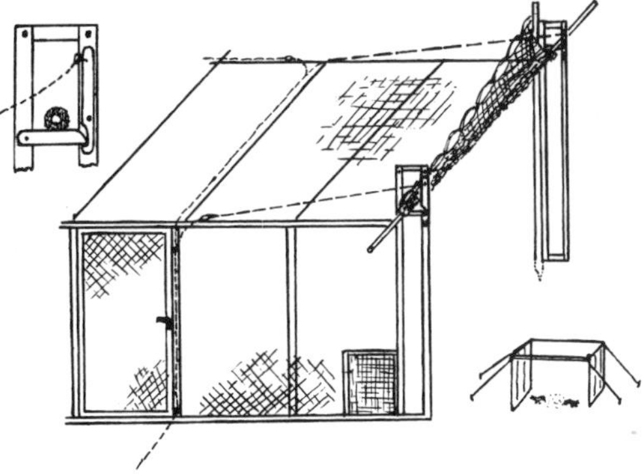

Fig. 138. Drop curtain trap. After LOCKLEY & RUSSELL, 1953.

5.7 Ladder entrance trap for crows, raptors and Collared Doves

So-called crow traps have popped up like mushrooms in recent years; not because of ornithologists, but because hunters want to get rid of crows. BICKEL (1951) first called our attention to the Scandinavian origin of this trap (Fig. 139). The crows move down a funnel-like top entrance and proceed into the trap between the rungs of a ladder. Only toward the center of the trap, for about 1.5 m, are the rungs 20 cm apart so that the birds can slip down between them. The rest of the rungs are spaced so closely that no bird can get in.

Now that we have discovered that this trap catches not only crows, Black-billed Magpies and Eurasian Jays, but also kites, buteos, Northern Goshawks, Eurasian (Common) Kestrels, Long-eared Owls, Tawny Owls, etc. banders also could use it to advantage. To some degree they have. In Switzerland, from January through

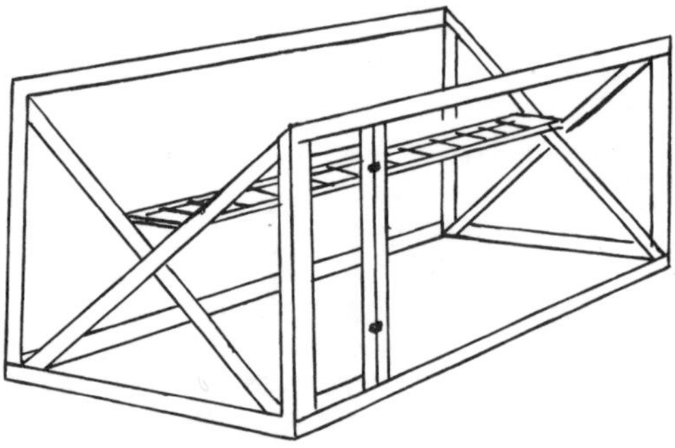

Fig. 139. Scandinavian ladder entrance trap for crows. After E. BICKEL, 1951.

March, 70 Common Buzzards were caught, many of them up to seven times. In January 1964, G. DOMEYER and coworkers captured 112 Rooks and Carrion Crows near Walkenried, Harz (Germany), in one day. Their trap was 5 m long and 4.5 m wide; they decoyed with a stuffed crow and baited with slaughterhouse offal. This same trap, without decoy but with snow cover, caught 25 Common Buzzards, 1 Rough-legged Hawk, 2 Northern Goshawks, 1 Red Kite, 2 Long-eared Owls, 144 Carrion Crows, 137 Rooks and 20 Eurasian Jays, from 1963 to 1968. In another trap they caught over 100 Eurasian Jays in a few winters. Their success would have been far greater if they had used one, or better two live decoys. The decoys should be color-banded or otherwise marked to distinguish between them and the wild birds. It is also advisable to clip their primaries.

Ladder entrance traps can vary in size: the minimum recommended is 3 m long, 2 m wide and 1.8 m high. As a rule, 5 to 6 meters is quite long enough, although JACK MINER's giant crow trap, 23 m long and 6 m wide at the top and 3.6 m at the bottom and 2 m high was obviously successful in Ontario (Canada) (LINCOLN and BALDWIN, 1929).

Recently new variations have been suggested: KEIL (1967 a and b) and HASSEL (1969) prefer deviations from the original Scandinavian model, though it is still a useful trap. Their models are more portable and more easily set up, so they are to be recommended. KEIL places his ladder entrance lower than the sides of the trap (Fig. 142), but HASSEL considers catching better if the ladder is even with the trap's top (Fig. 143). In fact, HASSEL even puts the ladder on top of the trap, thus raising it slightly. Furthermore, he places the ladder crosswise instead of lengthwise. This innovation works, but so does the old model. Large portable traps need frames so that they can be taken down or set up easily. The frames can be made of poles, cut for the purpose, or finished lumber can be painted to protect and camouflage it.

When setting a trap, predators that might injure the lure birds or the captured wild birds

Fig. 140. Big trap for crows. Measurements of the main part: 11 × 4.25 × 2.25 m. Length of funnel: 8 m; width at the entrance of the gathering cage: 1 m. The trap includes a second gathering cage, so that the first one can be exchanged if it is filled with birds. The cage containing the trapped birds is taken to the banding laboratory. Photo: K. GREVE.

Fig. 141. The same trap from the other side. The trap must be firmly anchored in the ground and has to be well braced, because it is exposed to wind and storm. Bread was mainly used for luring the crows near a refuse-pit. With 2 of these traps 5753 Rooks were caught during the winter 1984/85 and with 3 traps 7423 during the winter 1985/86. Of course other species have also been caught in the traps, such as Carrion Crows, Common Jackdaws, some Northern Goshawks, European Sparrow Hawks, Common Buzzards and Rough-legged Hawks. Photo: P. Eggerling.

must be borne in mind. Polecats, weasels, rats, etc. should not be able to squeeze into the trap, small birds, inadvertently caught, should be able to get out. Therefore, the lower ⅔ of the trap is covered with 15-20 mm wire mesh and the upper ⅓ with 30-40 mm mesh. One need not be so finicky about mesh size if one is not apt to run into either type of trouble. The crow trap at Wilhelmshaven (Germany) was simply covered with strong aviary wire which cut down

Fig. 142. Crow trap at Frankfurt-Fechenheim Ornithological Station. Photo: K. Lang.

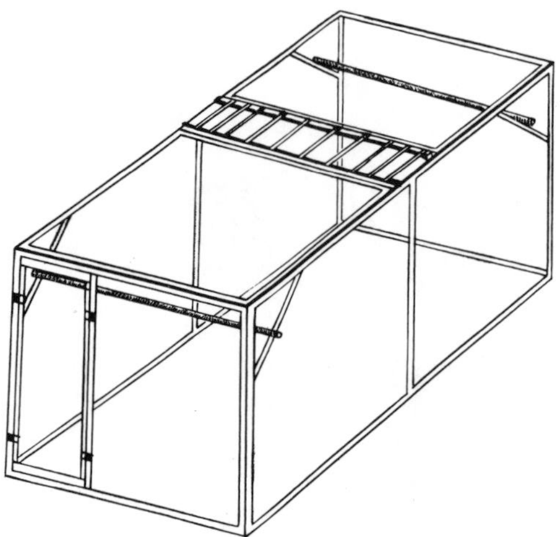

Fig. 143. Another type of a crow trap. Measurements 3×2×1.8 m. Measurements of the ladder 206×18 cm. After HASSEL, 1969.

rusting and kept foxes out. It was roofed with 40-50 mm poultry wire so that snow could fall through.

For crows, the rungs should be 20 cm apart; for Black-billed Magpies 20-15 cm, though magpies seldom escape between rungs intended for crows. HASSEL drives nails about 6 cm apart around the entrances to prevent escapes. This is wholly unnecessary and hardly in the spirit of bird protection.

It is important to have a shelter inside the trap for the lure birds to use during bad weather and at night. Two perches for the captured birds to sit on are desirable. It is not a good idea to have part of the trap sheltered from the weather as it makes the set too conspicuous.

It is sometimes necessary to unset the trap during bad weather (high wind, heavy rain, fog, sleet or heavy snowfall). To unset, the entrance door is left open and the ladder is removed.

When the trap is set, complete with lure bird and bait (garbage, slaughterhouse offal, dress-outs of game, ears of corn) it is usually checked twice a day: in the morning and toward dusk. But the probable catch must be considered: incompatible species may harass or injure each other in the trap. At Wilhelmshaven the crow trap can be kept under constant observation. KEIL, however, recommends not taking out the catch until after dusk so that any nearby crows will not become trap shy. This is true, but not to be recommended in all situations—especially not where other species, better released in daylight, are captured. Of course, it is not sensible to walk up to a trap when more birds are still trying to get in.

The birds are captured with a dip net and processed. Crows are first put into gunny sacks, whereas raptors are banded immediately and released at a distance.

Lure birds should always be supplied with waterers. Shallow tin cans set in the ground serve well. The birds are given fresh water and food: grain (including corn), bread, cooked potatoes or garbage.

Suitable trapping sites abound: in fields and meadows near woodlots, near forest edges, in large openings, near settlements. No food should be available except inside the trap, therefore a crow trap should not be placed in a dump, although exceptions occur.

The following hints should be taken into consideration:

1. If possible, be able to get to the trap by vehicle. This facilitates setting up the trap and checking it.
2. Be able to check from a distance with binoculars or a scope.
3. Keep the trap away from the public—not near roads or paths.
4. If it must be near the public, keep a close watch.
5. There should be at least one perch tree near the trap. If none is available, install one.
6. It is useful to move the trap 30, 50 or 100 m from time to time if opportune. Rubbish accumulates and has to be cleaned away occasionally.

H. HÜTTGENS used the Scandinavian type ladder trap for Collared Doves (Fig. 144). The trap

Fig. 144. Ladder trap for Collared Doves. Photo: H. HÜTTGENS.

Fig. 145. Trap for Collared Doves. Photo: H. Behmann.

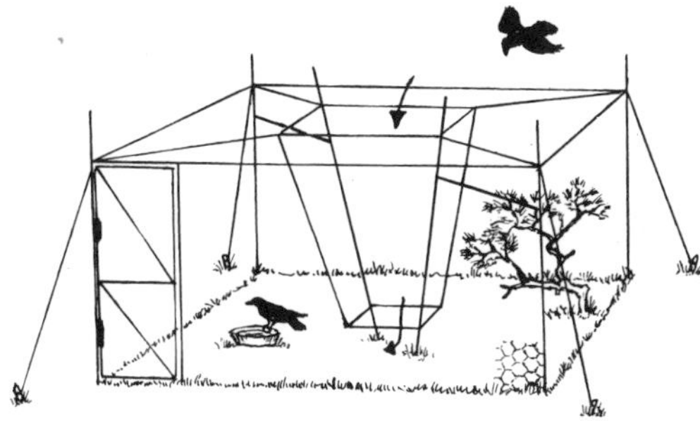

Fig. 146. Crow trap. After Hollom & Brownlow, 1955.

is 2×2 m, the side walls are 1.2 m high and the ladder is 0.85 m from the ground. The central rungs are 30 cm apart and those at the end 8-10 cm. There is also a funnel entrance at ground level. Hüttgens gets his best trapping done from the end of September to the end of November because the days are shorter and the doves are hungrier. Trapping hardly pays from December on as too many bird feeders are in operation and compete. It is important that sparrows can feed in the trap for then the doves trust it. From August 21 to October 16, 1967 Hüttgens caught 189 Collared Doves.

Another Collared Dove trap (Fig. 145) 195×195×65 cm is used by H. Behmann at Kiel (Germany) on the Baltic Sea. The ladder is 95×25 cm. As soon as the doves have finished feeding, they tend to flutter about. Therefore this trap is watched continuously and the birds are removed promptly with a dip net.

Fig. 146 shows a crow trap (Hollom and Brownlow, 1955), which is about 2 m high, 3.6 m long and 3.6 m wide. The top entrance funnel is square, 1.2 m at the top and 60 cm at the bottom. The bottom is 30 cm from the ground.

Crow trapping, traditional among Australian sheep herders, may in part have arisen from their American counterparts. Rowley (1969) describes several Australian traps in detail and illustrates the fundamental type (Fig. 147) which is apparently most commonly used. It is 2.5×2.5×1.8 m. The top entrance funnel is oval, at the top 90×30 cm, at the bottom 60×15 cm. It projects 30-40 cm into the trap. During hot weather leafy branches are put on the trap to give shade. Gunny sacking is tied to the trap corners to give protection from the wind. Rowley checks his traps twice a week (more often in hot weather). Such infrequent checks would be impossible in densely populated countries—who would be able to get a good night's sleep!

Another Australian crow trap is smaller (110×80×80 cm) and appears to be cut from one piece of poultry wire (Fig. 148). The entrance, again at the top, is 30×15 cm. Mobile hanging wires hinder escapes. This little trap, which would scarcely hold more than two crows in addition to the lure bird, is possibly intended for catching breeding pairs. The ornithologist takes the captured birds out through the entrance hole.

In addition to the Scandinavian type ladder trap, the smallest of which are 2.4×1.8×1.8 m, the Australians also use a larger tip treadle trap. Except at the entrance the treadle is roofed with wire so that the birds sliding down into the trap cannot raise their wings.

Rowley has some valuable comments on baiting. Crows are omnivorous. In Australia insects form a large component of their diet, but they also feed on grain, carrion, fruit, vegetables

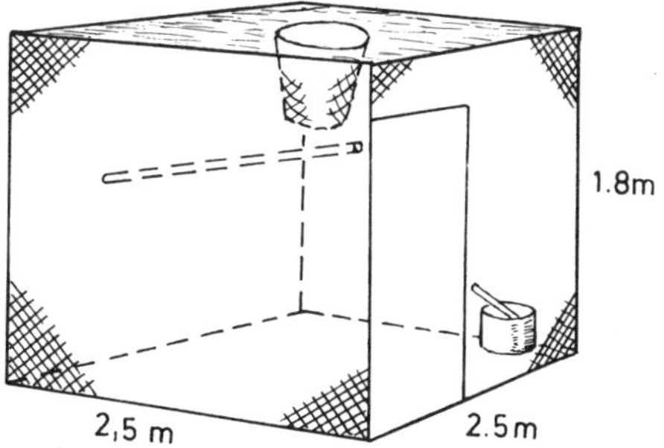

Fig. 147. Australian crow trap. After Rowley, 1969.

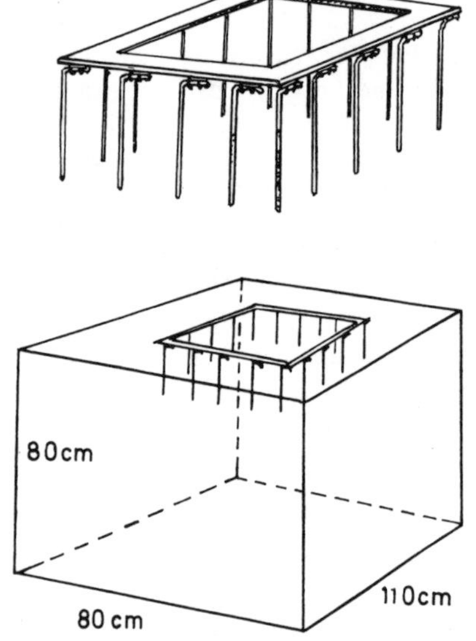

Fig. 148. Australian crow trap. After ROWLEY, 1969.

and garbage. Less frequently they take live quarry such as mice, small lizards, fledgling birds and weak lambs. That their menu varies according to time and place should be considered when baiting. Grain and meat are available throughout the year, but only in winter is meat good for bait. Grain is usually the only summer bait, but if obtainable, old bread and fruit can be added to advantage. The birds seem to prefer oats to wheat, and have been known to crowd into a trap having discovered ping-pong balls in it. Eggs, plastic or real, serve the same purpose.

Fig. 149 shows a vulture trap which can also catch kites, buteos, White-tailed Eagles and other raptors and should be suitable for gulls seeking carrion or meat remains. The size of the trap is not fixed, but for vultures and large raptors, it should be at least 15 m long, 8-10 m wide and 2-3 m high. The spacing of the rungs varies depending on the size of the birds to be caught. FRANCES HAMERSTROM has noticed that the rungs must be more closely spaced for owls; diurnal raptors tend to be less successful in flying upwards through a small space.

Fig. 149. Vulture trap. After MCILHENNY.

6. Installations for catching ducks and other water birds

"Duck decoys" may be massive installations designed for catching waterfowl or may pertain to live or artificial ducks.

The catching of waterfowl in special, permanent installations (often known as decoys) goes back to the Middle Ages and even earlier. According to VOLQUARDSEN (1933) an Act in the Netherlands specifically designated bird catching, especially the catching of swans, as a royal right in 1306. It is not known when the first duck decoys were installed. As there were already many duck decoys in Holland in the middle of the 16th century, they must have been built at least in the 15th or 14th centuries. It would seem that duck decoys spread from Holland into other middle and northern European countries. LINDNER (1940) shows a picture (Fig. 150) by the hunting artist WOLFGANG BIRKNER (1639). AITINGER (1653) also shows a well laid out duck decoy in his book (Fig. 151). PAYNE-GALLWEY (1886) also acquaints us with a duck decoy from earlier times. His special work on duck decoys is recommended to anyone who wishes to learn about duck decoys or how to install one.

Catching ducks, geese and other water birds with stationary nets, pull nets, double clap nets and rocket nets is described later.

May the instructions and representations offered here give wings to the catching of ducks, geese, swans, and other water birds! Banding of these groups, some of which make such long journeys, is wanted and needed for many reasons. Only the knowledge of their habitats and migrations paths will lead to their effective protection.

Fig. 150. Decoy for wild ducks from the Hunting Book by Wolfgang Birkner, 1639. After Lindner, 1940.

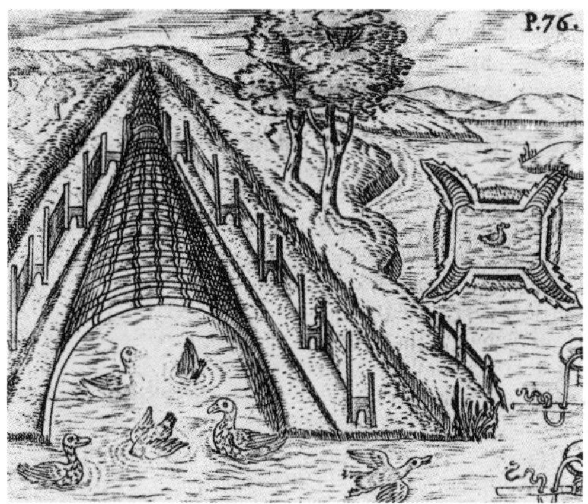

Fig. 151. Seventeenth century duck decoy. After Aitinger, 1653.

6.1 Duck decoys

Duck decoys from past centuries have become rare or have disappeared entirely from most European countries; a few are still to be found in Holland and there are several on the Island of Föhr off the North Sea coast of Germany where they are maintained for bird banding. In the lowlands of Lower Saxony the hunting associations near Celle (Germany) also maintain a decoy for banding.

In general we consider the duck decoy with four curving pipes consisting of narrow ditches, in which the catching actually takes place, extending from the main body of water.

It is desirable to have several pipes to take advantage of the wind direction. Ducks prefer to enter the pipe swimming into the wind. A diagram of such an installation is shown (Fig. 152). Payne-Gallwey (1886) considered a water surface of 4,000 to 12,000 square meters particularly favorable. But size is not always the critical factor as natural conditions have to be taken into consideration. If possible, the water depth should not exceed 1 meter, but here again, one must adapt to the site. Surface feeding ducks of course prefer shallow water where they can dabble. In general it is important for the shore to be level and for the water to be surrounded with trees and bushes.

Each ditch at its entrance under the first hoop is at least 6 m wide and in large decoys it may be two or three times as wide. Its outer side is about 70 m long. This measurement, like all the others, is not inflexible. As the pipe nears its end, it becomes narrower until finally it is only 60 cm wide. The depth of the ditch is also gradually reduced, tapering from 30-50 cm to only 10-15 cm.

Over each ditch arched iron hoops are erected about 1.5 m apart (wooden hoops were used in earlier times). The middle of the first arch is about 4 m above the water. The arches become

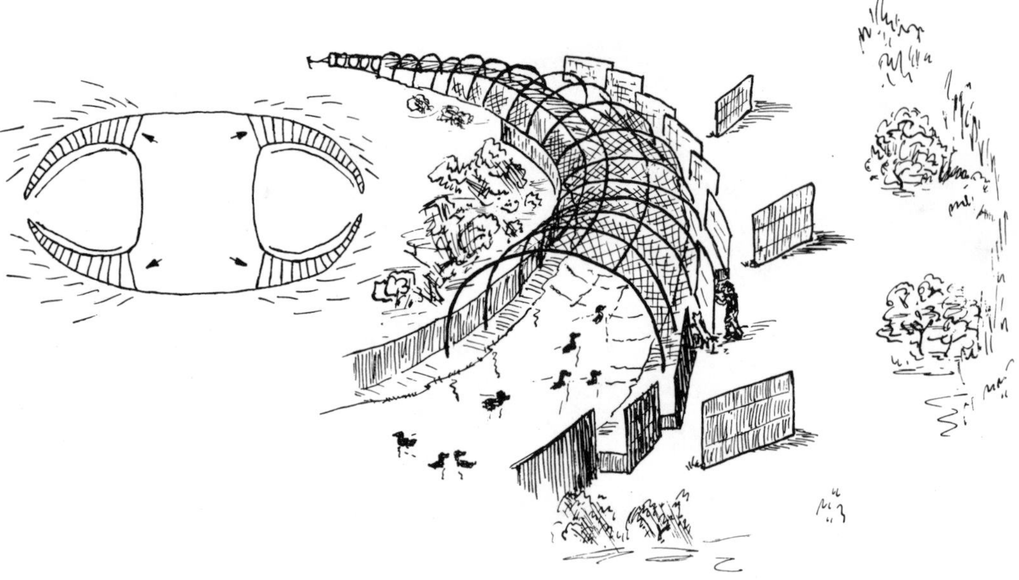

Fig. 152. Layout of a duck decoy. After HOLLOM & BROWNLOW, 1953.

lower and lower and, as they span the ditch, they become progressively narrower until at the end they are only 60 cm high. Netting is strung over the arches so that the whole looks like a giant snake-like cage. The ditches with their net-covered arches consitute the pipes of the decoy.

Along each pipe, for about ⅔ of its length, 10-12 overlapping screens of reed are erected. These are 60-90 cm from the edge of the ditch. In addition a far larger screen, known as the head-end screen, is placed where the net-covered pipe starts. There are two more screens known as breast-wall screens.

These screens are so arranged that the catcher can move along the pipe without being seen either by the birds or by birds out on the open water. The posts of the 3.5 m long screens are staggered in such a way that the screens overlap, but with a gap between each of them. The gaps between have only low screens (60 cm high), the so-called 'dog jumps'.

At various areas along the edge of the open water, the shores are flat. These are called 'landings'. The surface of the water itself and the entrances of the pipes have screens or 'bankings' so set about them that the catcher in the decoy can watch the birds through peepholes and can slip from one part of the decoy to another.

When catching, the trapper conceals himself behind the screens and often has a trained dog (preferably a Corgi or some other reddish-brown fox-like dog; a ferret has also been used with success). In order to entice the ducks into the pipe, tame ducks are induced to hang around the entrance of the pipe by the decoy-keeper who throws corn over the screens to them at particular times. Now the dog jumps from one screen to another, each time displaying himself to the wild ducks which are feeding together with the tame ones. A small, mobile animal makes the ducks extraordinarily curious; they follow it into the pipe. When the ducks are about halfway down the pipe, the decoy tender lets the ducks see him behind them. He shows himself at the small staggered gaps between screens and drives them, going from one low screen to the next. He continues to drive them thus until they flee into the narrow portion of the pipe. Of course, during this process, the decoy-keeper must not let any ducks still lingering behind detect him.

The details of the gathering cage at the end of the pipe of the Drielton decoy in England (Fig. 153) are shown by LOCKLEY and RUSSELL (1953). A dog is used only during the day. Sometimes ducks can be enticed into the pipe without a dog or decoys, simply attracting them with food. By day, a small decoy pond attracts the ducks; at dusk, the ducks leave to feed on

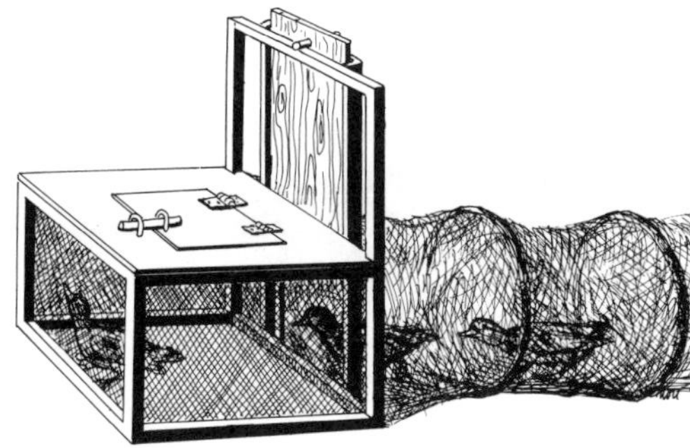

Fig. 153. Gathering cage at end of a duck pipe; its size depends on the number of birds in the anticipated catch. After LOCKLEY & RUSSELL, 1953.

plowlands, in marshes and along the coast whence they return to the pond at daybreak. Upon their return they sleep for a few hours. In the afternoon they start to move toward feeding grounds and it is then that they enter the pipe.

A short, vivid description of the catch is given by DIETRICH (1925): "I had the chance to become acquainted with catching in the decoys on Sylt, Föhr and Pellworm (Germany). From a distance a decoy appears to be a small woodland with crippled willows, alders, poplars, elders and other shrubbery; the vegetation seems strongly overgrown with lichens and mosses. In this little wood, which is enclosed by an earth wall for the most part thickly covered with vegetation, there is a rectangular pond from which the so-called pipes extend."

"In autumn, for a length of time depending on the weather, the tide flat is an arena for countless flocks of ducks and geese. At ebb tide the ducks feed upon eelgrass, mussels, snails, etc. on the dry flat. At high tide they like to seek out some quiet body of water for resting and digesting. The decoy pond with its peace and quiet looks just right to them and all the more suitable as there is already a gathering of ducks (live decoys) there. Some of the decoys, having unclipped wings, fly out onto the flat at low tide and then at high tide entice their wild cospecifics to fly to the decoy. The wild ducks settle down on the pond of the decoy in flocks and soon many hundreds are assembled there. Now the duties of the decoy-keeper begin. He throws a handful of barley through the reed wall into the water in the pipe. Immediately the intentionally not too well fed decoy ducks hurl themselves upon it and at least some of the wild ducks follow them. The decoy-keeper progresses softly behind the barriers, unseen by the ducks, toward the end of the pipe, throwing barley on the water from time to time. As soon as he has enticed the wild ducks far enough into the pipe, he doubles back and suddenly shows himself behind them—namely between them and the pond—so that their retreat is cut off. Fear drives them deeper into the pipe, whereas the lure ducks, accustomed to the decoy, swim back onto the pond. Finally the wild ducks get to the funnel where they crowd together, unable to go any farther. In this manner the catch is repeated 20, 30 or more times during the period of high tide. The decoy-keeper always used whichever pipe is upwind from the pond as the ducks prefer to swim into the wind. For the most part Mallards, Eurasian Widgeon, Northern (Common) Pintails and Common (Green-winged) Teal are caught; other species occur irregularly in small numbers."

A large traditional decoy, complete with small reddish dog, installed at the Delta Waterfowl Research Station on the shores of Lake Manitoba in Canada has proven highly successful.

6.2 Danish decoys

The Danish decoy (Fig. 154) has proven highly successful for a small body of water of 2,000 m^2

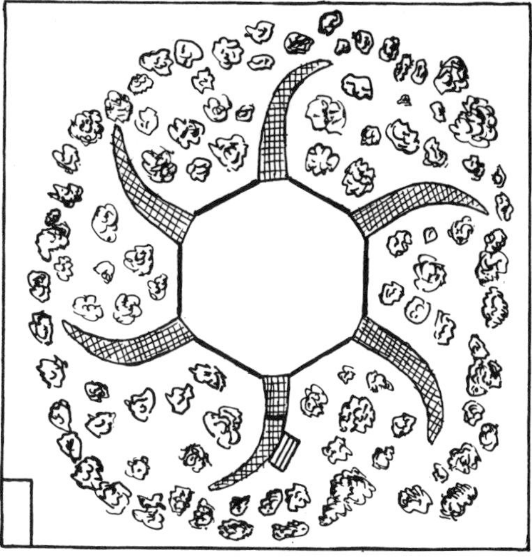

Fig. 154. The Danish duck decoy. Many duck decoys, also outside Europe, have more than four pipes. There are decoys with 10 or more. The pipes are often laid out asymetrically. After LOCKLEY & RUSSELL, 1953.

according to Lockley and Russell (1953). This method is fully described in the collected work of Mortensen (1950).

"Every autumn thousands of ducks are caught here; coming from the shallow-water stretches between Fanø and Jylland they alight on the pond of the decoy, which is carefully hidden in a small grove surrounded by bushes and low trees."

"After the ducks have settled on the pond to dring and bath themselves, they find some decoy ducks there and swim with them into a channel, which is covered with netting; here the decoy birds are accustomed to get their food (barley). From this channel the ducks are chased farther up into a trench, likewise covered by netting and ending in a trap which can be opened for the removal of those that have been caught."

"Every autumn some wild ducks are trained to become decoys, and this is done in the following manner. First one wing is clipped and they are placed in an enclosure ('taming-case') with room for a score of ducks or more. This is a high-roofed structure built out from one of the channels ('taming-channel') with a pool below and a window of wire-netting on one side of the roof. Through this barley is thrown to the ducks which are kept there from 10 to 20 days until they are no longer afraid to eat the barley when the decoy man is looking on. Then they are let free in the narrow end of the channel (which, as long as the ducks are in there, is shut off from the broader part by a sloping door, which can be surmounted by the wild ducks from outside, if they wish to get in, whilst the clipped ducks cannot get over it and out into the pond); there they become accustomed to search for barley, whether the grains fall on the dry parts or in the water and sink to the bottom. After two or three weeks they learn to desist from running up into the trap, where they would be caught with the wild ducks, and now they can be trusted out in the pond, which is surrounded by wire netting. Here they remain during the winter and next spring and on to the summer, returning several times daily into the channels for their food. A year after they have been clipped, in September/October, they grow new flight feathers and are then able to accompany the wild ducks visiting the decoy out to the shallow-water stretches, laid dry at ebb tide, between Fanø and South Jylland (the decoy men now call them 'flying

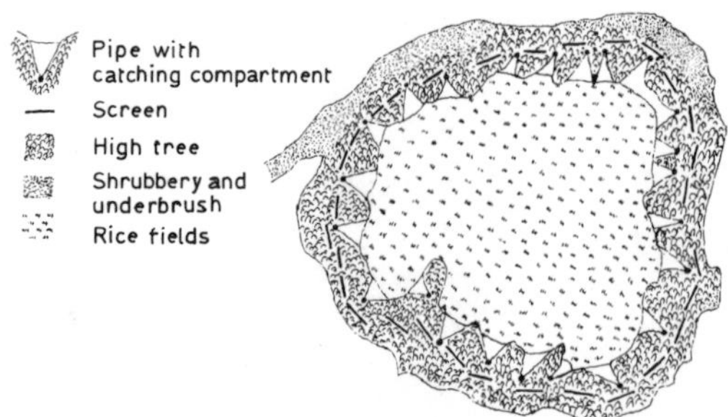

Fig. 155. A duck catching installation in Iran. After C. Savage, 1963.

ducks') but they return constantly to the good feeding grounds in the decoy channels, always followed by new ducks. When frost comes in October or November, most of the 'flying ducks' follow the hosts of wild ducks to milder climates."

According to Lockley (1953), this catching method was first developed in the Netherlands. Additional decoy types exist there, including one with a straight pipe.

Schüz (1957) describes a noteworthy method (Fig. 155) from Afghanistan. "In the Hindu Kush region geese and ducks are caught in pipelike bays. The catcher goes along these narrowing canals under cover of a wall and the ducks, suddenly frightened, struggle against a woven willow canopy that covers one of the water courses. It is said that at the right time of year, hundreds of waterfowl are bagged daily. In the valleys of Kohdaman and Kohistan, little huts are built over narrow water courses, leading back from the main stream or lake, where ducks assemble. The catcher opens a sluice at night so that the ducks, trusting the current in the dark, move through a narrow opening within the hut and now can be grabbed one after another. Two men reputedly can easily catch 150 to 200 ducks this way in one night."

6.3 Australian duck trap

For catching ducks on the Australian coast, several movable funnel traps (Fig. 156) have been constructed (H. O. Wagner, 1958). Each is box-like, with an iron frame covered with poultry netting, and they can be hooked together. The total set consists of several chambers. After the captured birds have pushed through several

Fig. 156. Duck trapping in Australia. Photo: H. O. WAGNER.

funnels into the first chamber, they slip through an opening into another, larger chamber to which the catcher obtains access through a door. Many of the birds caught are ducks. Only a few were Australian Wood Ducks or Chestnut Teal. The nightly catch in three setups is about 200-300 birds. About half of these are already banded. Ornithologists come from Melbourne each morning and remove the birds 10 to 15 at a time, placing them in a sack to await banding. As four men work together, the task is finished in two to three hours. Then barley is again strewn in the funnels; for which several Western Magpies, which get caught over and over again, are already waiting in the nearby eucalyptus.

6.4 Automatic duck trap

On the Orielton coast of South Wales, England, most ducks are caught in several automatic traps (Fig. 157) at the edge of a decoy pond (LOCKLEY and RUSSELL, 1953). The traps are placed $\frac{3}{4}$ in the water and $\frac{1}{4}$ on land so that the captured ducks can rest. The traps must be at least 3 m square and 1.5 m high. They are covered with 25 mm poultry mesh which is bent inwards under the water or buried in the bottom. A water depth of 50 cm suffices. There can be several funnel entrances at the sides so that shorebirds running along the water's edge can be captured too.

The traps are checked at dusk as soon as the birds have flown to their feeding grounds. Under some circumstances, checks are made by day as well. Barley or wheat is scattered in the trap and in front of the funnels. These grains do not float away as oats do, but sink to the bottom and lead the ducks to dive and so to get caught. When the ducks have been banded and the trap re-baited, a newly caught pair is left in the trap to decoy more. These decoys seldom escape if the funnel is so constructed that the ducks have to force themselves through an opening 8-10 cm wide and about 25 cm high. Shorebird funnels ordinarily taper to smaller openings. When the traps are not in use, the ends of the funnels are simply squeezed shut by hand or the door that the bander uses is left open.

The ducks are taken out of the trap with a dip net. A small funnel built into the inside of the trap makes it easier to seize the captured birds. Then the birds are taken to the banding place in baskets.

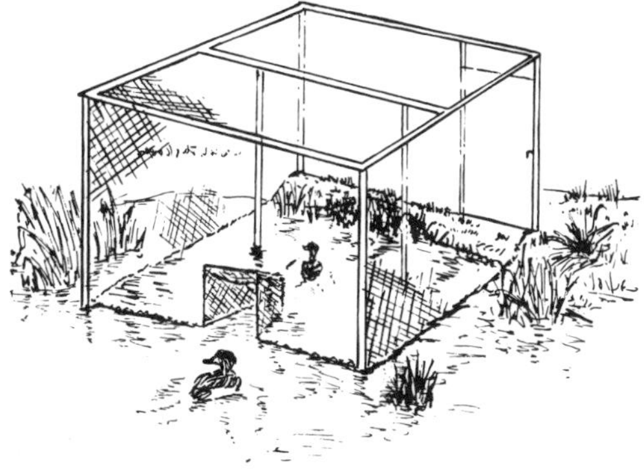

Fig. 157. Automatic duck trap. After LOCKLEY & RUSSELL, 1953.

Fig. 158. Floating duck trap. After LOCKLEY & RUSSELL, 1953.

6.5 Floating duck trap

Although originally designed for catching ducks (Fig. 158), this trap is also suitable for Common Gallinules (Moorhens), Common Coot and other water birds including sea birds. According to LOCKLEY and RUSSELL (1953) this trap was first used for Mallards. The trap, attached by a long cord, was blown out onto the water by the wind. It is more profitable, however, to anchor it in a selected place favored by the birds for feeding or for loafing. With the help of a rope running through two rings attached to heavy weights, one can haul the trap to and from the shore. The trap is secured against unwelcome visitors by locking the rope to a post on the shore. The water birds soon take to the float but according to HOLLOM (1950) they are mostly caught at night. When the trap is first put out, the funnels are stuffed shut with grass and their openings are baited with grain etc. until one sees that the birds are using the float. A wooden or stuffed decoy inside the trap in a sleeping position helps to lure the birds. The float has a rim 1-2 cm high to keep the bait from spilling away. The float's dimensions are 3.5×1.2 m and it is made of boards 5 to 10 cm thick. These are nailed to four cross pieces of the same material. The trap frame is of four strong wire bows over which poultry wire or netting is strung. The wire funnels taper to a height of about 22 cm and a width of 8 cm. The ends of the wire are bent inwards to make escapes less likely. Both ends of the platform have a 50 cm area free.

A small door is built into the top of the trap for taking the birds out. Smaller water birds can be induced to leave the corners of the trap by means of a thin bent stick.

U. ZWERGEL caught 130 Common Shelducks (of which 127 were flightless) on a 4×1.2 m float in Schleswig-Holstein (Germany) between 1968 and 1970. In addition, he caught 100 Mallards (but no Garganey Teal or Common [Green-winged] Teal), 120 Common Gallinules (Moorhen), 40 Common Coot, 45 Common Sandpipers and 30 Water Rails.

The trap itself was 3 m long and 0.6 high; the funnels 25 cm high, 20 cm deep and 8 cm wide at the opening, tapering to 5 cm. The mesh covering of the trap is large enough for small birds to pass through. ZWERGEL baits generously with grain (wheat and oats) inside the trap and has only a single plastic duck anchored 3 m from the trap.

According to LINCOLN (1926) a still larger floating trap, 11 m long, 3.6 m wide and 2 m high was used in North America. The trap was mounted on strong piles permanently driven into the bottom, but it can also be made into a floating trap. The trap is high enough so that the bander can walk into it. The funnels are closed when he is taking birds out.

6.6 Funnel trap, set in water, with leads

K. GREVE went duck catching with friends in the Braunschweig (Germany) area using a funnel trap (Fig. 159). The leads to the funnel were net walls 15 m long and the 'funnel' tapered from 8 m to the trap entrance. The height of the leads was 2 m of which ¼ could be under water. A box—in essence a trap (length 2-3 m, width 1.5-2 m, height 1.2 m)—received the captured birds. If the box cannot be held in place with

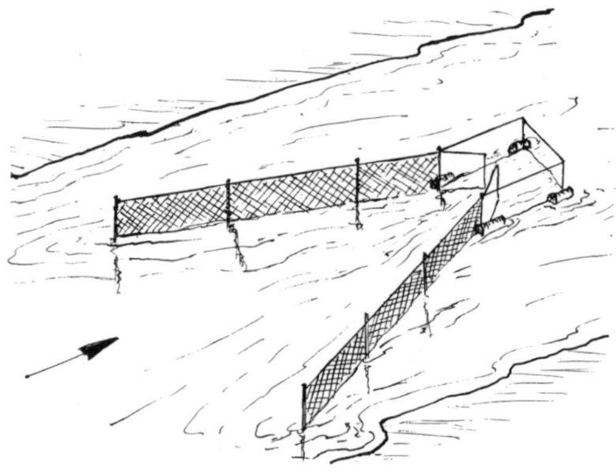

Fig. 159. Funnel trap, set in water, with leads. After K. GREVE.

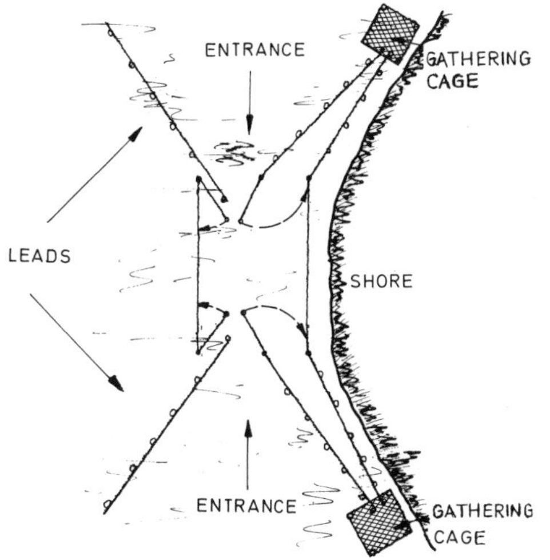

Fig. 160. The hurdle trap. After LINCOLN & BALDWIN, 1929.

stakes, it can be floated by fastening airtight cans at its sides. The bottom 30 cm of the box should normally be under water.

The ducks either swim into the funnel by themselves or are driven downstream toward it with a boat.

6.7 Hurdle trap

This trap (Fig. 160) is permanently installed (LINCOLN and BALDWIN, 1929) where the water level and terrain are suitable. The openings of the funnels should be movable so that the funnels can be closed easily when catching is poor. Bait and decoys improve the catch.

6.8 Garden duck trap

E. A. GARDEN (1964) of the British Wildfowl Trust designed a small portable trap for use in inland waters. It is 2.6 m long and 0.75 m high (Fig. 161). The funnel entrances must be on the landward side or parallel to the shore as the ducks generally make for the deep water. Leads 3-4 m in length and 0.5 m high are indicated. It is desirable to have a gathering cage inside the trap for taking up the birds.

6.9 Waterlily leaf trap

This trap (Fig. 162) was first used on the Illinois River in North America (LINCOLN and BALDWIN,

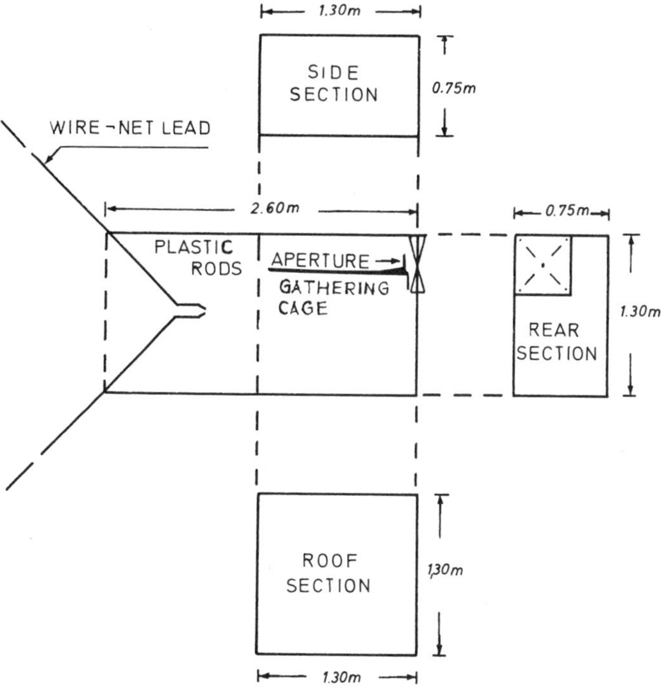

Fig. 161. The Garden duck trap. After GARDEN, 1964.

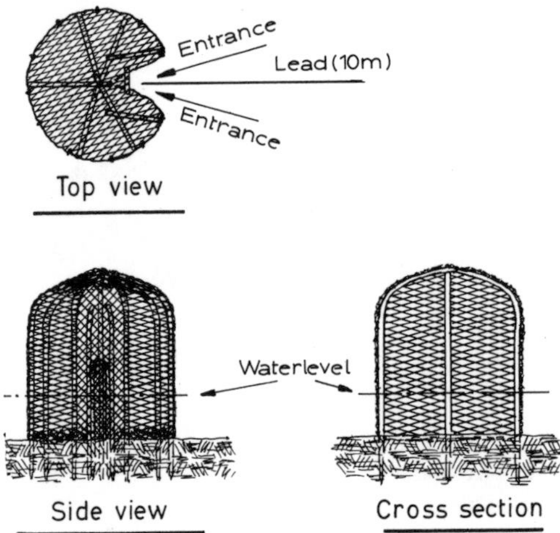

Fig. 162. Waterlily leaf trap for ducks. After LINCOLN & BALDWIN, 1929.

1929). Over 2,000 ducks were caught in one autumn in seven traps. The traps are easily built and not expensive if willow or poplar shoots, which must not be too thick, are available. In principle, the trap resembles many shorebird traps. Nine sprouts 3-3.6 m long are cut and shoved firmly into the mud so that they form the shape of a waterlily leaf or of a deeply indented heart. About 1.2 m above the water, the sprouts are bent to the middle and their tops are firmly wired together so that a frame is formed similar to that of the dwellings of certain Indian tribes. The two sprouts that form the entrance should be about 45-60 cm apart; from these the poultry wire covering is pulled to the middle of the trap until it is not more than 10-12.5 cm apart. If necessary, these parts of the wire netting can be stiffened with small willow withes which can be threaded through the mesh and stuck deep into the mud. The ends are then tied together up to 30 cm above the water. The opening is reinforced by an inverted U of strong wire if need be. The practical dimension for this trap is a diameter of about 1.5 m so that the trap walls can be covered with woven wire 5.4 to 5.7 m long. Two short pieces serve to cover the top.

The wire is pulled outwards about 45-60 cm at the base of the trap and stamped into the mud so that no ducks can escape by diving. Forked sticks driven through the wire can serve the same purpose; these should be rather close together as will be well put to the test if many ducks are caught.

It is best to place the trap with its back on shore so that the ducks can be taken out more easily. The funnel can be in water 20-25 cm deep. The trap, however, can also be placed in water up to 60 cm deep. The bander has to count on getting sopping wet when taking out the ducks so it is best for him to wear rain gear and waders or bathing trunks.

The entrance for the trapper can be on one side of the trap between the lower part of the wire mesh and the top of the trap. The wire for the entrance is simply hooked together; unlike that of the rest of the trap which is fastened firmly. Leads about 10 m long guide the ducks towards the entrance.

6.10 Diving duck trap

According to LINCOLN and BALDWIN (1929) this trap is a pen, closed on three sides (Fig. 163). It is erected on a natural feeding area in water about 90 cm deep. The sides are 2.4 m long. One side is a door which is raised upwards by a pull cord which in turn reaches to a blind. Four galvanized iron pipes form the corner posts. Two of these are 2.4 m long so that they can be driven 60 cm deep and firmly into the ground. These are used as the front posts. Holes are bored about 65 cm from the bottom matching holes on the door. Small bolts through these holes serve as hinges. The two other pipes are 2.7 m long.

The door, 1.8×2.4 m in size, is of thin iron pipe covered with poultry wire. The door is fas-

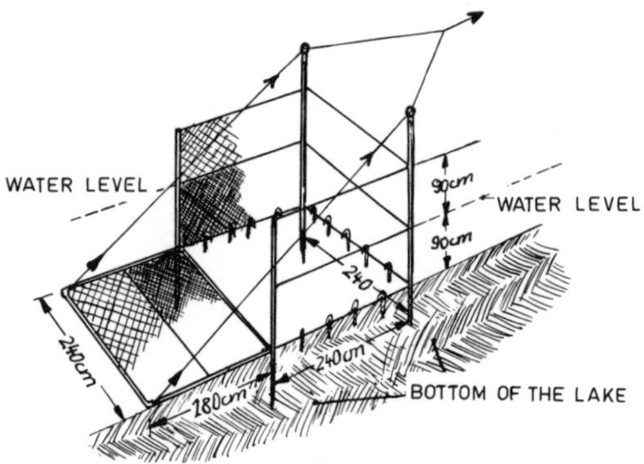

Fig. 163. Diving duck trap. After LINCOLN & BALDWIN, 1929.

tened in position before the posts are driven into the ground. The two 2.7 m poles are similarly driven into the ground until they project 1.2 m above the water. The wire mesh (15-20 mm) must be anchored to the ground with forked sticks for the underwater portion of the trap. The above water section can be of wider mesh (25 mm).

Two small rollers at the top of the rear posts, through which two wires pass to the pull cord, insure smooth operation. These wires are attached to the top corners of the door. The bander pulls the cord when enough ducks are in the trap. The catches are better if the trap is baited with grain.

6.11 Concerning duck catching in the Camargue (France)

The Tour du Valat Biological Station in the Camargue in southern France catches several thousand ducks annually, especially during the winter. According to R. Lèvêque, several possibilities are taken advantage of and the traps are adapted to the existing water courses and tamarisk shrubs. The height of such a trap may be two meters and the width 6 to 7 m. The shape of the sides is adapted to the tamarisk vegetation; thus the trap is camouflaged in a highly natural manner, awakening no distrust in the ducks. But on all sides, there are funnel entrances leading into the trap, which also has an entrance for the bander and a compartment into which the captured ducks can be driven. The trap is covered with poultry wire which is burried 30 to 40 m at the bottom and simultaneously curved under. Rice is used as bait. Daily catches of up to 300 ducks are not uncommon.

As soon as these traps become dry in the summer heat, similar traps are baited with dead fish for catching Black (Common) Kites, Marsh Harriers and Herring Gulls, especially when food for these species is relatively scarce. Also in summer, when there is little standing water in the traps, the various heron species enter them, attracted by fishes and frogs. The entrances need to be relatively larger for herons. On the other hand in winter, after a freeze-up, many Common Jackdaws get into the traps.

6.12 The tent net

In Russia S. A. Postinikov (1958) uses a special tent net for catching ducks and game birds. The net is 8 × 8 m and made of 0.5 mm thick twine (mesh width 55 mm). The center of the net is fastened to a 16.5 cm ring made of 20 mm thick iron rod. Eight 1 mm cords run from the ring to the edge of the net so that it forms an octagonal tent 3.2 m high (Fig. 164 a). The eight cords project and are staked so that the lower edge of the tent is suspended 0.7 to 1 m above the water.

The release mechanism is a stick 23 cm long and 2 cm thick, fastened to a ring with a strong 20 cm cord. The midpoint of this stick rests on a groove cut into the tip of the center pole and its tip is placed in a notch at the midpoint of a second stick. The upright stick (Fig. 164 b) has a simple pull cord tied to its top and leading to

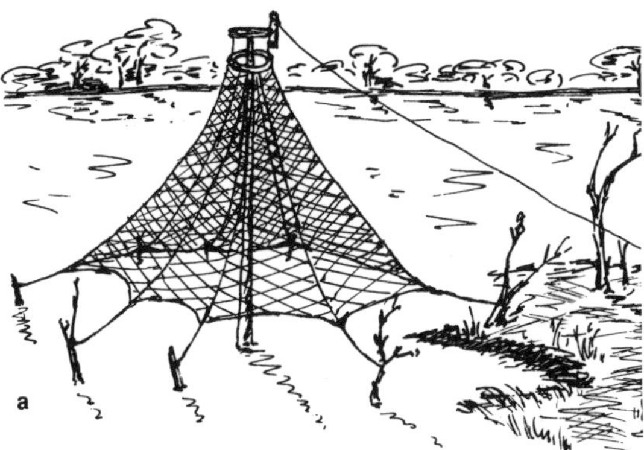

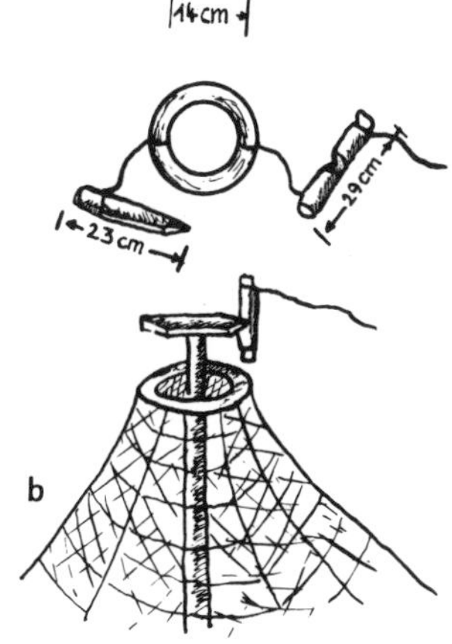

Fig. 164. (a) Tent net for ducks; (b) Release mechanism. After Postinikow, 1958.

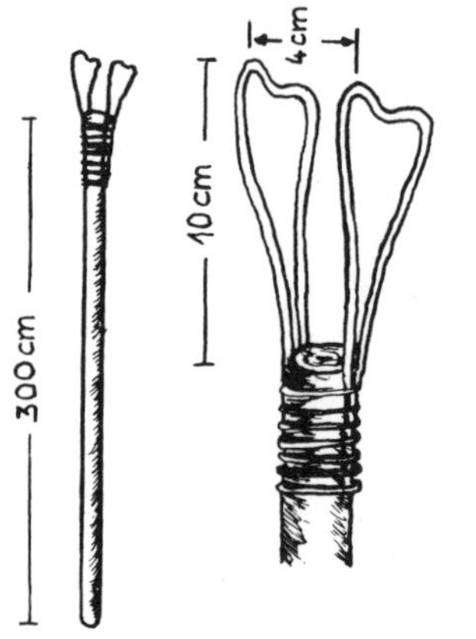

Fig. 165. Rods for setting up a tent net. After POSTINIKOW, 1958.

the trapper's blind. The pull cord releases the sticks so that the net drops.

The center pole is 4.5 m long and its diameter tapers from 10 cm at the bottom to 4.5 cm at the top. The groove at its tip is 2.5 cm wide and 2 cm deep and lined with thin sheet metal to receive the horizontal trigger stick. The bottom of the center pole is pointed and driven about 50 cm into the ground under water.

For catching drake Mallards, swamps with tall timber and brushy vegetation near flooded meadows are best.

It is easy to set up the net where water is 40–50 cm deep. The brush and tall timber help to conceal it. Catches are made at dawn and at dusk. Hen Mallards hide from the drakes during these cooler parts of the day and drakes become more active. The lure duck is tethered by a foot under the middle of the net, but so that the net ring will not fall on her. A drake is similarly tethered on a nearby shore, but where the duck cannot see him. This causes her to call more often and the voice of the drake encourages other drakes to hurry close and to approach the duck more boldly. POSTINIKOV observed this on numerous occasions.

For successful trapping, the net site must be changed for the drakes go trustingly under the net only during the first twilight and obviously shun it on the following days. The explanation for this appear to be that the drakes have become trap-shy either because they were caught or had seen the catch.

A fork, mounted on a slender 3 m rod, is helpful in raising the net and spreading it. The two fork prongs are indented loops of wire 4 cm apart (Fig. 165). With the help of this device, passed through the ring, the trigger mechanism is gripped in two places and with the attached net is raised and set in the notch on top of the pole.

It is advisable to accustom the ducks to the net early in spring long before catching starts. Experience has taught us that the lure duck must be conditioned in advance or she gets

Fig. 166. Duck trap on the east coast of Sweden. Photo: R. DE VRIES.

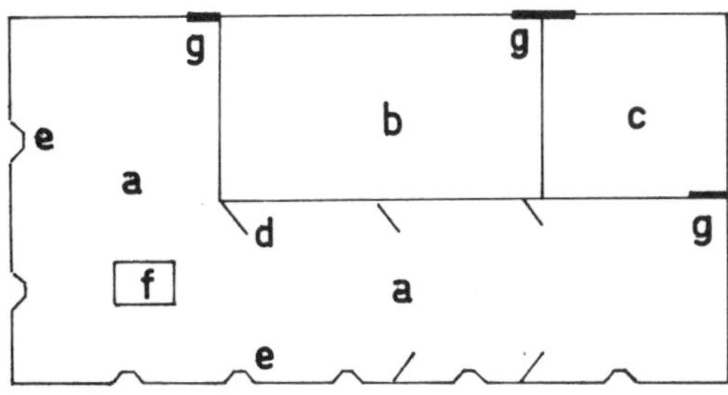

Fig. 167. Ground plan of Fig. 166. (a) trapping area, (b) lure bird compartment, (c) banding compartment, (d) leads, (e) funnels, (f) feed, (g) doors. After R. DE VRIES.

badly entangled when the net falls, becomes shy and ceases calling.

Drake Common (Green-winged) Teal have been caught together with Mallards in this type of trap. Undoubtedly, a teal hen could be used with a net of smaller mesh to catch teal drakes.

6.13 Swedish duck trap

On the east coast of the Island of Oland a funnel trap (Fig. 166) is used by banders of waterfowl and shorebirds. A small dam near the sea keeps the water in the trapping area about 20 cm deep.

The frame of the trap, according to R. DE VRIES, is made of iron pipes. It is 25 m long, 12 m wide and 2 m high and covered with 20 mm poultry wire. The seven funnels are 8 cm wide and 35 cm high and face the ocean (Fig. 167). The decoys—a few ducks and shorebirds—live in a closed compartment. In addition, a raft with food is placed inside the trap to attract ducks. The trap is checked regularly and the birds are caught with a hand net. Leads, about 50 cm wide and 2 m high, facilitate driving the catch into the banding compartment.

6.14 Eckernförder duck trap

In 1970 U. ZWERGEL had good results with a small (4×1.5×1 m) funnel trap (Fig. 168). It caught 78 Mallards, 4 Common (Green-winged) Teal, 2 Gadwalls, 1 Garganey Teal, 2 Common Shelducks, 7 Common Coots and 4 Common Gallinules (Moorhens). The funnels, however, should have longer leads and not all be at the front of the trap. A board just above the surface of the water gives the birds a chance to rest in the trap.

Fig. 168. Eckernförder duck trap. Photo: U. ZWERGEL.

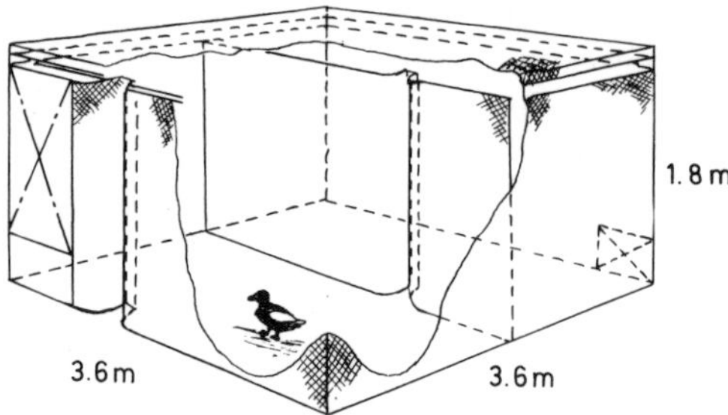

Fig. 169. Ythan-Abberton funnel trap. After GARDEN, 1964.

6.15 Ythan-Abberton funnel trap

The distinctive feature of this trap is that after the birds are caught, they pass through another funnel leading to a second compartment, from which they are far less likely to escape (GARDEN, 1964). The internal funnel should not be opposite the entrance funnel (Fig. 169). This trap is primarily for Tufted Ducks, Greater Scaup and Common Goldeneyes. Barley and wheat are the best bait.

6.16 Funnel trap for swans

GARDEN (1964) illustrates an all purpose trap for waterfowl including swans (Fig. 170). The funnel is usually 1.2 m high and 20 to 30 cm wide—wider for swans. The trap is placed below high tide with funnels on the landward side. Ducks are most apt to enter the trap at high tide when its bottom is flooded. On the Ythan River on the east coast of Scotland, the trap is set about 25 m from the high tide mark. At normal flood tide the trap's bottom is about 1 m under water, but sometimes the tide rises unexpectedly so that the captured birds must be able to find a way to escape. A 30 cm gap between the

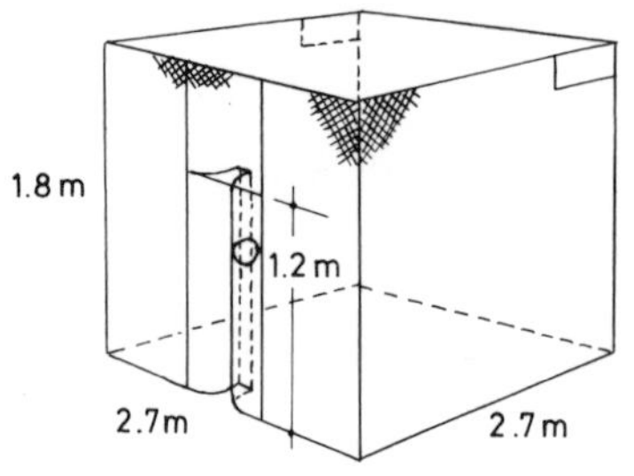

Fig. 170. Swan trap. After GARDEN, 1964.

sides and the top of the trap is provided for such eventualities. In windy weather, small diving ducks sometimes climb the walls. To prevent them from escaping through the gap, a wire deflector pointing inwards is fastened to the bottom of the gap. The mesh at the lower part of the walls must be big enough so that marsh birds and other nonswimmers can escape drowning by passing through it.

Common Eider, Tufted Ducks, Greater Scaups, Common Goldeneyes, Mallards and Common Pochards, as well as Mute and Whooper Swans, were caught in such traps baited with barley and wheat. Common Goldeneyes find their way out rather easily. Plastic tabs 0.4 cm in diameter slanted at an angle in the funnel help prevent escapes. Such tabs can be used in similar traps.

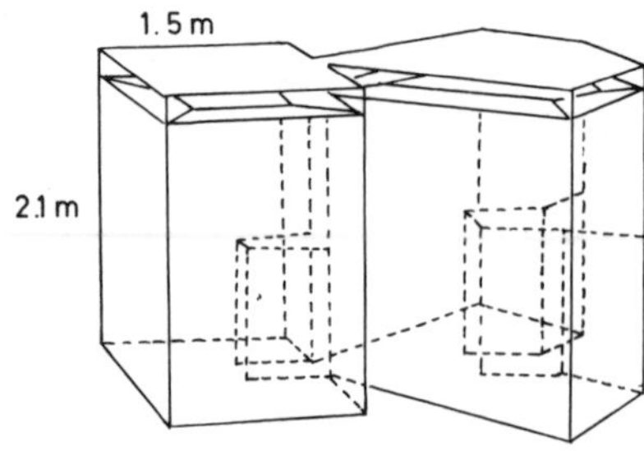

Fig. 171. Funnel trap for shelducks. After YOUNG, 1964.

6.17 Funnel trap for Common Shelducks

At the mouth of the Ythan, Scotland, a special Shelduck trap (Fig. 171) was made to catch the birds when they returned from their winter quarters early in spring (YOUNG, 1964); over 200 Common Shelducks were caught within two years. The techniques for catching Shelducks are different from those for other ducks. For example, they congregate in large numbers only in tidal waters so care must be taken to keep captured birds from drowning.

At the mouth of the Ythan the traps are at the very outer edge of a great tide flat where they must withstand inundation and even the crush of winter ice. The location is of prime importance; where the birds put in to feed, but where they will find only bait (barley and wheat) in the traps to feed on.

The funnel is wide, for Common Shelducks are wary of narrow entrances. The meshes of the trap's covering are large so that the ducks are not apt to injure their bills and also so that non-swimming species, caught incidentally, can get out.

The double roof (Fig. 171) with its 20 cm head room offers an escape hatch when the water rises too high. The inner edge of the lower roof is also 20 cm wide.

7. Cage traps

Historically the smaller cage traps evolved before the large cage and funnel traps. For example, the idea of a catching cage was known in British Guiana in the form of a trap covered with nooses (Fig. 172, ROTH, 1924). An even simpler method paralleled this beginning in the Congo (LINDBLOM, 1926). These were not cages in the true sense; they contained a lure bird, but the trapper had to lurk nearby so that he could quickly cover the opening (Fig. 173) with a leaf as soon as a wild bird slipped in.

A further development can be seen in the grouse trap of the Yakuts in Siberia (Fig. 174). The bird catches itself by landing on a tip top. The European tit trap (described below) is also a simple cage trap (Fig. 175), but with a trigger.

The French sparrow trap is still another development (Fig. 176). It contains a lure bird and several narrow funnels serve as entrances. This trap dates back to 18th century France (QUANTZ, 1941) and was already illustrated in 1712. Although it may have been an original invention, it is also possible that primitive societies provided the model for the trap. MÈRITE (1942) shows a woven basket with a single funnel used in the New Hebrides. The French sparrow trap, at any rate, should produce massive catches.

These may not only have been the traps from

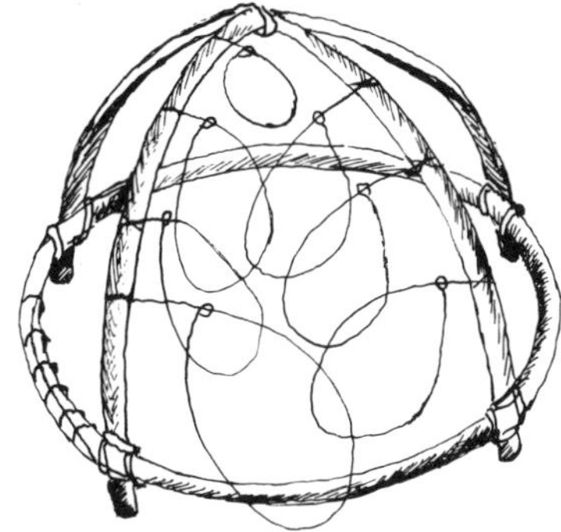

Fig. 172. Noose trap in British Guiana. After ROTH, 1924.

Fig. 173. Trapping a small dove *(Ptilopus fasciatus)* on the Samoa islands. The dove in the basket is tethered. As soon as a wild dove jumps into the basket the catcher leaves his hut, covers the opening with a banana leaf and grabs the bird. The tribal chiefs vied to catch the greatest possible number of these small doves, which were not used for food. After KRÄMER, 1903.

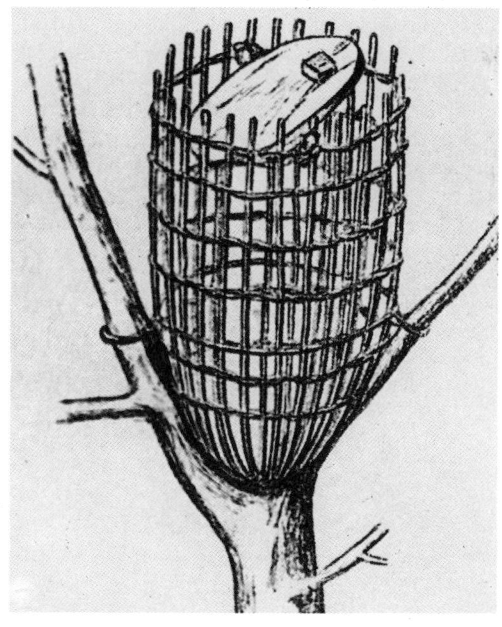

Fig. 174. Yakut grouse trap. After PFITZENMAYER, 1926.

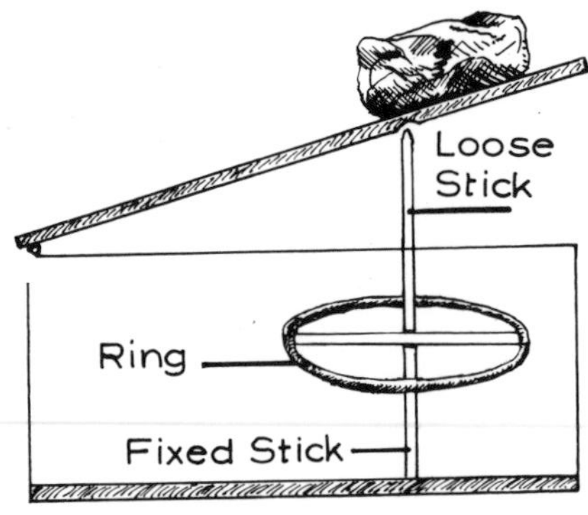

Fig. 175. Two types of the European tit trap. After KRACHT, 1952.

which modern traps evolved, but they had attained considerable sophistication by the 17th and 18th centuries (BECHSTEIN, 1797).

The origin of lures and baits is not known. Live decoys may have been used first or baits such as seeds, berries or worms may have antedated them. The question is worth pondering; but in either case, the trapper needed to know the birds and to be a good observer. Bub suspects that the earliest trappers used live decoys: nestlings, fledglings or adults that could not fly. Wild birds visit no place more regularly, and therefore more obviously, than their nests. If the parents are not wary, this could readily be observed. Furthermore, it is often easy to capture an incapacitated adult.

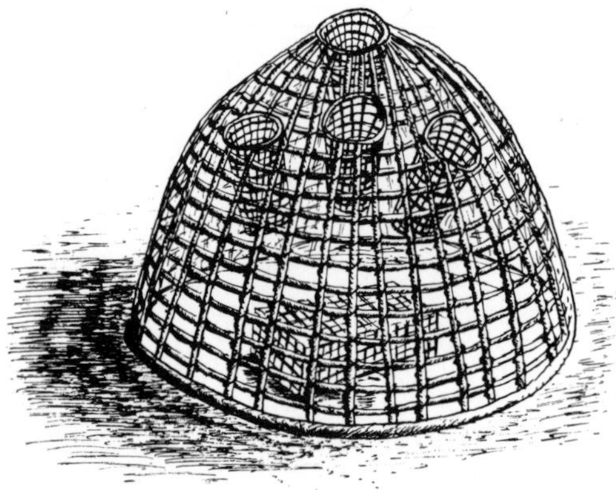

Fig. 176. French sparrow trap. After QUANTZ, 1941.

On the other hand, feeding behavior of some birds is equally conspicuous and easy to observe as, for example, they go after berries. Berries may have been the first bait. At any rate, the earliest trappers must have understood attractants.

Pit traps are somewhat related to cage traps, but we recognize a distinct development, which continues to the present day. They will be treated separately.

7.1 Cage traps with a lure bird compartment

Cage traps will never make the big catches that funnel traps and stationary nets do, but we cannot get along without them, especially for catching songbirds. They are useful at all times of year for both granivorous or insectivorous birds. Eurasian Siskins, *Acanthis* finches, European Goldfinches and Common Serins react especially well.

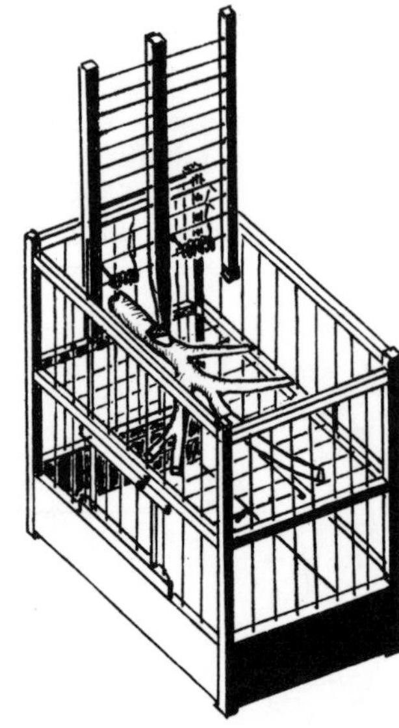

Fig. 177. Cage with catching compartment above the lure bird. After SUNKEL, 1954.

Fig. 178. Trap with 8 cells. Very successful for catching crossbills at the Serrahn Biological Station. Photo: H. Bub.

There are several types of these traps (Fig. 177-179), one with the lure bird below, another with the decoy on the side, and still another with the decoy surrounded by catching compartments. Fig. 177 illustrates a typical top entrance trap; the lure bird's position in a compartment below has generally proven most practical. No set rule applies as some birds prefer to stay on the ground and to enter the trap at that level. The trap illustrated is 30 to 40 cm long and 20 to 25 cm high. The trigger is released by a twig upon which the bird alights. Its weight pushes the twig down, in turn releasing the catch of a spring-loaded door.

E. Franz at Magdeburg, Germany, preferred a 4-cell trap for finches, etc. Two of the cells abutted the lure bird compartment. Franz erects these traps on poles; hangs them in trees, placed on a board, or sets them on the ground, singly or in groups. Between October 4 and November 10, 1962, he caught 154 Eurasian Siskins; in nine mornings he caught 300 Twites, and in addition he captured Common Bullfinches, Red Crossbills, European Goldfinches, Eurasian Linnets, Hoary (Arctic) Redpolls, Common Serins and Bramblings. Insect eaters such as *Sylvia* warblers, pipits, Old World Flycatchers, larks, Dunnocks, Icterine Warblers, and Coal Tits (these especially in February and March) can be caught with live lures. In autumn we bait the traps with elderberries. Nieselt-Lausa (1924) states that the Icterine Warbler reacts to other species used as decoys during the breeding season, but the lure bird must give voice.

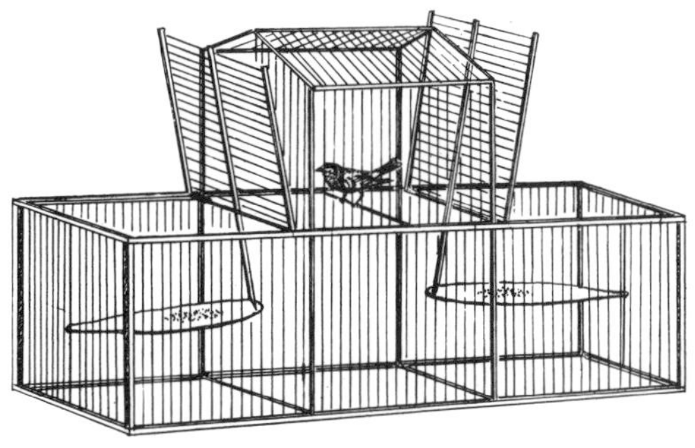

Fig. 179. Two-cell trap used in the Balearic Islands. After Salvator, 1897.

Lure birds, of course, are also used in lands outside of Europe. This is reported by Voss (1967) from a trapping expedition in the primeval forests of Southern Brazil to capture especially tanagers.

7.2 Cage traps with tip-top entrances

The many commercial sparrow traps sold since 1930-1935 (Fig. 180) have given wide distribution to this method. Mansfeld (1950) gives us an idea how efficient it is. Between August 8

Fig. 180. Sparrow trap with tip top entrance. Today there are many traps with wire floors. Photo: A. PRÄKELT.

and December 10, 1935, 1,104 House Sparrows were caught in two traps set in a threshing machine. Such examples are numerous.

The birds enter from the top. VIANDEN (1955) reports that individuals sometimes escape when entering sparrows make a momentary gap. Others take the bait with outspread wings, without entering the trap. This demonstrates the sparrow's ability to learn from experience.

The sparrows first fly toward the sparrows they see in the trap. Grain must be available too and can be glued to the treadle. In addition to sparrows, tits, nuthatches and finches (for example, Twites) get caught. G. DIESSELHORST (1968) even had success with these traps in his Greater Whitethroat study: a number of breeding birds slipped into his trap as soon as he used the same species as a decoy.

In one autumn, W. STÜRMER caught 180 Dunnocks in seven traps hung 1 m high and 1-2 m apart in his garden in Hannover (Germany).

H. LÖHRL finds this trap suitable for Spotted Flycatchers when hung under eaves against the wall of a house: the flycatchers investigate the opening as a nesting possibility and get caught. Seven were caught at one place in one spring.

P. FISCHER built a very useful trap for Common Bullfinches, Eurasian Siskins, other finches, tits and other species, near Hohegeiss in the Harz Mountains (Germany) (Fig. 181). The entrance is in the center, and the birds slip off the treadle into the holding compartment at the left which may be covered with unobtrusive cloth. The lure bird compartment is at the right.

VAUK and GRÄFE (1962) built a tip-top entrance in the top of a 3×2.5×2.5 m aviary (Fig. 182 and 183), and caught essentially all the Collared Doves that visited the catching garden on Helgoland. Two Collared Doves were kept in

Fig. 181. Cage trap from the Harz Mountains. Photo: H. BUB.

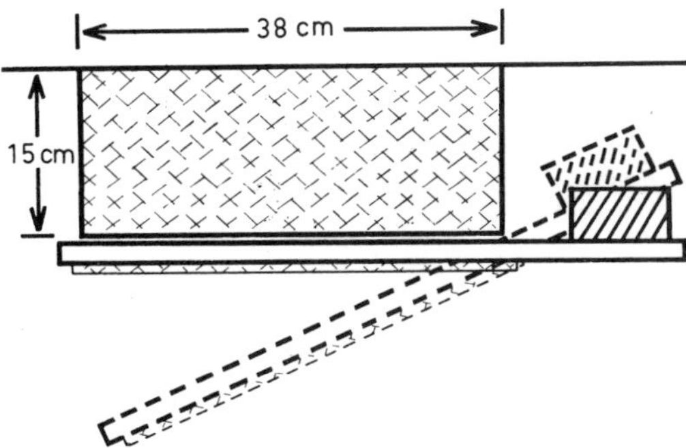

Fig. 182. Cross section of wire basket top of aviary. Treadle trap for Collared Doves. Solid line = treadle when closed. Dotted line = treadle when tipped. After VAUK & GRÄFE, 1962.

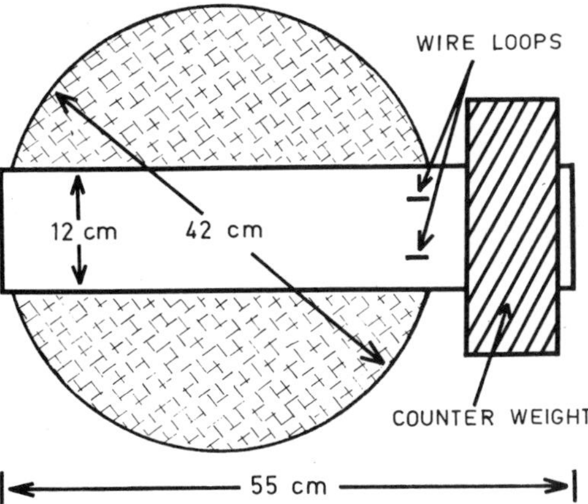

Fig. 183. Wire basket and treadle seen from above. After VAUK & GRÄFE, 1962.

the aviary as decoys. Whenever possible, a pair was used, as pairs get along better with each other and besides the mating display of the male causes him to be an especially fine lure bird. The aviary was supplied with perches, some protected from wind and rain, a bath and—in plain sight—a feeder. We refer to their description:

"The catching device consists of a basket resembling the mesh of the aviary. It is 15 cm deep and 38 cm in diameter. The basket is open at the top. Its circular floor is of wire mesh extending 2-5 cm beyond the basket's sides. Part of the floor consists of a treadle which is attached to its edge by wire loops which serve as hinges. The treadle is longer than the diameter. A counter weight of scrap metal or something similar keeps the treadle up, thus closing the bottom of the basket. On the other hand, the treadle tips easily to let the bird into the trap. Feed can be glued onto the treadle as an additional inducement. The construction of the trap can be altered but to insure success three points are essential: (1) the lure doves must be well adapted to the aviary; (2) the top of the basket must be flush with the roof of the aviary and the basket should be deep enough so that a dove cannot quite open its wings as it slides down the treadle; (3) the treadle must be nicely balanced so that it gives with the weight of a dove, yet will swing back into position automatically.

The set can be improved by placing dead trees at the corners of the aviary. For best results, a fairly stout branch should extend over the treadle entrance.

7.3 Wilhelmshaven gull trap

About 1950 A. PRÄKELT built a highly successful gull trap near the buildings of the Helgoland Ornithological Station at Wilhelmshaven (Germany). It was adjacent to a sea wall by a wharf where gulls loafed. In one winter 100 Herring Gulls, some Greater Black-backed Gulls and one Glaucous Gull were caught.

This gull trap (Fig. 184) is 1.2 m high and the frame is of 3 cm diameter iron pipe. The doors and sections are framed with wood. A lure bird chamber, 1.2×1.2 m, is in the middle for the decoy gull. Three catching chambers are grouped around this; two more can be added. Fig. 184 shows the set without additional catching chambers.

Whenever gulls have entered one of the baited compartments, its door falls, released by pulling the appropriate pull string. Each compartment has a pull string leading to a hiding place or to one of the windows of the institute's buildings.

7.4 Raptor cage traps

There are various methods of catching raptors. Stationary nets, bow nets and nooses will be covered later. This section deals with raptor drop traps or Swedish goshawk traps. These traps are not difficult to build with wood or metal frames, but it is probably no longer possible to purchase them.

Fig. 184. Gull trap at Wilhelmshaven. Photo: A. Präkelt.

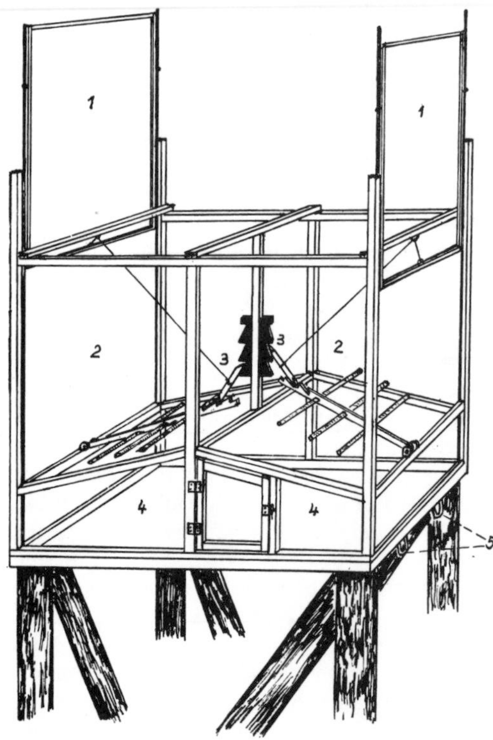

Fig. 185. Two compartment raptor trap. (1) Drop doors; (2) Catching chambers; (3) Releases; (4) Pigeon compartment; (5) Eyes for ladder. After Glasewald, 1927/28.

The raptor trap (Fig. 185) has two compartments (Glasewald, 1927-28). The trap is about 1.4×0.7 m. The lure bird compartment is 30 cm high, the catching chamber 60-70 cm. Each drop door (1) is held up by a string fastened to a prop (3) which is hinged to a false floor of sticks at the bottom and rests in notched racks at its top. A bird landing on the slanting false floor (hinged at the bottom), releases the prop and the door falls down. Each compartment catches independently. The lure bird chamber (4) contains feed and water.

Such a trap can also be built with a single catching compartment (von Schwerin, 1934).

7.5 Cage traps without lure birds

Many useful cage traps do not require lure birds, but are baited with grain, berries, worms or live insects.

The prototype is probably the tit trap (Fig. 175) described in detail by Brehm (1855). In Central Europe at least, it is surely the forerunner of these cage traps. Various authors (Brehm, 1855; Tenner, 1892; Neunzig, 1927; Kracht, 1952) prefer the sides of the cage to be made of natural sticks (usually elder twigs) and to use old, grayisch, weathered boards for the top and the bottom. The tow types of setting are shown in Fig. 175.

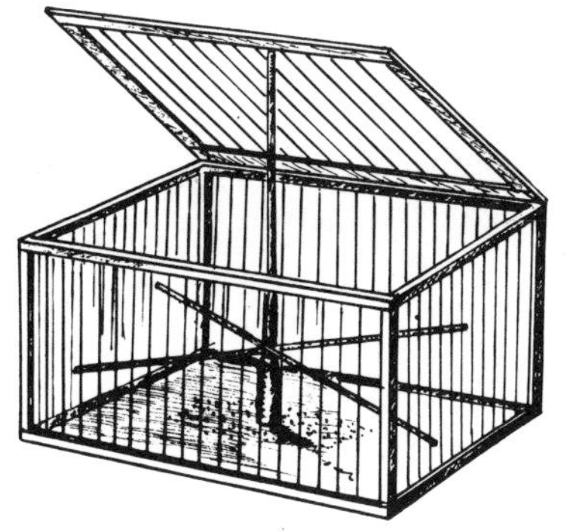

Fig. 186. A novel type of a tit trap. After NEUNZIG, 1927.

NEUNZIG (1927) gives dimensions for a tit trap: 28×16×18 cm. He describes the setting device of this somewhat modern version (Fig. 186). Two small perches are placed crosswise on a 4-5 cm upright block. The trigger stick rests at the cross point of these perches and props up the roof so that about 9 cm stays open. Bait is placed on the floor of the trap. A bird on its way to the bait must stop on one of the perches thereby tripping the trigger.

SUNKEL (1956) considers a tit trap 35×40×30 cm practical for modern banders and it is easily and cheaply built. These traps are useful on expeditions as they can be set immediately before getting up the mist nets. SUNKEL adds that one can place a cage with a lure bird under a trap or several traps can be placed around the lure bird's cage for catching Chaffinches, Icterine Warblers, Black-cap Warblers or others. The placing of the traps is simplified

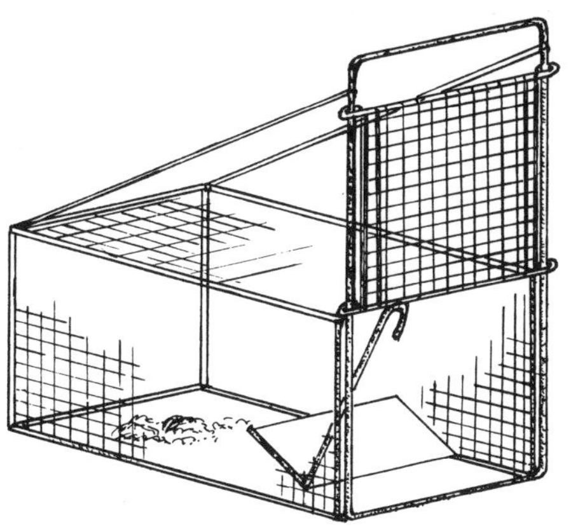

Fig. 187. Potter single cell trap. After LINCOLN & BALDWIN, 1929.

if they are constructed of boards instead of wire mesh. This is usually done when the 'cage' is used as a ground trap and buried precisely to its top for catching European Robins, Common Blackbirds, Northern (Winter) Wrens, Northern Wheatears and other ground-feeding species.

Potter trap. HOLLOM (1950) considers this the best of the perch trigger cage traps. This may be single cell or two-cell as CHAPMAN recommends. The earlier 4-cell types are not particularly useful. The measurements of a one-cell trap are 40×25×30 cm (Fig. 187). Two-cell traps are best made with entrances on opposite sides to insure that one trigger does not inadvertently set off the other (Fig. 188). CHAPMAN's version of the Potter trap is twice as wide at the rear as the front in order to give a bird the impression that

Fig. 188. Improved two-cell Potter trap. After HOLLOM, 1950.

the trap is open at the rear. Therefore the inner wall of his two-cell trap is diagonal. The back and roof of each cell are transparent—preferably of plastic rather than glass to prevent breakage. The sheets of plastic need not be built into the trap as shown in Fig. 188, but can be attached with small hooks. Instead of the usual wire mesh treadle, it is more effective to use a low perch across the inside of the entrance. Hopping birds will bump against it; running birds will step on it and in either case the trigger is sprung. The trigger can be fashioned by twisting a flexible twig (for example, lilac) over a wire. The wire is arched slightly and its ends lie on the ground, with the arch projecting back

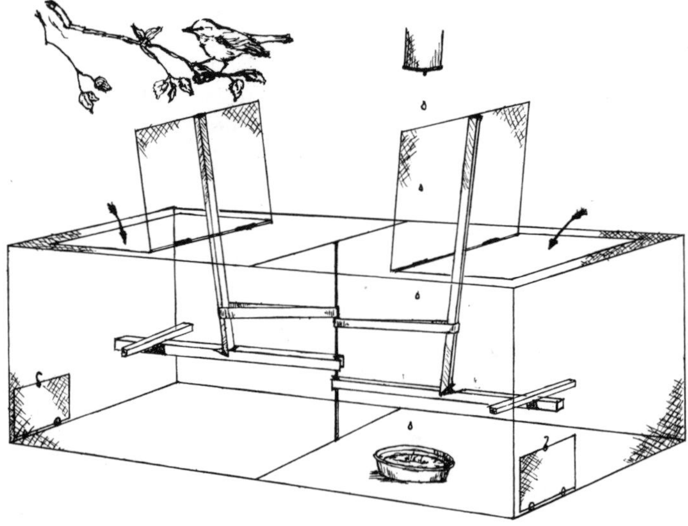

Fig. 189. Two-cell water-drip cage trap. After HOLLOM, 1950.

into the trap 30°–40° above the horizontal. One end of the wire extends upright so that the door rests on it. A bird moving the perch sets the trap off.

HOLLOM (1950) describes a trap used by the English especially for Old World Warblers. It can be built with one or two cells. A thin stick which projects down into the trap is permanently fastened to the middle of the top entrance door. Its bottom rests in a notch in a trigger perch. One end of the trigger perch is also notched to sit in the partition between two cells. (The trap can be set without the latter notch.) The perch falls free when the trap is sprung. The perch sticks must be light enough so that the weight of the door does not spring the trap prematurely. If necessary, the set can be reinforced with a rubber band (see Fig. 189). The captured bird is removed through a small door near the floor.

Dripping water attracts especially *Sylvia* and *Phylloscopus* warblers to this trap. One bowl with water dripping into it suffices for two cells. G. R. MOUNTFORT (HOLLOM, 1950) makes the entrances as large as possible and does not reinforce the edges of either doors or entrances, believing that the least framework produces the best results. The doors are neither weighted, nor spring-loaded and simply fall by their own weight. The speed of closing is regulated by the angle at which they are set.

MOUNTFORT considers moving water by far the best attractant and his waterers are so arranged that a drop falls every three seconds. Care must be taken so that the birds cannot get a free drink from the hose or elsewhere nearby or they may fail to enter the trap. MOUNTFORT usually fastens the hose 20 cm above the trap. Before the days of electric refrigerators, a number of American banders fastened the hoses to their ice boxes. As the ice melted, their traps were supplied with dripping water. It is also important to set the trap near bushes or twigs so that the birds can view the set before entering. Last, but not least, the set should be within sight of trees as warblers and flycatchers often feed high up.

The trap should be painted dark brown or dark green. The floor should be of wire mesh giving the natural appearance of the ground. For several years MOUNTFORT caught up to 250 *Phylloscopus* and *Sylvia* warblers, Old World Flycatchers, *Regulus* warblers, European Redstarts (*Phoenicurus* sp.) etc. annually.

A smaller (20×15 cm) single-cell model (Fig. 190) has four 2.5 cm thick uprights. The rest of the frame is of 2.5×0.5 cm laths. It is covered with wire mesh except for the entrance half of the top. The door should be about 13×11 cm. If it is too big, Common Blackbirds can get in and if is too small, it is hard to get one's hand out after setting the trigger. The door pivots on a strong knitting needle or on a length of motorcycle wheel spoke. The pivot passes through two screw eyes attached to the door.

A 1 cm wide rubber band, run through the

door and fastened to the frame (see Fig. 190) snaps the door shut. When set, the chisel-shaped projection from the middle of the door rests in a notch on the pencil-sized perch.

The English bander CAMPBELL caught more than 400 Old World Warblers and Flycatchers in 12 such traps in one trapping season. He placed them in favored spots such as a row of peas, a lettuce bed, a blackberry patch and at the edge of a potato field. For bait, he used a broad leaf covered with lice, caterpillars, raspberries, currants and other soft fruits. Good baiting and trap function were essential (HOLLOM, 1950).

JOHOW (1961) also caught hummingbirds in the Cordillera of South America with a pull string cage trap. The traps were baited with flowers commonly visited by the hummingbirds. JOHOW als describes a catching expedition to Juan Fernández, the island of ROBINSON CRUSOE: "But back to our Juan Fernández Firecrown. With the help of our cage traps and a particularly stout butterfly net that my son wielded in the crown of a tree in full bloom, we succeeded in catching 23 hummingbirds which we brought down from the mountains alive and adapted to living in our room in the fishing village. There were 11 pairs and an extra female. Of these I released 10 pairs on the coast of the continent. Perhaps they are breeding there. After three months we saw some again at the Zapallar Spa. Now I will take this opportunity to describe transporting the birds from where they were caught to the adaptation cage. At the start we caught one or two and carried them on foot over the island, then along the mainland by car in a small covered cage. Nevertheless, the birds fluttered almost to the point of exhaustion and arrived tired out. The metabolism of flying hummingbirds is so rapid that they cannot withstand such overexertion for more than a few hours without food. As I happened to know that hummingbirds go into a stiff coma when it is very cold or at night and that they can stay alive so for some time and be awakened by warmth I tried to immobilize the little creatures. I wrapped them in gauze like mummies, bound them with thread and laid them in the open box. As an experiment I left one wrapped up without nourishment for 24 hours. It had become stiff and cold and its eyes were closed. I thought that the experiment had taken a bad turn. After a few minutes in my warm hand, its heart beat strongly again, it opened its eyes and it sucked honey water greedily. When first unwrapped it staggered a bit, but soon it whizzed about in the cage and was fit as a fiddle. Now we routinely wrap each hummingbird and set it aside as soon as it is caught so that we can catch more. From time to time we offer them honey water which they refuse once or twice but then suck up with their long split tongues. Thus we can let the captured hummingbirds lie quietly and give our attention to catching more throughout the day."

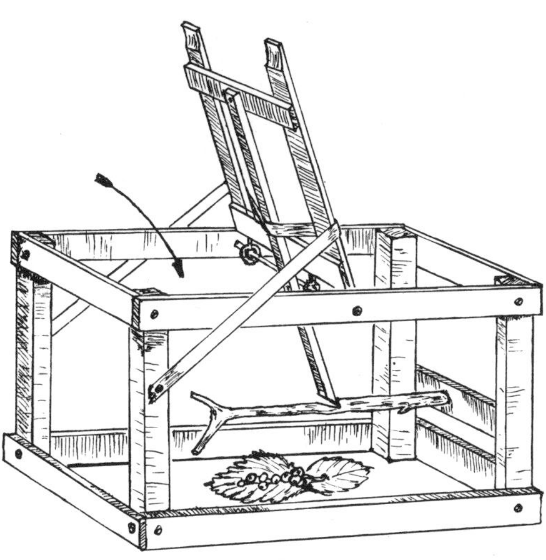

Fig. 190. One-cell cage trap. After HOLLOM, 1950.

The 'dwarf' with its simple construction is a modern tit trap. After the bird has sprung the trigger the door is shut with lightening speed, usually by means of two springs. The door part of the frame is padded with strips of felt so that the bird will not be frightened by too loud a snap.

The dwarf (Fig. 191) is primarily for songbirds and the size varies according to the size of bird sought: for small birds 16×14×14 cm and for larger ones 25×18×18 cm. In the standard model (SUNKEL, 1947 and 1956) the door closes upwards from the bottom of the cage. Doors of other types close downward or even from the side, which was convenient for BUB when catching Northern Wheatears on their breeding grounds. Wheatears, like Eurasian Nuthatches and Common Blackbirds, tend to run into the trap. SUNKEL (1958), and BUB concurs, warns against general use of other models that have a 270° door closure. The doors of the standard

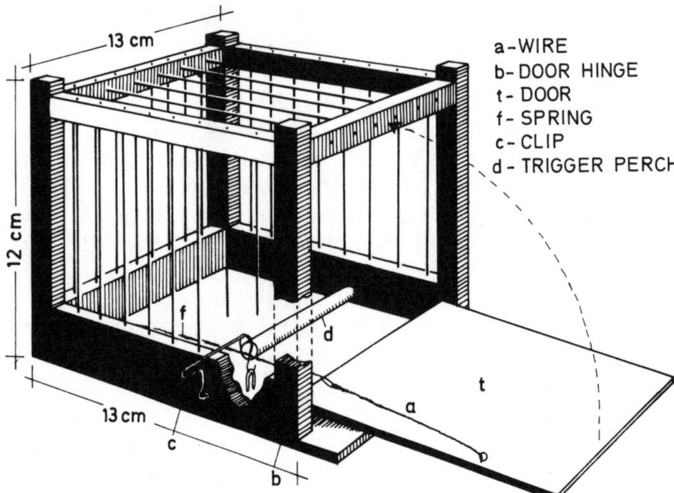

Fig. 191. Dwarf cage trap. After SUNKEL, 1947.

model only move 90°. The perch trigger of traps with doors that close in a wide angle should be no closer to the entrance than the middle of the trap. If there is no handy perch near the dwarf, one can be put there.

Not only seed eaters but insect eaters as well can be caught in the dwarf. W. BRAUN (1942) caught many—particularly European Robins, Dunnocks, Black-cap Warblers, Greater Whitethroats, Lesser Whitethroats, Pied Flycatchers, Black and Common Redstarts, European Willow Warblers, Northern (Winter) Wrens and *Regulus* Warblers. He baited with a mealworm or sometimes with a caterpillar or larger spider. He states, "It is good to have moving bait. These creatures are not put in the usual feed tray but are fastened to small splints well within the trap. The slightest pressure from a bill releases the trigger.

Dunnocks, European Robins and Black and Common Redstarts are caught on the ground, in hedges, etc. and redstarts in particular near posts out in the open. I placed the dwarf higher up in hedges, bushes and trees for other species. It isn't always possible to wedge the trap up in a hedge or in a fork in such a way that the mechanism works. Besides these very species prefer the leafy outer twigs of the tree tops. Therefore I nail a weathered board on a pointed stick; the board is somewhat larger than the trap. This little 'table' can be put anywhere among the twigs with the trap on it. Pied Flycatchers are easily caught thus in trees with holes."

K. GREVE caught migrating Pied Flycatchers in dwarfs on the Isle of Neuwerk (Germany) in the North Sea. He placed the traps on 1 m poles among oaks and maples and baited them with mealworms.

SUNKEL (1948) lists further species that are caught with live insects as bait: Whinchats, Northern Wheatears, Common Nightingales, Bluethroats, thrushes, Common Starlings, shrikes, Crested Larks, *Phylloscopus* and *Acrocephalus* warblers, etc. At the beginning of the nesting season W. NOLL baited with light colored chicken feathers and caught Long-tailed Tits and other species. If seeds are used as bait it is best to glue them in lest they fall out.

Especially for catching Common Bullfinches, Common Serins and Eurasian Siskins, SUNKEL (1948) used several dwarfs in conjunction with a lure bird. The traps should not touch each other or the springing of one may set the rest off. If one has several lure birds, they are placed so that they can hear, but not see, each other and dwarfs are placed around each. This technique increases the catch.

It is worthwhile to mention CREUTZ's (1942) experiences. He recommends that the side walls be of wire mesh or glass rather than of wire uprights, and that at least the lower third of the walls be covered with wood to keep the bait from falling out. But perhaps this last recommendation would keep birds from entering the trap: a 1 cm rim should suffice.

"It is undesirable to build the dwarf too low. Skillful, adaptive birds soon learn to take the bait without getting caught. They do this by grabbing the upper edge of the entrance and simply bending over to feed. A European Greenfinch once sat on the top edge looking in, pressed the treadle perch with its tail and the door caught it by the neck. Such casualties can only be prevented by building the trap high enough and placing the trigger perch well down. I have had good luck gluing the feed on a small board on the floor; the board can be lifted easily and is fastened to the trigger perch so that it sets the trap off.

"The hanging of the trap is also important: Common Bullfinches, European Greenfinches—in fact all finches—must have a chance to jump into the trap from above. Therefore, top entrance traps, if placed near a windowsill, should have their tops on level with the sill; or a platform, such as a box, can be baited and placed at trap height."

7.6 Cage trap for rails

In the Ismaninger Reservoir region of Bavaria (Germany), M. Sumper developed an ingenious method of catching rails. He cut 1.5 m alleys 150 m apart in the reeds and placed five to ten cage traps (Fig. 192) in each slot. The width of the reed margin here is up to 50 meters. The traps are 50×40×30 cm. The birds push against a string which causes two props to tumble and the door falls. The rails are removed from above through a flap (Fig. 193). Leads guide the birds to the entrances. Sumper has caught hundreds of Water Rails and Spotted Crakes in this manner.

Adapting Sumper's rail trap, M. Dornbusch used traps measuring 50×33×33 cm and 50×50×33 cm. The smaller traps were more satisfactory. His traps were collapsible (iron rod

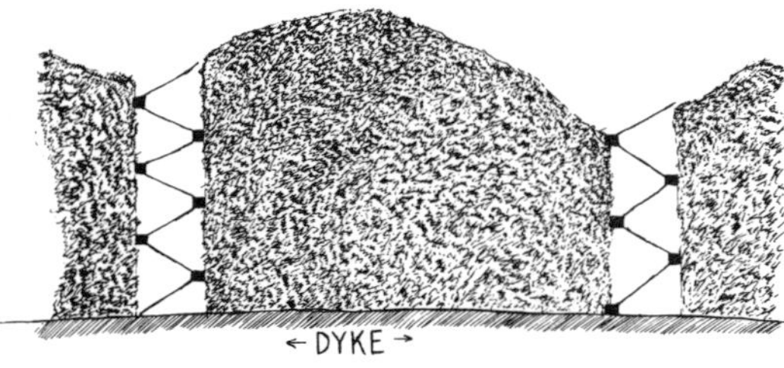

Fig. 192. Alleys cut through the reeds and leads for catching rail. After M. Sumper.

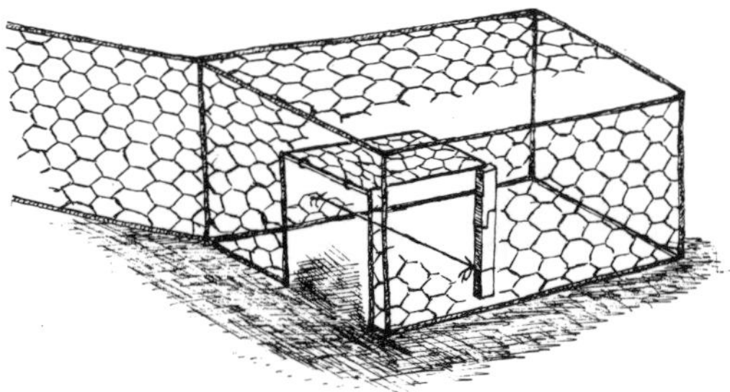

Fig. 193. Cage trap for rails. As soon as the bird bumps the trip cord, both props fall and the door closes. After M. Sumper.

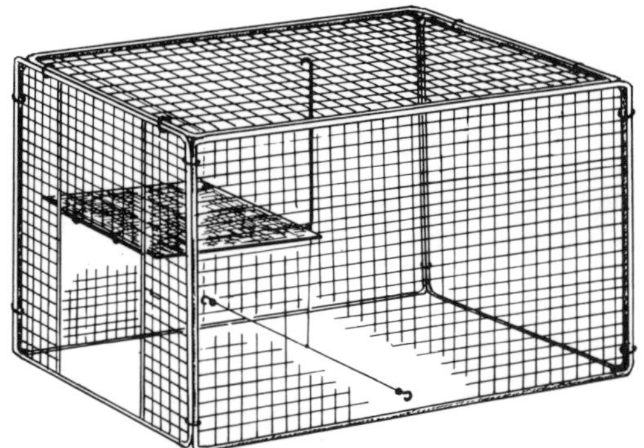

Fig. 194. Cage trap for rails. After M. Dornbusch.

frame covered with wire mesh). The entrance was at one short end of the trap (15 cm wide and 20 cm high). The drop of brown fiber board was slightly shorter and wider than the entrance; it is pulled up to a horizontal position inside the trap and held in place by a hook hanging from the roof. A cord runs from this hook to a trip cord somewhat behind the middle of the trap. The trip cord is attached to the side walls by two small S-hooks. The tension of the cord may be adjusted by pulling the hooks through the wire and fastening them elsewhere on the mesh. As soon as a bird bumps the trip cord, the hook on the roof is dislodged and the door falls (Fig. 194).

For catching rails and other reed and brush dwellers that frequent paths, box traps built on the principle of martin or cat traps work well. For bird catching it is better to have walls of netting rather than of wood. The trigger must be adapted. Inconspicuous leads of netting or wire mesh increase the catch.

7.7 Tree trunk traps

Creepers, nuthatches and woodpeckers are led into this trap (Fig. 195) by narrow hoops of netting spiraling toward the entrance. Once inside the trap, they fly against the window and glide into the lower compartment whence it is harder to escape. It is often desirable to fashion the rear of the trap of thin, bent wood. If the side walls are nailed to the trunk, one can forget about the rear wall. If the trap is to be moved often, the rear wall is best made of wire netting. The bottom of the funnel is open. It is about 38×30×30 cm and it is supported by frames of stout wire, rather than wood. To unset the trap, the box is removed.

Fiske (1968) describes a 26×18×36 cm woodpecker trap (Fig. 196) in which he caught small woodpeckers and nuthatches. A hardware cloth trough in the upper part of the trap is baited with suet. As shown in the figure, the trough is hinged and a projection from the hinge rests against another projection from the door which holds the door open. When a bird picks at the bait the trough swings, the abutting projections are jostled apart and the door is shut. The mechanism operates without friction. The holes must be drilled with precision and there should be no play between the working parts. The trough needs about 1.3 cm play so that it does not bump at the sides and it should be able to swing back 1.3 to 2 cm or the birds can climb up behind it. It is hard to get them out from behind the trough. The bait should be placed so that the birds cannot get at it from outside. A small door at the top (not shown in the figure), secured with a hook, facilitates baiting the trap.

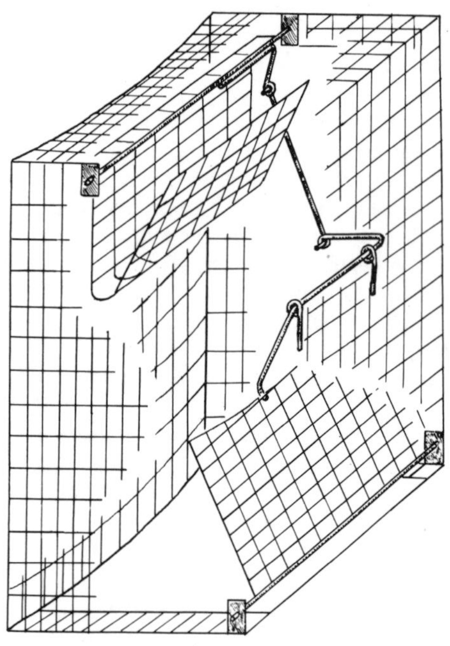

Fig. 196. Woodpecker (and nuthatch) trap. After Fiske, 1968.

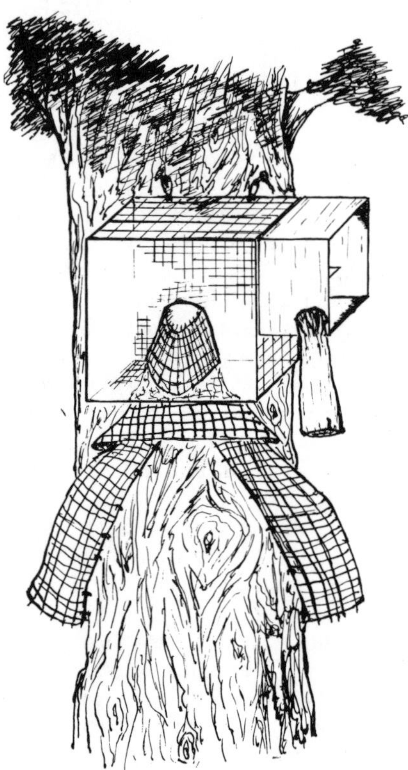

Fig. 195. Tree trunk trap. After Lockley & Russell, 1953.

Fiske sets his trap on an elm near his house so that he can check is easily. To unset, he props the door open and the birds can feed freely.

The rear of the trap can be shaped to trees or it can be straight to fit housewalls or fences. In any case the rear of the trap must fit its substrate rather well. Sometimes it pays to add an unfinished board or bark so that the birds can climb up easily.

8. Pit traps

Pit traps are particularly primitive in origin. They are related to ancient pits used by our ancestors and by some primitive people to capture both large and small animals. All early pits were constructed to render the captives as helpless as possible and to crowd them into a small space so that they could be killed easily. Furthermore, pits are related to primitive traps in that the captured mammals or birds were out of sight. Other traps such as funnel traps and stationary nets are a departure from pit traps in that they leave the animals in their natural habitat.

Pit traps are a separate branch of bird catching that has been adapted to scientific banding.

The web trap is described by MARINKELLE (1957) from an island near Kalimantan, Borneo, and we include it as an insight to Borneo catching methods (Fig. 197). The islanders dig two holes, one broad and one narrow, about 10 cm apart. The web lies in the larger and the trigger in the smaller. The holes are connected by a tunnel. A 'bell' of woven bamboo, weighted with stones, is hung from a frame in such a way that when a bird alights on the sunken web it releases the trigger. The trigger is made of polished wood or a piece of bone and it extends through the tunnel and is fastened to the middle of the web with a thin piece of cord or bast. The 'mesh' of the web is about 5 mm and the web is set about 5 cm from the bottom of the pit

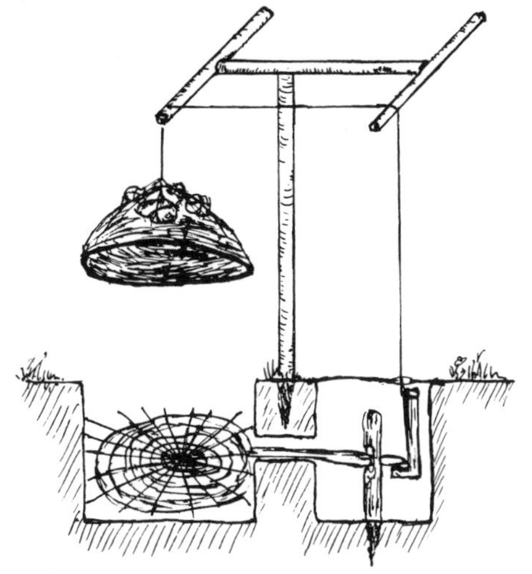

Fig. 197. Web trap used in Indonesia. After MARINKELLE, 1957.

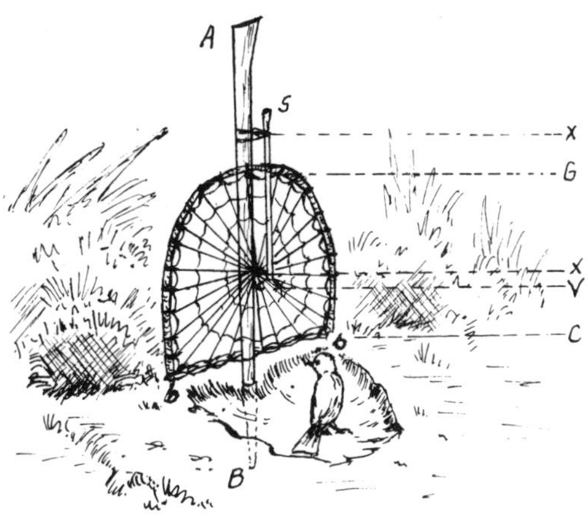

Fig. 198. Deadfall trap used in Tunisia. (AB) supporting stake; (bb) stick bent into semi-circular frame; (CG) tension cord; (V) bait; (XX) upper and lower loops to hold the trigger stick. After J. STEINBACHER, 1957.

and about 20 cm below ground level. The web is carefully set in position with small bits of bone set in the sides of the pit. Its diameter varies from 20 to 60 cm. It is essential that the string running from the web's midpoint and the trigger be exactly horizontal. Bait is strewn around the larger hole and the web. When a bird goes into the hole, the web tips, releasing the trigger and the bell falls. A variety of species can be caught in this carefully constructed trap. An experienced trapper can vary the weight needed to release the trigger or catch several birds at a time.

J. STEINBACHER (1957) reports a successful pit trap from Tunisia (Fig. 198) which he saw used in the Cap Bon region and—in greater quantities—farther south in Gabes. He writes, "Here they use various sizes and the better they are made the better they work. Nevertheless any halfway adept Arab boy can make them quickly from materials at hand from Nature. He needs a few sticks and a few twisted plant fibers that grow anywhere and a flexible stick 70–80 cm long and about as thick as a thumb. He gets the fibers from the hard, resilient grasses of the hills, as well as two sticks, one a stake, the other thinner. In the Gabes Oasis pomegranate branches, palm fronds and plant fibers are used. The flexible stick is bent into a half circle and held in position with a twisted cord; the enclosed area is converted to a sort of a spider web by means of concentric lattice work. Live bait,

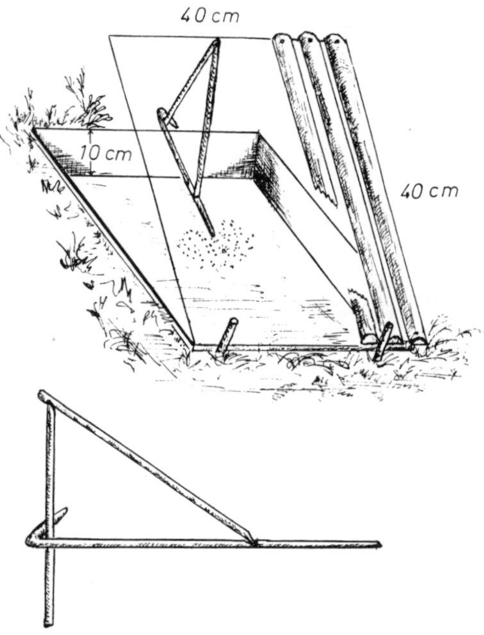

Fig. 199. Jay trap. After Fiess, 1969.

such as a grasshopper or beetle is attached to the midpoint of the web by means of a noose. At Gabes crickets, abundant there, are most frequently used. To rig the trap the stake is forced through the cord that holds the semicircle and the cord is twisted together several times like tightening the blade in a bucksaw. Thus the cord—now fastened—has the tension of a spring and the half circle, with its mesh, presses firmly against the stick, or, if the bow is raised, it slams to the ground when released. To set the trap, the stake is driven into the ground, the bow is lifted, the thinner (trigger) stick is slipped through a loop at its upper end and its lower end is simultaneously slipped into the loose noose at the midpoint of the lattice. The tension on the bow holds the trigger and holds the lattice in position until the bait is seized from below by a bird. Then the noose that held the bait slides off the trigger, freeing the bow which—with its lattice work—slams to the ground.

"By this method every bird is caught, but if the lattice is strung tightly it is injured or even killed by the blow. If, however, a hollow is dug beneath the bow's fall few birds are injured. Of course, either negligent or more or less careful lattice weaving may prove decisive. The catcher can place the set so that it catches by chance or in relation to the bird and the bait."

The jay trap, built of wood (Fig. 199), is somewhat like the tit trap. Fiess (1969) caught 50 Eurasian Jays in seven of these traps in March and April 1968 and 20 more in November. The traps were often unset by wind and rain. Sets have to be placed with this in mind. Few materials are necessary for this trap: the frame—about 20 cm high—is set into the ground and no floor is needed.

The trigger consists of three smooth hardwood sticks. The upright (22 cm) is pointed and pressed into the floor of the trap so that it is not quite as high as the closed trap. The second stick, about 30 cm long, is hooked at one end and laid at right angles to the first. The third stick (20-25 cm), slightly pointed at the top, fits into a notch on the second stick, rests on the point of the upright and holds the door open. It takes dexterity to set the trigger to catch jays rather than smaller birds.

The trap is baited with up to two handfuls of grain. Fiess scouts before setting out his traps to see where many Eurasian Jays are found. His best sets were in small woodlots or by single trees near the edges of woods. Sunny wood edges were good, as were pheasant feeding places. Used with board floors, the traps were also successful on the roofs of pheasant feeding places. For catching in pheasant habitat (near feeding places or on the roofs of them), the traps should be at least 50×35-40 and 30 cm high and floored. The walls are built high so that pheasants, feeding on the ground, won't see the bait. It is also a good idea to bore two 3 cm

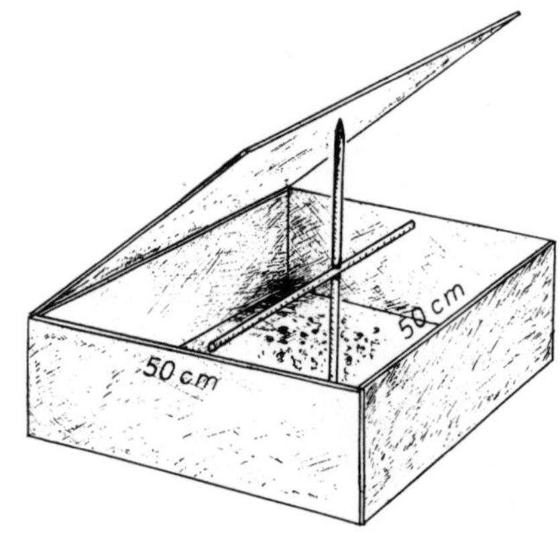

Fig. 200. Jay trap. After Fiess, 1969.

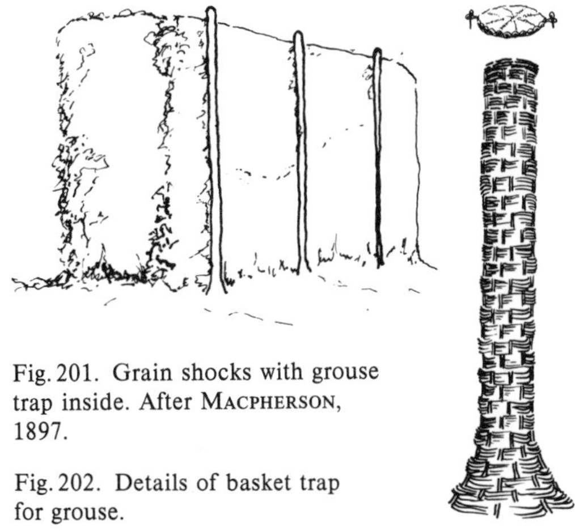

Fig. 201. Grain shocks with grouse trap inside. After MACPHERSON, 1897.

Fig. 202. Details of basket trap for grouse.

holes in the back wall of the trap so that small birds, inadvertently captured can escape.

Another jay trap (Fig. 200) consists of a hole in the ground with a cover and trigger. The cover of wire mesh can be laced with fresh spruce twigs (FIESS, 1969).

9. Methods for catching grouse

9.1 Grouse catching in Siberia

For catching grouse in Siberia, tall funnel-like baskets were set up near grain shocks (Fig. 201). These were slightly higher than the shocks. The covers of these baskets are somewhat smaller than the tops of the baskets (Fig. 202). They are fastened with a cord so that the lids can tip in either direction. A bird alighting on the lid tips the cover and slides down into the basket whence it cannot escape.

According to PALLAS (1773) similar traps are set in birch woods for catching Black Grouse (Fig. 203). Forked sticks are set up where the birds feed in winter and a stick is placed across the fork. A few heads of grain are tied to the tops of the forked sticks. A basket about 3 m high is fashioned just beneath by pushing thin birch branches into the ground. The opening is 50 to 60 cm wide at the top. The lid is made with two half hoops, sharing an axis. The hoops are covered with straw and heads of grain. The lid is free to pivot. At first the grouse tend to sit on the stick between the forked sticks, but when

Fig. 203. Basket trap for grouse. After PALLAS, 1773.

they fly to the grain on the pivot lid, they have no place to alight except on the lid, which turns and causes them to fall into a funnel. Pallas sometimes found the trap half full of Black Grouse.

9.2 Finnish and Russian grouse traps

The Finnish zoologists Ilkka Koivisto and Esko Anderson supplied much material for this section. Since long ago hunters in northern Europe have obtained grouse in great numbers. For the most part, traps were used in early times but less so recently with the development of firearms. Grouse trapping, except for research, is no longer legal in Finland.

In the course of the research conducted in Evo in southern Finland many methods of live-trapping grouse were tested. Some of the earlier traps were used and new ones developed. For the most part Black Grouse were caught, but also some Capercaillie and Hazel Grouse. Unfortunately we have no experience in Willow Ptarmigan catching as this species does not occur in our study area but the game manager, J. Lahtinen, told us of his experiences.

Under the conditions in their study area the big grouse—Capercaillie and Black Grouse—can only be caught during the display season or in autumn on cultivated fields. Attempts to catch Hazel Grouse gave a paucity of catches as these birds seldom flock. With the proper bait one can catch throughout the year except in summer. Willow Ptarmigan can be caught in winter when they flock.

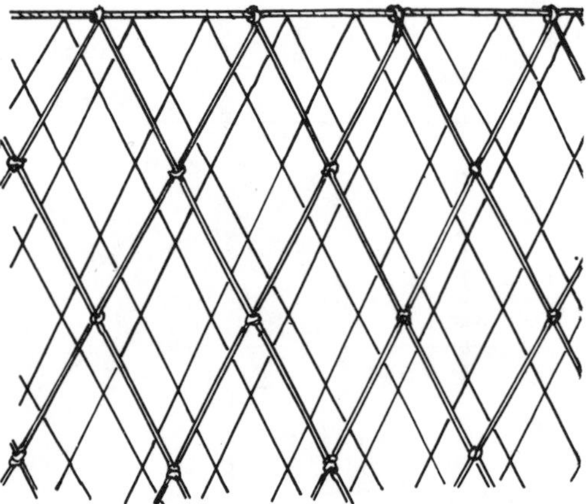

Fig. 204. Finnish grouse net: length 10-50 m. Height: for black grouse, 1.2-2 m; for capercaillie, 1.6-2 m. After I. Koivisto.

When Black Grouse assemble on cultivated fileds or on leks they can be netted. The nets are attached to light poles which fall at the slightest touch. The mesh should be rather large—about 7 to 10 cm. It is advantageous to fortify the net with another of even larger mesh (Fig. 204). Black Grouse get into the nets—mostly during twilight—when they are chasing each other on the lek. This method of catching has one disadvantage—any captured bird scares the rest away from the lek so that usually only one is caught. Often one catches none as the bird escapes from the net. Moreover, one has to stay near the net, otherwise these strong birds may injure themselves.

Grouse can be caught in open country or on leks with rocket nets such as are used for geese. This net has the disadvantage that it tends to freeze to the ground in this terrain. Black Grouse like to display on frozen ponds or ocean bays. A length of plastic under the net cuts down the likelihood of freezing somewhat. The Hamerstroms found rocket nets highly unsatisfactory for winter prairie chicken trapping in Wisconsin, U.S.A. Freezing, thawing or drifting snow kept altering the weight of the net. Each change in weight altered the trajectory. Either the net spread too high and the flock escaped, or it spread too low and birds were injured. The drop net (Fig. 278, 279) proved far more satisfactory.

Kirpitshev (1962) describes methods of catching Capercaillie in Russia. The birds are caught in 20 m long and 1 m high nets with a mesh of 80 mm. The net is hung from a rope stretched across a lek that is also frequented by hens. The rope is stretched tightly between two trees and the ground beneath the net is cleared of branches and debris. The net hangs free on both sides. When a bird flies into it, the net pulls together at the top and wraps around the bird, which promptly receives a hood to cover its eyes. This is a new use of a stationary net, suitable only in special cases. Some American biologists have used mist nets on display grounds of prairie chickens and Sharp-tailed Grouse.

Trapping of grouse on display grounds tends to catch more cocks than hens. In some studies it is important to observe normal behavior and those methods that result in newly banded

Fig. 205. Box trap for grouse. Length 60-80 cm, width and height 30-45 cm. Measurements for capercaillie: 80-100 cm, width and height 50-80 cm. After I. Koivisto.

cocks returning promptly to their territories are to be recommended. Grouse often appear to be disturbed by rocket netting; after bow-netting they may be back on territory within 20 minutes.

Box traps with wooden frames covered with netting are good for catching Black Grouse (Fig. 205).

Koivisto devised triggers with a rat or mouse trap (Fig. 206). Hamerstrom and Truax (1938) used a simple trigger. An ear of corn was placed on a small (70-120 mm) stick. A string ran from this stick up through the top mesh of the trap and was fastened to a long (about 30 cm) stick, the other end of which held up the door. Any bird moving the ear of corn released the small stick which flew up as the door fell. The small stick was of wood, but later, when plastics came into general use, they found that old ballpoint pens made better trigger sticks as they are less likely to freeze down.

The birds can hurt themselves even when the trap is covered with fish net. To cut down injuries, a compartment can be added to the trap, so built that any bird entering it finds itself in total darkness. Such traps do not need to be watched constantly. For Black Grouse, berries—their favorite food—are the best bait. For some species of grouse it is worthwhile to use leads even with single catch traps. Aggressive cocks can sometimes be lured into the traps with mirrors. Ruffed Grouse also react to their mirror image and are caught in this way. Half tame lure birds used during the display season were the best bait of all. Mostly cocks were caught.

Black Grouse are easily caught in grain fields with so-called 'bird baskets', according to Sirelius (1934). The baskets were originally made of thin sticks. Nowadays we make them of rather thin boards or of thin sheets of fiber board. Here, too, a darkened chamber into which one can reach easily, can be added to reduce injuries (Fig. 207).

The methods here recommended for Black Grouse are also suitable for Capercaillie, except that because Capercaillie seldom use open

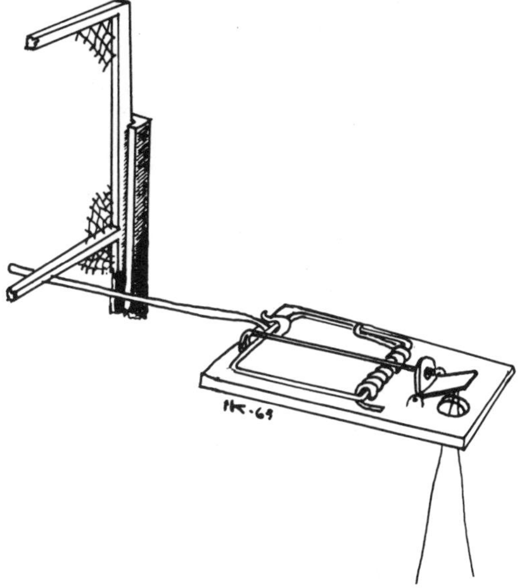

Fig. 206. Release mechanism for a drop door trap for grouse. After I. Koivisto.

Fig. 207. Basket trap for black grouse. Height: 1.5-2 m; diameter of top, 60-70 cm; bottom, same diameter or slightly smaller. An artificial decoy in the rear. After I. Koivisto.

country, rocket nets are impractical. Traps for Capercaillie must be correspondingly larger than those for Black Grouse. From our experience traps with doors are best for Hazel Grouse and Willow Ptarmigan.

All four species just mentioned—and probably Rock Ptarmigan as well—can be caught in summer with a pointing dog. Old birds cannot be caught this way, but it is easy to catch young ones. A very well trained dog is essential for the 'hunt'. A net or net basket is thrown over the birds while the dog is on point. Pointers are also used for finding young American Woodcock for banding in the United States. Quite often the tiny young are simply picked up by hand.

Hens caught while incubating usually deserted the clutch; but John Toepfer has successfully caught and banded incubating prairie hens without causing desertion.

9.3 Tip-top trap for Sharp-tailed Grouse

R. E. Farmes (1955) describes a trap for Sharp-tailed Grouse used in stubble fields and brushy country in Minnesota, U. S. A. It consists of a wooden box about 50×60×45 cm, buried in the ground with its top at about ground level. Buckwheat or other grain is spread on and near the traps. It is best to use several traps. The birds are caught by means of four tip-treadles which take up most of the top of the trap. The treadles

Fig. 208. Catching quail with a stationary net; from illustration from a Theban grave. After Lips, 1927.

Fig. 209. Stationary net for catching Emu in Australia. After ROTH, 1897.

are made of smooth wood so that the birds find no footing while sliding down. All the treadles tip toward the middle of the trap and are brought back into position by means of springs.

10. Stationary nets

The setting of vertically spread nets meets a deeply rooted desire to render creatures with wings or with long necks immobile and so convert them to possessions (Fig. 208 and 209).

The stationary net category includes vertical single or triple-layered nets of various sizes.

Stationary nets may be raised or lowered, but as a rule, they are not moved from place to place.

Where did this wide-spread method originate? Which devices for catching animals are inherently stationary nets? The first primitive beginnings seem to be snares placed close to each other. This method appears in the hunting techniques described by SMYTHIES (1968) in Kalimantan (Borneo). The Indonesians erected fences of bushes for catching gallinaceous birds and wherever there was a gap in the fence they placed a snare.

A further step in the evolution of stationary nets is the Mexican snare for game (Fig. 210). LIPS (1927) made a pertinent statement, "Sometimes the snare also evolves into a net to facilitate the closing of the snare and to increase catching success." Such an arrangement of snares increases the potential catching area and the catch is less limited to pure chance.

Stationary nets have been in use for over 2,000 years. Already in SOPHOCLES's tragedy "Antigone", a strophe of the chorus begins, "Nothing is worse than man ... he catches merry migrant birds with nets."

As for the existence of nets themselves, we may well go even farther back into the past. Quoting LINDNER (1937), "It is probable that trapping techniques of the Early Stone Age were widely adapted by Middle Stone Age men; men of the latter period understood pitfalls, nets and snares, which were used, not only alone, but also in combination with other devices and hunting methods such as game fences and drives."

In this connection—at least regionally—fish catching methods may have made significant

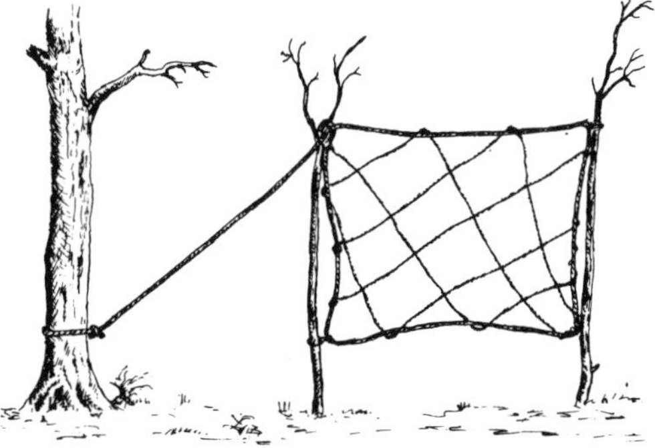

Fig. 210. Mexican snare-net for game. After LUMHOLTZ, 1903.

contributions. But the degree to which these inspired our early ancestors evades our critical judgment. It is not impossible that fish catching methods just made stationary nets possible. Oldsquaws (Long-tailed Ducks) and sometimes other avian species are not infrequently caught deep under water in nets set for fish.

A notable refinement of materials was required for small bird catching.

A number of authors—and most particularly an anonymous author writing in 1802—describe the art of making nets.

10.1 Trammel nets

The trammel net is used for catching birds in almost all European countries: in those benighted parts of Southern Europe where songbirds—some of them rare—are roasted on spits (Fig. 1); and also at the stations of banders striving to increase scientific knowledge, as well as in other parts of the world.

Trammel nets, which tend to be about 2 m high and of various lengths, are inconspicuously colored black, tan or dark green and consist of three layers: a fine-meshed net on either side of which two coarse large-meshed nets are hung (Fig. 211). To set the trammel net, the inner, fine net is gathered upwards—especially the top quarter—and the fine net is fairly evenly distributed between the two layers of coarse netting. Thus the fine net is loose enough so that any bird flying against it will push on through the coarse mesh, pulling the fine net through and so land, captured in a pocket. The meshes of the coarse net are large enough so that most songbirds 'think' they can get through—and the meshes of both coarse nets are nicely lined up to give passing birds exactly this idea.

Trammel nets will always be useful. Mist nets, to be sure, are used throughout the world, but trammel nets make surer catches. Essentially no birds can free themselves and escape.

The top cord of the trammel net should be strung tightly. Special poles are not necessarily needed; the net can be spanned between branches, roots or tree trunks. Sunkel (1954) published on the particular usefulness of roots in setting up trammel nets. Strategically placed hooks make it quicker to reset the net where catching is particularly good. A path about 1 m wide is cleared of branches, tall weeds, twigs, etc. to keep the net from getting entangled.

Trammel nets should be placed in the shade. They are too visible against the open sky. Conspicuous nets do not make good catches. Woods, fence rows and hedges are ideal. Nets in such locations are usually sheltered from the wind as well. Wind shakes the inner netting down. Pastures and parks with pedestrians are poor sites. The nets tear all too easily.

Birds should be taken out of the net promptly. Many, especially most small insect eaters, stay quietly in the pocket as though lying in a hammock, but others, particularly tits, can become badly entangled; and so can starlings. Mass catches of such species in trammel nets are best avoided.

The first step in removing a bird is to find out from which side it went in. Next hold the bird in one hand and slide a finger of the other down its pathway (Fig. 212). The hind part of the body

Fig. 211. Trammel net. After Sunkel, 1927.

Fig. 212. Removing a sand martin from a trammel net. Photo: R. STRUWE.

is freed first, then the legs and finally the remaining meshes are pushed over its head. Special pains must be taken in freeing the wings. Patience and gentleness are essential. Practice makes perfect. Only under dire circumstances are any meshes cut. Even old hands have to resort to the careful cutting of a mesh or two.

The nets are taken down during rainy weather. We take any birds, caught in a sudden shower, out of the nets as quickly as possible. RUTTER (1962) places any wet birds in separate small bags to give them a chance to dry out before letting them go. Care must be taken not to place them too close to a stove.

Birds in stationary nets are vulnerable to predation from both mammalian and avian predators, so frequent checks are essential. Shrikes can kill or maim netted songbirds in a matter of seconds. SUNKEL (1954) pointed out that Eurasian Nutcrackers can be equally dangerous. Small avian predators at least get caught in the net too, whereas larger mammals tear or bite great holes in the netting. It is hard to know what to expect. In South Africa an ichneumon *(Herpestes pulverulentus)* killed a bird hanging in a mist net a meter above the ground. In Austria, at Neusiedler Lake and elsewhere, Water Rails have frequently killed birds (ASCHENBRENNER et al., 1957). PYLE (1964) in North America saw cicada killer wasps *(Shecius speciosus)* swarm around netted birds sometimes stinging them so that they were temporarily paralyzed or even died.

Before taking the net down, all foreign matter should be removed. Then one side is detached and the net is pulled to the other, taking care to gather up the lower half. The top cord may be wound around the net or around a pole. Wet nets should be hung up to dry. Dry nets are easier to put up. Nylon nets do not suffer from moisture; they can even be stored wet, but they should be kept out of direct sunlight as much as possible. Cotton nets are best dried quickly in direct sunlight to keep them from rotting. Nets can be dried in a dryer. Nylon is so heat sensitive that even a cool dryer may ruin the netting.

SUNKEL (1950) and his colleagues caught about 100 species in trammel nets. In addition to many songbirds such as *Acrocephalus, Phylloscopus* and *Sylvia* warblers, finches and other birds of brush, woods and reeds, they caught also Eurasian Jays, woodpeckers, European Nightjars, Tawny Owls, Water Rails, Common Gallinules (Moorhens) and Mallards. GREVE trapped with a single trammel net on the Isle of Neuwerk (Germany) during the spring and fall migrations of 1958. In addition to others, he caught 132 Chaffinches, 61 Pied Flycatchers, 108 European Willow Warblers, 49 Greater Whitethroats, 566 European Robins and 102 Dunnocks. In the fall of 1957 he caught 50 Redwings and 165 Goldcrests.

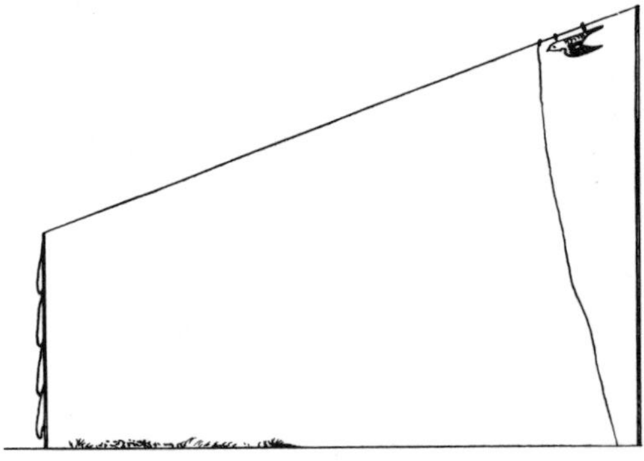

Fig. 213. Sparrow hawk decoy used for trapping near a feeding spot—in this case near a mist net. After WÖHRMANN, 1959.

WÖHRMANN (1959) used a sparrow hawk decoy to scare feeding birds into trammel or mist nest (Fig. 213). The decoy is a silhouette of a sparrow hawk sawed out of plywood. It is attached to a tightly stretched mono-filament line by means of two almost closed screw hooks. One end of the line is fastened high—to a tree or wall of a house—the other is fastened to a pole near the middle of the net. The net is so placed that escape cover (a bush for example) is on its far side. The angle of the line determines the speed of the artificial hawk's flight. The release is a split cork pushed onto the line to hold the decoy up. When the cork is jerked free with a pull string the hawk stoops. The flock of feeding birds flushes en masse toward the escape cover and against the net wall. It flushes with sufficient force to get well caught. This method works particularly well for Bramblings. If the birds notice the 'hawk' only at the last moment though they will disperse every which way rather than towards the escape cover.

Trammel nets are especially good for catching Northern (Winter) Wrens. During the breeding season, RUSCHKE (1963) found them useful only within a given wren's territory. Less precision in placing the net was necessary at other times of the year.

Sometimes it seems that wrens fly into the net on purpose as though expecting to fly right through the meshes. Others whizz closely past the net if there is good cover at its end.

BRAMBLINGS, using a winter roost in Switzerland, were so wary near the ground that catching was difficult. MÜHLENTHALER (1952) raised the net 15 to 20 m above the ground (Fig. 214) and then made remarkable catches: 304 in 17 days.

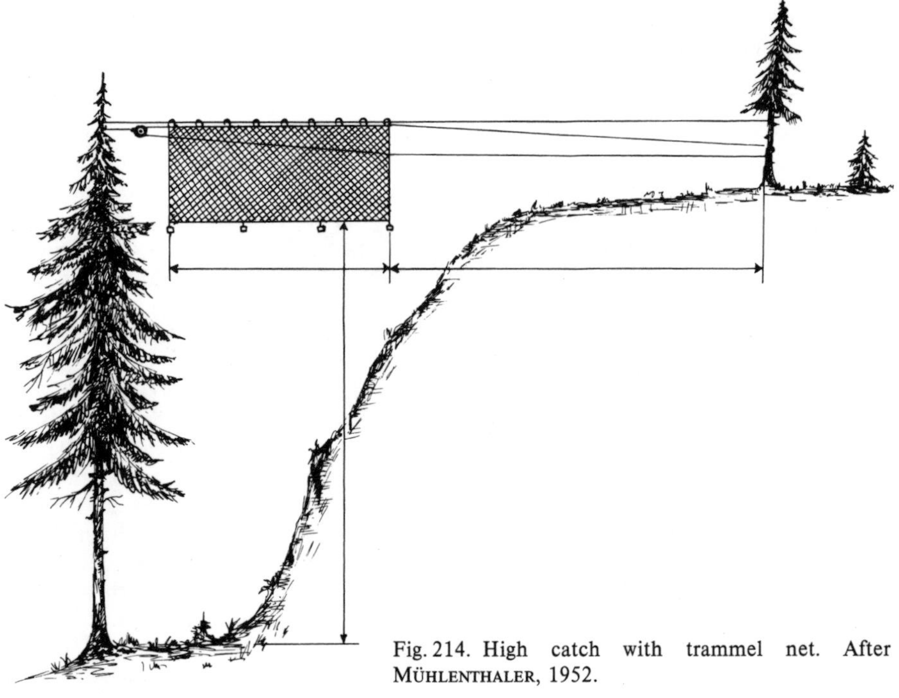

Fig. 214. High catch with trammel net. After MÜHLENTHALER, 1952.

10.1.1 Catching in various habitats

a) Catching in woods, gardens, and other tree and bush covered places

Hedges and fence rows are particularly propitious as birds are reluctant to leave such good cover. Nets are placed where the vegetation has been cut away for their protection and birds are easily driven into them. Each net is set across the hedge and should project 1 to 2 m beyond the vegetation so that birds flying along the edge are caught too. Skillful driving augments the catch. Drives are made toward the net, first from one side and then from the other. When catching tapers off the set can be moved to another location, or it may be worthwhile to wait a little and then try again. Migrants can often be driven long distances toward the net. Several sets can be placed in one fence row and sometimes it is advantageous to place several nets at the end of a fence row so that the birds are driven into a pocket of nets. The Sempach Ornithological Station (Vogelwarte Sempach, 1935) had good success with a high net and a low one 3 m apart.

A Little Owl—either alive or stuffed—placed about 1.5 m from the net increases the catch appreciably. BOLEY (1933) points out that the owl should be far enough from the net so that it cannot reach it either by running or flying. Two hedges intersecting at right angles with the owl within the apex of the angle and the nets along the hedge edges form a good set. Bare branches half the height of the net are pushed into the ground between the hedge and the owl. Birds, sitting on them, tend to flush toward the hedge for cover if startled by the bander or by a sudden movement of the owl, and to end up in the net. One may wish to enlarge the set by adding nets within the hedge, behind it, or all around the owl about 5 to 10 m from it.

Other species make good decoys too. BOLEY (1933), catching Eurasian Siskins, spanned the net across a watering place—usually a flat, pebble bottomed part of a brook. And a good lure siskin is put in a cage nearby. The cage is placed low enough so that any siskins intending to perch on the cage tend to get caught on their way there or upon departure. It is best to put nets on both sides of the lure or all around it. As siskins and other finches lie quietly in a net without becoming entangled, many can be netted without having others become suspicious. Furthermore, the siskins netted early in the game act as lure birds too; they attract others by their calls.

BOLEY (1933) also noted that one Long-tailed Tit in the net gradually brought in the whole family of seven or so. A lure bird is also effective for Common Bullfinches, especially if the net is set up in shady shrubbery. The same is true of Common Redpolls, crossbills, Bramblings, Common Serins and Hawfinches. Catches can be increased by bait such as twigs with berries, etc.

b) Catching in meadows, cultivated fields and reeds

Even cultivated fields present opportunities for netting. Push nets are used here. The net poles are simply pushed into the ground about 1 m apart. The poles are usually rather short (Fig. 215).

Quail catching in the Netherlands, described by MOOIJMAN (1955) cannot be overlooked. His push net is 18 cm high and can be just about any length. Nets of 10 m are the most practical. The nets are fastened to 25 cm rods set 60 cm apart in grain fields used by quail. When a singing spot of a quail is found, the net is spread silently and without any unnecessary movements. Then the catcher kneels a few meters behind the net and waits for the quail to call, whereupon he imitates the hen's call twice on his whistle. Whenever the bird is silent, he is silent too. But after each sequence the cock comes nearer until he gets entangled in the net on his way to the fake hen. Then the catcher quietly walks around the net in order to approach the bird from the rear.

The catching period is from early May until early August. Up to 24 cocks were netted in a single night by this method. KRUIS considers morning and evening twilights good for catching and has had some success in broad daylight. Hot weather favors good catches; cold, wet weather does not.

Meadows as well as cultivated fields are suitable for catching. Corncrakes are most easily caught in mowed meadows (C. L. BREHM, 1855). A strip about 5 m wide is left uncut; the push net is placed at its end and the birds are noisily driven into the net. All other rails are easily caught in low push nets too. The nets need not

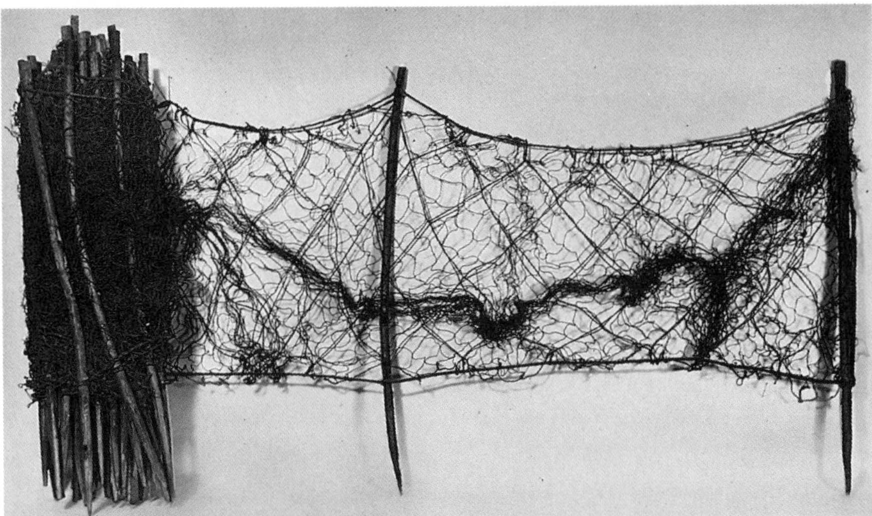

Fig. 215. A push net used for quail in Italy. Suitable also for dwellers of high grass or rushes. Photo: A. Toschi.

be higher than 50 cm, Kruis used 30 cm high push nets for Common Gallinules (Moorhens), Water Rails and quail.

The outer trammel netting mesh was 140 mm stretched and the inner 60 mm. The push sticks, about 15 mm in diameter, are set 1 m apart. For catching flightless ducks Kruis used larger mesh; 180 and 100 m stretched. Mist nets and ordinary trammel nets are unsuitable as they are too conspicuous. Push nets project little or not at all above the vegetation. Kruis used only dark green or brown, but never black, nets in rushes (Fig. 215).

According to Ghigi (1933) the first quail arrive on the Italian coast of the Tyrrhenian Sea from North Africa at the end of February. The migration reaches its peak in April and may continue into May, June or even July depend-

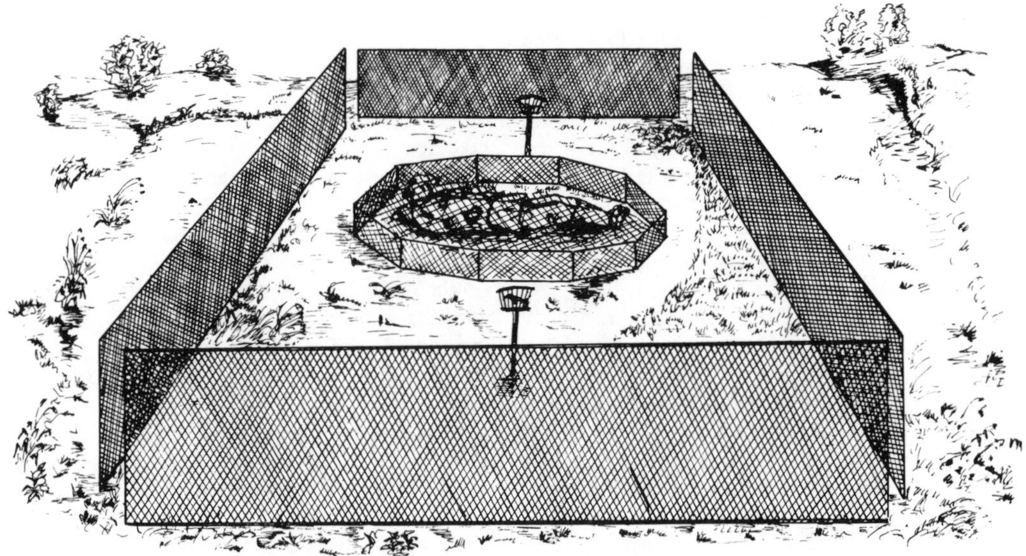

Fig. 216. Quail catching with trammel nets and push nets. After Bechstein, 1797 (Pernau, 1720).

Fig. 217. Banders setting up a trammel net at the edge of a pond. Photo: H. BEHRENS.

ing on the weather conditions. The quail tend to leave the African coast in the early evening and arrive at dawn or earlier. The morning flight is good until 09:00 but may continue to 15:00. Some of the birds put in directly along the coast; others fly inland. Vertical 3 m trammel nets are placed on the ridges of the dunes parallel to the beach. Many quail are not caught: some pass over the nets and others squeeze beneath the narrow space below them. The main nets on the first dunes behind the beach are 400 m long. Other nets, 500 m long, are strung along dune tops farther inland. Not only quail, but other species migrating from Africa are caught too.

As quail make a tasty meal, comprehensive descriptions of the catch have been given by earlier authors, for example, AITINGER (1653), BECHSTEIN (1797), BREHM (1855), WINCKELL (1878). BECHSTEIN describes an original method (Fig. 216). Higher nets are first placed in a square. The lower nets are placed about 2 m away to conceal the trapper who lies on the ground with his whistle. Live lure hens that cannot see each other are in nearby cages. With this set catches of 15 to 20 quail per morning or evening were not unusual.

WINCKELL (1878) describes the hunters' methods of catching Hazel Grouse, Black Grouse, Eurasian Woodcock and geese with push nets and a lure whistle: the hunter, well hidden, calls; the young birds rush toward the sound, get entangled in the nets or are shot. Some old hens are also fooled by the lure whistle, mistaking it for the call of lost young. USINGER (1963) describes a number of modern methods for luring various gallinaceous birds.

WINCKELL (1878) praises push nets for catching Eurasian Woodcock. The nets are set zigzag or at angles in propitious cover. Several drivers herd the birds into the net by stamping their feet from time to time. Push nets also serve well to catch molting ducks and geese or their young.

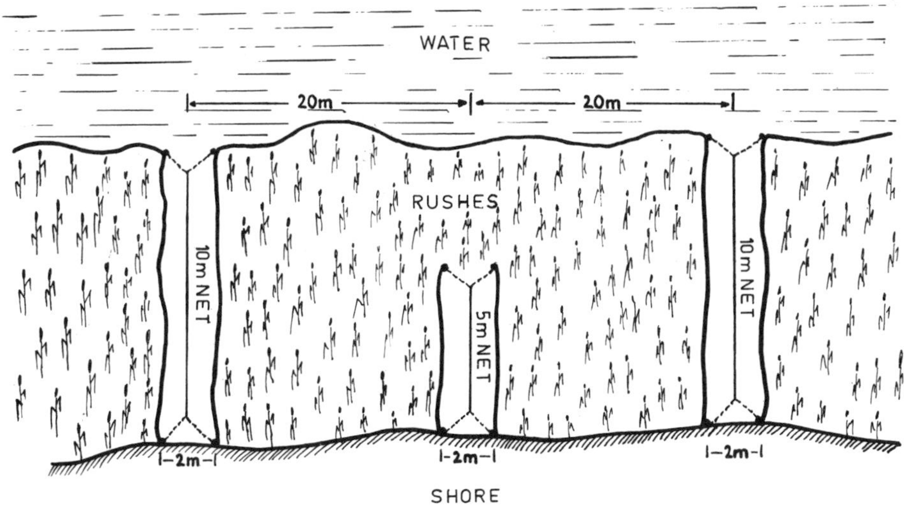

Fig. 218. Arrangement of trammel nets in rushes. After K. GREVE.

The largest catches of miscellaneous birds are in rushes (Fig. 217) Relatively narrow rush borders produce the best results. Fig. 218 shows K. Greve's set of three trammel nets which caught 50 European Reed, Great-Reed and European Sedge Warblers in a short time. The bottom of the net should be 15 to 20 cm above the water to keep birds caught in the bottom shelf from drowning. The most successful trammel net operation was T. Samwalds at the Neusiedler See (Austria) (Fig. 219). He and his wife utilized over 200 m of net, mostly on weekends and holidays, and caught up to 8,000 Bearded and Penduline Tits, *Acrocephalus* warblers and other species annually. Paths—not too wide—were cut in the rushes. The paths changed direction frequently. Permanent boardwalks resembling piers made it possible to tend the nets along the paths and short nets at the ends of the boardwalks increased the catches. The Samwalds technique included lure birds and the spreading of sand on the boardwalks for the Bearded Tits.

Night roosts in rush beds along migration routes are apt to be rewarding. The net poles must be firmly anchored. Masses of swallows that have tipped a net in their onrushing flight are a bit hard to disentangle. At the Radolfzell Ornithological Station on Lake Constance (Germany) several tens of thousands of Barn Swallows and Bank Swallows (Sand Martins) were netted during migrations in lanes cut in the rushes (Fig. 220). The bottom shelf of the net was kept at least 50 cm above the water, not only to keep the birds from drowning, but also to prevent the annoyance of catching Common Coot. Swallows often flew into the nets at the onset of twilight. Boats were used to take the birds out where sets were in deeper water. Even small roosts sometimes produced substantial catches. Careful driving—if need be with a light behind the net—increases catches. The trapping area should not be obtrusively disturbed.

Peltzer (1967) caught Water Pipits in Luxemburg in night roosts. He had no success in a 10 to 15 by 100 m sedge marsh (*Carex* sp.) next to a ditch because there was no place for the trappers to hide. He had better luck in lanes cut in the rushes and caught four Water Pipits on January 16 with five 10 m nets. Although there was cover on both sides of the nets, here too he ran into difficulties. The first was to get the Water Pipits to go down into the rushes, which he sometimes accomplished by driving them from their habitual haunts, but they also showed a certain distaste for stands of rushes. If, however, the birds were already in rush stands, he could not drive them toward the nets, as they flew upwards instead. Experience showed that it did not work to simply drive pipits. Instead, they were more or less herded toward the nets from distances of up to 1 km. Forcing the birds to put

Fig. 219. Trammel net set up at Neusiedler Lake. Photo: A. Bellingrath.

Fig. 220. Trammel netting swallows near night roosts in Lake Constance. Photo: G. GRONEFELD.

in near the nets first brought success. As experienced birds that had already been caught shunned the locality of the nets and drew away the unbanded birds, PELTZER was forced to keep moving the nets. He varied his sets, sometimes placing them on the edges of rush borders (with cover on one side only), in sedge marshes (where dead stalks offered camouflage), and across ditches.

Autumnal lark catching as described by NAUMANN (1824 and 1905), BREHM (1855) and VON KRIEGER (1876 and 1878) deserves special mention. WOLFGANG BIRKNER, the painter of hunting scenes, portrayed this method as early as about 1639 (Fig. 221). The money expended was so great that this method was reserved for princes. Nevertheless, many larks may still be caught in this way, for example on the steppes of Southern Russia where they are abundant.

The set consisted of eight 200 m long, vertical nets—namely about 80 nets 20 m long and 2.5 m high—placed specific distances apart (Fig. 222). The materials—nets, poles, etc.—were transported in a horse-drawn wagon. The big drives started toward evening when the larks had just gone to roost. But according to NAUMANN, catching itself was not begun until the first stars showed in the sky—in particular a medium-sized star low to the south known as the 'Lark Star'. Not many birds fly into the nets earlier, but to drive too late is even worse.

It was all over at nightfall. NAUMANN considered 1,000 to 2,000 Common Skylarks a normal catch.

c) *Catching next to flowing water*

Some, but not all, have the delightful opportunity of catching over brooks and streams, narrow enough so that nets can be strung across.

Fig. 221. Autumnal lark catching. From the Hunting Book by WOLFGANG BIRKNER, 1639.

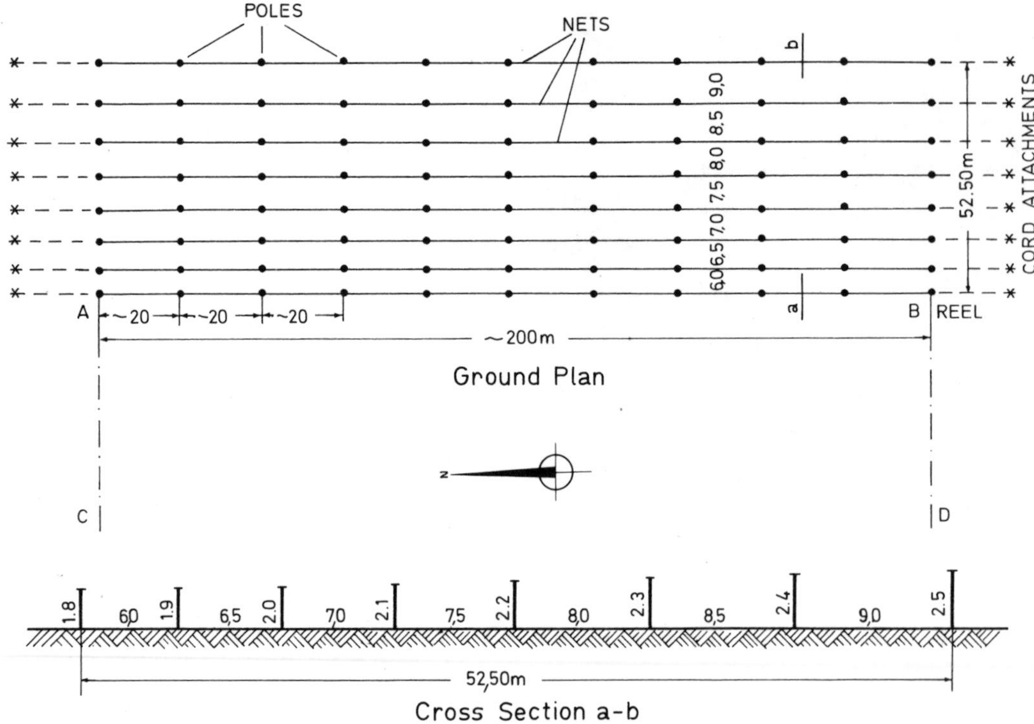

Fig. 222. Arrangement of nets for lark catching from a description by J. F. NAUMANN. After PÄTZOLD, 1963.

Species such as White-bellied Dippers, Common (River) Kingfishers, Gray Wagtails and Common Sandpipers can be netted for general banding or for area studies. Other species such as swallows, *Acrocephalus* and *Phylloscopus* warblers and some shorebirds can be caught by the same method.

Nets are set in several positions—often just one across the brook, or a supplementary net parallel to it. Whether or not to erect nets along the stream's banks depends on how visible the net across the stream is.

Some banders put a conspicuous net across the stream on purpose, planning their set so that birds wishing to evade the big net are caught in inconspicuous nets along the banks. As usual, care must be taken to keep the bottom shelf well above the water. SUNKEL devised a pull net. Its cord could be moved like an endless belt through pulleys attached to trees on both banks. This is handy for keeping the bander out of the water in winter and practical, too, if one plans to net at the same spot repeatedly. If two banders are working together one operates on each bank; the bander who has the net throws an end of the top cord (to which a stone has been attached) across the brook; his assistant fastens the end and ties it taut and horizontal. The net is suspended from this top cord. Now another cord, the same length as the net, is thrown across the brook. The net is fastened to this (care must be taken not to let it drag in the water) and pulled into position. Hooks at various heights can be used to adapt the net to changing water levels.

The nets are often set across the brook at about 100 m intervals. When they are in position, the drivers proceed noisily along the banks—preferably along both banks simultaneously for 1 to 2 km. The drivers keep moving until they have come to the last net (unless someone has to stop to take out a bird hanging so low that it might drown). The birds are removed on the return trip, which serves as another drive. It pays to have other teams netting simultaneously over adjacent brooks or in nearby cover. The drives can be coordinated by blowing a whistle or by starting each at a set time.

To keep the nets inconspicuous, they should be strung in the shade. Dusk is a good time for catching, especially for shorebirds.

10.1.2 Catching hole and house nesters

Trammel nets are indispensable for catching Bank Swallows (Sand Martins) at their colonies. The nets are put up either before or during the day, but catching is better in the very early morning when young are in the nest holes. Later in the day the adults tend to be away catching food. We plan to band the adults when the young are small to avoid nest desertion. It is also profitable to band the young when they are free flying but still spend the night in the nest hole.

The nets are either hung from above the colony or erected from below. If the bank is too high or there is too much vegetation in the way, we fasten the nets to poles and hang the set in front of the colony. This necessitates at least two helpers at the top of the bank to receive the net and fasten it with 20 to 30 cm long hooks. The trammel net is pulled into position like a curtain. It must be pulled silently in order not to spook the birds out of their holes. Some individuals may spook even in total darkness. Coughing, conversation and scuffling are to be avoided. Hand signals should suffice for setting up the net.

Large colonies require four to six workers. Sometimes ladders are needed. Some banks are so high that they present the danger of landslides and are best not tackled at all. We often carry shovels for cutting steps to reach the high nests. If possible, we cut away in advance rootlets and any other protruberances that might interfere with catching.

When we arrive at daybreak, we remove any birds already caught, except for those that it will take too long to disentangle in poor light. Next we stamp our feet about 1 m from the top of the bank to get the birds moving, but take care not to bounce around so hard that falling sand jeopardizes the young and eggs.

As the birds are taken out of the nets, they are put into loosely woven white bags—white, so that no bag lying on the ground will be overlooked. We start banding as soon as it is light enough to read band numbers. We never leave the nets up for more than two to three hours at a time. We stuff shut any holes that our nets cannot reach and then shift the nets to catch the birds within these too.

Trammel nets are not the best for catching in daylight. It is better to use mist nets (see next chapter).

The trammel net technique as described here can be adapted for other bank nesters such as Europen Bee-eaters and Common (River) Kingfishers.

Trammel nets are also useful for catching birds such as Barn Swallows or owls that inhabit barns, sheds or buildings. Just hang the net in doorways or windows, or cut up an old trammel net and fasten it on a frame to fit the opening.

10.1.3 Italian trammel netting

The Italians have special sets: the bresciana (Fig. 223) and the roccolo. These have been described by many (for example, by VALLON, 1882; STEINFATT, 1931; GHIGI, 1933; SCHÜZ, 1931; WITZIG, 1952; BOSSINI, 1958). DE BEAUX'S

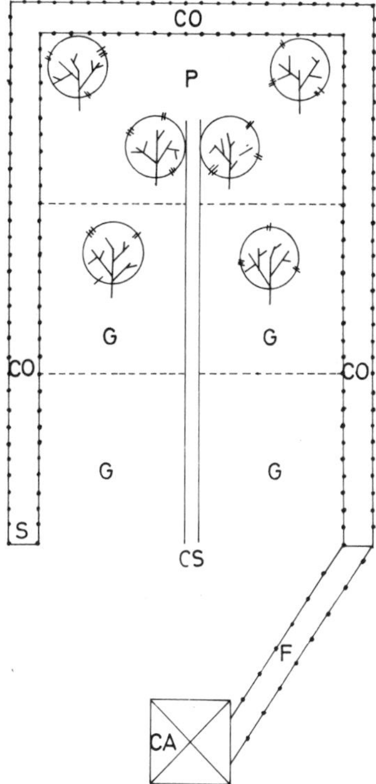

Fig. 223. A bresciana showing groundplan. (CA) blind; (CO) path between nets; (G) lure bird positions; (P) bait for thrushes; (CS) path without nets. After BOSSINI, 1958.

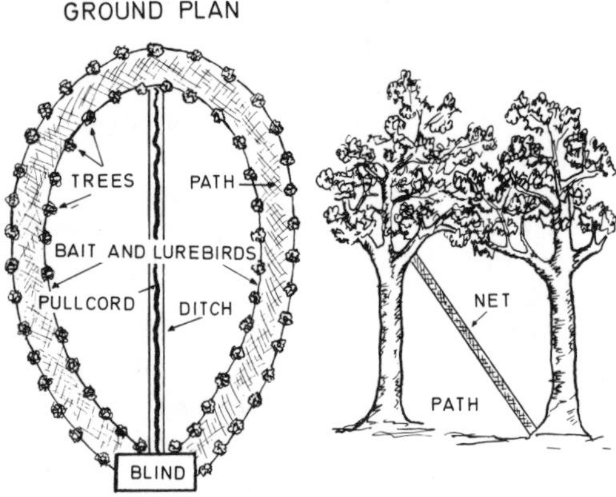

Fig. 224. The brescianella. After DE BEAUX, 1903.

(1903) description of the so-called 'brescianella' (Fig. 224) follows: "An egg-shaped hedge of trees is planted. It must be at least four years old. There is a blind at the pointed end for the trapper to hide in. The hedge is trimmed and bent into shape each spring so that no branches interfere with the path of the nets. The branches are trained to form a shady canopy to make the nets inconspicuous. There are scattered tall single trees near the path. The trammel net runs all around the inside of the egg-shaped hollow hedge from the blind until it meets the blind again and it is set at a slant from the inside toward the outside, rather than perpendicular. A pull cord is strung from the blind through the longitudinal axis of the 'egg'. The pull cord, fastened at its far end, is concealed in a small ditch. A pull on this cord presents the scare devices: it elevates bits of tin, rags and sometimes many small bells are set a-jingling. But first the birds are drawn in; small cages with lure birds, held over from the preceding year and sometimes blinded so that they will sing better, hang in the branches of the trees. Some of the lure birds, however, need to be able to see well for they spot the new arrivals and twitter—a signal for all lure birds to twitter and lure a passing flock to settle first on the tall trees and then on the catching paths. At this point they spot the shady canopy and others of their own kind feeding on favorite foods and apparently free. Lured by their own species, they fly down into the space between the hedges. The moment is ripe. The trapper gives the cord, lying in the little ditch, a strong pull, chunks of tin tinkle, rags float aloft, bells jangle, and terrified, the birds seek to escape between the tree trunks and low branches. They fail to see the nets and fly into them with full force.

"The roccolo is used particularly in mountain passes. It is a large brescianella that sometimes encompasses several kilometers of woods. The nets are like those just described, but ever so much bigger, many more of them and they are set concentrically—not within a specially prepared planting, but under existing trees. The blind is in the center, covered with ivy and concealed by tall trees. It overlooks the entire woodland, enlivened with the twittering, singing and trilling of lure birds. As the birds pass over the Alps in dense flocks, selecting passes to conserve energy, they are enticed to put into the trees in the roccolo by the songs of their own kind. Suddenly the scare devices emerge from the blind with full force. Some trappers produce certain vocalizations of the falcon. The birds fly low to cover to save themselves under branches and trees and end up in one or the other of the concentrically placed nets. In the evening the birds are collected (Fig. 224) and the numberless hawk decoys—thrown during the day—gathered up.

"The brescianella and especially the roccolo are the devices that catch thousands of birds daily."

10.2 Mist nets

10.2.1 General information

There is probably no catching method that has had such a strong appeal to banders as the mist net. Catches mounted significantly with the introduction of mist nets. In the early days of banding it was mostly nestlings that were banded. Now there is a pronounced reversal. These statistics need to be evaluated.

Mist nets are not only practical, but cheaper than trammel nets. Trammel nets last longer and under certain conditions they are better for catching given species.

Mist nets, called 'kasumi-ami' by the Japanese, have been used for over 300 years. They may have been introduced to Japan by the Chinese (AUSTIN, 1947). They essentially attained their present perfection by the last half of the 19th century. The feudal system terminated in 1853; armies were disbanded and many soldiers suddenly found themselves without a means of existence. It is thought that many turned to bird catching. From 1873 on only those individuals that had permission and paid fees were permitted to catch birds. Meanwhile, mist-netting had spread throughout Japan.

Originally the nets were made of silk, spun from cocoons by the trappers themselves. Specialization set in before long and net makers in the villages became the source of supply. They were replaced by industrialized factories about 70 years ago, but even today some trappers prefer to make their own nets.

For the most part Japanese netters still use two mesh sizes, one for thrush-sized birds and one for smaller birds. They used twoshelf nets in the rushes and three-shelf nets elsewhere. New variations have been developed.

Nets with five shelf strings are even baggier and there is less chance of having birds fly against flat net walls.

Until recently the Western World was not aware that bird-catching gardens and special layouts had been developed in Japan many years ago. These 'toyabas' are comparable to the Italian layouts and will be discussed under 'catching in the mountains'.

The mist net principle differs from the trammel net in that birds are not captured in individual pockets. Shelf strings are substituted for the pocket principle. One fine, almost invisible net is strung loosely so that each shelf is essentially a long hammock in which the captured birds lie (Fig. 229). They get into this position by flying against the net, falling or fluttering down, and becoming sufficiently entangled in the net so that they cannot escape (Fig. 225). Any bird striking the net just below a shelf string is apt to bounce right back out, but if the net has sufficient bag a high proportion of birds flying against it will be caught.

Fig. 225. A Sand Martin hanging in a mist net. Photo: H. BEHRENS.

Modern mist nets are made of nylon rather than silk or cotton thread. Our usual net weighs 80 g, is 6 m long and 2.5 to 3 m high. We sometimes use nets that are twice as long and others 9 to 15 m in length. Nets longer than these have seldom proved practical. There are exceptions; we have used 40 m nets for shorebirds. The mesh was 50-60 mm stretched and the shelf strings were stronger. The mesh for small birds is normally only 24-30 mm stretched.

Black is the best color for any standing net. Black masks other colors, absorbs light rays rather than reflecting them. A black net up in the air against any background, even against snow, is the least visible color. Furthermore, black netting blown about by wind near bushes or trees takes on the appearance of shadows.

When unpacking a new mist net, save the packing string; most companies use shelf string and it can come in handy for mending later. Before ever using the net, may banders extend the length of the shelf string loops so that if a side wind blows the fine netting toward a pole, it will not get caught in hooks and such. Some use strong rubber bands to extend the shelf strings and thereby insure better tension. Some save much time by coloring each loop with string or spray paint the first time the net is put up. Each self string should have the same color at both ends and all nets at a station should be color-coded alike. It takes only a moment to tell a new assistant, "All the top loops are red and then we follow the colors of the rainbow." The self strings are about 60 cm apart if there are four shelves and 45 cm if there are five. Fig. 226 shows various possibilities for attaching loops to poles. Remember that if you select the dog leash type of snap swivels or any sort of metal hooks, they are apt to fall through the meshes and create a miserable tangle.

Net poles are usually about 2.7 m long and may be of bamboo, wood, metal, etc. and should

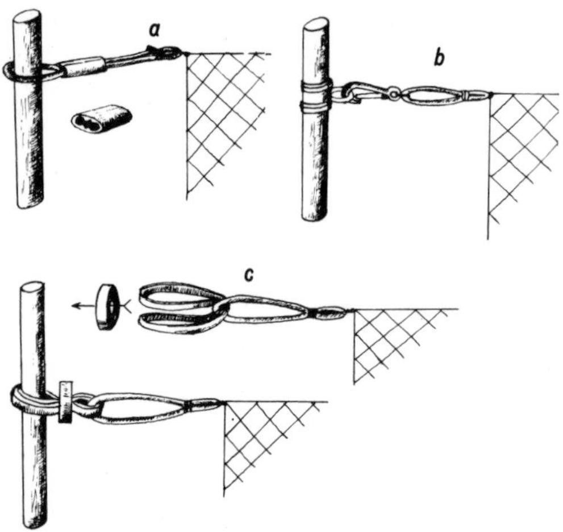

Fig. 226. Three methods of fastening shelf loops to poles. After SANTOS, 1965.

be incospicuous. Some prefer jointed poles because they are easier to transport. Thinwall or light weight pipe make good poles and have the advantage that they can be slipped over short rods driven into the ground to erect them. The rods can be left in the ground if the site is to be used again. Very often, poles are just pushed into the ground and sometimes do not even need to be braced.

Sometimes the terrain is difficult. Some banders, for example GRISEZ (1965) have resorted to earth borers. Others have used 45×30×8 cm cement blocks. A pipe is cemented in so that it projects 8-10 cm and the poles are slipped into the pipe.

As a rule, the bottom of the mist net should be close enough to the ground so that no bird can fly or run under it. It is important to clear away vegetation so that the bottom of the net rests on a good, clean path. A machete is useful for cutting mist net paths.

Wind is the bugaboo of all mist netters. It is best to place the nets where they will be sheltered; if this is impossible, they should be placed cross wind so that the nets are not blown along the shelf strings. SPENCER (1962) devised a good solution in England. He ran a second string along the top shelf string and knotted the two together at about 15 cm intervals with simple half-hitches (Fig. 227). It does not matter if the knots slide a little—they won't move far after the taske is completed. The knots are most easily made with a small shuttle. A similar knotted line along the bottom shelf string proved unnecessary. SPENCER's device makes mist-netting possible with a wind force of up to 5 or 6 and the nets can even be erected on steep slopes.

One manufacturer puts out nets with 3 to 4 vertical spacer strings already knotted in. These raise the price of the nets considerably and are only a partial solution to the problem. Some banders try to keep the nets from sliding by means of small (2-3 cm) and equally spaced clothespins. At Radolfzell rubber discs (25 mm in diameter and 3 mm thick) are slit so that they can be pushed onto all shelf strings.

Nets that have been taken down carelessly and untidily stored are difficult to put up again. PATTERSON (1962) and YUNICK (1967) made canvas aprons to store the nets in (Fig. 228). Each net is put directly into a pocket as it is taken down. Each pocket holds three 6 m nets or two 12 m nets. One pocket is reserved for rubber bands and small equipment.

Before starting to take any bird out of a mist net one must determine which way it went in. It is best to take the birds by their legs, free their toes of netting, and little by little disentangle the rest of the bird from the net. Some birds struggle and bite, making it hard to take them out. Chickadees and tits are particularly exasperating in this respect. Occasionally a bird gets a twist of netting stuck behind the forks of its tongue. Gentle persuasion with a common pin can solve this nasty little problem.

Mist nets do not last forever. SCHIFFERLI (1956) stated that 8 to 10 weeks of constant use was all that they were good for. We count on using our nets for one, two or more years, but not with daily use. K. GREVE found that his 40 m

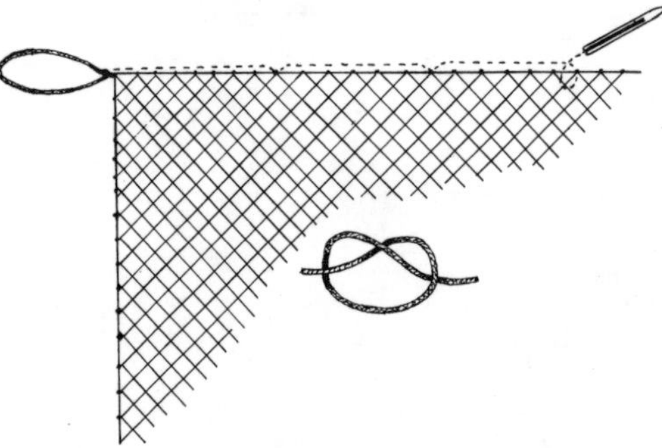

Fig. 227. Knotting of the net to the top shelf string. After SPENCER, 1962.

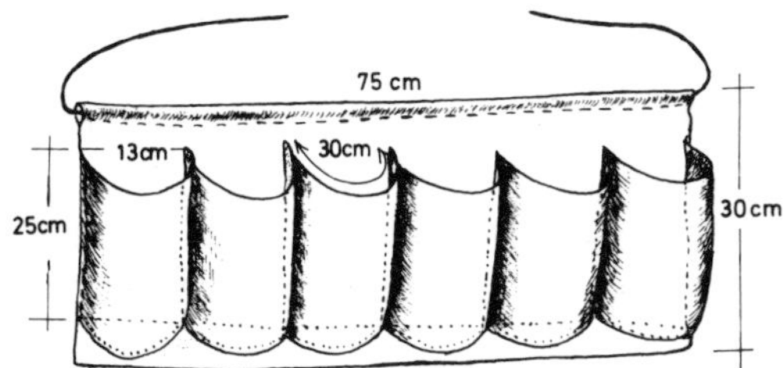

Fig. 228. The mist netter's apron.

shorebird nets were essentially ruined after 1,000 hours of actual catching, as ducks and pheasants have made too many holes. A deer can ruin a net in moments.

The list of species captured in mist nets is huge. In addition to songbirds, captures have included woodpeckers, pigeons, owls, hawks, jays and other corvids, and more. Even pheasants have been found hanging in songbird nets, but stronger mesh is to be recommended for the larger birds.

In Australia, Stewart and McKean (1962) netted small raptors including the Boobook Owl, Little Falcon, Brown Falcon and the Collared Sparow Hawk. Hundreds—perhaps thousands—of Sharp-shinned Hawks and Northern Pygmy Owls have been mist netted on flyways in the United States of America in recent years. The Australian banders felt that it was best to hood the birds as in falconry. Americans have not found this necessary, even for the larger species such as Great Horned Owls.

When bats get caught in mist nets, they start immediately to free themselves by biting through the netting. Many parrots do the same. According to K. Preywisch, Common Bullfinches also sometimes manage to bite through threads. It is advisable to wear gloves when taking bats out of mist nets—these animals are frequently carriers of rabies. F. Hamerstrom frees bats from the net bare-handed, but admits it is not one of her favorite occupations. She seizes them at the back of the neck so that they cannot bite her and little by little works them loose.

Large beetles also try the netter's patience. As long as they are still mobile, all six legs tend to keep moving, gathering up more and more netting for the bird bander to disentangle.

Before we close, let us mention various horrors that pass through mist nets. Among the uninvited guests are horses, sheep, dogs, deer, wild pigs and people. When our luck is really bad, the latter tear through in cars, with farm machinery, on horseback or bicycle. At any rate, the net is done for. But perhaps we can take consolation in the experience of a South Aus-

Fig. 229. Mist net set with loudspeaker and lure bird. After Johnston, 1965.

tralian bander, JOE MACK, whose net was demolished by a donkey while his own dog simultaneously demolished another net. Portions of MACK's (1966) account follows: "On a quiet day, I set up my nets near Berri to increase my mist net catches and I was able to drive away the horses grazing nearby. But soon things started to go wrong. Just after I had checked a net strung among trees and bushes, a horse began to play with the net. Then, horror of horrors, a donkey occupied himself with the other side of the net. Suspecting what was up, I sneaked toward the net and induced the horse to depart. Not so the donkey! With bouncy jumps he leapt into the net and through it, leaving only fragments behind on sticks and stones."

10.2.2 Other methods of putting up mist nets

The simple placement of mist nets along hedges, in paths and between trees (Fig. 229) does not always suffice. Trappers all over the world have exercised their ingenuity. Two nets, placed at right angles to each other with bait, water or a Little Owl decoy within the angle (Fig. 230) are successful. Banders in Czechoslovakia for example caught Eurasian Golden Orioles, Common Hoopoes, Eurasian Rollers and Crested Larks with an owl decoy.

Nets set parallel to each other also work well. P. LOKIETSCH tried to drive Eurasian Nutcrackers into two nets and failed until he placed small pieces of meat in the netting. Australian banders hung bright red or blue plastic jar tops in the net to attract birds such as the Flame Robin and the Blue Wren.

Mist nets can be placed to form funnels for catching Water Pipits, Common Sandpipers and other species that run along beaches or mud flats. Instead of being set vertically, the net is at a slant. Its front edge is held up by thin 50 mm twigs and its back edge rests on the ground. The birds are slowly driven into the 'funnel'.

R. FÜRL erected mist nets without poles between two trees (Fig. 231) and fastened the loops to the edge line with hooks made of 1 mm copper wire.

A mist net mounted within a frame (Fig. 232) is useful for catching Bank Swallows (Sand Martins) (BRUHN, 1965) and can also be hung high in trees. EVANS (1965) describes improvised methods of anchoring mist nets (Fig. 233). Tree top netting is well underway, but no bander has yet succeeded in stringing a long mist net between two balloons. The Serrahn Biological Sta-

Fig. 230. Right angle mist net set.

Fig. 231. Mist net set between trees. After R. FÜRL.

tion in Mecklenburg (Germany) prefers high nets. Two are hung 2 to 3 m apart and parallel above a small lure bird aviary in a woodland opening. Pedestals, 2 m high, make it easier to take the birds out. The catching area is camouflaged with a few bushes and trees. H. WEBER has conducted his 'count' in this manner for 15 years. The immediate vincinity of the trapping area was cleared of vegetation. At the highest point a small group of bushes 3 to 4 m high entices the birds to alight. A somewhat higher tree, purposefully defoliated, nearby adds to the effectiveness of the set. The birds are caught in strategically placed mist nets. Double decker nets, one above the other, work very well, but these should not rise above the enticing perch tree.

In order to 'count' migrants from the north, or sporadic invaders, an aviary or a large cage of lure birds of various species is placed inside the small woodlot. According to WEBER, many species, especially finches and particularly the first arrivals, tend to put in to such places. But other birds, lured by the many different call notes, are also attracted.

Baiting such places with spurce cones, or berries of mountain ash, viburnum or elder or sunflower heads, etc. adds to the catch. At Serrahn, the sets are placed in small (20×20 m) openings in beech-spruce forests.

It is necessary to make a distinction between single nets erected at a height and double deck er nets. Even two nets directly above each other present a very formidable net wall. Up to 8,000 birds—mostly songbirds—were thus

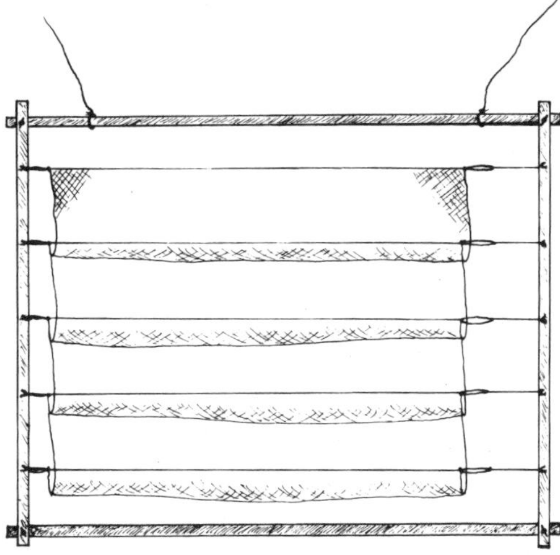

Fig. 232. Mist net mounted on a frame. After BRUHN, 1965.

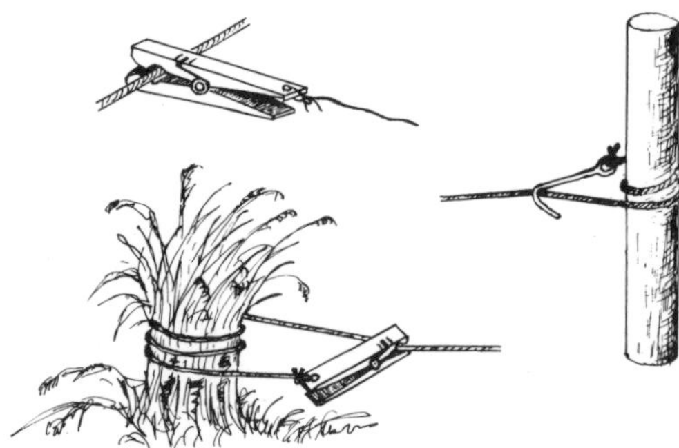

Fig. 233. Methods of anchoring mist nets. After EVANS, 1965.

caught at Cedar Grove Ornithological Station in Wisconsin (U. S. A.) annually. Of course, there must be a way of lowering the upper net quickly and simply to remove captured birds. At Cedar Grove, a cord runs through a ring at the top of each pole and back down to the ground. The shelf string loops of both nets are fastened to this cord so that both nets can now be hauled up or down simultaneously (Fig. 234).

A. GIESE and his colleagues evolved a new and impressive method of erecting high nets at their permanent trapping station near Hilden (Rhineland, Germany) (Fig. 235-237). They found that 2 m high nets, erected in alleys cut in 8 m vegetation permitted most of the birds to bypass the nets. Therefore they erected tall masts made of pipes brazed end to end. Up to four tiers of nets could thus be raised by means of pulleys. Masts 8.5 m high supported up to four tiers of nets. Each net was raised by its own pulley and each net could be raised or lowered independently. The heavy masts themselves were seated in larger pipes sunk in the ground so that they could be taken down when need be. The tall masts had to be braced with guy lines. Nylon cords were used to raise and lower the nets and these proved practical. The best catch (on October 4, 1961) was 155 birds banded, including 120 Eurasian Siskins.

Patterned after the Hilden set, G. PSCHOWSKI erected a somewhat similar one at the edge of the town of Burgdorf near Hannover (Germany). The site was in a small woodland opening and close to the trees. As at Hilden, cages

Fig. 234. The lower part of a pole supporting two mist nets to be raised one above the other. These cords raise and lower only the left hand nets.
Photo: H. BUB.

Fig. 235. Mist net erected up high at Hilden, Rhineland.
Photo: K. STORSBERG.

Fig. 236. Pulleys at the base of a tall mast. Each pulley serves one net. Photo: K. Storsberg.

with lure birds were raised by cords run through pulleys.

The desirability of adding the top net to the set is clear. For example in 1962 the catches were as follows (high net catches in parentheses): Coal Tit 11 (11), Fieldfare 3 (3), Redwing 27 (21), European Goldfinch 12 (7), Eurasian Siskin 227 (190), Chaffinch 60 (40) and Brambling 34 (25).

An inexpensive high set with wooden masts 8.3 m tall was built by Saemann and Stötzer (1968). The three tiers, however, could not be raised or lowered independently.

In Fig. 238 and 239 we see J. P. Binnert's innovation in France for attaching shelf loops on poles up to 10 m tall, which, according to A. Schierer, were constructed of 2 to 3 m long steel pipes 22 mm in diameter. The masts bend a bit at this height, but this does not hinder the catch.

Atop each mast is a cap 4 cm long (inside diameter 25 mm) to which four rings (diameter 25 mm) are brazed. Two rings carry the guy line cords and one carries the net-raising cord (Fig. 239). The cords are all of 2 mm nylon; the rings are of 3 mm iron rod.

In order to raise or lower nets easily, the French use guides (Fig. 238 and 239). These guides are movable and can be slid up or down to adapt to any shelf size. It was found practical to fasten a 'stop' by winding string around the mast 1 m from the ground to keep the guides from gliding all the way down. The 'stop' also makes it easy to furl the nets overnight. The sliding guides remain in proper sequence and can be pulled into position quickly the next morning. A. Schierer suggests that if the same mist net is to be used repeatedly, the end cords

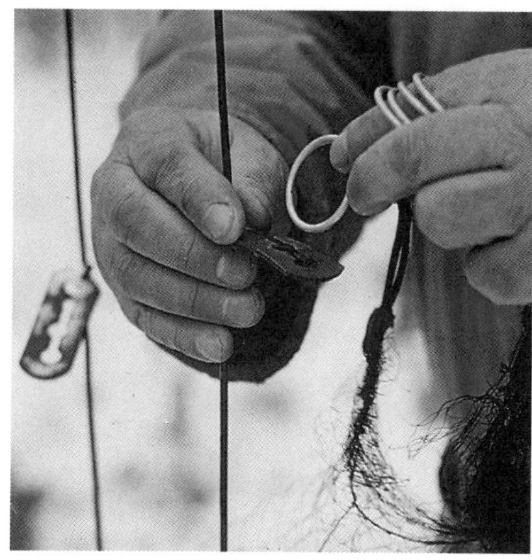

Fig. 237. Attachment for shelf string loop. Photo: K. Storsberg.

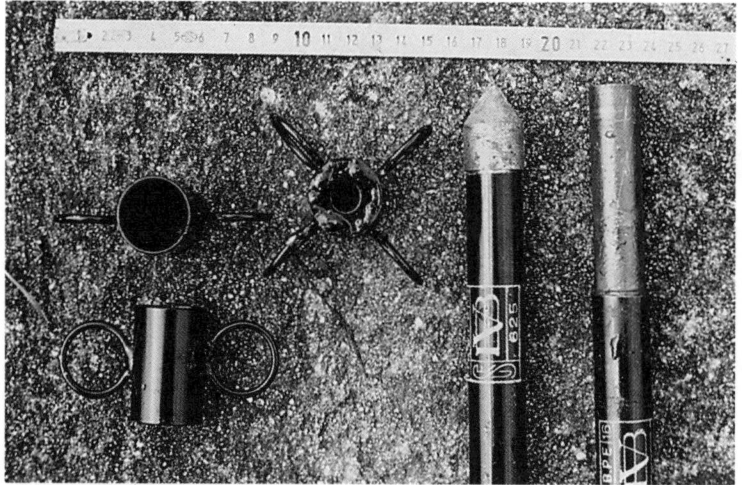

Fig. 238. Pole section, cap and guides. Photo: A. Schierer.

Fig. 239. An upper section of a mast showing the net-raising cord (a) and the guy line (b). After A. Schierer.

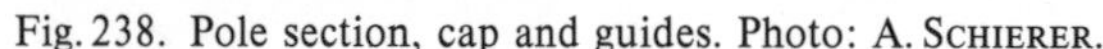

(between the loops) be strengthened (Fig. 239). Thus the net can be set up more quickly.

For summer catching of American Woodcock, Sheldon (1960) used two 9 m sections of aluminum television aerial poles. Two were fastened together to make each pole. A cord, run through a pulley at the top of each pole enabled two men to raise 2 to 3 nets simultaneously.

McClure shows a simple rig using trees (Fig. 240). Still another is shown by W. Böhme (Fig. 241).

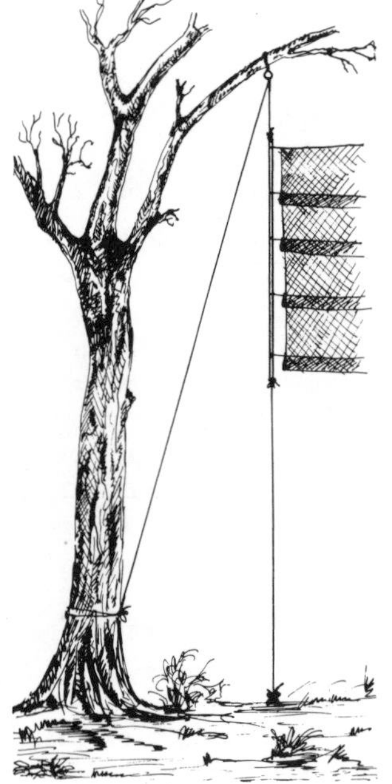

Fig. 240. Method of raising a high mist net. After McClure, 1966.

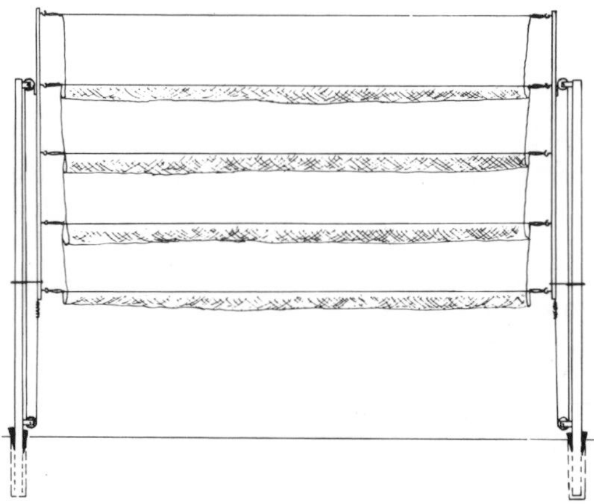

Fig. 241. Simple rig for raising a mist net. After W. Böhme.

10.2.3 Catching in woods, thickets, groves, hedges and fence rows

Mist nets work best in shade cast by trees, in narrow alleys cut for them, in hedges and in fence rows. Birds thus captured by individual banders and by banding stations total into the hundreds of thousands or millions. Migration paths and areas where birds fly about a lot produce the largest catches. Anyone wishing to catch birds had best use at least several nets. The Radolfzell Ornithological Station (Germany) used 250 m of nets and bands 5,000 to 7,000 birds on the Mettnau Peninsula, Lake Constance, in a single summer and fall. The banding station at Hilden, Rhineland (Germany) has 300 m of net included in its high sets and bands 5,000 to 6,000 birds annually.

A good example of short-term success is the catching of 900 Bramblings with only five nets, in the beech woods near Constance (Germany) in the winter of 1962-63. P. Berthold and two coworkers accomplished this in a few days!

To assess the population and compare net shyness of Wood Thrushes in a forest, the North American ornithologist Swinebroad (1964) set his nets in a pattern, using twenty-four 12 m nets (Fig. 242). Each 100 m^2 contained one net. Half the nets were moved systematically every third day. There were no catches on rainy days. In general, there were behavioral differences, but Swinebroad concluded that further work on net shyness was needed.

Lovejoy (1967) and Humphrey, Bridge and Lovejoy (1968) describe catching with mist nets

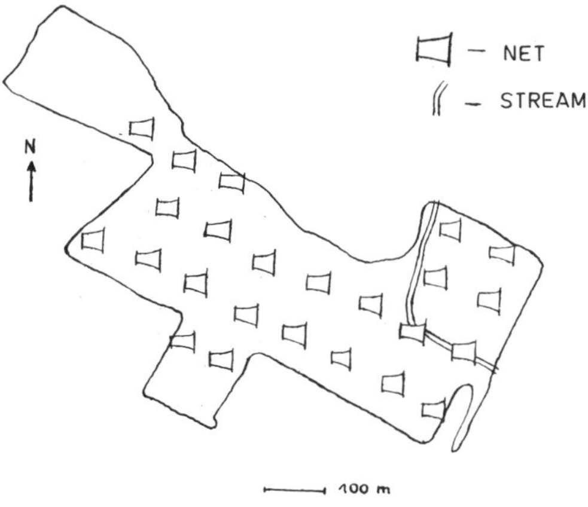

Fig. 242. Mist nets systematically set in a wood. After J. Swinebroad, 1964.

Fig. 243. A high net (made up of 3-4 nets) in the Amazon's primeval forest in Brazil. Photo: T. E. Lovejoy.

25-40 m high in the primeval forests of the Amazon (Fig. 243). Probably no one had netted at such heights before: of 47 species caught, 14 were caught only in the high nets.

Humphrey and Bridge worked out a system for hanging nets at practically any height within a forest—provided one is a good tree climber (Fig. 244 and 245).

A top support rope stretched between two tall trees supports the net rig. Two loops in this rope and attachments at ground level support a system of metal rings which act as pulleys. Through these metal rings a continuous nylon line is strung so that one vertical line at each end will move in the same direction and at the same speed. The mist nets are attached to the nylon line and can be raised or lowered by one man from either end.

10.2.4 Catching in plowlands, meadows, steppes and deserts

It is difficult and sometimes impossible to catch birds where there is little or no vegetation as

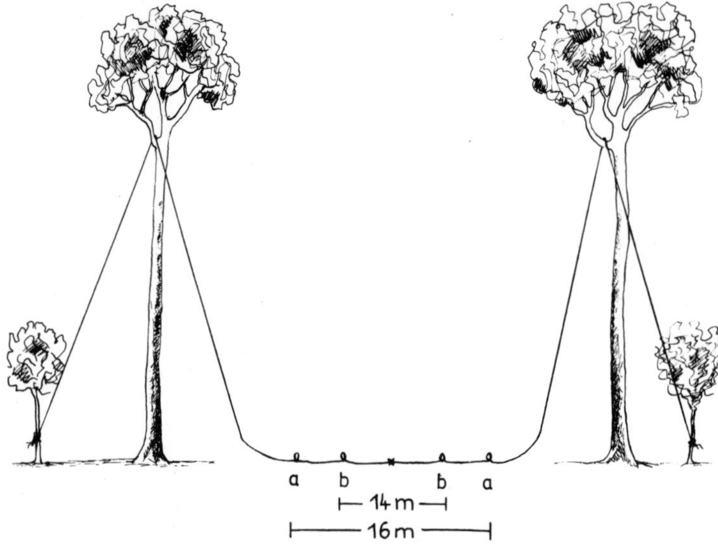

Fig. 244. The net cord before the net is fastened on. (a) loops for the net cords; (b) loops for attaching the net. After HUMPHREY, BRIDGE & LOVEJOY, 1968.

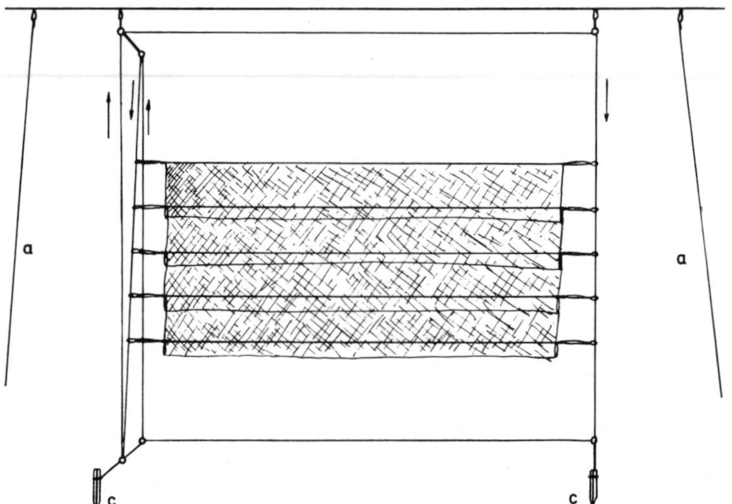

Fig. 245. The net, raised for catching. The arrows indicate the direction that the cord is pulled. (a) top support rope. (c) stakes in ground. After HUMPHREY, BRIDGE & LOVEJOY, 1968.

they can detect the nets too easily. Twilight or when the sky is overcast are the best times to try. SCHIFFERLI (1956), however, caught Ortolan Buntings, Northern Wheatears, swallows and Yellow Wagtails on bright sunny days and attributed his success to the 'invisibility' of his nets.

Catching in open country is often facilitated by using only two shelves of the net—a set reminiscent of push nets (Fig. 215) which have been used in open areas for ages. KRUIS catches European Quail in mist nets when these birds are abundant.

Low nets are used for catching Aquatic Warblers during the breeding season. G. SOHNS and H. WAWRZYNIAK wrote that "except for the display flight this species almost always flies low over the sedge and approaches its song perches from below. We utilized this trait to advantage by positioning a low net near the sedge but slightly rising above it.

"Three low nets can be made by cutting up a 3-shelf mist net, restringing the edges of each piece, and adding loops at the four corners. The vertical cords should be shortened to form sufficient bag. A well strung low net does not advertise itself by fluttering in the wind and it is

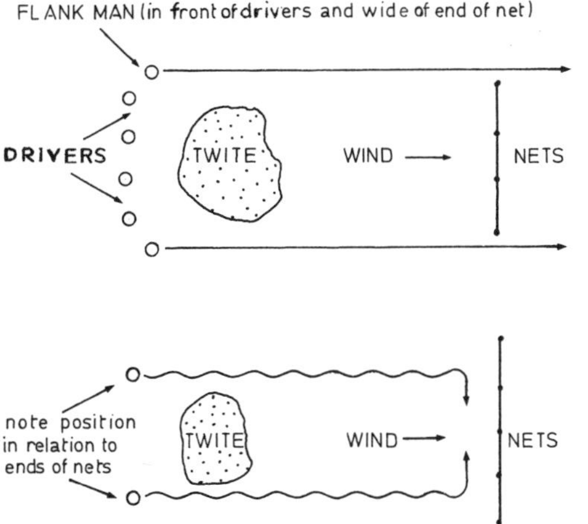

Fig. 246. Twites drivers. After ORFORD, 1967.

quickly set up. One person can erect three nets in four or five minutes. The net rises at most 50 cm above the sedge.

"The singing display of the males increases toward dusk and it is only then that the males keep shifting position. The best catches are usually about sundown. If the song perches were not too far apart, up to four males were caught in 30 to 40 minutes. As a rule the net had to be moved to catch each one. We tried to set the net between two regularly used perches but closer to one than the other.

"Sometimes we had to shorten the singing perches so that they would not project above the top edge of the net. At twilight it usually sufficed to set the net about 20 m from the singing bird and then to drive it slowly toward the net persuading it to take one short flight after another. Often, when the song broke off at deep dusk, silence denoted that the catch had been made."

The English ornithologist ORFORD (1967) experimented with the catching of flocks of Twites in open fields. The birds were driven toward the net from about 200 m away (Fig. 246) and the plan for each drive depended upon the number of available drivers. One person led each drive, keeping track of the others and directing them with hand signals to speed up or slow down depending on the situation. After the catch, the birds that evaded the net are driven toward it again.

HOESCH (1958) describes his difficulties in mist netting in Southwest Africa. Setting up at the right time and in suitable places took much time and planning. Most bushes were thorny and some of the grasses had prickly awns so that just getting the net up was fraught with many a misadventure. Straight sticks for net poles were in short supply. The preparations were often not proportional to the catch, which mostly consisted of long-toed, wildly fluttering, self-entangling weaver birds. Their removal from the meshes of the net took much time. In this land, where shade is scarce, the nets had to be checked constantly lest the captured birds met their death from the blazing sun. "In short, when one has stewed in the sun for an hour or more to disentangle a flock of 20 or more Vitelline Masked Weavers from the meshes of a mist net, enthusiasm for this method of catching wanes appreciably."

10.2.5 Catching in populated areas

There are many opportunities for mist netting in villages, towns and cities, not only in gardens great and small, but also in parks of various sorts. The available birds may be breeders, migrants or winter visitors, for example, Bramblings, Eurasian Siskins, Common Redpolls, Twites, Bohemian Waxwings, crossbills and Eurasian Nutcrackers.

In Bangkok (Thailand) ornithologists strung nets high over the street (Fig. 247) between 12 to 20 m high buildings to catch swallows (KING, 1969). The birds roosted in the city. They were driven into the nets by waving a long-handled

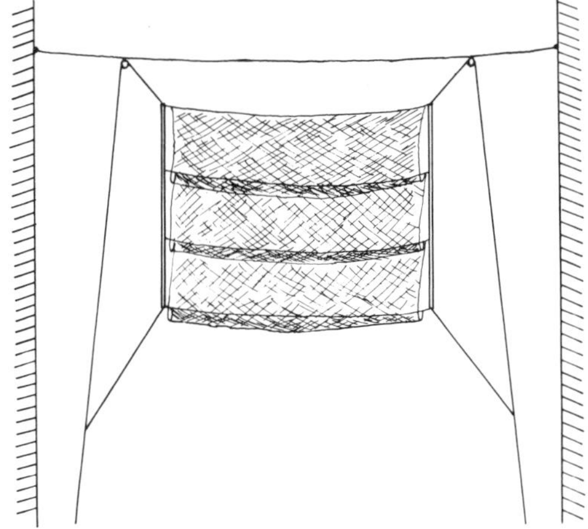

Fig. 247. Net setup used in downtown Bangkok for catching swallows. After KING, 1969.

flag and up to 500 were caught in a single drive. Disturbance by the public was resolved: up to 300 people watched the action and many wanted to help take birds out of the net. Most of them meant well but were a bit rough; some took birds home to supplement their diet. As soon as the banders realized that they could not turn down these many offers of help, every effort was made to get people to handle the birds with care. Special thanks were extended to three police stations in Bangkok for "nightly supplying policemen to direct traffic and keep law and order."

Swallows and swifts in the breeding season offer many chances for studies and banding.

Catching in gardens has been discussed under funnel and cage traps. The planting of berry bushes, hazel nuts and conifers (as well as food patches of corn, millet, buckwheat or sunflowers) attracts many birds. Backyard banders in the United States of America are highly successful. Margared Morse Nice's classic study on the Song Sparrow was essentially a backyard project while she was a housewife and bringing up her children. Most backyard banders in America use mist nets. In Europe catches are often enhanced with regularly tended trammel nets and a small aviary of lure birds.

W. Jahnke spread fresh manure over a 1 m^2 surface in his garden and became successful in catching White (Pied) Wagtails. He set two nets at right angles 1.5 m from his 'bait' which consisted of countless insects congregating on the manure. By appearing suddenly and throwing a weighted white cloth into the air, he could usually prevent the birds from flying upwards and scare them into the nets instead.

S. Schöne nets over his swimming pool, setting his poles in cement sockets.

J. Sadlik catches migrating Coal Tits in his garden. At the midpoint of a 12 m net he sets an additional 6 m net; one of the corners contains two lure birds out of sight of each other. These, unlike other lure birds, are on the set all day as the Coal Tit migration may continue well into the afternoon. The lure birds work in shifts, but are not released. A bare branch near the lure birds served as an attractive perch. The birds flying 10 to 20 m above the ground were easily caught. Sadlik banded 76 in 1969.

10.2.6 Catching on seacoasts, lakes, rivers, brooks and in marshes

As we know, the catching on bare surfaces along the seaside and near water is not without difficulties. On bright days, even with long net walls, one usually has little success. The best catches are during twilight or on darker nights. The coast dwellers knew this long ago when they set long, high stationary nets for ducks and geese. Nocturnal expeditions are not always successful, especially if the nights are very bright. Thus, during a night on the shore of Neuwerk Island (Germany), Bub did not catch a single bird although countless shorebirds were flying over the tidal flat and three 40 m nets certainly presented a fine net wall.

Knorr (1963) has had considerable experience with the difficulties in shorebird catching and gives useful advice which is presented here in detaii. These suggestions no doubt will have to be augmented under certain circumstances.

Nets that are set on open coasts or on damp tide flats generally present a silhouette against the sky that the birds see all too easily. The nets are least conspicuous just before dusk and on cloudy days.

Winds are always stronger along the shore. Strong, shifting winds are best coped with by setting the nets independently, rather than fastened to each other. Nets that are attached to each other cannot be moved quickly. Single nets, on the other hand, are easily moved to reset them cross-wind.

Tides require frequent resetting of nets. One must be especially careful when the tide is coming in. Whenever a net fills up with birds while the water is rising, the net poles in the deepest water should be pulled up first and carried toward land resetting them to keep a semi-circular set. Nets are less threatened by the ebbing tide.

Shorebirds often fly in flocks. Surveillance of over-water nets cannot be discontinued for even a minute. The weight of a flock of shorebirds caught simultaneously can cause a net to collapse and be the death of many or all of the birds. The bottom of the net should be at least 40 to 50 cm above the water. Along the coast it is often not necessary to set over water, for the birds tend to fly over their favorite feeding places at high tide level.

Soft mud and wet sand do not offer firm footing for net poles. The nets can tip even when

Fig. 248. Long net wall for catching Brant Geese on the North Sea island of Jordsand. Photo: M. Fog.

they are empty. Longer poles can be set deeper, and for best results they should be set at an angle so that the tops are farther apart than the bases.

Flotsam is a further danger. The bander should be thoroughly familiar with the terrain on which he traps and avoid all firmly embedded obstacles such as posts, anchors and markers sticking in the mud. The netting site should be selected at low tide when such things are visible.

If a net collapses because of the weight of captured birds, it should be taken down quickly and carried to solid ground; this is one of the reasons why at least one helper should be on hand. The net is laid full length on land that is free of mussels, driftwood and other beach debris. Dry sand should be avoided because it gets into the birds' eyes. Moist sand is ideal.

Along the coasts and tide flats, we set primarily for shorebirds, ducks, geese and gulls. The Danish zoologist Fog (1967) started studying the movements of the Brant Goose—a species that had scarcely been banded before—on the North Sea island Jordsand. For this purpose he set up nets 30 m long and 6 m high on a tide flat (Fig. 248), and caught 56 Brant Geese in three autumns and 110 in three springs. Fog used fish nets with a 1.5 mm diameter thread with a 19 cm distance from one knot to another. Nylon nets of about 200 mm mesh stretched are also suitable and serve to catch birds less strong than Brant Geese as well. He netted in dark nights and checked the nets at half hour intervals.

French ornithologists erected long connecting net walls of several hundred meters on the Atlantic coast and caught shorebirds, terns and gulls.

Yunick (1965) developed his own method of setting nets (Fig. 249) and used a large number of artifical lure birds. Hardman (1965) adds some good suggestions. The catches of birds on tide flats can be substantially increased at dusk or during moonlit nights by putting lure birds on the windward side of the net. They cause

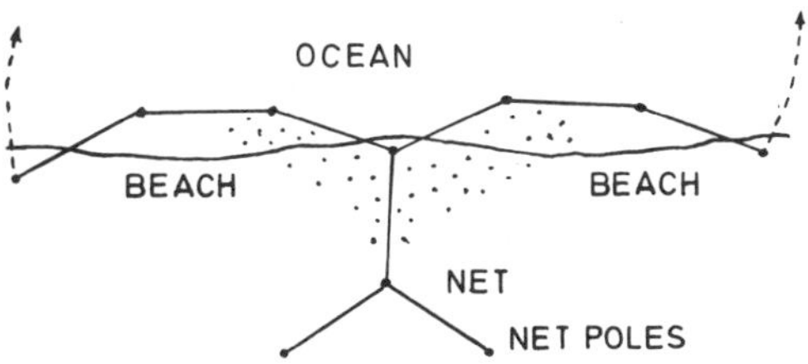

Fig. 249. Arrangement of mist nets on an exposed coast. The dots represent artifical decoys. After Yunick, 1965.

Fig. 250. Mist net on the Dutch coast. Photo: J. TAAPKEN.

birds passing over to lose altitude and to fly low into the wind to land among the lure birds—whereby they end up in the net.

Silhouette decoys made of wood or cardboard can be used, but they are not as successful as stuffed birds. The latter are expensive and not easy to set out; they do not stand well and are not handy with wet footing.

A simple way of converting any dead birds into decoys is to inject them with concentrated formaldehyde. This dries them out or mumifies them if one keeps them for a few weeks in a warm place. These have an advantage over stuffed birds in that they are more rugged.

The dead bird is first placed in a wire (mesh) holder that keeps its head, body and legs in a natural position. The head, each muscle and the internal organs are injected with formalin (10 to 20 cm^3 suffice for an entire bird). Skin coming in contact with formaldehyde should be washed thoroughly afterwards.

When the bird has dried hard, the net frame is removed. The lower 7 to 14 cm of the legs are wound with wire to give the decoy greater stability. Similar decoys can be used for other types of trapping. Plastic decoys are serviceable for rainy or foggy weather; some of these can be bought and look remarkably life-like.

Although catching by day is generally less fruitful, there are places where this is not so true. CARRUTHERS (1965) is among those that catch by day. He captured 70 shorebirds on the southern coast of Australia in one afternoon: ducks, kingfishers, bee-eaters, and Zebra

Fig. 251. Catching swallows from a bridge. After LUESHEN, 1962.

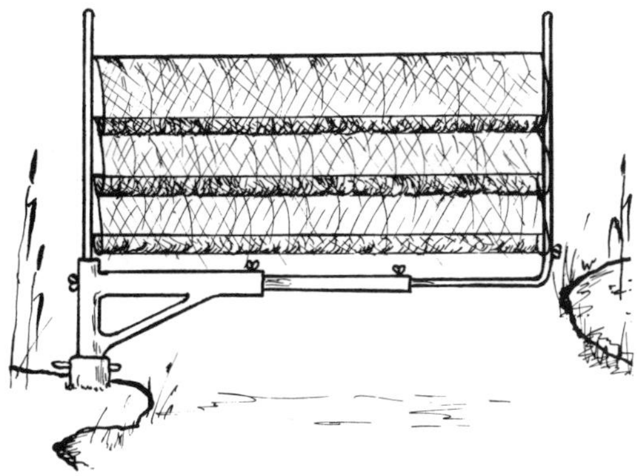

Fig. 252. Slewable mist net. After G. WEPLER.

Finches also landed in his nets. Once he set the nets in a triangle and another time in a great U. He watched the flight patterns of the birds carefully.

MINTON (1965) described a method of keeping birds hanging in nets from drowning. He makes small m-shaped supports of galvanized wire to keep the bottom strand of the net above the water. Thus the net does not droop down when carrying the weight of many birds. The height of the support frames depends on the depth of the water. The dimensions MINTON uses are: height of the sides 1.2 m, breadth 1.2 m, indentation at the top in the middle, 15 cm. After use, the support frames may be folded.

J. TAAPKEN shows a method (Fig. 250) of diurnal catching, suitable also for songbirds, on the Dutch coast. In the daytime it was useful to set only the top third of the net along bare beaches and coastlines. Shorebirds, pipits and Northern Wheatears, flying to and fro are less apt to notice them. J. TAAPKEN catches Horned Larks by driving them carefully into nets set only half high. In the same manner Twites can be captured in the beach aster *(Aster tirpolium)* zone. For Common Gallinules (Morrhens) which show little fear of man, TAAPKEN scatters bread and drives them into the half set net. It does not always work, however.

Now we come to catching along smaller watercourses: brooks and streams, where success has been good with one or more nets; mist nets are under discussion here.

WILLETTA LUESHEN (1962) caught swallows from a bridge with a 9 m net in Nebraska (U. S. A.) (Fig. 251). Once she caught 76 Cliff Swallows in two hours.

SELLICK (1960) describes catching Harlequin Ducks in Iceland. He erected nets at strategic points along a stream.

G. E. WEPLER devised a noteworthy method. He fastened a mist net to a rectangular frame mounted on a hinge (Fig. 252) so that he could catch where access was difficult: primarily at brooks and stream shores. The hinge facilitated adapting the net to utilize the direction of the wind as well.

A method of setting mist nets in marshes or over brooks was devised by R. DE VRIES at Wilhelmshaven (Germany) (Fig. 253). He fastened the net between two thin bamboo rods with a hook at the top of each. These in turn are hooked into 4 mm wire fastened between two stationary poles. All the tiers of the net are immovably fastened to the bamboo rods. The length of the rods determines the height of the net. The distance between the poles must be at least twice the length of the net (and, of course, can be greater) so that the net can be shoved over to where the bander has good footing for removing the captured birds. A cord running from the right hand bamboo rod to the right hand pole and then close to the ground from the

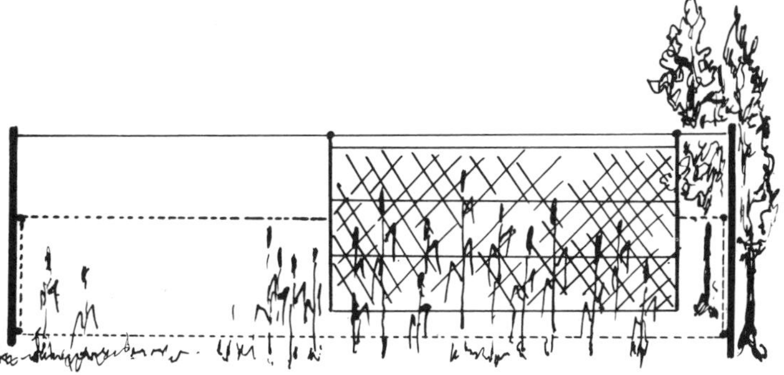

Fig. 253. Movable mist net. After R. DE VRIES.

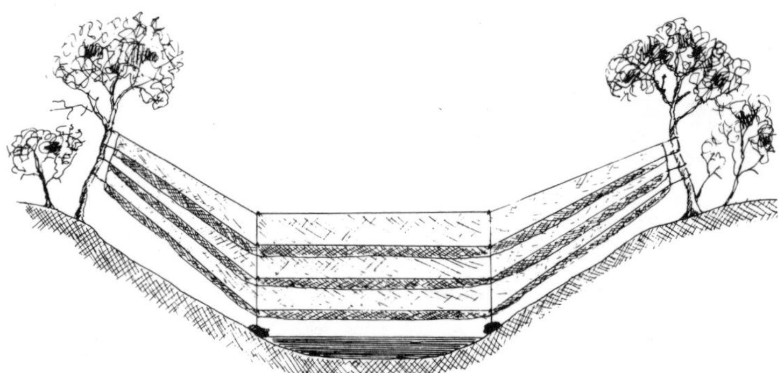

Fig. 254. Catching by streams with steep banks. After A. SCHIERER.

right pole to the left pole and from there to the left bamboo rod makes it possible to pull the nets into the desired position. The cord runs through rings or channelizers, and always in such a manner that the net remains stretched between the bamboo rods.

For catching over brooks 2 to 8 m wide, W. JAHNKE utilizes a fringe line to close off the area between the bottom edge of the net and the surface of the water. Gray Wagtails, White-bellied Dippers and Common (River) Kingfishers otherwise fly easily under a net having the necessary safety factor against drowning. The distance between fringes is 10 cm. JAHNKE has never seen a bird fly through the brown colored fringe.

For catching by streams with steep banks, A. SCHIERER devised a set (Fig. 254) in Alsace-Lorraine (France). This set is suitable only for streams shallow enough to wade across. Perpendicular cords are shortened to 70 cm. This not only prevents thrush-sized birds from slipping out the ends, but also enables one to adapt to variable tie points on trees, shrubs or rocks (with saddler's twine or other strong black cord), and poles are not needed at all.

Three or four small hooks (paper clips will serve) are fastened to a strong cord 50 to 70 cm apart and these are hooked into each shelf string. The strong cord is fastened to the ground so as to lower the net into optimum catching position. It is generally better to have two sets of such hook strings for each 6 meters of net. Hook strings are best stored wrapped over cardboard to keep them from getting tangled.

There are excellent catching possibilities at ponds and lakes, especially near their shores, marshes and fens as well as in irrigated fields near cities and villages.

Shorebird catching, in the Middle European inland, was inconsequential before 1950, probably because of the high price of trammel nets. With the advent of Japanese mist nets, the situation altered: many sets were made for shorebirds with great success. Inland catching stations in Middle Europe have become an essential completion of the important northern stations such as Ottenby (Sweden), Amager (Denmark) and Revtangen (Norway).

The model for many European banders and banding stations was surely Tour du Valat Biological Station founded in 1947 in the Camargue in southern France (Fig. 255). In addition to many fyke nets a great number of mist nets were utilized. Some of these were set in irrigated fields to catch shorebirds as MCCLURE (1956) already described in his publication from Japan (Fig. 256).

Before 1969 the following shorebirds were caught—mostly in fyke nets however: 4,831 Wood Sandpipers, 3,231 Snowy (Kentish) Plovers, 2,878 Little Ringed Plovers, 2,264 Common Redshanks, 1,714 Little Stints, 1,625 Pied Avocets, 1,566 Common Snipes.

SCHÜLER's (1960) report on shorebird catching near Marburg/Lahn (Germany) is of general interest. Common Snipe were captured during the misty hours of daybreak. The flocks were still together then. They permitted the catcher to come nearer than in good daylight and they failed to see the nets as quickly. The nets were set in a zig-zag line about 150 m long across a marshy meadow; mist nets and trammel nets were set alternately. SCHÜLER found this arrangement productive as rails and Common Gallinules (Moorhens), as well as Common Snipe, often run before taking flight. These running birds were most frequently caught in tram-

Fig. 255. Flocks of birds in the Camargue. Photo: A. R. JOHNSON.

mel nets which can be positioned to touch the ground. Of course, it is possible to catch with mist nets only, as has been demonstrated elsewhere, but one will miss some of the running birds—in total, indeed, a small percentage of the catch. For catching rails push nets or trammel nets are more efficient.

With other shorebirds SCHÜLER's experiences were different. These birds preferred marshy banks of streams for feeding. Of these the simplest catches were Common Greenshanks and Common Sandpipers as these both habitually fly low over the water at dusk.

Eurasian Jacksnipe can be mist netted in spring and fall. K. KLIEBE warned against trammel nets as they are too conspicuous in open country. Before setting the 12 m mist nets up, it is important to seek out favorable spots within the large roosting area. The rule of thumb is that Eurasian Jacksnipe put in at dusk where other snipe do. This is not a social bond; it is the joint utilization of a feeding area.

After sizing up the catching area, the nets are set up in zig-zag. Two to four nets are not sufficient. On favorable days KLIEBE utilizes at least 10 to 12 nets covering 100 to 120 meters, shorter because of the zig-zag set, but more efficient. A sort of maze increases this type of catch. If, when getting the nets up, snipe are already flushing, set up in that very place, for in all likelihood, the birds will return to the same area at dusk. The only good time for catching is dusk for then Eurasian Jacksnipe are active enough and easily caught in the maze of nets. The nets should be set an hour before the onset of twilight and then remain undisturbed. The advantage of this method is that usually several are caught at a single setting. The disadvantage is the time and effort in setting up.

KLIEBE uses this method only during peak migration periods (March-April and October-November). In winter he uses another method (suitable also for migration) requiring only one undamaged 12 m mist net. One, or better still, several helpers are required. One tries to flush a Eurasian Jacksnipe, marks down where it has put in and then covers it with the net. The observes should be placed so that the flushing

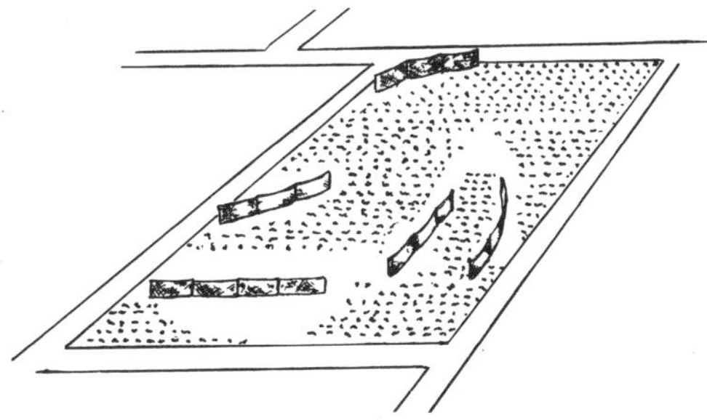

Fig. 256. Mist nets set in a rice-field. After MCCLURE, 1956.

snipe can be followed with binoculars. Whichever observer has marked the bird down most strategically, approaches the spot quietly taking care to keep his voice lowered and not to get so close as to flush the snipe prematurely. If possible, he should have taken note of the exact spot—by a thistle stalk our rush clump—otherwise the net may miss the bird. Sometimes the bird must be flushed repeatedly, but there is a chance of catching it each time.

When the exact spot where the bird has put in has been marked down, it is then covered with the spread net. All members of the party participate, communicating only with nods and prearranged hand signals. The bird by no means always flushes when the net covers it. Sometimes one has to keep searching for it, as it hides in a depression scarcely recognizable against the surrounding terrain. Once spotted, it is simply grabbed from above through the net. If the bird just cannot be detected, the net is slid to and fro so that the snipe, alarmed either by the sound of the net or by its touch, flushes and is captured unharmed.

Inland shorebird catching was highly successful in the sewage irrigated fields north of Braunschweig (Germany). From 1962 to 1970 K. Greve and helpers used the customary 12 m mist nets and 40 m long shorebird nets (50 mm stretched mesh of strong thread). They banded 9,028 marsh and water birds: 1,833 Common Snipe, 1,626 Common Sandpipers, 1,554 Ruffs, 1,349 Wood Sandpipers, 276 Northern (Common) Lapwings, 242 Common Greenshanks, 131 Spotted Redshanks, 127 Common Redshanks, 116 Ringed Plovers, 98 Curlew Sandpipers, 76 Little Stints and 49 Eurasian Jacksnipe. Here, too, most of the catches were during twilight or at night. Daytime sets succeeded only on overcast or drizzly days. Such weather favored catches at night as well. Moonlit nights or clear, starry skies usually produced less than average catches.

For several years, banders have trapped successfully in the sewage irrigated fields near Münster (Germany). M. Harengerd, W. Prünte and co-workers captured from 1969 to 1980 4,032 Ruffs, 2,994 Common Snipes, 2,286 Lapwings, 1,045 Common Sandpipers, 968 Wood Sandpipers, 957 Greenshanks, 512 Green Sandpipers and 428 Black-tailed Godwits in addition to other birds.

One can catch swallos on or near pond or lake shores in autumn. Some banders even string nets over the surface of ponds.

Hocheder (1956) had varied experiences in swallow and swift catching near Regensburg (Germany). His first try failed. The nets had been set in shallow water between two islands of reeds, and side by side as well as at right angles. For the most part, the birds flew to within 20 cm of the nets and then, suddenly, swung up over them. Brinkmann, too, noted that he caught mostly young birds and that adults often evaded the net. But beside the open water, a marsh with man-high willows was often hunted in low flight by the swifts. A net was placed between two willows where the flight was most intense. At first, the success was meager. At the third try, two nets were set at right angles to each other and the catch improved, for when the birds spotted one net, to evade it they usually fell prey to the other. It was important to string the nets between willow bushes as the swifts flew among them utilizing the gaps between them. Cold, nasty weather was a prerequisite for good catching. Catches predominated with light to heavy rain from 17 June to 3 July and amounted to 58 Common Swifts and 38 Common House Martins. Good trapping was limited to approximately 6:00-7:00 and again roughly to 18:00-19:00. For House Martin trapping W. Sunkel prefers sultry weather preceeding a thunderstrom—at such times, the swallows gather near water or edges of woods in dense swarms and nip through the air in such a lively manner that if a mist or trammel net is rightly placed, one has one's hands full taking them out of the nets. Recaptures, too, are often netted.

Catching with trammel nets in rushes has already been mentioned. Without extensive catching activities, which nowadays are provided by mist nets, one cannot do adequate research on a larger *Acrocephalus* warbler population according to Bezzel (1961). This statement presumably stems from the fact that about 70% of the color-marked birds were reidentified through mist netting. As Springer (1955) already pointed out, the cutting of catching alleys does not disturb the establishment of territories; the alleys are included within them.

Buysee (1968) cut 60 cm wide lanes in the great rush marshes of the Netherlands. All the lanes were interrelated. He captured *Acrocephalus* warblers, Bearded Tits and rails.

10.2.7 Catching in the mountains

In the hills or mountains we usually have to observe the same rules as in the lowlands. Two catching methods offer special possibilities in the high country. Japanese mist netters have developed a special system, independently of the Italian netting arrangement. A toyaba (bird-blind-place) is always set up according to the situation, but it follows a certain pattern. Attempts over a period of years found the best locations—usually on ridges. At an ideal site, the slope falls off sharply from the toyaba on three sides and rises sharply above it on the fourth.

The nets are set up in the shape of a horseshoe so that three sides of a coppice are surrounded with nets. An elevated platform by the open end of the 'horseshoe' is the center of operation—this must be erected in advance. The nets are always parallel to a contour line, rather than running uphill and downhill, and they are never set at right angles to each other, except on level ground.

The size of the toyaba is determined by the terrain. The center of the set, where the lure birds are positioned, may vary in diameter from 15 to 100 m. Generally 35 10 m nets are utilized, having a periphery of 350 m. The better the site is, the fewer nets are needed. For the most part, one row of thrush-sized mist nets at the edges of the set suffices, but sometimes the catcher sets three or more rows of nets below the 'plateau'. Small mesh mist nets, on the other hand, are placed near the center of the toyaba between the lure birds.

In lower, less steeply cleft country more labyrinth-like sets with up to 250 nets are put up.

McClure (1956) reported on mist netting in Japan (Fig. 257) and recorded a variety of sets. Those with hundreds of nets are generally on slopes that present their sides toward the main direction of migration. The catcher sets his nets parallel and two or three meters apart in the vegetation. The nets range between 30 and 60 m in length. McClure also mentions the utilization of plenty of lure birds.

Great success has been attained by the Swiss in their mountain passes. M. Godel and G. de Crousaz (1958) describe their set (Fig. 258). "After our preliminary attempts in 1952 and 1953, Col de Bretolet struck us as the best catching place for two reasons. It is better protected from the wind than Col de Cou so the nets can be erected already spread for catching. Also, the alders, scattered in the pass, attract various birds and offer some cover for the nets. The best catching technique was developed in the fall of 1954 for these local conditions and in 1956 we became even more efficient by using nylon nets."

"The set on Bretolet is 2.5 to 3 m high and about 100 m long, essentially extending across the notch of the pass. In part it crosses open country and in part it passes among low alders. A few supplementary nets which may be replaced with small Helgoland fyke nets to facilitate removal of the birds are set in the bushes below the pass on the Swiss side of the border. And finally, two high nets, strung between two 5 m bamboo poles, are very useful for capture on windless nights. Most of the diurnal captures were close to the bushes; most of the nocturnal ones in open terrain."

"Four banders can barely cope with this spread of nets, and sometimes—especially if there is a strong tit migration—some of the nets

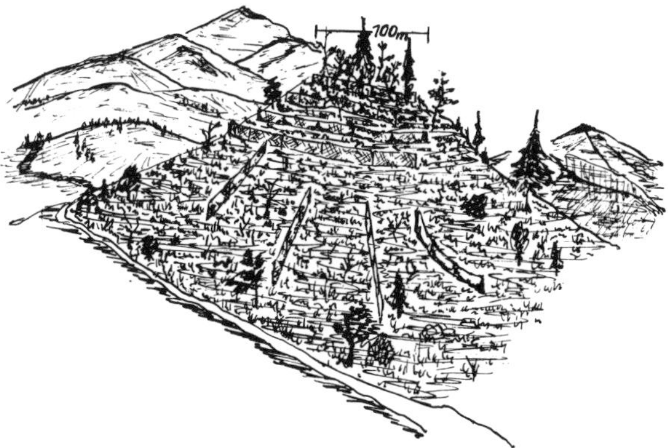

Fig. 257. Mist nets systematically arranged on a hill in Japan. After McClure, 1956.

Fig. 258. Arrangement of mist nets on the Swiss Alp pass Col de Bretolet in the fall of 1957. Right, the bander's blind. After M. GODEL and G. DE CROUSAZ, 1958.

have to be furled. Of course, the set up can be expanded if there is adequate manpower."

The catches have been fabulous: in 1957, between October 2 and October 26 a total of 5,982 birds were caught and banded. They were mostly finches and tits.

Groups of three to six banders worked in shifts and a total of 20 co-operators took part in these catching expeditions.

10.2.8 Catching colonial birds breeding in embankments

In general, trammel nets are best suited for this purpose because the birds can be removed more quickly. Also, as the net is strung on rings it can be pulled into position more quickly and quietly. These can be decisive factors at daybreak. As many ornithologists do not own tram-

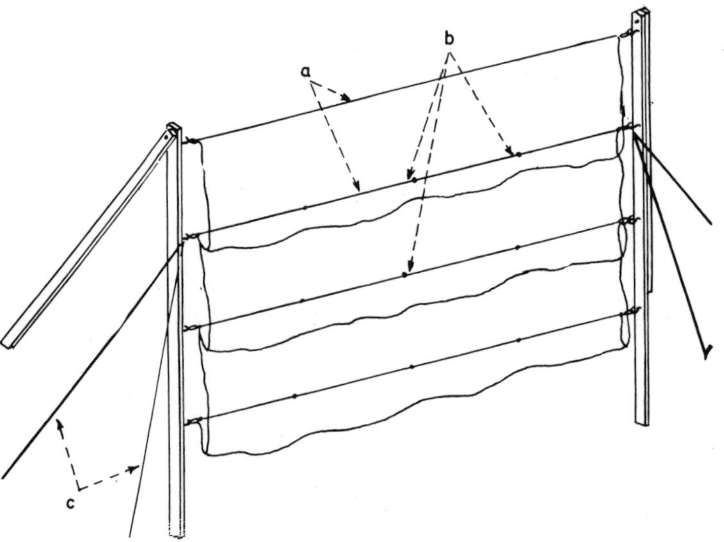

Fig. 259. Set for catching at an embankment. (a) the shelf strings are somewhat shortened to give more bag; (b) the net is fastened to the shelf strings here and there; (c) the net poles are held in position by guy cords as there is often no footing for the poles. After MEYLAN, 1966.

mel nets nowadays, we will describe some successful mist net techniques.

Meyland (1966) suggested catching in the evening after the birds were in their holes. We fear that this may be too great a disturbance during the breeding season. Mist netting at daybreak is less convenient but justifiable if the birds are banded and released promptly.

Meylan's nets were up to 10 m long. He, like Schierer, shortened the shelf strings at the ends to cut down the number of escapes (Fig. 259). Meylan also suggested fastening the net to the shelf strings here and there so that wind cannot shift the net too far to one side. The net poles are braced; where this is not possible on high banks, one fastens still another pole to the ground up above the colony. In special cases Meylan set one net above the other (Fig. 260). Birds that are missed by the upper net are often caught in the lower. As already stated, trammel nets are better than any of these sets. Bub has caught 10,000 Bank Swallows (Sand Martins) with trammel nets so far.

Leijs (1963) has described two net arrangements on level ground close to a Bank Swallow

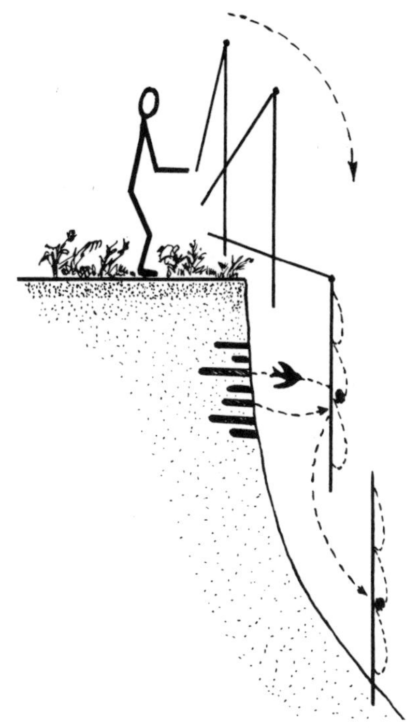

Fig. 260. Catching with two nets propped against a bank. After Meylan, 1966.

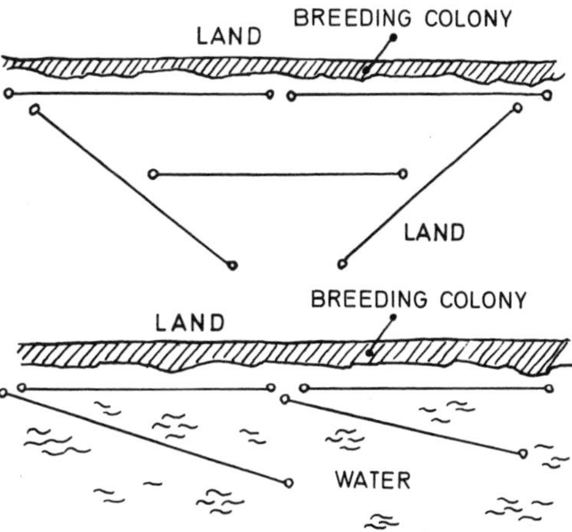

Fig. 261. Two arrangements of nets in front of a sand martin colony. After Leijs, 1968.

(Sand Martin) colony (Fig. 261). The net should be placed at least one or two meters from the nesting bank so that the birds can fly between it and the net. Catching should be limited to an hour or two to avoid excessive disturbance of nest-brood events. Persson (1965) and Yunick (1970) described similar methods from Sweden and the U. S. A. respectively.

10.2.9 Catching at night

Catching at roosts of swallows, wagtails, pipits and other birds has already been covered under trammel nets. Mist nets, of course, can be used as well. Peterson (1963) mist netted 23 Saw-whet Owls—maximum number 10 in one autumn— for the most part by setting mist nets over a brook that the birds used for hunting.

Liddy (1964) banded more White-cheeked Honeyeaters late in the evening when they were seeking food than in the early morning. First he listened to the weather forecast and then he set his nets up on Friday for the weekend catch. He checked the nets with a flashlight after dark and then left them set overnight. Only one bird in 50 weekends was caught after darkness set in. Nonetheless, nets left open overnight should be checked regularly during the night—too many things can happen!

Jenkinson and Mengel (1970) made comparative studies of the Chuck-will's-widow and the Whip-poor-will. They succeeded in catching males of each species by luring them in with tape recordings of the appropriate species. They

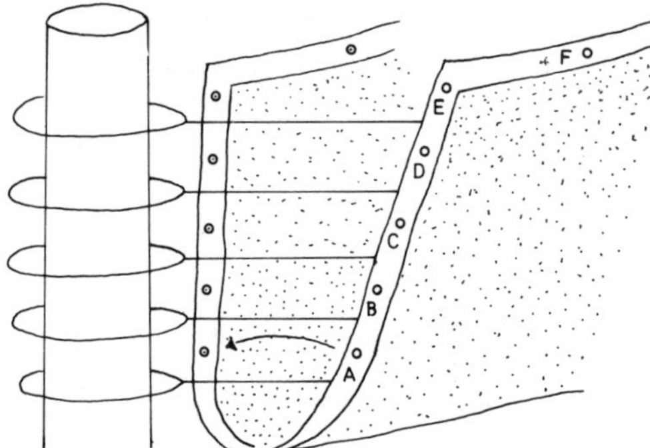

Fig. 262. Chiffon cover for noctural mist netting. After JENKINSON & MENGEL, 1970.

drove the gravel roads of their study area until they heard a bird calling. Then, at some distance from their car, they set a net across the little traveled road, placed a loudspeaker under the net and hid themselves close by with a playback-recorder.

The most critical part was getting the net up quickly and quietly before the bird shifted to another song post. Although JENKINSON and MENGEL generally caught when there was some moonlight, they strove to set their nets in shadows of bushes, branches and twigs. The nets often snarled in these as a result of a hasty set up in the dark; a flashlight tended to spook the bird. Sometimes the net became so entangled that they had to give up for the night.

The problem was solved by a net cover of dark chiffon, which protected the net at all times except when it was set (Fig. 262). Snaps A through F and every 50 cm along the length of the net could be opened quickly as soon as the (dark) net poles were in position. The cover was as long as the net and 30 cm wide for a five shelf net. After a catch, snaps A through E are closed and then those along the top. Next one pole is pulled up and the encased net is wound around it toward the other pole.

Finally, the poles are tied together with a man's handkerchief. Unsetting takes about five minutes. To set the net, the process is reversed and two people can get the net into position in one minute. An additional advantage of the cover is that poles and net can be carried on the luggage rack of a car without getting tangled.

10.3 High placed nets for water birds, woodcock and pigeons

Simple net walls were used in past centuries in many European countries—apparently mostly in north, middle and eastern Europe. OLAUS MAGNUS (1555) even left us an illustration. At any rate, according to RADDE (1884), SIRELIUS (1934) and LAGERCRANTZ (1937) this method was also known in northern Asia (Fig. 263) and in the Near East and was certainly independently known to some degree by the primitive

Fig. 263. Duck net in Western Siberia. After SIRELIUS, 1934.

people of southeastern Asia as is demonstrated by the bat net (Fig. 286).

This method was primarily intended for ducks and geese and therefore largely limited to coasts or edges of lakes and rivers. Such catching is hardly practiced anymore having for the most part been replaced by gun hunting. There are good netting possibilities, however, for scientific purposes at suitable sites. The material needed is simple: an ordinary net, its strength dependent upon the species to be caught. Ancient and modern techniques follow.

10.3.1 Catching ducks and geese

Pallas (about 1773) describes a method of catching geese that he learned on his travels in Siberia. The geese preferred to land on lakes that were half protected from the wind. On the side favored by the geese for taking off and landing, a lane about 30 m wide was cut through the birchwoods which was soon utilized by the birds. Where lakes were not far apart, such lanes were cut from lake to lake. At some distance from the shore two tall birches were left uncut to string the net between. The nets were of strong double hemp, 20 to 25 m long and 4 to 5 m high. To quote Pallas: "The top edge and the short sides are hemmed with a string, and at two corners a thin line of about 50 to 60 m is tied on. To both trees, from which branches have been trimmed off, a long pole, forked at the top, is firmly fastened. When one wishes to use the net, someone climbs the tress and lays the lines to which the nets are tied over the forks and thus the net is pulled up and spread so that the corners reach seven or more meters above the ground to the forks, but the lower edge is only 2 to 4 m from the ground. This is then fastened to stakes in the ground with five or six pieces of string. Now the catcher goes as far back from the net as the lines (that hold the net) will permit. Holding these tightly in his hand, he lies in the grass awaiting his booty at dawn. Together, the geese leave the lake an hour before sun-up, and as they cannot see the net in the twilight, they fly into it with outstretched necks. The net must be released at that moment when the geese pull it down with their own weight and catch themselves. If the net is taken down in the morning, new threads are fastened to the pull strings so that one does not have to climb the trees again the next night."

Radde (1884) speaks of large mesh standing nets set up by hunters. Snow is cleared from part of a meadow and so, when snow is deep and winter hardship has set in, one attracts Red-breasted Geese. It is said that early in March 1880 over 200 were caught in this manner at a single netting in one of the Burani Islands in the Zaliv Kirova on the Caspian Sea (U. S. S. R.).

Albarda (1885) reported that in Friesland (Holland) great standing walls of net were set on the meadows nearest the coast and on the south side of the islands. For the most part, they were 25 m long and 3 m high. They were suspended from a cord between two posts so that their lower edges were about 1 m from the ground. Intermediate poles were not driven into the ground so that they could be put down at the side. The nets were rather wide meshed but very baggy so small shorebirds as well as curlews and geese were captured. To catch birds migrating low on dark, stormy nights, the net was set across the beach. Von Droste (1869) found rainy, windy evenings (wholly unsuitable for mist nets), or snow flurries at dusk the most propitious. Nothing was caught when the moon was bright or by daylight. Dietrich (1911) said that in earlier times 4 to 5 m high nets were set on the North Frisian Islands, North Sea, against which geese flew at night and in which they became entangled. The geese were gathered up at the next low tide. "But the wary Brant Goose soon caught onto this trick and the islanders had to give up this catching method."

The coastal inhabitants of the Caspian Sea

Fig. 264. Stationary net from Volhynia, Russia. After Lagercrantz, 1937.

put up stationary nets for waterfowl and shorebirds. Schüz (1957) saw rows of wide-meshed nets about 10 m long erected on tall poles over the water from February until April 1956 in the South Caspian province near Pahlavi (Iran). They extended 2 m over the surface of the water and were 1.5 m high. Ducks settling in at night were caught; coots appeared to ricochet (see also Radde, 1884).

Lagercrantz (1937) shows a net in Volhynia (Russia) (Fig. 264). Ducks and geese flew into a stationary but baggy net and became entangled in the meshes without causing the net to fall. We may assume that the hunters remained nearby.

10.3.2 Catching woodcocks

Eurasian Woodcock up to a few decades ago were sought as a special sort of game especially on the North Sea coasts of Germany and Holland. Nowadays only a few such catching spots remain in Holland and they are used for banding. Mörzer Bruijns (1960) and Bontekoe (1967) have described the catching. The stretched mesh of the net is 100 mm. The net is 28 m long and 11.5 m high and it is elevated with pulleys. Two 13 m high poles are spaced 30 m apart. Their bases are buried 1 m deep and the tops are held firmly in position with long guy lines (Fig. 265). The set is made in an opening of about 33×20 m. If possible, the bushes and trees leading toward the opening and the broad side of the net are cut down. The birds usually appear shortly before sun-up over a period of 20 to 45 minutes. Gamekeeper Visser has noted that catching is over as soon as the human eye can detect grass blades on the ground.

The catching period lasts from mid-September until after the middle of December. As soon as frost gets into the ground, there is no further point in trying for woodcock. Wind influences success strongly; when the wind is too strong, the birds often fly over the net instead of flying at tree crown level. When a woodcock is almost at the net, the catcher, hidden in a small blind, releases the lines so that the bird flies into the falling net. With quick, deft movements, the bird is removed from the net and is put into a strong, but not airtight bag. The light weight net is immediately raised again. According to Visser's experiences, most woodcocks are caught when they are seeking a landing spot. They appear to come mostly from north or northeast and seem to have had a long journey behind them. In a favorable little woodland near Rijs in Friesland (Holland) Visser banded 287 Eurasian Woodcock from 1955 to 1966, of which 42 were subsequently recovered.

Woodcock catching was conducted similarly but more simply on Helgoland (Germany) some time ago (Gätke, 1891; Rohweder, 1900; Buse, 1915). The birds were so numerous that special measures were hardly necessary. A net, generally about 15 m long and 4 to 6 m high was elevated between two tall poles. Both upper corners were attached to a line that ran through pulleys and was weighted at both upper ends with a stone. The set was placed where Eurasian Woodcock tended to fly by at dusk: on the upland, near shrubbery or a row of houses; on the lowland near cliff faces. The catcher stood close to one of the poles, if possible taking advantage of any available cover, with the pull string in his hand.

Thrushes also got caught in the 20 cm

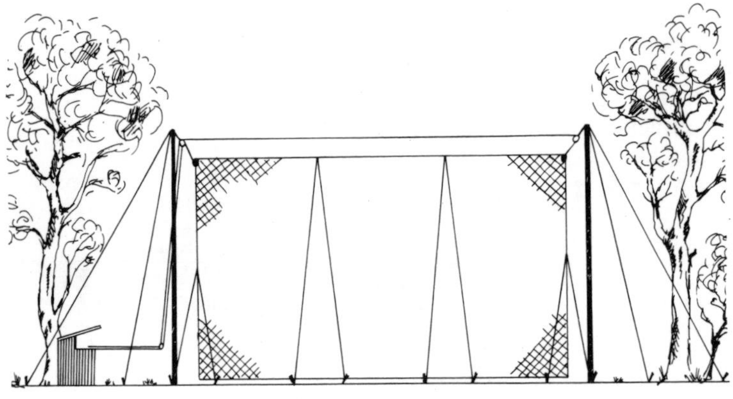

Fig. 265. A Dutch net for woodcock. To speed the falling net stones are fastened to the upper corners of the net. On top of the right hand pole there is a single pulley, on the other one a double pulley. After Mörzer Bruijns, 1960.

stretched meshes. On good migration days, the Helgolanders caught 20 to 30 Eurasian Woodcock in two nets in one morning. Gätke (1891) has more to say about this. "Long before the first grey of daybreak, the catcher stands near one pole with the pull cord that releases the net in his hand. He must pay great attention so as to release the net at the moment that the woodcock flies into it. To speed the falling of the net, chunks of flint about twice the size of a fist are fastened to the upper corners of the net, and with careful watching and if the lines run smoothly through the pulleys, few woodcock that have hit the net fail to be captured. Although they are clearly visible by day, oncoming woodcock still fly into them for the most part without fear during the morning hours."

10.3.3 Catching Wood Pigeons

The netting of Wood Pigeons as a form of hunting has been especially highly developed in the Pyrenees—in the land of the Basques—and stems back to the times of the Templars.

When, as Chaigneau (1961) points out, about 30,000 to 40,000 wild pigeons are caught yearly during migration periods of May and June and September and October one can only begin to suspect what multitudes must have passed through the area.

The 80-100 mm stretched mesh nets are 15 to 18 m long and 6 to 10 m high. They are strung as simple net walls (trammel nets are used in certain places). The catching area must be specially prepared so that the net 'walls'—up to 14 of them—can be arranged somewhat slanted one behind the other (Fig. 266). Each net is tended by one man who sits in a blind beside it. As soon as a flock of pigeons appears, men sitting in high blinds toss chunks of wood into the air, wave white cloths and shout to frighten the pigeons into flying low and into the nets; then one by one the nets fall. The net cords, as is customary, run over pulleys. Burdett-Scougall (1949) describes the procedure.

Fig. 266. Wood pigeon catching in the Pyrenees. After Chaigneau, 1961.

Fig. 267. Wood pigeon catching in the Pyrenees with large net cage. After Chaigneau, 1961.

Chaigneau (1961) acquaints us with the following catching method (Fig. 267). The catchers cut an alley and then place a net 'cage' about 20 to 30 m long, 8 m high at the front, 9 m high at the back and about 15 m wide.

This cage is raised with the help of pulleys on nearby trees or poles set up for this purpose. The bottom of the net is firmly fastened to the ground, but by releasing the ropes that pass through the pulleys the whole cage can be made to collapse. It is impossible for the birds to escape.

The set is carefully camouflaged on both sides by trees; the hunter sits in a blind 25 m above ground and 50 m in front of the entrance of the cage well armed with countless chunks of wood about 30 cm long, some of which resemble sparrow hawks. When the pigeons approach he throws these into the air a few at a time as soon as the main body of the flock has passed him. Others of the 12 to 24 hunters join him and throw their chunks too. The first man, however, has the most important part to play. If he throws chunks too soon, the pigeons rise and fly over the net; if he throws them too late, the pigeons do not alight until they have passed it. Therefore he always throws his chunks after most of the flock has passed.

The hunters on the ground carry large sticks to which straw or feathers are attached and it is up the them to see to it that no pigeons fly back out of the net.

Dark, overcast skies produce the best catches. Few pigeons are caught on clear days. Wind is also a detriment to catching, but winds from the south or west have been more favorable. All in all, a multitude of factors must be understood; the natives know the routes that the pigeons take and on which days catching will be good. All the hunters must be experienced and seasoned. One false move and the catch is in jeopardy! Watchers spot the incoming flocks of pigeons and alert the hunters when the birds are still several kilometers away.

10.3.4 Catching Cassin's Auklets

Attempts to mist net Cassin's Auklets in the Farallon Islands (37.40 N 123.00 W) west of California (U.S.A.) where they nest abundantly, essentially failed. Most of the birds flew right through the nets at about 35 miles per hour and those that did get caught, tore the net badly with their feet. Therefore, RALPH and SIBLEY (1970) devised a new method using ordinary nylon or perlon fishnet 18 to 20 m long and 6 m high. Detailed descriptions of the method are given by the authors.

10.4 Capture of raptors

10.4.1 Dho-gaza

For catching Northern Harriers (Marsh Hawks) during the breeding season, FRANCES HAMERSTROM (1963) uses a method that undoubtedly would be suitable for other harriers (Fig. 268). This catching method had a forerunner in earlier centuries (Fig. 269).

A black nylon net 2.4 m wide by 1.8 m high is stretched between two 5 m poles close to the harrier's nest. The mesh is 120 mm stretched and the thread diameter is 1 mm. A loop is tied to each of the four corners of the net. Four hooks of fine soft wire are permanently taped to the poles to receive the loops at the net corners so that the net is stretched tight. A Great Horned Owl is tethered to a 1 to 1.5 m high block perch so that it cannot reach the ground, for it has been known to brood nestling harriers on the ground. The owl wears jesses and a swivel so that it cannot get tangled up. FRANCES HAMERSTROM uses hand-reared owls. Tawny Owls and Long-eared Owls may also be used.

The net is placed crosswind preferably 3 or more meters from the nest. Catches have been made with the owl 30 m from the nest. The harrier stoops at the owl and flies into the net.

All four wire hooks straighten out and the harrier falls to the ground with the net. Lest the bird fly away with the net, a 6 m long nylon cord is fastened to one of the lower corners of the net: at the other end of the cord a nylon stocking weighted with about 250 g is tied firmly. The stocking also facilitates finding the captured bird in deep vegetation.

A light wind (3-5 mph) is advantageous to determine flight direction. Hawks tend to stoop into the wind or downwind. When there is no wind the net is set across the path of the sun's direct light. If there are eggs or small young in the nest the best trapping time is during the first hour after sun-up. If the young are already well feathered the adults may not come until two hours

Fig. 268. The dho-gaza for catching harriers. After F. HAMERSTROM, 1963.

Fig. 269. Falconstoop. After King Modus manuscript in the Copperplate Cabinet, Berlin, first half of the 15th century. After LINDNER, 1940.

after sunrise. Individual adults often react differently. Some hit the net without getting caught, some show no inclination to stoop at the owl, or do not even arrive on the scene for two or three hours. Females are usually more easily caught than males, and males tend to be caught more easily when the young are small.

Fig. 270. Osprey catching with a dho-gaza. After F. HAMERSTROM.

When the young are half grown, the males sometimes no longer come at all. The most successful catching is shortly after the hatch. In such cases it is better to use a stuffed owl. FRANCES HAMERSTROM caught 14 harriers with a stuffed owl and a total of over 130 by this method, including 'foreign' individuals—i.e. not one of the mated pair.

It is best to have two trappers; one hides in the grass near the nest and the other—partly hidden—watches from a distance of 150 m, prepared to blow a horn when a bird is caught. At this signal the hider in the grass rushes out to pick up the catch.

The dho-gaza also works for Osprey catching (20 catches up to the present time) (Fig. 270). The set is made about 30 m from the nest. The net poles are erected on a cross made of 2×4' construction timbers, which is anchored; the owl is tethered to a low perch (Fig. 271). The procedure resembles harrier catching. The banders wait in a boat ready to retrieve the captures Osprey immediately. Owl lures have not proven successful at migration time.

10.4.2 Japanese raptor nets

According to MACPHERSON (1897) the Japanese netted raptors with a dho-gaza long ago. Be-

Fig. 271. Dho-gaza with tame Great Horned Owl at an Osprey nest in Wisconsin. Photo: F. HAMERSTROM.

tween two bamboo poles about 2.2 m high hangs a square net of the same dimensions (Fig. 272). Beneath the net a fyke-like container about 4.5 m long and 1.5 m in diameter holds the lure birds. For larger raptors, pigeons are put in the container; sparrow hawks and small falcons are attracted by smaller birds.

10.5 Stationary nets for starlings

For catching Common Starlings Swiss banders use a 7 m long boat with a 5 m long and 2 m high net wall of fine mesh poultry wire or tightly strung net (Fig. 273) (A. SCHIFFERLI, 1932). This rig is pushed into position at night and drivers cause the birds to fly against the obstruction among the reeds. Shores with a reed border more than 10 to 15 m wide are not suitable. Catching is at night—for best results on quiet, dark, rainy nights. The boat is staked across the reed border. A driver gets the starlings moving with a stick which he brandishes slowly and carefully not to flush the birds, but to cause them to become uneasy so that they move through the reeds towards the boat. Most of the birds fly over the net, but many fly against it and hang on for a moment or two. Three or four catchers stationed nearby with small dip nets cover the birds as they hit the net and take them out quickly with the other hand and put them into a bag fastened to their waists. These containers are topped with a 30×30 cm board from which a stout cloth sack about 50 cm deep hangs. A spring-loaded self-closing cover on the board permits the bander to slip the birds quickly into the sack. The set can be repeated along the shore. Top catch at Sempach was 430 Common Starlings in one night.

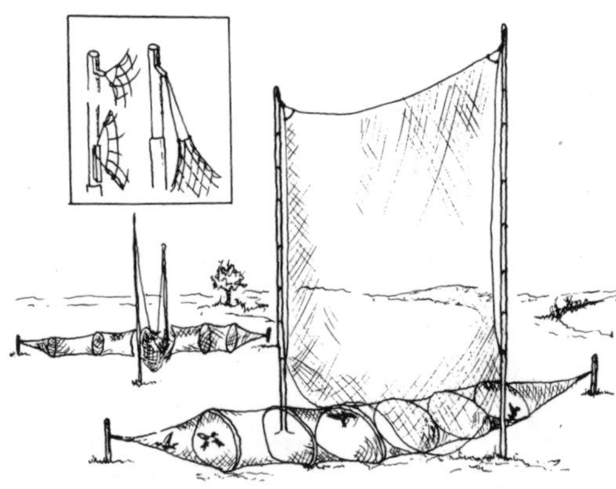

Fig. 272. Japanese raptor net. Set in the foreground and after a catch in the background. After MACPHERSON, 1897.

Fig. 273. Net wall in a boat for starling catching. After A. Schifferli.

10.6 Stationary nets for larger birds in the reeds

The Dutch ornithologist Koridon (1958) caught Great Bitterns and Water Rails in the winter by hacking a hole in the ice of a ditch or pond and spanning a 10 m long and 2 m high net in V formation on one side of the hole. As soon as the birds had found the open water and had finished drinking and feeding, the catcher came out of his hiding place and herded the birds into the net. Although the set was obviously open at the top, they did not fly away, but tried so find escape holes in the netting. Several Great Bitterns and Water Rails were caught thus in a short time. After banding, the birds revisited the water hole regularly.

10.7 Quick-rise nets and curtain nets

Quick-rise nets may be especially useful for catching swallows and swifts (Fig. 274). We use netting of trammel net strength or mist nets; at any rate, they must not be too coarse. G. Walther (1935) had a net 5 m long and 1 m wide and bound it with a 2 m cord to the top of a 1.25 m stake. The pull cord measured 5 m and when pulled tight the net hangs down between the two cords. The catcher holds the pull cord loosely enough so that the net lies on the ground. When a swallow is about to fly across the net, the catcher pulls the cord and then lets it drop quickly together with the approaching bird. This method takes practice for split second timing is required. On a favorable day—windy, rainy weather helps, but is not essential—practiced catchers can net a good number. For example, on May 27, 1934 three catchers netted over 200 swallows, primarily Common House Martins, near Wörlitz (Germany).

Kruis uses a mist net for catching swifts with good success near Prague (Czechoslovakia). But he catches only in the evening at the edges of ponds.

Lockley and Russell (1953) used a related method (Fig. 275). Two poles and a crosspole are erected. Curtain rings on the crosspole facilitate drawing the net shut. This set is used over brooks and woodland lanes and is for catching birds flying low over the water or along the ground such as swallows, kingfishers, woodcock, etc. The net is swiftly pulled shut at the moment that the bird is about to pass under the frame. It is even simpler to string the curtain net from tree to tree without erecting a frame.

Scheithe (1956) used a 5×2 m net and with a helper caught 315 Common Swifts and 40 Barn Swallows in six days. The catching site was especially propitious as it was on a tongue of land about 20×20 m which projected into the water of a dredge ditch. The swallows passed over at only 30 to 100 cm height. Scheithe states that a lower temperature than normal for the time of year is of primary importance. The mornings of cooler days and after nights cooler than +10° C were favorable. On the best days the catch was good from 17:30 to 20:45 in the evenings; thus 114 swifts were caught on one June 25.

Also worth mentioning is what Pallas (1773) saw in the Ob region of Western Siberia (Russia) in the 18th century.

"In order to make wildfowl more trusting and to decoy them they put ducks and geese stuffed with hay on the water near the fieldworks. It is astonishing that wild swans and geese so gladly fly toward these stuffed decoys and straightway start to bite at them. At dusk, later in spring, many of these waterfowl are caught in swaying air nets called 'perewessi'. Yes, one has also devised a way to fish geese and ducks out of the air in broad daylight."

"The bird catcher makes a blind out of bushes where he can hide and watch the passing birds. The net, which is called 'kyskan' lies on the ground in readiness on the lines. As soon as the hidden Ostyak sees the oncoming birds near

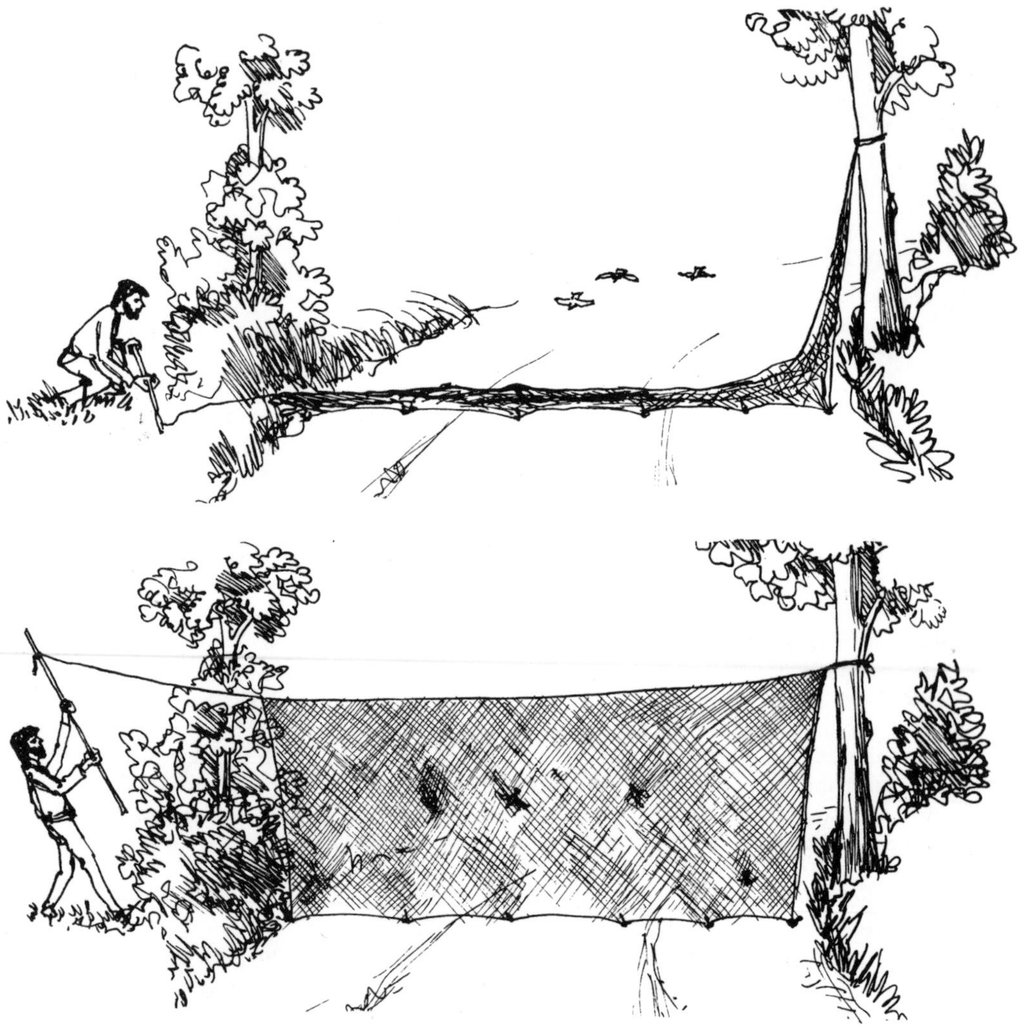

Fig. 274. Quick rise net in "lying-in-wait" position and in catching position. After Ornithological Stations Helgoland, Rossitten and Sempach, 1953.

Fig. 275. Curtain net. After LOCKLEY & RUSSELL, 1953.

enough he spreads the net in the air by means of the lines which must be highly movable. The heavy-bodied birds, which cannot rise quickly, seldom escape him. As soon as they reach the net it is dropped and they become entangled in it and are caught. But if it happens that the geese wish to fly over the net, the Ostyaks have already put out live lure geese and they know so well how to mimic the flock call of the wild geese on a piece of birch bark that the flying birds forget the net, put down near the lure birds and become the booty of the bird catcher."

LAGERCRANTZ (1937) describes a method that was apparently used only until the middle of the

Fig. 276. Large curtain net of Swedish farmers of the coast. After LAGERCRANTZ, 1937.

19th century (Fig. 276). Where two small islands were not too far apart the Swedish skerry farmers erected two strong masts. A rope ran through pulleys over the tops of the masts to which a net was fastened with brass rings. As soon as a flock of ducks came in sight the net was spread quickly with the help of a boat. It had to be pulled across the entire strait before the flock arrived. A catcher on the shore released the rope as soon as the ducks reached the net. Apparently Oldsquaws (Long-tailed Ducks), Velvet Scoters, Common Eiders and also Razorbills were caught in this way. Such nets were made of 3-stranded bottle twine. Their length varied between 50 and 180 m and their height between 6 and 7 m.

The Soviet ornithologist SKOKOVA (1960) describes cormorant catching on the sandbanks and spits of the Volga Delta on the Caspian Sea. The birds are primarily caught in August and September before they fly to roost, when they are resting on the open beach and drying their feathers. They are captured in a 'catching yard' (Fig. 277)—a method we have not heard of before. W. D. TREUS and A. A. NESTEROV first tested this method in 1954 in the Damtshik Division of the Astrakhan Nature Preserve (U. S. S. R.) The principle is simple: to keep birds that need a long take-off before flying from getting airborne. It is suitable for cormorants, loons (divers), grebes, swans and others.

The construction—in this case for cormorants—is simple. Fish netting with a 120 to 140 mm stretched mesh is fastened to 1.5 m poles which are kept upright above and below by stout cords (Fig. 277). There must be an uneven number of poles, for the side walls (of equal length) extend toward shore from the key pole, out on a land spit. The lower ends of the poles are tied in a movable manner to pegs in the ground (Fig. 277 b). They are attached to stakes in the water. The pull cord is strung through the top of the net and fastened to the top of each pole and leads to the trappers blind.

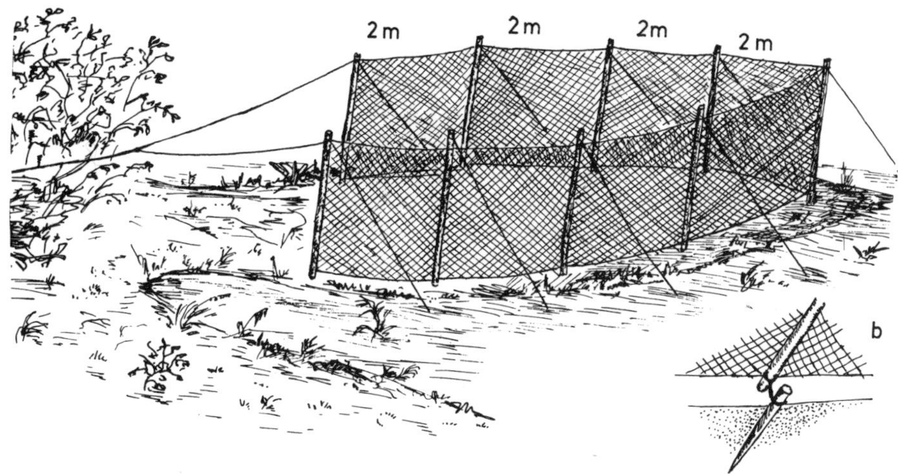

Fig. 277. Cormorant net from the Volga Delta. After SKOKOWA, 1960.

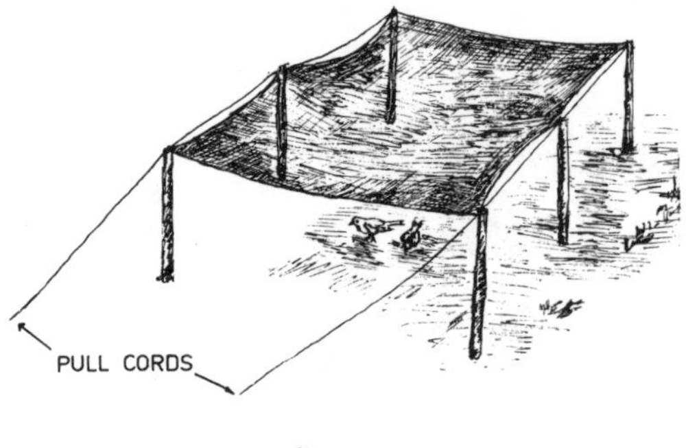

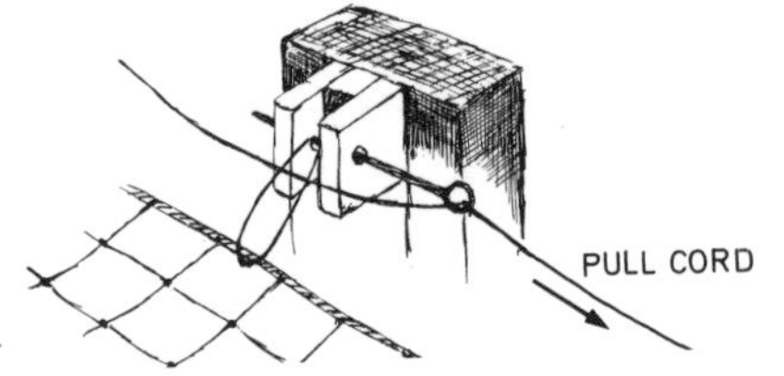

Fig. 278. Drop net with trigger method. After K. F. JACOBS, 1958.

It is best to use wire for the part that leads to the blind. To set the trap, each part is carefully laid out on the ground so that the trapper can pull the trap up into catching position with one pull.

It is best to lay the net in a shallow trough on the beach or to cover it with water plants so that the birds will not be wary of it. As soon as the birds are within the 'yard', the trapper pulls the net up, ties the pull cord to a stake and, with his companions, rushes out to grab the birds quickly. His helpers should really emerge first so that no moment is wasted.

The Soviet ornithologists used stuffed cormorants as well as live lure birds for decoys. Many individuals were caught in this manner; sometimes several dozen with one pull.

11. Drop nets

Drop nets cover the birds form above. Some types are permanently positioned and others are portable.

11.1 Permanent drop nets

Substantial catches are made by baiting the birds under the drop net with grain (Fig. 278). They can reach the grain pile from any side. The net is released by a pull cord (sometimes $\frac{1}{4}$ mile long) leading to the trapper's hiding place. JACOBS (1958) uses two different sizes for catching Greater Prairie Chickens in Oklahoma (U. S. A.). The smaller net, 9×9 m, needs three poles on each side; the larger net, 13×13 m, needs 4. The net is hung horizontally 1.4 m above the ground. The pull cords from each side are fastened to a single long pull wire about 22 m from the net.

The net mesh is large enough so that captured birds can stick their heads through easily, but small enough so that their wings cannot pass through and become entangled. JACOBS found prairie chicken trapping was far better when snow covered the ground, but caught quail in mild weather. FARREL DOPELIN also had good catches, even in summer (Fig. 279).

A special innovation for catching Wilson's

Fig. 279. Drop net in the North American prairie. Photo: F. COPELIN.

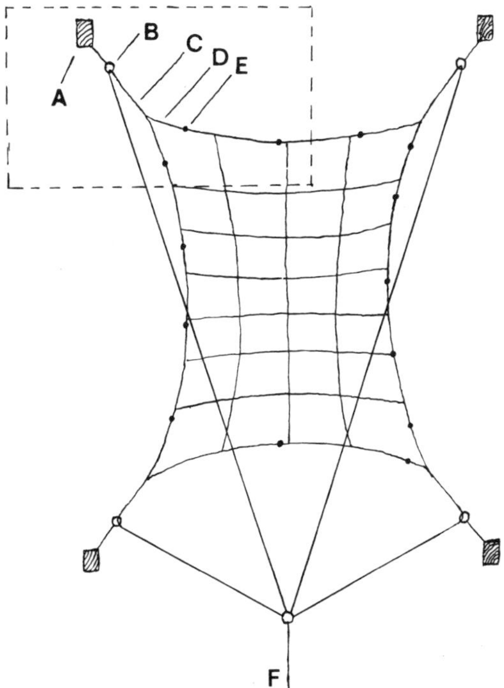

Fig. 280. Mist net stretched horizontally for catching Wilson's Phalaropes. After J. E. JOHNS, 1963.
A. 5×5 cm stake driven into mud of a pond bottom in selected site. At least 1 m protrudes above water surface.
B. Metal ring attached by cord to stake (A) through which cord (C) from corner of mist net is passed.
C. Line attached to corner of mist net passed through metal ring (B) and attached to line (F).
D. Nylon mist net held parallel to and approximately 70 cm above pond surface.
E. Lead split shot attached to all edges of mist net (D).
F. Single line attached to four corner lines (C) and leading to place of concealment.

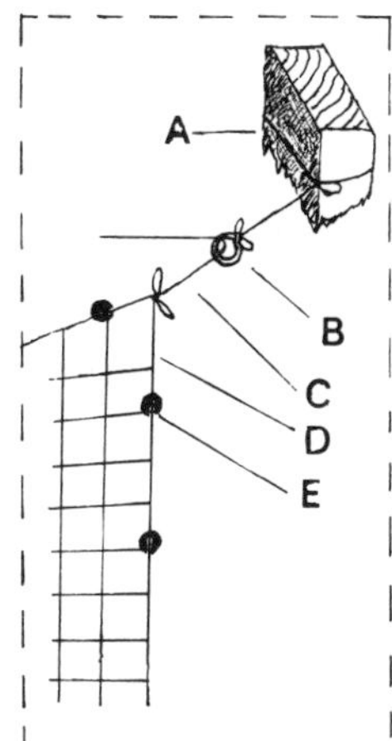

Fig. 281. Detail of a net corner. After J. E. JOHNS, 1963.

Phalaropes was developed by JOHNS (1963) after only a few individuals had been caught by nets placed in the normal manner. The nets were strung parallel to the ground (Fig. 280 and 281). Therefore the shelf cords were lengthened to flatten the net, getting rid of the pockets. Big nets are successful, but nets about 12 m long are easier to handle. The net is stretched tight about 70 cm above the water: a height that does not seem to disturb the birds. For best results part of the net should be over the shore where shorebirds often feed. Thus one can catch them on land or in the shallows. Special attention should be paid to see that the lines attached to the corners of the net (C) pass freely through the metal rings (B) so that the net drops quickly when line (F) is released. In addition, 'split shot' lead weights (E) are attached along the edges of the net at regular intervals. These cause the edges to fall faster than the center and prevent escapes. Sometimes one can drive the birds under the net especially with a helper. Once caught, the birds should be removed from the net quickly as the weight of the net may force their heads under water. The trapping site should be essentially devoid of vegetation. No doubt this method is also suitable for other shorebirds.

11.2 Portable draw nets—the tirasse

In old bird books we generally come upon an engraving showing catching with a tirasse: two men are pulling a large net over a flock of Common (Gray) Partridges which is 'holding' to a crouching dog (Fig. 282). C. L. BREHM (1855) describes the tirasse as a square at least 6×6 m. Other authors describe a more elongated rectangle. Mesh size is adapted to the size of birds to be caught. The tirasse is used in two different ways: either the net is pulled over the birds, or it is set with the front elevated on forked sticks and the rear fastened to the ground in which case the partridges or quail are enticed under the net by a whistle or by a decoy bird. Next they are suddenly frightened so that they will fly into the net and get caught. Generally, however, the catchers drag the net across a field or mea-

Fig. 282. Catching with the tirasse.
After AITINGER, 1653.

dow with the leading edge about 60 cm above the ground. A rope, running through the mesh of the leading edge and projecting about 3 m at each end keeps the net stretched and facilitates pulling (Fig. 283).

More recently WARGA (1929–1930) has described quail catching with a tirasse; a mist net will also do as KLIEBE has demonstrated with Eurasian Jacksnipe. The net WARGA used in Hungary was 8 m square with a stretched mesh of 8 cm. Smaller nets are used as well. "The catching takes place in cultivated fields, in alfalfa or millet, or on meadows, primarily in May and June during morning or evening twilight or late in the afternoon. We spread the net loosely and lightly over the crops or over the grass, where we have seen quail flocks. We lie down quietly, with our heads under the net facing the calling quail. After a while we let the whistle sound. Three times, in even rhythm we imitate the natural call of the hen quail: 'tri-tri, tri-tir' or 'prü-prü'. Then we wait for a little while for the reaction of the cock. The calling must be practiced in advance as faulty calling or any mistakes will simply scare the cock away. Generally, the cock reacts to the first call, approaching on foot right away to seek the luring hen while uttering a soft 'wa-wa, wa-wa'; later he starts calling loudly. This calling is individual and is not connected with the breeding season. The more cautious, less temperamental cocks walk under the net, the more fiery cocks, on the other hand, often come flying. Frequently they fly into the net from above and become entangled in it. As soon as one has lured an approaching quail well under the net, one startles it by hand-clapping or beating on the ground; usually the startled, flushed bird remains hanging with head or wings pushed through the meshes of the net. Therefore, if a smaller net is

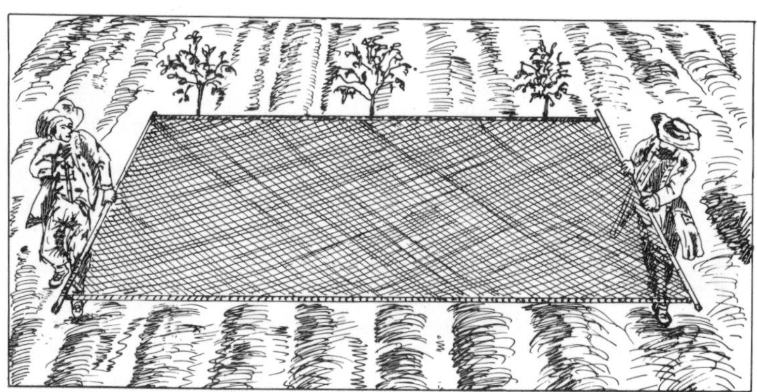

Fig. 283. Catching with a larger draw net (tirasse).
After CHOMEL, 1743.

used (1×1.5 m) it must be laid lightly over the tops of the crop for even if the push of the rising bird raises the net its edges fall together under him. It is inadvisable to place the net on a crop standing more than 60 to 80 cm high. When catching with small nets we lie about 5 m from the net."

The importance of á pointing dog used with a draw net has been mentioned. Such dogs are useful, not only for holding partridge and quail, but also for rails, snipe and other birds.

Nocturnal catching with portable draw nets was, or still is, practiced in many countries. This presents good possibilities for the bander as well. In addition to 'skylarking' by day, NAUMANN (1824, 1905) describes catching by night with a long rectangular net (22×7 m with a 70 mm stretched mesh). Smooth poles at each end make it possible to carry it about ready for catching. After sundown the catchers walked the fields methodically covering the sleeping larks with the net. Larks that flushed were pulled out through the netting. The method, though painstaking, was fruitful. Some nights produced 200 to 400 or even more birds; others scarcely ten. "On bright, moonlit nights no lark will put up with this nor will it when the ground it too cold or wet. It works best on inky black nights and if they (the larks) sit too tight or let the net be pulled past them, small straw 'awakeners' are tied to the back of the net on 1.2 m strings and towed behind." Strong winds and rain are not suitable for catching.

This so-called 'night-lark-net' is recommended for many other ground roosting species such as pipits and buntings.

More recently HOLLOM (1950) described a drag net. A group of English banders worked an old over-grown field for 45 minutes one night and caught 17 birds—among them Meadow Pipits, Common Skylark, Corn Buntings, Redwings and Common Snipe. The banders carried the leading edge of the 18×4 m net at shoulder height. Terrain with thorny bushes and shrubbery is unsuitable and the night must be dark.

12. Aerial clap nets

To the best of our knowledge these nets are not used by banders in Europe. Aerial clap nets should prove useful and in some instances they may be a last resort.

As has so often been the case, these nets stem from the fertile minds of hunting peoples. SALVATOR (1897) watched such nets (Fig. 284) in use on the Balearic Islands, Spain, in the Mediterranean. The hunter selects a spot at the edge of a wood where the trees form a narrow lane. At dusk he places himself in this path. As soon as the birds using the travel lane had flown against the net the hunter either dropped it to the ground or clapped the net poles together always clapping the lighter right hand pole against the heavier left hand pole.

Generally the hunter sits with his back to the approaching thrushes and tries to hide his net-poles and himself under branches. This hunt, dearly loved by the peasants, occurred regularly between November and February.

In North America, banders have caught birds in this way for decades. LINCOLN and BALDWIN (1929) show a longish oval net with a center pole. Three 5 to 6 m bamboo poles and light weight fine netting are used. The slender tips of the poles are permanently pulled into a bow with strong cord and the tips of the poles are hinged together with deerskin or some other soft leather so that the tips can move freely in any

Fig. 284. Aerial clapp net in the Balearic Islands. After SALVATOR, 1897.

Fig. 285. Two North American quail catchers. After NELSON, 1928.

direction. The upper 4 m of the poles are covered with netting to form a bag at least 2 to 3 m wide. It is advisable to dye the poles and netting light green. Extension poles can be attached with metal sleeves for catching at higher elevations, but the poles should never be more than 7 to 8 m long. The bander wears a heavy leather belt with sockets at each side and another in the middle of his back to support the ends of the poles. The belt must fit well to prevent chafing. The rear pole remains at the bander's back; the other two are clapped together when birds fly into the spoon-like net. After a catch, the bander bends over so that an assistant can remove the birds.

NELSON (1928) shows two quail catchers (Fig. 285). Such nets are good for catching on roosts, by day or at twilight. MCCLURE (1967) describes nocturnal catches in the Philippines.

We would like to suggest this method for individual birds that come close to people in defense of their nests.

13. Catching bats and flying foxes

We will only touch upon this subject. Many banders have caught bats in mist nets. If they are left in too long they may bite holes in the netting. We do not know whether this is an individual trait or characteristic of certain species. K. GREVE noticed that bats hanging in the net squeaking attracted more bats.

REINER (1965) describes a method of catching flying foxes in New Guinea—a method used extensively elsewhere in Southeast Asia. The natives south of Wewak (Australia) cut lanes through the forest on the mountain ridge closest to the coast, especially where fig trees abound. The lanes are 4 to 8 m wide. A net, woven of twisted vines with meshes 6 cm on a side, 15 m long and 3 to 6 m wide is suspended from the branches on either side of the lane much as a sail is hoisted. It essentially shuts off the lane (Fig. 286). One pull on a cord and the net collapses.

At night the catchers of flying foxes take to their lanes, raise the net and wait until the large flying foxes (kalongs) go forth to feed. When the fruits of the fig ripen, these animals appear in greater numbers. The alleys in the forest cause them to fly low to reach the fig trees. The moment a flying fox touches the net the trapper

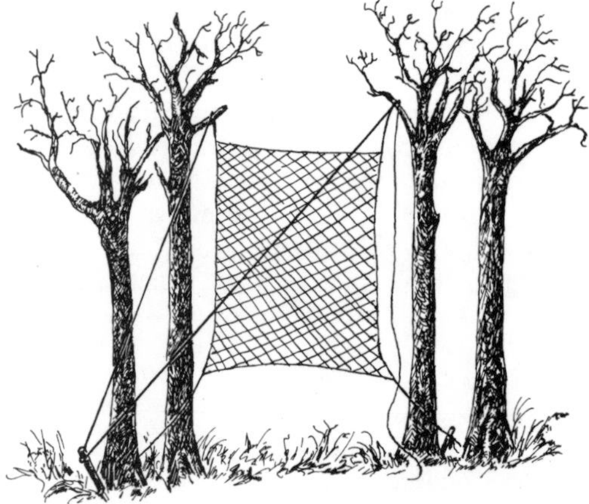

Fig. 286. Bat net in New Guinea. After REINER, 1965.

lets it fall. The captured animal is swiftly grabbed and taken from the net so that it will neither escape nor damage the net too much. Four or five flying foxes are often caught in one night. This is considered rich booty.

14. Catching with bow nets

14.1 History and development of bow nets

When we use bow nets today in ornithological field work for catching and banding, we normally do not think of the fact that this device has its own special place in the long history of hunting. This type of trap seems to have been developed after drop traps and snares.

Bow nets operate by torsion, or the turning around an axis. Elastic energy is stored, to be converted into motion at any moment as the trap closes. The source of the elastic energy is different from that in a shooting bow. Use of the elasticity of a bent rod and a therefore tightened bowstring which converts the elastic energy into motion—via the arrow—corresponds in a bow net or 'torsion trap' to the gathering of elastic force (physical energy) via twisted and thereby shortened elastic material.

This energy gained by twisting elastic material is retained as long as the trap is not released.

Elastic material twisted by a crossbar wants to return to its original shape and can have considerable force. In torsion an animal sinew, root or fibre is shortened and the stored energy becomes greater as the shortening process is increased with appropriate aids. "This torsion energy is transmitted via a lever to the effect carrier, whether this be a wooden slat fitted with sharp bones or iron nails which drives down on the animal with great force or a frame, often combined with a net, which pins the animal down or catches it." (Lips, 1927) The opposing force which prevents shortening of the tension carrier also often determines the shape of the

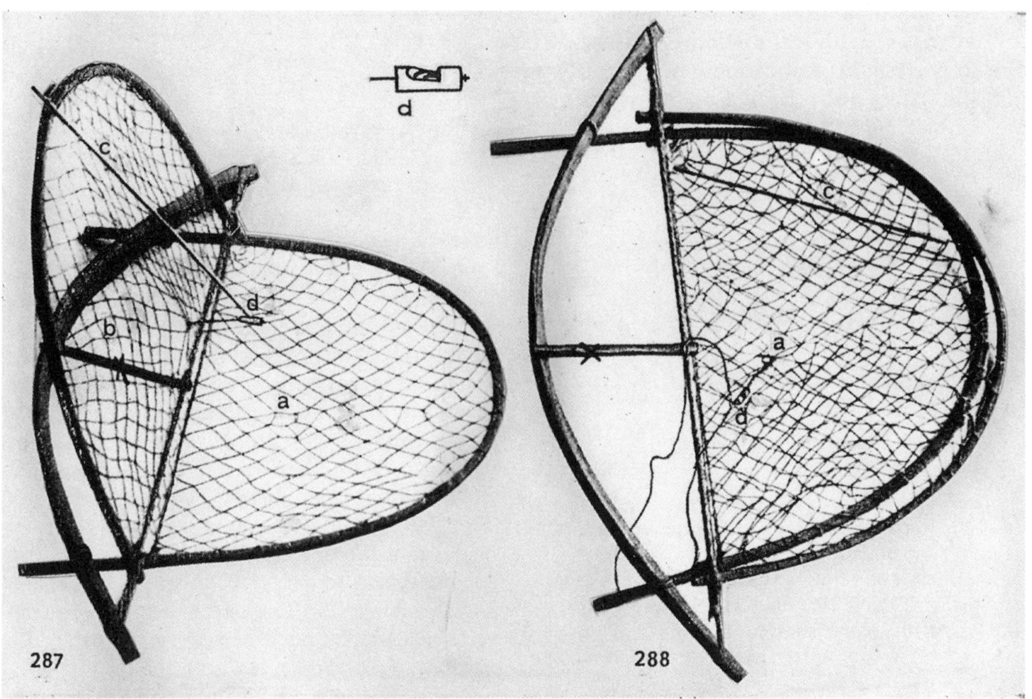

Fig. 287. Chinese bird trap (before release). Length of the torsion cords 48 cm; width of net frame 35 cm, length 25 cm. Material of net: silk, of frame: palm ribs. The bird is caught uninjured. From Lips, 1927.

Fig. 288. Chinese bird trap after release (see Fig. 287). From Lips, 1927.

traps. Lips shows a number of different traps mainly intended for the hunting of mammals.

All bow traps should be operated from close by. Success can only be obtained within a narrow circle around the trap.

According to Lips, torsion was used only in traps for small animals and birds in China and its area of influence as well as in Eastern Turkestan and Egypt. Their appearance and construction have many parallels in the traps still used today by banders. Only the materials used are different. Instead of single slats as effect carrier we use a frame which hits another frame upon release and thus becomes a catching compartment which holds the captured animal (Fig. 287 + 288). Lips also mentions the Egyptian trap (Fig. 289) described and built by Schäfer (1918/19) which is closely related to our bow nets of today and which is further described by Lagercrantz (1937 a). The wooden platform that forms one part of the bow net was purchased in Thebes in 1905. Nothing further is known about the origin of this platform, but a similar flat wooden piece was found in a grave in Thebes dating from the beginning of the XVIII dynasty. With this platform Schäfer was able to prove in his reconstruction that the bow (i. e. the frame) was activated by two sources of force, that it had a net, and that it must have been set with the usual Egyptian release mechanism.

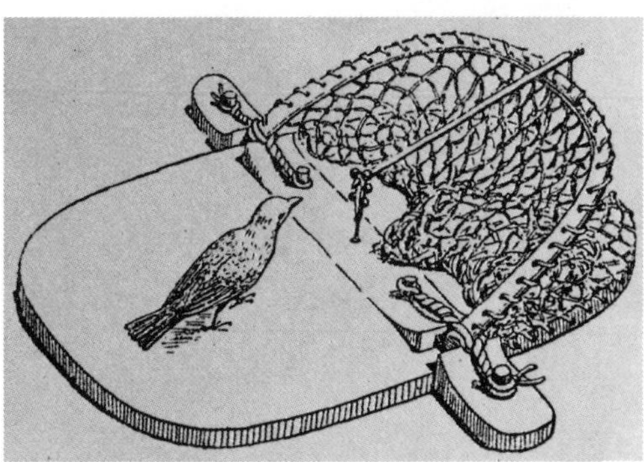

Fig. 289. Reconstruction of an Egyptian bird trap from the time of Amenophis I (app. 1550 B.C.). Similar in principle and trigger mechanism to the Chinese trap in Fig. 287 and 288. This trap differs in the construction of the power source, as the ends of the net frame are set in **two** separate and independent torsion devices which are held in place by wooden plugs in the board. It is typical of these torsion devices that the frame should not lie flat—contrary to our bow nets of today—as the torque apparently is not sufficient to force the frame around its axis entirely to close the trap. From Lagercrantz, 1933.

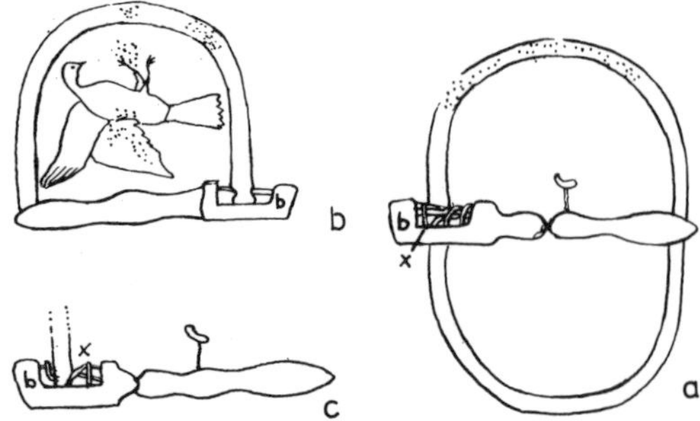

Fig. 290. (a) Egyptian bird trap after a picture in Benihasan (grave of Achtoj, XII dynasty). (b) Egyptian bird trap. (c) Part of an Egyptian bird trap. (b) and (c) both after pictures in said grave. After Lips, 1927.

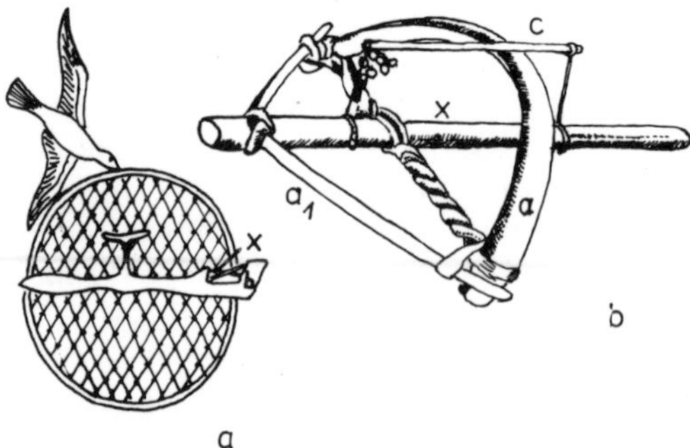

Fig. 291. (a) Egyptian bird trap after a picture in Benihasan (grave of Beket). (b) Present-day Egyptian bird trap (before release). After Lips, 1927.

Paintings in graves at Benihasan show another series of old Egyptian traps and a different way of constructing bow nets (Lips, 1927). The peculiarity of this construction (Fig. 290) is that the net frame is only propelled on one side b by torsion cords which are fastened in a depression on the cross beam of the trap. The wooden stick x (Fig. 290) indicates that this served to twist the strings and thereby regulate the release energy. A comparison to the mechanism of Fig. 288 is worthwhile. In Fig. 290 x even seems to be connected to the frame which serves to catch the pressure of the torsion cords on the tension bolt.

In Fig. 291 we see a bird trap used in present-day Egypt. According to Lips it corresponds in construction to the traps we use for bird catching with the difference that our traps use metal springs and the frames are covered with netting for live catches. The twist of the string, made of the inner bark of palm and mixed with camel or goat hair, is counteracted by the bow (a) made of palm ribs which presses the bird's neck against the lower frame a_1 after release. Today steel wire is sometimes also used.

LAGERCRANTZ (1937a) devoted a special study to the question of the origin of torsion traps. LAGERCRANTZ believes the age of the Egyptian torsion trap to be difficult to determine. Some of them are credited to the VIth dynasty of Sakkara and even to the pre-dynastic times of Hieraconpolis. LAGERCRANTZ believes too that the torsion technique originiated in the Far East. The bow traps of those regions are older than the ones from Africa or Europe. They came to Egypt at some early time and they can still be found today at oases. Only the trigger system has been altered.

Presumably the traps typical of Benihasan (and Sakkara) were not the usual bow traps and must have proven to be rather impractical (from a purely technical point of view). They were then modified into the trap with a wooden platform and two points of energy (Fig. 289). The Asiatic trigger system was not derived from the Egyptian bow nets.

This circumstance increases in importance considering that French and German bow nets have retained the system of the Far East thereby proving that they were imported directly from Asia (and not via Egypt). Even as far as the torsion traps combined with nooses or snares are concerned, no connection exists between European and Egyptian areas of influence.

The Egyptian bow trap has counterparts in Asia, but as far as ascertainable, none in Europe. The trap is found in Dolan (Iraq) where it is known as gïsmag, in Cherchen (Sinkiang, China), in a specialized form in Taiwan and a more antiquated form in Korea (Fig. 292): The Asiatic trigger system is used throughout the area stretching from Cherchen to Taiwan and Korea. Its characteristic is that the pressure of the clap trap bow comes from below. The longer trigger is thus held in a position determined by the shorter one with bait.

All traps mentioned here—except the ones combined with nooses or snares—are similar in that they all have two bows and that the strings themselves are not twisted in a stretched state. Very interesting is a clap bow trap from Northern China (Chihli) now in the possession of the Swedish Ethnographic Museum and which is different from the other traps. Whether set or not, it shows exactly the development of the torsion technique (Fig. 293). Presumably this is the oldest form of a torsion trap that ever existed. A detail worth mentioning is that the shorter trigger is kept in the proper position by upward pressure of the bow—a construction detail as rare as it is old.

LAGERCRANTZ concludes in his informative publication—despite the scant material available—that this technique was developed in an advanced culture. As the Egyptian torsion trigger system obviously follows the Far Eastern one, Egypt cannot be the origin of this technique. Moreover, the distribution of the torsion trap in Africa (Fig. 294) indicates clearly that the trap came from Asia, especially as we can trace the bow trap to the Far East and the development of the torsion technique can be followed back into earliest antiquity.

In the Far East the regular bow trap was developed from the Chinese clap bow trap. Combined with a net, it became our bow trap of today. Both kinds of trap continued using the

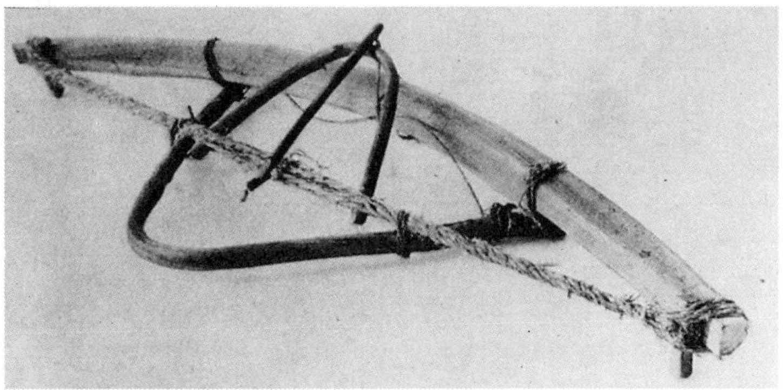

Fig. 292. Clap-bow-trap, Korea. After LAGERCRANTZ, 1937a.

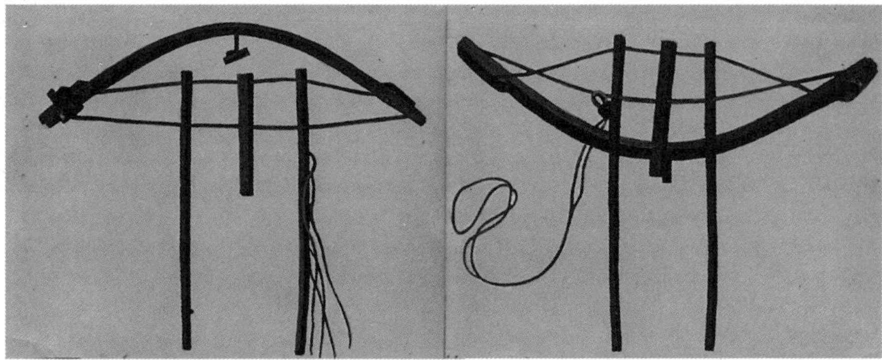

Fig. 293. Clap-bow-trap. China, Chihli. After LAGERCRANTZ, 1937a.

short bait trigger. But using a second lower bow the two triggers had to be rearranged. This trigger system of the Far East has spread together with the concept of the bow trap. There is some possibility that the torsion trap combined with a noose or snare descended from traps similar to the Chinese bow trap, although in the absence of intermediate forms, a few other possibilities for their independent development exist.

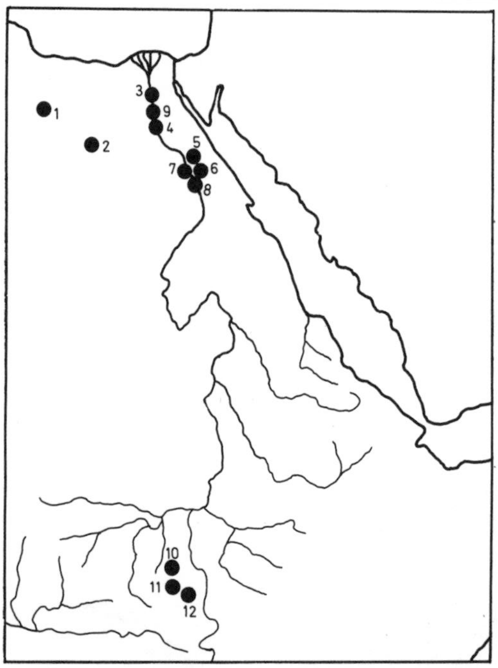

There seems to be enough evidence that all main torsion trap types have their origin in the high culture of the Far East, where the original trigger system was developed. It is impossible to determine the exact age of these main types of torsion trap, but we do know that they were introduced in the pre-dynastic era of Egypt. Everything seems to indicate that the torsion trap concept spread and reached Egypt by one route and Europe by another. This theory also explains why the two distribution areas do not lie adjacent to each other. Several special types of torsion traps were developed in Egypt. LAGERCRANTZ mentions in this context that the main component of the clap bow trap was made of the wood of the wild fig *(Ficus sycomorus)*. Of greater importance is the fact that the short bait trigger of the Far East was replaced by a "loop" to which bait was attached.

In Europe the regular clap bow trap never established itself. The reason is probably that it was no more effective than the widely used dead-fall trap which killed the prey with a heavy board or woven mat and which had long been used for bird catching. One torsion-type clap bow trap was indeed adopted, but it was imported from Egypt and modified soon under the influence of early Eurasian trigger systems.

The greatest importance of the torsion technique is probably that the well-known bow nets were derived from it and in turn from earlier hunting and catching methods. LINDNER (1940) deals with torsion traps in his basic work on hunting methods of the medieval age. He even includes those wooden clap traps which he de-

Fig. 294. Distribution of torsion traps in Egypt in antiquity. Explanation of discovery sites: (1) Siwah; (2) Farafra; (3) Sakkara; (4) Benihasan; (5) Kuft; (6) Thebes; (7) Luxor; (8) Hieraconpolis; (9) Lahun; (10) Bongo and Beli; (11) Mittu; (12) Moro. After LAGERCRANTZ, 1937a.

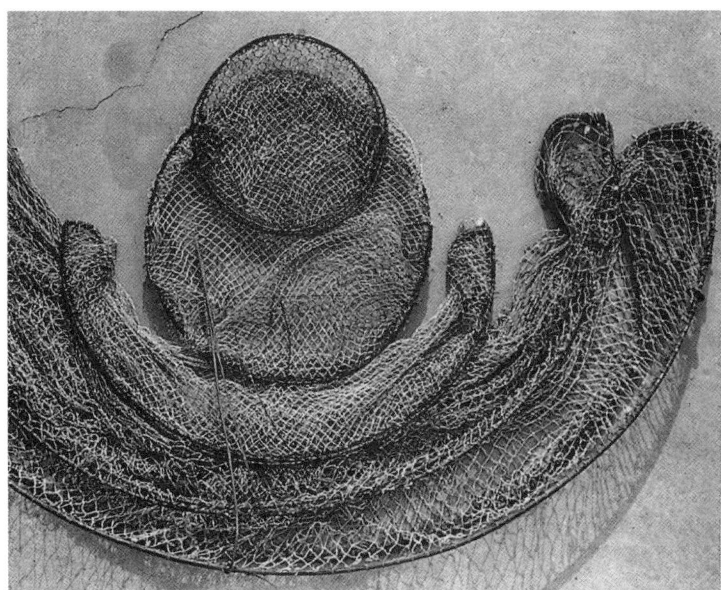

Fig. 295. Bow nets of various sizes.
Photo: I. SCHUPHAN.

scribed under neolithic hunting methods in this group. According to LINDNER, the double clap net too must be included in this group of trap-type devices operating with torsion. The double clap net only differs in being operated by the human hand.

The double clap net could be regarded as an oversize clap bow trap which does not operate on the pure torsion system. It derives its kinetic energy needed to close the net walls quickly from thin bent tree trunks, or originally through hard pulling on the pull cords by human hands. But some ethnologists may not count the double clap net among the true torsion traps.

14.2 Catching with terrestrial bow nets

Bow nets have long been used for scientific bird banding. The most common bow nets—for single catches—are round ones, but a number of square models also exist (Fig. 295).

Two springs—usually one on each side—quickly close the set part of the frame upon triggering. Small bow nets are generally triggered by the bird itself; larger ones are set off by the catcher via a pull-string.

Round bow nets (Fig. 296) should have a diameter of at least 30-40 cm. This size is suffi-

Fig. 296. Set bow net. The mealworm holder may also be fastened with a wire to the ground.
Photo: I. SCHUPHAN.

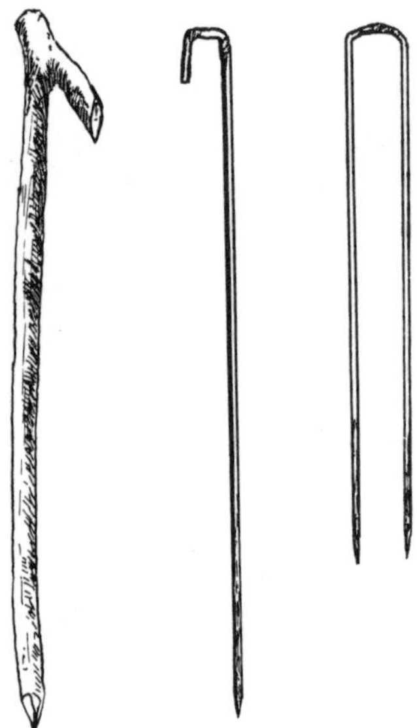

Fig. 297. Various hooks for staking bow nets to the ground. Diameter and length of hooks depends upon the nature of the ground. Wooden stakes are suitable for soft ground (soggy meadows, etc.)

cient for small birds: smaller nets can injure the birds as the distance between bait and the closing frame becomes too short. The frames are covered with cotton or nylon netting. Nylon is more durable and does not soak up moisture. Wet cotton netting should be dried in the shade. Mesh size for small birds should not exceed 12 mm so that the captured bird cannot stick head or wings throught the net. It is important to use plenty of netting. A tight net presses the bird to the ground, can injure or even kill it. The color of frame and netting should be inconspicuous—olive or gray are best. In special cases the ground color itself must be taken into consideration.

It is also possible to cover the bow net with some inconspicuous material. The captured bird then cannot be seen by members of its own or other species. A cloth-covered bow net should be used only in those banding programs where it is necessary that the captured bird in the clap-net not be seen by its own species.

The following example shows that birds can learn from the experiences of conspecific individuals. Klopfer (1957) trained ducks to avoid a water container which gave electric shocks to any bird trying to stick its beak in. Other ducks in the same cage which had not yet received an electric shock also began avoiding that water container.

During catching the bow nets should be fastened to the ground with two or three 15 to 20 cm long wire or wood stakes (Fig. 297); the length depends on the ground. To save time—or if the ground is either too soft or too hard—nets with a wire floor can be used. Usually a wire frame of the same diameter as the frame of the bow net itself is covered with wire mesh (Fig. 298). Kruis believes a floor of netting to be more practical. The rectangular bow net (Sunkel, 1948) has proved useful for very loose (for example, dunes) or very hard ground (for example, concrete, window sill, roof). The net is mounted on a board or a stable rod frame. The dimensions are: for small birds 30-35 cm, for thrushes 35-40 cm, for jays 40-52 cm. On the narrow side a small box (with hinged lid) restrains the movable net frame. The mechanism is triggered when a bird touches the swiveling food tray in the middle. Food such as seeds, nuts etc. has been glued to the tray. This net can be used in a more or less slanted position as well as the horizontal one.

Mealworms are particularly good bait. The can be fastened with their bodies in the holder specially developed by J. Winkler, (Fig. 26). The great advantage of this holder is that the mealworm is not damaged. In bright sunshine, however, the worm will soon have to be changed. Kruis ties the mealworm on an olive colored

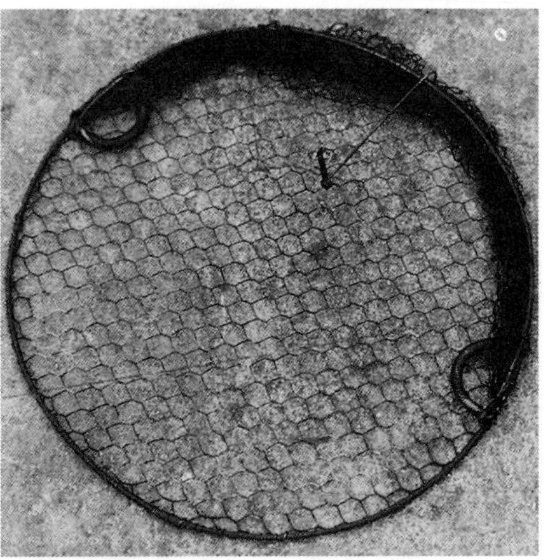

Fig. 298. Set bow net with wire floor. Photo: A. Präkelt.

thread and adjusts the trigger mechanism accordingly. The worm hangs close to the ground.

According to SUNKEL, earwigs are well suited as bait because they are rather tough and attract not only insect eaters but also bullfinches, finches, and buntings. Shrikes like beetles and melolonthids. According to KRUIS, earthworms cannot be fastened well in a holder and quickly dry out. Sometimes dragonflies, big flies and similar insects can be used successfully. SUNKEL baits large bow nets with hazelnuts for nutcrackers. He glues the nuts to a thread which he then attaches to the trigger in such a way that it releases the mechanism as soon as a bird tugs at a nut. Acorns can be used for jays.

When catching bullfinches SUNKEL does not hitch the free end of the trigger to its loop but puts a cluster of mountain ash berries over it. If a bullfinch nibbles at one of these berries, the cluster falls off the trigger, thus releasing it and catching the bullfinch under the bow net.

When setting up bow nets one needs to check whether they close perfectly and whether captured birds can escape under the frame if the ground is uneven. Sometimes it is advisable to conceal the net by covering it lightly with leaves, grass, sand, etc.

Which species can be caught with bow nets? First, all birds responding to the baits mentioned, for example many insect eaters such as Nightingales, European Robins, wheatears, pipits, European Redstarts, accentors, shrikes, warblers, Reed Buntings, Yellowhammer, Blue-headed or Yellow Wagtails; but Winter Wrens usually are not captured.

ANZINGER (1902) caught two Crested Tits in nets set for European Robins. KRUIS compiled a list of 75 species which may be captured with bow nets. There may well be others that could be included.

Mealworms are not always the ideal bait. According to BERGER (1967) Thrush Nightingales are not easily captured with a mealworm during times of abundant insect life. This also applies to other passerines. H. PRÜNTE emphasizes that sandpipers usually avoid mealworms while Ruffs like them; Rock Pipits likewise ignore them while Water Pipits respond.

WINKLER tells us that Blackcaps and Corn Buntings scorn mealworms; Tree Pipits and Icterine Warblers do not get caught too often either. On the other hand, KRUIS works with mealworms when catching Nightingales and

Fig. 299. Wheatear on a perch near the bow net.

captures Blackcaps, Barred Warblers, and Whitethroats as well as Lesser Whitethroats. Garden Warblers take mealworms only in the fall and if the net is under an elderberry bush, Tree Pipits take them only in spring and fall and Winter Wrens only in spring. Warblers of the genus *Phylloscopus* are only rarely interested.

Other species responding to mealworms are: rails, occasionally a kestrel (something KRUIS never observed, however), European Cuckoos (the net must be set up under the tree the cuckoo calls from!) and others.

One day in March BAUMANN (1905) set up a bow net at a hole which had been dug into an anthill and concealed it well. He caught a Three-toed Woodpecker, apparently the digger of the hole. K. GÜTH warns that bow net sites must be checked for ants beforehand as large numbers can kill the captured bird.

K. ROTHMANN piles up small earthmounds besides each net when catching field and meadow birds such as wheatears, whinchats and others. The birds like to perch on such "high" points. R. HELDT sticks a twig or branch into the ground next to each net (Fig. 299). During 40 days in the fall of 1964 HELDT trapped with 50 nets on the meadows near the outer dikes at St. Peter (Schleswig-Holstein, Germany) and caught 82 Skylarks, 137 Meadow Pipits, 66 Blue-headed or Yellow Wagtails, 4 White Wagtails, 513 Wheatears, 49 Whinchats, 6 Lapwings, 2 Grey Plovers, 20 Golden Plovers, and 7 Common Sandpipers. These were all migrants. The mealworms were generally changed twice a day.

14.3 Catching with bow nets at windows and on flat roofs

During the colder months of the year trapping at a window feeding station can be very successful. In order to catch Herring, Black-headed and Common Gulls, M. Riegel attaches a board with a bow net (diameter 80 cm) to the outside of the kitchen window of his third floor apartment at the outskirts of Wilhelmshaven (Germany). Bait is kitchen scraps and seeds. The net is triggered from the apartment. Although only one gull at a time can be captured, the total results for the winter were 36 Black-headed, 22 Common and 40 Herring Gulls. Success depends upon the weather. During high snows 126 European Starlings were caught in 6 weeks. During the snowy winter of 1962/63 Riegel caught over 300 Twites at this feeding tray alone. There were no lure birds except Twites.

In seven years at this site Riegel banded 69 Herring Gulls, 79 Common and 298 Black-headed Gulls, 69 Collared Doves (42 in 1964), 180 European Starlings, 488 Greenfinches, 400 Twites, 158 Great and 43 Blue Tits, 18 Chaffinches, 11 European Robins, 5 European Blackbirds, 3 European Tree Sparrows, 2 Hawfinches, 2 Great Spotted Woodpeckers, and 1 Dunnock, Brambling, Bullfinch, Linnet and Siskin each.

Fig. 300. Bow net on a roof. Beside it the lure bird whose cage has been protected with wire mesh against raptors. Photo: H.-W. Nehls.

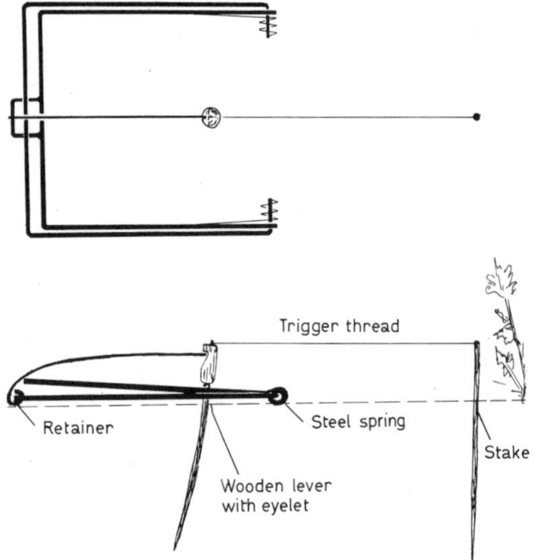

Fig. 301. Bow net for shorebirds (upper: top view; lower: side view).

Riegel mounted his net on a flat board which had a wooden rim of only 1-2 cm height. Kruis of Prague (CSSR), on the other hand, mounted his bow net on a wooden box 50 cm long, 35 cm wide and 6 cm high. He too obtained remarkable results. H. W. Nehls of Rostock (Germany) set up a bow net on a roof directly below his living room window and released the trigger from there (Fig. 300). Nehls caught many birds in the center of the city of Rostock, similar to the results in Wilhelmshaven. Snowy winters produced better results here also.

14.4 Shorebird and rail trapping with bow nets

Shorebirds are caught mostly in mist nets, but bow nets still have useful applications.

In the region around Dessau (Germany) H. Kolbe uses 70×50 cm nets for shorebird trapping (Fig. 301). The frames consist of 4.5 mm steel wire, the springs are made of 1.2 mm steel wire and have 20-25 coils. For better utilization of the net capacity Kolbe triggers the larger frame. Therefore an extension section must be welded to the smaller frame. Netting of 10-12 mm mesh is used as covering. Also needed are a holder of steel wire 2 mm thick and approx. 12-14 cm long, a prop and the stake

Fig. 302. Set bow net for shorebirds. Photo: H. Kolbe.

(15-20 cm long depending upon the ground). The trip cord (strong black twine) is fastened to the prop. When dismantling the trap the twine is wound around the stake. Two wire hooks (length depends upon ground conditions) hold the lower frame in place.

Kolbe sets up his clap net on muddy banks (Fig. 302). The moving half of the frame closes toward the water and actually reaches the edge of the water. The net itself must be on "dry" ground so that captured birds will not drown. The trip cord is set 2-5 cm high depending upon the species expected. A bird moving along the water's edge runs into the trip cord and thus triggers the closing of the trap. Camouflage of the net is not always necessary but generally advisable. In order to increase catches small obstructions can be constructed from plant material in front of and behind the nets. According to Kolbe at least 5-10 nets are needed. The minimum distance between nets on mud banks should be 1 m. The nets are checked every half hour or hour. Catching at night is very advantageous for Jack Snipes and nets must be checked at similar intervals. With skill one can drive birds close to the nets into them. Clap nets are also useful in catching ducks and rails, but should then be set near the edge of the reeds. A thick vegetation barrier was very useful in this case.

On Norderoog (Germany) Schuphan quickly caught 15 turnstones with bow nets mounted onto wire platforms (diameter 35 cm). The bird triggered the net via a trip cord. A sprung trap was never empty.

In the fall of 1971 in Northern Bohemia (ČSSR) a bander caught 100 Moorhens, 37 Spotted Crakes and 11 Water Rails plus 38 Bluethroats with a bow net mounted on a wooden board and baited with mealworms. The wooden platform was left in place at the trapping site, while the net itself was set up each time and then taken away afterwards. Thus bow nets mounted on solid surfaces give good results in environments other than on a roof or at a window.

Large shorebirds—with the exception of Redshanks, Wood Sandpipers, and Ruffs—rarely are caught in funnel traps. Therefore Fredga and Frycklund (1965) made 8 clap nets with square bottoms and triangular sides (Fig. 303). Inside the trap a thin black cord is strung, about 10 cm above the ground. Small shorebirds can walk underneath the cord unmolested, while the larger birds touch the cord walking through and thereby trigger the mechanism. The thin cord leads to a set mouse trap whose frame is connected to a fork via a strong wire. The fork holds together the two sides of the clap net, which snaps shut quickly when triggered. The closing

Fig. 303. Bow net for large shorebirds, used by FREDGA and FRYCKLUND in Sweden. The net should be hitched up better when trapping. Photo: G. HANSSON.

is accelerated by two pieces of elastic attached to each side between the ground and the frames, as shown in Fig. 303. HEYER (1968) used a remarkable method of triggering to catch three Sanderlings at a dammed lake near Weimar (Germany). Behind the slightly camouflaged net the trigger cord leads through an eyelet on a stake in the ground. Trigger cord in hand, HEYER drives birds into catching area of the net, marked by clearly visible small sticks. The net itself can be triggered from any direction. W. PRÜNTE came up with quite an unusual way of "baiting" near Münster (Germany). As soon as a Ruff gets caught in a bow net, his "relatives" nearby come in to have a good close look at the captured bird. At that moment PRÜNTE releases a pull net and thus captures the entire party.

NØRREVANG and MEYER (1960) and SCHILDMACHER (1965) show trapping with bow nets at shorebird nests. We need to point out strongly that catching with bow nets on nests can cause great damage. The captured bird necessarily damages or destroys some eggs or even the entire clutch under the net and is frightened excessively by the sudden capture.

14.5 Mouse and rat traps in the service of bow nets

SUNKEL (1938) mentioned that mouse, or better yet, rat traps can be converted into bow nets. He used a 50×50 cm wooden board as the base. The deadfall frame is replaced by a larger one which is connected to the spring. A new trigger wire of appropriate length is connected at the place where the back of the opened frame touches the board. A larger but light board or cardboard tray is fastened on the original small food tray. We now attach some bait onto this new tray. The mouse trap with attached bow net is now ready to be set.

PARRY (1968) reports on the catching of 88 Kookaburras during two years in Australia. The trap consists of a circular net 75 cm in diameter. The rim is reinforced with wire and wooden strips and is soldered to the deadfall frame of a large rat trap. Two hooks keep the trap in place on the ground. The trap is set as usual (Fig. 304). The trap is released as soon as a bird tugs at the bait. Sometimes three birds who all had rushed to the meat bait got caught at the

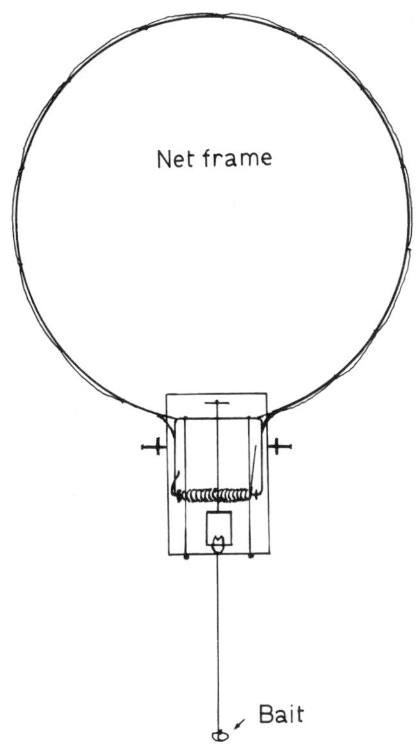

Fig. 304. Bow net attached to a rat trap. After PARRY, 1968.

same time. Young birds occasionally tried to take the meat while the trap was still being set up. If on the other hand a bird witnessed another bird being caught, it would take a while before it in turn was caught. One old male refused to be caught despite many attempts.

ENGELMANN (1928) talks about a Red Kite which got caught by a dozen breast feathers in a rat trap set at a hamster's burrow. As the bird had just molted, the feathers did not come loose and the wild attempts at freedom were unsuccessful. The bird was captured.

14.6 Push bow nets

Many bird species like to sit on the tops of bushes or trees. They almost always prefer any kind of perch raised above the surroundings that will give them a good view to all sides. Insect-hunting flycatchers, European redstarts, shrikes and others use them as starting points for a hunt; European Robins, thrushes, pipits, and buntings use them for territorial song and courtship display.

History shows that these preferred song perches were always used by bird catchers. The perches were coated with bird lime, or nooses, hooks or snares were attached. A bird catcher observing the behavior of the birds to be captured soon knew which were their preferred perches. This is as true today as it was then.

But today we use the so-called "push net". This type of bow net (Fig. 305) has two frames covered by netting which clap shut above the perch as soon as a bird alights. Additional details are described below. These nets have proved very valuable even in places where we could capture birds with other traps. There are situations in which they are indispensable. Their advantage is a quick set-up.

In some instances they are particularly use-

Fig. 305. A set push bow net.
Photo: I. SCHUPHAN.

Fig. 306. Set push bow net for Twites. Photo: H. BUB.

ful. In Norway, BUB used these push bow nets together with a lure bird to capture Twites (Fig. 306). He attached holders for two push bow nets to the roof of his car. At each rest stop he set two nets and put the lure bird in its cage on the car roof. The Twites came without much ado to investigate the lure bird and got caught in the nets.

As we know from experience, catching with push bow nets is not without danger to the bird, especially if the nets are too small or the springs too weak to close the frames quickly. For small birds each frame must be at least 30-35 cm long and 20-25 cm wide and high. The measurements for small raptors are 60 cm and 45 cm respectively.

For many years the two upper bars were made of steel wire. If quick birds got caught between them, they often were injured or killed. H. WEBER replaced the steel with stiff rubber and thereby eliminated all danger for the bird. Of 1,000 catches at the Biological Station Serrahn (Germany) only one bird was injured. It is also possible to use two pieces of rubber hose opposite each other on part of the frames so that they do not slam shut hard.

The nets should be set up 1-2 m above ground. For some species such as crossbills it is necessary that the nets have the highest perches in the vicinity of the lure birds.

Nets and frames should be painted an olive color for best camouflage. Berries and mealworms work fine as bait although often no bait is needed.

14.7 Prairie Chicken trap for single catches

O. MATTSON developed a trap for single catches of gallinaceous birds (Fig. 307). The trap is usually released when the position of the corn cob is changed by the bird. The net frames are fastened to 7 cm wide boards. At the opening

Fig. 307. Prairie chicken trap for single catches. Photo: H. BUB.

the wooden frame is 85 cm wide, in the back, 17 cm. The boards slanted towards the rear are 60 cm long. The height of the net frame is 50 cm in front, 29 cm in the middle, and 15 cm in the back.

15. Raptor trapping with bow nets and stationary nets in North America: The falconers' camps of the present

15.1 General

The falconers' camps of olden times, such as existed in Europe and the Orient over many centuries, have probably become only "hunting history" over the last few decades, at least as far as Europe is concerned. The last European falconers' camp near Valkenswaard in the Netherlands ceased to exist sometime between 1920 and 1930. More recently, for example in Scandinavia, there may occasionally be some simple falconer's camp even today.

Continuity has been provided through scientific bird banding. Over many years catching raptors was only a "side line" for ornithologists. Banders used only various raptor fall traps or sometimes bow nets. J. WINKLER used nets with a diameter of 80 cm; the frame was made of 5 mm thick steel wire, the springs of 2.5 mm thick steel wire.

Suitable bait for buteos are parts of a hare or rabbit, chicken or other animal. The trapping place is marked with scattered feathers and pieces of fur. Sometimes rags can be used as bait for kites during their nesting season. R. LEVEQUE catches mainly Marsh Harriers, kites and buteos in the Camargue (France). He uses mostly dead ducks as bait.

H. KLAUENBERG, a successful raptor catcher in the area between Hannover and Braunschweig (Germany), baits his camouflaged trap with a stuffed young hare. In addition he fastens a bit of fur or wool on a length of twine to a stick. The wool moves in the wind and attracts attention to the hare. Also successful is a clean piece of rabbit fur, without head or feet, in which KLAUENBERG wraps a fresh piece of meat, heart, lung or liver. If he attaches a bit of fur and splashes about a bit of lime so that it looks like a fresh kill, he catches all harriers and buteos. If the meat spoils in the sun, only kites will come.

According to KRUIS, banders in the CSFR use 70×40 cm bow nets to catch Eurasian Kestrels, European Sparrow Hawks, Long-eared Owls and Little Owls. An Eurasian Tree Sparrow (a very active bird), a Greenfinch or a Brambling in a very small cage serves as the bait. Eurasian Kestrels and Little Owls respond well to a mouse, especially a field mouse. It will remain active for a long time in the small round cage. A twig attached above the cage serves as the trigger. The same bow net baited with white bread catches Rooks and Jackdaws in the snow.

M. RIEGEL of Silesia (Germany) set a cage with a dove in a hole in the ground that had been dug so the cage top reached only slightly above the ground level. Two strings ran over the cage to the trigger hook and released the trap upon a slight touch of the strings. Common Buzzards, Rough-legged Hawks and Northern Goshawks were captured.

Some raptors also have been caught in mist nets or double clap nets. Many are banded as nestlings, a fact that proves important because we know their birthplaces.

Many European Sparrow Hawks have been caught in the large funnel traps of the Ornithological Research Station Helgoland (Germany)—1,260 birds from 1926 until 1965—and several hundred Rough-legged Hawks in the pull nets of the crow catchers of Rossitten (Germany) (SCHÜZ 1942).

In 1938 the HERDEMERTEN expedition (HERDEMERTEN 1939) to Greenland had the objective of observing Gyrfalcons and catching a number of them. They did not establish a regular falconers' camp. They attempted to catch the birds with one bow net and several nooses, but caught only a few. The catching attempts did not give the impression of a well planned operation. MATTOX (1970) planned better, and we will describe his catching of Gyrfalcons later.

Recently raptor catching has been revived in Europe, where it had almost completely ceased, by two methods imported from North America. One is the bal-chatri trap, an East Indian raptor trap, which has already fascinated many a bander. In East Germany this method, which does not endanger the bird, has already been officially approved.

The other method is the combined catching

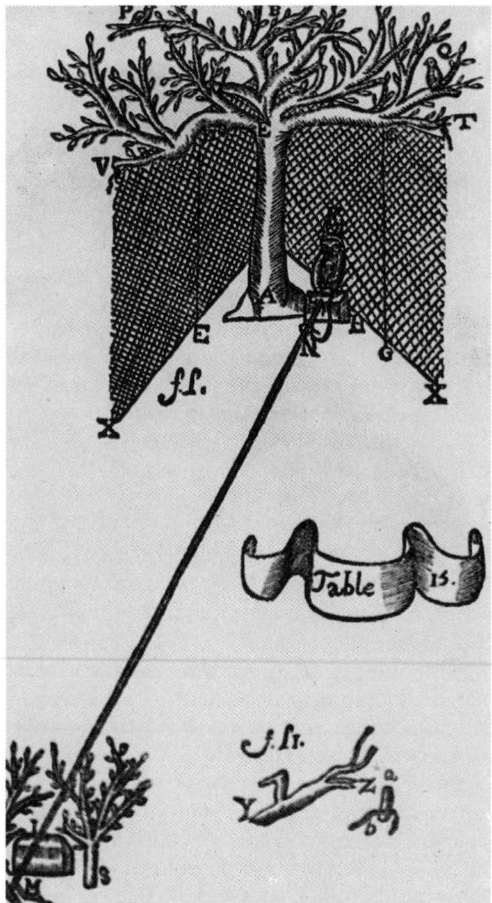

Fig. 308. Two stationary nets set at an angle to catch raptors. Inside the angle an Eagle Owl as lure bird (Les Ruses Innocentes, 1660). After MACPHERSON, 1897.

with bow net and stationary net or catching with a stationary net alone, as described by HAMERSTROM (1963).

This last method has a precedent in a trap dating from the 15th century (LINDNER, 1940). MACPHERSON (1897) shows two nets set at a sharp angle to each other with a Northern Eagle Owl inside the triangle (Fig. 308).

The stationary nets used at Cedar Grove (USA) are similar, but no owls are used as lure birds.

15.2 Raptor catching at Cedar Grove (USA)

In the fall of 1962 BUB visited the Cedar Grove Banding Station on the western shore of Lake Michigan in North America and observed the raptor catching operations. H. C. MUELLER and D. D. BERGER built the set-up themselves.

First the historical background of this station. H. STODDARD and C. JUNG first "discovered" the raptor migration along this coast of Lake Michigan in 1921 (Fig. 309), similar to the "discovery" of other favorite places to observe migrations. In 1930 M. DENSING and JUNG attempted to catch raptors with little success until a falconer advised them in 1936. The Milwaukee Public Museum then built a trapping station and manned it with the help of the Civilian Conservation Corps. Work was suspended in late 1941 when the United States entered World War II. After the war cooperators of the museum only occasionally caught raptors. From 1936 until 1949 a total of 961 raptors were captured.

In 1950 H. C. MUELLER and D. D. BERGER resumed regular trapping. They built an entirely new station as well as new catching set-ups. In 1962 there were five raptor trapping sites which had only been completed two years earlier. Raptor catching here has set the example for other stations in the USA.

The observation figures for the fall migrations of 1952–1957 give an impression of the numbers of raptors passing through (MUELLER and BERGER, 1961): 15,965 Broad-winged Hawks, 8,524 Sharp-shinned Hawks, 1,407 Red-tailed Hawks, 1,115 Northern Harriers, 798 Merlins, 370 American Kestrels, 268 Cooper's Hawks, 187 Ospreys, 150 Peregrine Falcons, 72 Red-shouldered Hawks, 39 Rough-legged Hawks, 19 Goshawks, 17 Turkey Vultures, 7 Swainson's Hawks, 6 Bald Eagles, 2 Golden Eagles, 1 Prairie Falcon, 115 unidentified raptors. Total: 29,061. This count does not seem to reflect the entire migration as later observations indicate. For example on one day (Semptember 18, 1962) over 5,000 Broad-winged Hawks migrated overhead.

Raptors caught during fall migration between 1950 and 1971 number 6,263. An additional 872 birds were banded during spring migrations. In spring the migrations are not nearly as heavy and banding operations were not as regular.

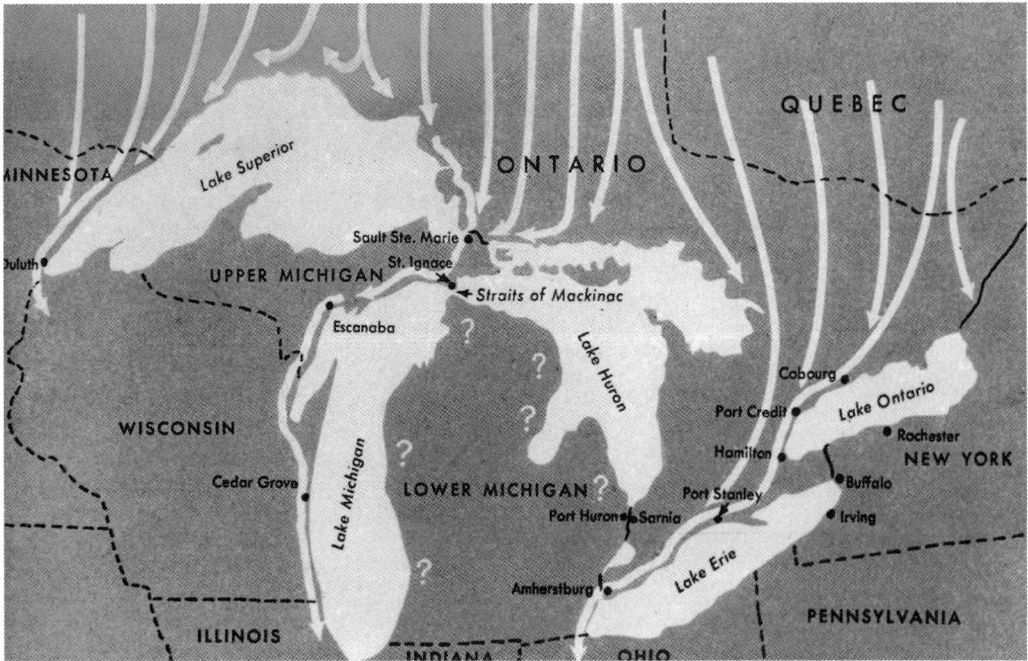

Fig. 309. The Great Lakes region with migration paths indicated. After PETTINGILL, 1962.

BERGER and MUELLER indicate their best spring catch for April 4, 1956: 11 Marsh Harriers, 7 Goshawks, 2 Red-tailed and 2 Rèd-shouldered Hawks, as well as 23 American Kestrels.

DAN BERGER compiled the following list of banded raptors. Fall catches are mentioned first, followed by spring bandings in brackets. This means that between 1950 and 1971 over 7,000 raptors have been banded.

Turkey Vulture *(Catharthes aura)* 1
Northern Harrier *(Circus cyaneus)* 265 [S 337]
Cooper's Hawk *(Accipiter cooperii)* 201 [S 186]
Red-shouldered Hawk *(Buteo lineatus)* 35 [S 37]
Broad-winged Hawk *(Buteo platypterus)* 18 [S 11]
Swainson's Hawk *(Buteo swainsonii)* 7 [S 1]
Rough-legged Hawk *(Buteo lagopus)* 30 [S 4]
Peregrine Falcon *(Falco peregrinus)* 143 [S 1]
Merlin *(Falco columbarius)* 380 [S 3]
American Kestrel *(Falco sparverius)* 152 [S 148]
Golden Eagle *(Aquila chrysaetos)* 4
Sharp-shinned Hawk *(Accipiter striatus)* 3,361 [S 50]
Northern Goshawk *(Accipiter gentilis)* 171 [S 2]
Red-tailed Hawk *(Buteo jamaicensis [harlani])* 1,008 [S 91]
Bald Eagle *(Haliaeetus leucocephalus)* 2
Great Horned Owl *(Bubo virginianus)* 3
Common Screech Owl *(Otus asio)* 28
Short-eared Owl *(Asio flammeus)* 3
Long-eared Owl *(Asio otus)* 49
Barred Owl *(Strix varia)* 2
Tengmalm's (Boreal) Owl *(Aegolius funereus)* 1
Saw-whet Owl *(Aegolius acadicus)* 396 [S 1]

Owls were generally caught in mist nets.

Especially interesting is the relationship between birds ovserved and birds captured. In the fall of 1962 the following birds were caught (numbers in brackets are birds observed): Northern Harrier 7 (185), Sharp-shinned Hawk 168 (1,485), Cooper's Hawk and Goshawk 37 (75), Red-tailed Hawk 72 (392), Broad-winged Hawk 0 (8,217), Red-shouldered Hawk 5 (50), Rough-legged Hawk 0 (49), Bald Eagle 1 (5), Golden Eagle 0 (5), American Kestrel 11 (44), Peregrine Falcon 2 (15), Merlin 18 (93), Osprey 0 (21), Turkey Vulture 0 (3). That amounts to 330 bandings. The best day in 1962 was September 17 when 30 Sharp-shinned Hawks, 20 Red-tailed Hawks, 4 American Kestrels, 3 Merlins and 1 Bald Eagle were captured. On October 13, 1955 a total of 85 Sharp-shinned Hawks were banded.

Fig. 310. Part of the trapping area of Cedar Grove (Wisconsin, USA), taken from the blind. In the middle is an area for catching with dho-gazas and bow net (see also Fig. 312). Catching is done only with bow net at the high pole in the background. This high pole serves to "top" the lure pigeon. Photo: H. Bub.

The catching area (Fig. 310) is 200 m distant from the shoreline of Lake Michigan. Hidden in the bushes lies the little hut which serves simultaneously as catching blind, banding hut and living quarters. One side (Fig. 311) faces the open inconspicuously. A small slit provides a view of the area with the 5 specific catching locations and it is from here that the catchers operate the bow nets—with the aid of powerful batteries in the hut and a solenoid at each net.

Catching equipment: At 4 of the 5 catching locations a large bow net with a diameter of

Fig. 311. The side of the catching and banding hut facing the catching area. The small slit and the two long narrow windows on top serve for observation purposes. The round openings are for the strings to which the lure birds are attached at 4 locations. The strings for the fifth location are on the left side of the blind. Photo: H. Bub.

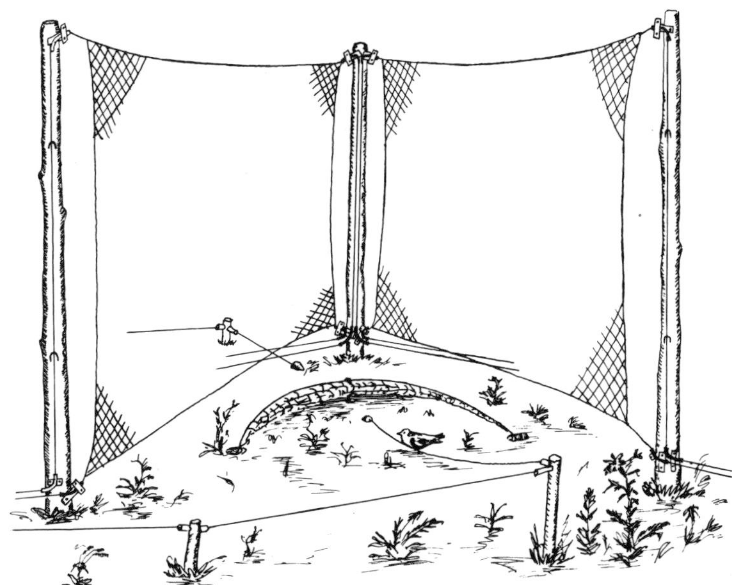

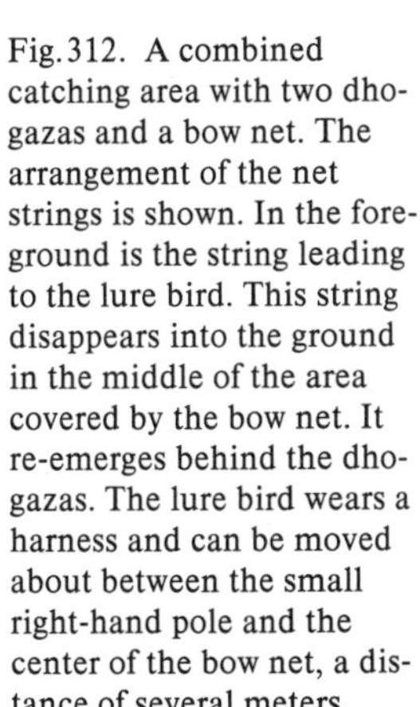

Fig. 312. A combined catching area with two dho-gazas and a bow net. The arrangement of the net strings is shown. In the foreground is the string leading to the lure bird. This string disappears into the ground in the middle of the area covered by the bow net. It re-emerges behind the dho-gazas. The lure bird wears a harness and can be moved about between the small right-hand pole and the center of the bow net, a distance of several meters.

1.6 m complements two dho-ghaza nets or stationary nets (Fig. 312). The 5th location has one bow net only. The combination of both methods is favorable. The nets catch those raptors that try to grab the lure bird in flight, while the bow nets catch those that will drop onto the lure bird. Each lure bird—an European Starling, House Sparrow or Rock Dove—wears a harness and is fastened to a nylon string which leads from the hut to the bird and back again in a loop. In this way it can be pulled in the desired direction.

The dho-gaza nets are 2.4×1.8 m. The mesh measures 100 to 120 mm stretched. The mesh twine is 0.5 mm in diameter and the string framing the net as well as that leading from net corners to the poles is 1 mm thick. It is a flat single-walled net. Berger and Mueller make their nets with the aid of a 1 m high wooden frame. Care must be taken that all sides are equal. Net-

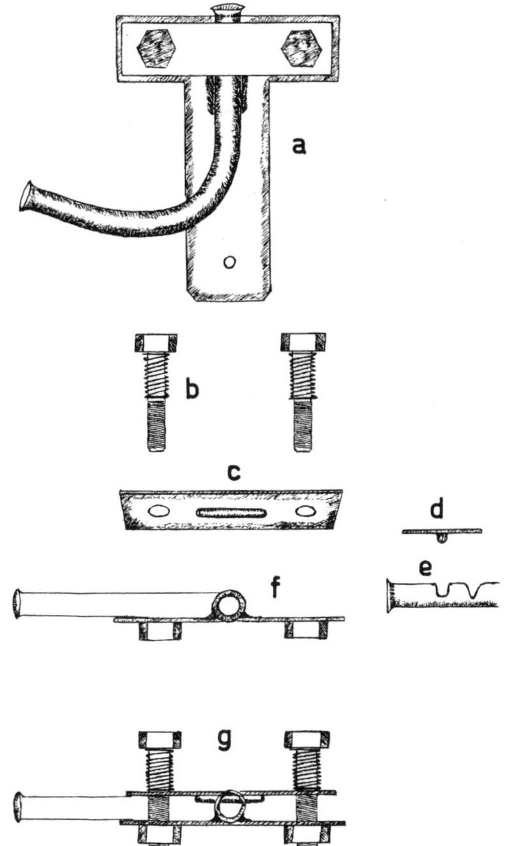

Fig. 313. Copper tubing with base and components. (a) view from above; (b) restraining screws (length 47 mm, diameter 8 mm) with springs; (c) brass strap (47 × 14 mm) with release (length 18 mm, diameter 2 mm) bottom view; (d) side view; (e) side view of copper tubing with filed groove and center-punched indentation; (f) front side view of brass base with soldered tubing and nut; (g) the same as (f) but with brass strap [see (c) and (d)] and both restraining screws, but without swivel (see Fig. 316).

ting and strings are of nylon and are soaked for 2 days in liquid black shoe dye. This will protect them against chewing by mice. The three poles which support the nets are 2.2 m high and 12-15 cm thick. They form a right triangle with the outer poles 2.6 m from the middle one. The nets are set between these poles. The distance between the two outer poles is about 3.6 m. This side remains open.

A very practical way of fastening and tightening of the nets has been devised. Copper tubing which has been soldered to brass plates is fastened to the poles at the top and bottom (Fig. 313). The arrangement of the tubes at the middle pole can be seen in Figs. 314 and 315. The end poles have only one tube above and below as they have to hold only 2 net corners each.

The tightening of the nets is important. A long black nylon string 1 mm thick is fastened to each net corner. At the upper corners it must be 8-9 m long, at the lower corners 6-7 m. At the appropriate distance from each net corner (see Fig. 314) a swivel (Fig. 316), such as anglers use, is attached. The string must be cut and fastened to each end of the swivel. Now the string with the attached swivel is threaded through the copper tubing but only until the ball of the swivel lifts the "stop", which sticks into the tub-

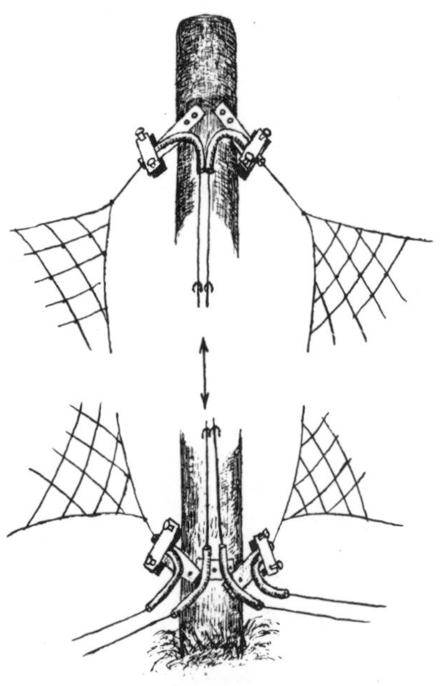

Fig. 314. Center pole showing arrangement of tubing and net strings.

Fig. 315. Arrangement of tubing at the bottom of the center pole. Photo: H. Bub.

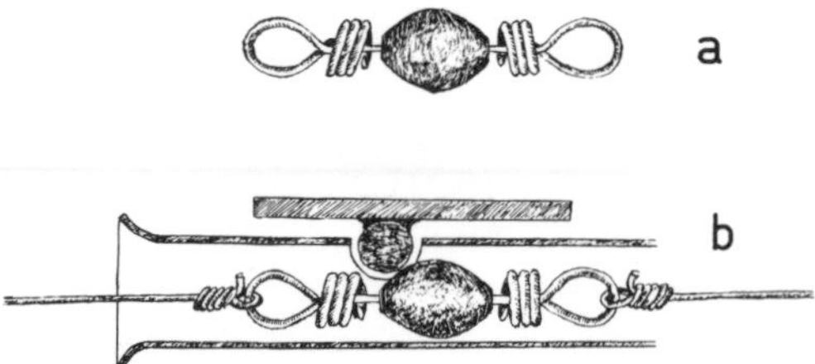

Fig. 316. (a) The swivel, total length 21 mm, length of "ball" 6 mm, diameter of ball 5 mm; (b) the swivel inserted into the copper tubing, at the right and left the attached nylon cord. Above the tubing is the strap with the release (see also Fig. 313 c, d, g). Approximate enlargement: 3 x.

ing, and is under tension. The swivel ball is held in place by the stop (Fig. 313 and 316).

This method in marvellous insofar as this stop can be adjusted very delicately (Fig. 313) with the aid of two upright screws (Fig. 313) which have a spring and whose nuts are soldered in place. In other words the resistance to pull of the ball can be adjusted via the two screws and nuts with the special springs; the ball can be set "soft" or "hard".

Each raptor flying into the net can therefore pull it out of its four holding devices and then fall to the ground together with the net. Small raptors can also hang in the net (Fig. 317) if their stoop is not too swift or the ball has been set rather "hard".

The thin black nylon strings are guided down the pole via some staples. Before leaving the pole they run through simple copper tubing and are laid out on the ground for another 6 m. The same is done for the strings on the lower net corners (Fig. 312). This considerable lengthening of the strings is supposed to make resetting the nets easier after a catch. Only very rarely does a large raptor make his escape with the net, losing it in flight.

Catching with the bow net. BERGER and MUELLER built big bow nets of 1.6 m diameter at four of the catching places described. A fifth bow net is used alone. The springs of the net are of 2 mm thick steel wire, are 4.7 cm long and have 12 coils. The wire of the frame is 6-7 mm. The nylon mesh of the net is 80 mm stretched and the individual strands are 1 mm thick. The nets are fastened to the ground in 3 places, 2 of them under the springs.

Release of the net is transmitted from the blind via five 6-volt batteries connected to each other and a solenoid at the net (Fig. 318). The connecting wires are laid underground. After release the frame of the net hits an iron hook (Fig. 319) which positions itself over the frame making it impossible for a bird of any size to open the net.

No catch is possible without lure birds. In order to excite the curiosity of as many raptors as possible, BERGER and MUELLER use Rock Doves

Fig. 317. BERGER (left) and MUELLER (right) take two birds out of the net: a Sharp-shinned Hawk and an American Kestrel flew almost simultaneously into the net. Photo: H. BUB.

Fig. 318. Solenoid for the release of the bow net. The trigger for the closing bow is released from the blind: the cables are underground. Photo: H. BUB.

Fig. 319. Catch which locks the bow closed. Photo: H. BUB.

(Fig. 320) as well as European Starlings and House Sparrows. Only doves are used on the farthest away spots because they are more readily noticed by hunting and migrating raptors. Once they have come closer they see the smaller lure birds as well. The distribution of sparrows and starlings at the other catching sites is not fixed. It also depends on the lure birds available (only the three species mentioned above may legally be used as lure birds in North America). The birds are not tied to one spot but rather they are tied to black nylon strings coming from and returning to the blind. The strings for the starlings

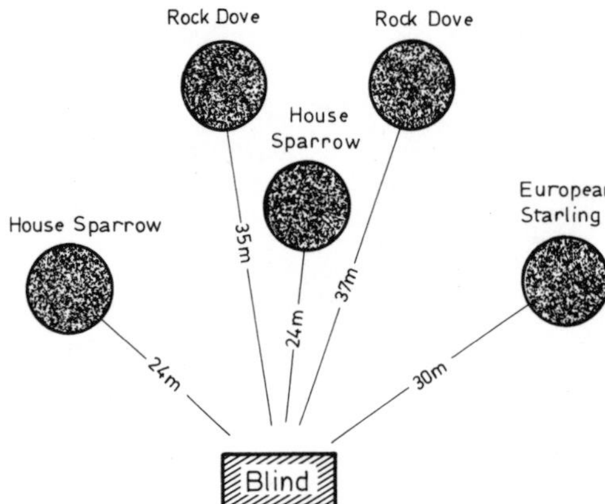

Fig. 320. The catching areas with the lure birds used.

are 20 lbs. weight and those for the doves 50 lbs. For the first 10 m these strings (Fig. 312) run along the ground held in place by periodically being guided through copper tubing attached to vertical stakes. About 3 m from the center of the bow net the string rises 80 or 100 cm into the air and then leads diagonally to the center of the bow net where it disappears briefly into the ground. It is very important to have the string rise because now the lure bird can be 'topped' to the height of 80 or 100 cm in order to make him visible to the arriving raptors. If one stoops the bander pulls the lure bird quickly back to the center of the bow net and releases the net.

The lure birds rarely become the victims of the raptors. At sites 3 and 4 Berger and Mueller raised the string up to 6 m in order to top the doves properly. From this height of 6 m the distance to the center of the bow net is 6.5 m. The buried string reappears behind the poles and leads back to the blind. Perfect operation of the lure bird strings requires cutting "lanes" in the grass. The lure birds are attached to the same swivels as shown in Fig. 316, which are fastened to the underside of the harness. Doves wear a guide for the pull string on their backs. The guide is also fastened to the harness.

The various families and species of raptors do not all react equally to the lure birds. H. C. Mueller specified this:

Accipiters. All accipiters want to see the lure bird move. So do American Kestrels.

Buteos. Top the dove at short intervals. If a Buteo approaches, quickly pull the lure bird to the net; the attack usually happens like lightning. If the hawk grabs the dove in front of the net, the dove and the raptor are generally both lost. This also applies to other raptors.

Northern Harrier (Hen Harrier). These birds prefer starlings in the fall, but in the string they will also attack doves. Move the starling only very little. At Cedar Grove this raptor usually appeared suddenly near the lure bird which had previously been sitting quietly.

Peregrine Falcon. Top the dove vigorously. Peregrines are usually caught in the dho-ghazas; only 1 in 10 or 20 is caught in a bow net.

Raptors must be taken out of the nets as quickly as possible so that new arrivals are not repelled. If a catcher is busy taking a bird out of a net when another raptor arrives, the catcher in the blind gives a short warning call upon which the man at the net freezes, avoiding even the slightest movement regardless of position taken. Many a raptor has thus not noticed the catcher and flown into the net next to him.

Hints on the temporary 'storage' of raptors have been given previously. At Cedar Grove long empty sausage cans are used for Sharp-shinned Hawks. Two empty coffee cans—the bottom of one can having been removed—were tape together with insulating tape for buteos. Sometimes 20 and more raptors were kept lying in semidarkness without their getting excited and ruffling their feathers. Keeping raptors in bags and sacks must be discouraged strongly.

Falconers of all ages have been very solicitous about keeping their newly captured birds. Frederick II (1194-1250) already mentions the 'falcon bag'. This linen bag was cut to size for the various species and it was open at both ends. One opening was only big enough to let the bird's head pass through. The other end had draw strings and the feet, tail, and wing tips remainded outside.

15.3 Raptor catching at Cape May (USA)

Clark (1969, 1970) began catching raptors at this favorable migration spot in 1967. The catching location is at Cape Point, within sight of the lighthouse, on the west side of tilled field approximately 40 m wide (Fig. 321). The blind,

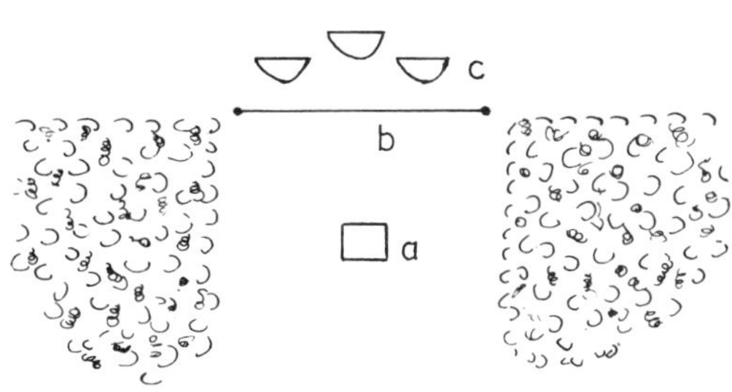

Fig. 321. Raptor trapping area at Cape May, N. J., USA. (a) blind of bander; (b) dho-gazas; (c) three bow nets. After CLARK, 1970.

facing east, is 7-10 m from the edge of the field. The catching set-up consists of one large bow net and two smaller ones left and right. The smaller nets have House Sparrows as lures, the large one a Rock Dove. Catching protocol is essentially the same as at Cedar Grove. The bow nets are released by hand here.

As many raptors are suspicious and only fly low over the lure birds, CLARK set up two large nylon mist nets one above the other between the bow nets and the blind, giving a total height of over 4 m.

The bodies of the lure birds are protected by "leather vests". Thus no doves were killed by raptors in 1968, and only a few House Sparrows.

From 20 Sept. until 1 Nov. 1969, 271 raptors were captured: 157 in the 2 small bow nets, 105 in the nylon nets, and 6 in the large bow net. Almost all Sharp-shinned Hawks (112), American Kestrels (121) and Merlins (16) were captured in the two small bow nets and the nylon nets. CLARK and two cooperators captured 1,152 raptors at two locations in the fall of 1971.

For historical reasons we must mention a raptor trapping method used by IMLER (1937) in Kansas (USA) with which the captured 108 birds of prey. IMLER set up several steel mammal leg traps whose jaws were wrapped heavily with cloth, around a dead rabbit. Strangely enough the birds of prey did not react much to live and strong chickens. As the traps were not under constant observation a small number of birds sustained injuries while trying to free themselves. The use of these traps is out the of question for modern banders trapping raptors.

15.4 Other methods of capturing birds of prey

We have already dealt with historic raptor trapping, especially with bow nets. Some other trapping methods remain to be mentioned. We will not discuss in detail those methods—some of them in use until the beginning of this century—which easily injured birds of prey or even killed them. Hunting literature shows many examples of such raptor traps (i.e. RAESFELD, 1921; ENGELMANN, 1928). The frames of such raptor traps must be large enough and be wrapped thickly with material or rubber on two or three spots so that the frames do not close to hard on each other after release. Only in this way can trapping in such clap bow nets be justified. Even if the raptor has not been injured, but only his primary tips have been broken off by the closing trap, this constitutes an irresponsibility towards the welfare of the bird.

Fig. 322 shows a bow net for catching birds of prey up to Goshawk size. The cage under the net is intended for the lure bird or animal. A sparrow or Greenfinch is sufficient for catching small birds of prey. In setting the net the perch must not point too far upwards so that the closing net frames do not injure the wings of the captured bird. BUB has used this trap outside breeding season (winter half year) to capture interfering raptors (Eurasian Kestrels) for deportation from the vicinity of a banding operation.

K. SCHWAMMBERGER built an ideal raptor trap (Fig. 323) for the Ornithological Research Station in Ludwigsburg (Germany). He built this model because some birds of prey (Goshawks, European Sparrowhawks, Tawny Owls) occa-

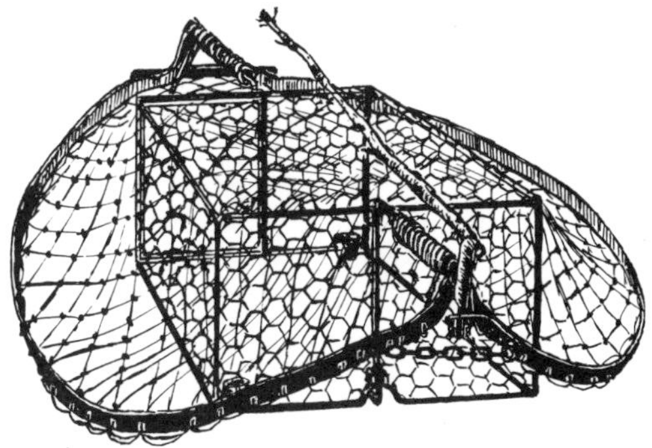

Fig. 322. Raptor bow net with lure bird cage.

Fig. 323. Raptor basket trap with mounted simple bow net. The set bow net slams on the wooden block at left as soon as the bird has triggered the release mechanism. Photo: K. SCHWAMMBERGER.

sionally injured themselves in the regular taps. This raptor trap is very simple. It consists of a frame covered with thin wire mesh. The lure bird is kept in the lower half. A regular bow net corresponding to the size of the basket frame is mounted on top.

The basket is elliptical, 80×90 cm. The false bottom is slightly dome-shaped and triggers the net when weighted down. A sturdy piece of wood catches the closing bow. The bottom is round with a diameter of 40 cm. The catching compartment is 15 cm high, the lure bird compartment 20 cm. A door on one side serves to provide the lure birds with food and water.

In general this trap sits on the ground and does not need to be fastened. The trigger is bent downwards and its length is selected so that it can neither be held back by the bird nor can it hurt it. The iron frames used are 8 mm thick. The springs for the bow net are 2 cm above the rim of the frame so that the bird's wings cannot get caught.

In addition to Goshawks, European Sparrowhawks and buteos, SCHWAMMBERGER also captured Eurasian Kestrels, Black-billed Magpies, European Jays, shrikes, woodpeckers, Tawny Owls, Long-eared Owls, Tengmalm's Owls and Little Owls. Apples, meat, corn cobs and mice served as bait, while doves, European Jays, European Blackbirds, sparrows and Greenfinches were used as lure birds. LÖHRL thinks the Brambling is the most suitable lure bird for bird-catching raptors. Bramblings are even restless during the night, when several together will flutter and even call.

LÖHRL (1962) reports specifically about the capture of European Sparrowhawks. European Blackbirds were used as lure birds only when female Sparrowhawks ignored Bramblings or sparrows. In such instances a female Sparrowhawk usually was caught within a few hours. "European Blackbirds are not suitable as lure birds for longer periods of time as they quickly begin to remain immobile during each alarm or visible approach of a Sparrowhawk. Sparrows also sometimes do this after several catches and must be relased and exchanged for new ones. Most useful by far are the Bramblings as they tend to fly up at the approach of a Sparrowhawk and thus attract its attention."

This trap, which is used in various sizes at Ludwigsburg, should be used more widely. Its manufacture is easy. Besides the basket part a bow net is needed.

16. Capture with nooses

16.1 General

Catching with nooses has been practiced for millenia and has remainded with us until today. The variety of its uses is large. The aim is always to capture the mammal or bird. MACPHER-

SON (1897) recounts an instance which illustrates the adventures possible when catching with nooses. A native hunter in Australia who had set a noose at the nest of a cassowary was killed through a singular accident. The unhappy man fell asleep while awaiting the arrival of the bird. In the meantime the bird came to her nest and laid her egg. Leaving the nest, one of her legs got caught in the noose. The hunter who had tied the rope around his body was dragged along the ground and hurled against a tree trunk which stunned or killed him. The bodies of man and bird—still tied together—were found several days later at a considerable distance from the nest. Normally hunters quickly tied the rope around a nearby tree after the bird's capture and waited until it got tired and was helpless.

The reader wishing to learn more about catching with nooses is referred to various publications by BREHM (1855), LINDBLOM (1925/26), SUNKEL (1927), LIPS (1927), LINDNER (1937, 1940), and LAGERCRANTZ (1938), all of which deal in greater or lesser detail with the development and use of nooses and snares around the world.

Although the need to hunt with nooses and snares may still exist for primitive peoples, it has not existed for Europeans from the last century on. Catching with snares became so unpopular, especially because of the use of lines of snares, that it was finally forbidden altogether in 1908 in Germany. The practice had suffered continuous criticism and attacks in the second half of the 19th century. In Europe many millions of songbirds probably were caught in these lines of snares in past centuries. We must keep in mind, however, that despite these trapping methods one cannot necessarily say that most of the species captured also declined during those times. In many cases individual numbers only began declining when man started to alter the environment to ever increasing degrees and thereby decreased or even eliminated the habitat for many species.

If we now speak again of catching birds with nooses, there is no need to be alarmed. We deal amost entirely with raptors—or else with larger birds—which get caught in these loops by the feet. In addition, capture of these birds should be carried out in such a way that they are held only briefly by the noose. Should a bird get its head caught in a noose, there is no danger to it, as the bander must have the trap in view at all times in order to be able to free any bird immediately. All birds should only receive their official bands, be measured if necessary for studies of sex, age, etc., and be released immediately again to the wild. The ornithologists themselves have the greatest interest in returning the birds safe and sound to the wild.

16.2 The East Indian raptor trap or bal-chatri

East Indian falconers have long captured their birds with a trap which MACPHERSON (1897) called the "shikra" trap. CRAIGHEAD (1942) called it the "bal-chatri". The Shikra, an accipter, was the most commonly trained hawk in In-

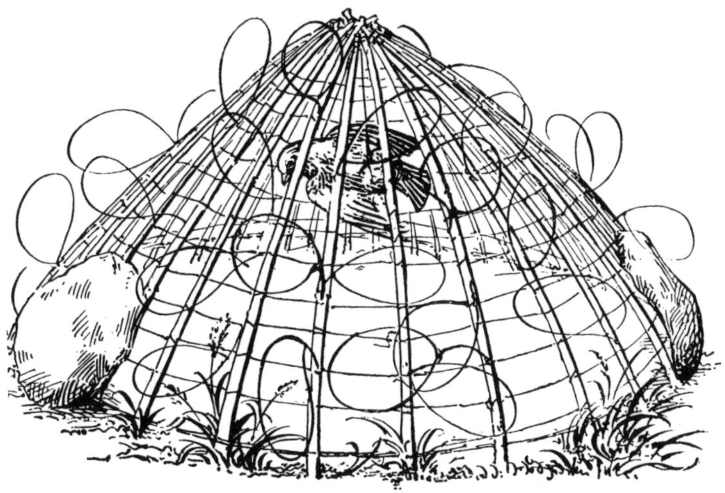

Fig. 324. East Indian Goshawk trap (Shikra trap), an ancient form of the bal-chatri trap. After MACPHERSON, 1897.

dia in former times. Fig. 324 clearly shows the original trap. It consists of about 15 thin split cane sticks, each about 45-50 cm long. These sticks are spaced evenly and are tied on top at the center so that they form a conical basket. The individual sticks are tightly connected with strings. Two wet lumps of clay on each side weigh the trap down. The nooses measure approximately 8 cm in diameter when open and are made of strands of black horse hair. Inside the trap, one or two lure birds serve as bait. The catcher hides close by.

American ornithologists Frances Hamerstrom, D. D. Berger and H. C. Mueller (1959, 1962) began catching raptors in this manner in 1953 and have the honor of having tried it in an altered form and of having introduced it to scientific bird banding. The bal-chatri provides a way of catching raptors which is ideal for a number of species. The traps are easy to make and inexpensive. Injuries during trapping are almost impossible.

The main attraction to the raptor is the lure—either bird or small mammal—which, in its small cage, has only limited room for movement.

The authors mentioned above used these traps at regular catching locations as well as during trips through the countryside. During the trapping of gallinaceous birds during the winter semester, it was necessary to trap raptors in the area and to deport them as they would otherwise have attacked the birds in the traps.

Description. The bal-chatri is a cage made of wire mesh having nylon nooses on top. Berger and Mueller first used three sizes which are listed below. Of course, sizes can be varied, as shown in Fig. 325. Lately the trap shown in Fig. 326 has been widely used.

Type 1. This trap is a cylinder, 15 cm in diameter and 8 cm high. It is made of wire mesh or hardware cloth with 8 mm wide mesh. A little door on the bottom permits the lure animal to be put in. The "roof" has about 40 nooses of 4 cm diameter. An iron ring is attached to the bottom with wire. This ring increases the total weight to 230 g.

Type 2. This type is the same size as type 1, but it receives a heavier metal ring which increases the weight to 750 g. On top are 35 nooses of heavier nylon. The diameter of the opened nooses remains at 4 cm.

Type 3. This model is a half-cylinder, 30×25×15 cm. The roof and sides are made of 12 mm wire mesh. The bottom is sturdy hardware cloth. This trap can have several hundred nooses of 4 cm diameter. The bottom of the trap also has an iron ring to which the wire mesh has

Fig. 325. Four different sizes of the bal-chatri trap for raptors. Photo: H. Bub.

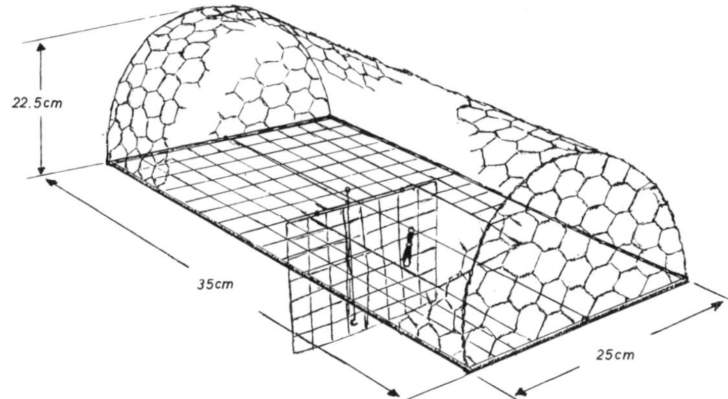

Fig. 326. Raptor trap without nooses. After BERGER & HAMERSTROM, 1962

been attached. The total weight should not be more than 700 g.

Chicken wire should be used for the larger traps (possibly also for the smaller ones) as raptors are less wary of it than of hardware cloth. BERGER and MUELLER noticed this especially for accipiters.

For raptors up to the size of a Great-horned Owl, BERGER and HAMERSTROM used a slightly larger trap (Fig. 326) and for eagles a semi-cylindrical one 50×45×38 cm. The mesh of the strong wire is 25 mm; the door of this trap is on one side. It is very important to weigh the trap down by additional pieces of iron or iron rings. For the smaller traps it also serves the purpose of making the trap land bottom down when being thrown from a slow-moving car. For the larger traps they serve solely as weights. The total weights mentioned are large enough so that a raptor may be able to drag the trap for a very short distance but cannot escape with it. If the trap is too heavy, the nooses can be torn by the bird taking wing.

The nooses. It is necessary for catching success to make the nooses large enough. For birds the size of Great-horned or Eagle Owls and Snowy Owls the diameter should be 11 cm; for the smallest raptors 4 cm; for accipiters, buteos and harriers 8-10 cm. The trap top should be thickly covered with them. The nooses can be knotted in a variety of ways. BERGER and MUELLER (1959) show a single noose (Fig. 327). As soon as each noose is loosely attached to the trap, a pencil is inserted in the noose and pulled strongly away from the trap. This procedure fastens the attaching knot firmly. The noose tied to the pencil is then opened again and arranged so that it stands up as straight as possible. A drop of epoxy holds the attaching knot in place and makes the noose stand upright. A # 3 knitting needle makes things easier, especially when nooses have to be opened again after a catch.

The trap hardly needs care. Attention should be paid that nothing lies on top of the nooses. From time to time they have to be arranged anew. After a number of catches the nooses may tangle or break. They must then be renewed.

BERGER and HAMERSTROM (1962) used a noose-pair (Fig. 328). They developed a method of making them quickly in large quantities.

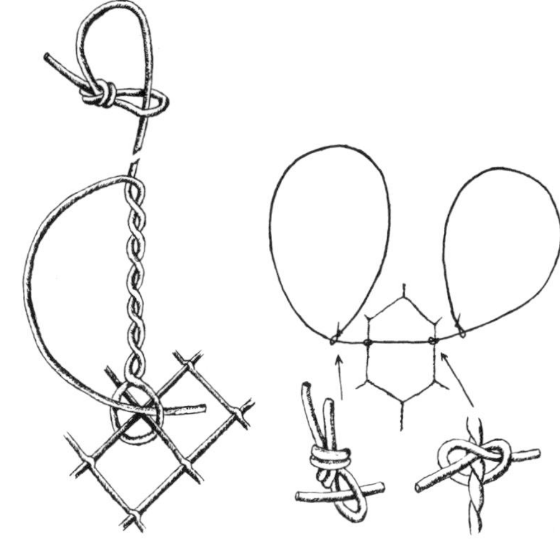

Fig. 327. A single noose for catching raptors. After BERGER & MUELLER, 1959.

Fig. 328. A set of 2 nooses and their method of attachment to the trap. After BERGER & HAMERSTROM, 1962.

They suggest the following procedure: Double the end of a long piece of nylon monofilament line and make a knot, but do not tighten it. Slip the knot on a # 3 knitting needle and tighten the knot so that the small loop has the diameter of the needle. About 54 cm from this knot, repeat the procedure, then 8 cm further on again. Repeat this about 40 times, alternating the distances between 54 and 8 cm. Leave all knots on the knitting needle. Finally pull on the lines forming the larger distances and immerse the whole for a moment in boiling water. This will set the knots. Cut all 8 cm segments and take all pieces with a small loop at each end off the needle. Slip each of the 2 small loops through the other so that each piece of string has 2 nooses. These double nooses are fastened with 2 knots on top of the bal-chatri, opend and erected. The trap is ready for use with approximately 40 nooses.

The strength of the nylon monofilament is important. It is classified according to the weight it can carry. Weaker line should not be used. An alert bander who can quickly reach the catching location can use a 7 kg nylon line.

	Weight of nylon line			
Eagles, Snowy Owls, Eagle Owls	20	to	23	kg
Accipiters, buteos	13	to	18	kg
Falcons, small accipiters	4.5	to	6.5	kg
Small falcons	3	to	4	kg

Lure animals. Berger and Hamerstrom (1962) decided that Rock Doves and European Starlings are the best lure birds for the larger traps. Starlings were very active. They must be watched so that they do not pull nooses into the cage as they can strangle themselves on them. It is best to protect them with another (false) roof. In winter starlings must be exchanged often. Rock Doves are more resistant and can remain outside 24 hours a day. Berger and Hamerstrom changed birds twice a day. It is important to keep the lure birds strong and in good health. Starlings need dark corners in the cage in which they live so that they can hide. They want bath and drinking water and can be kept well on dry, protein-rich dog food.

Berger and Mueller (1959) used mice for the small traps. The Common House Mouse *(Mus musculus)* was the best. Other kinds often do not attract the attention of raptors, though fieldmice (*Microtus* sp.) can be used with some success. House mice were very active and therefore the best attractant. They did not cease activity upon seeing a raptor and they are easy to care for in captivity. Some seemed even to be getting better after some hunting experience, contrary to other mice. H. F. Arentsen in the Netherlands had good experience with the Wood Mouse *(Apodemus sylvaticus)*.

Berger and Mueller (1959) keep most kinds of mice during the catching season in large metal cans with holes in the cover. The can is filled two-thirds with dry shredded paper or coarse wood shavings. The mice are fed corn or commercial mouse food (available in pet stores). Every 2nd or 3rd day some greenery is added. They do not get water. The can must be cleaned regularly and the shredded paper or wood shavings must be replaced.

House Sparrows can also be used. Some learn quickly to remain quiet when a raptor appears. They are then unsuitable for further use as lure birds.

Catching. The trap can be set up in any location where raptors are expected. It must be checked regularly. All authors mentioned here always carry traps and mice along in their cars and attempt a capture as soon as they see a raptor. Falcons, and particularly American Kestrels, usually sit on wires and poles. If a bird has been spotted, drive to the shoulder and decrease the car speed to about 10 mph and drop the trap out of the door opposite the bird while driving by. The door should only be opened as far as necessary to drop the trap and ascertain that it landed right side up. The door remains open until the car is far enough away from the bird. Each movement with the door or inside the moving car can cause the bird to fly away. Sometimes he does so anyway. In such cases it is best to approach from far away, put down the trap and retreat inconspicuously. The car should be at least 60 to 100 m from the trap. To catch buteos an even greater distance is necessary.

Some birds react immediately or within a few minutes. A capture is not always immediately noticeable. If the road is somewhat busy, the first passing car answers this question by causing the captured bird to flutter. The captured animal should be approached from the direction of the pull of the nooses.

Resetting the nooses is much easier if the bird is freed quickly. When catching in areas where birds are numerous, BERGER and MUELLER set a carton upside down over the trap so that the nooses do not get squashed and the trap is ready immediately. The mouse can remain in the darkened trap, and is then also more active. If trapping is poor, the lure animal is put into the trap only upon sighting a bird.

BERGER and MUELLER caught hundreds of American Kestrels, once 22 in one day. During two North-South trips through the USA—the first one 13 days long in January 1955, the second one 37 days in December 1955/January 1956—227 American Kestrels were banded. During the first trip 45.8 % of all birds which had seen the trap were captured, and during the second trip the total rose to 57.8 %. The total of birds observed was 1,258 of which 16.7 % were captured. Every other attempt at catching was successful. But success was not equal in all locations. On January 7, 1955 in a semi-arid stretch in Texas, 22 of 25 birds were caught. In summer sometimes entire families could be trapped in the northern U. S. Good catching results can also be obtained with Harris' Hawk, which often appears in groups of 3 to 6 individuals. BERGER and MUELLER constructed a trap specifically for this species so that 2-3 birds can be captured simultaneously.

BERGER and HAMERSTROM (1962) provide an overview of raptors captured in bal-chatri traps from 1959-1961:

Species	Road catches	Other catches
Northern Goshawk	-	6
Sharp-shinned Hawk	-	1
Cooper's Hawk	1	22
Red-tailed Hawk	30	56
Red-shouldered Hawk	4	5
Broad-winged Hawk	5	7
Swainson's Hawk	2	-
Rough-legged Hawk	8	10
Northern (Hen) Harrier	2	78
Prairie Falcon	1	-
Great Horned Owl	3	6
Snowy Owl	52	2
Barred Owl	1	2
Short-eared Owl	-	5
Total	109	200

No American Kestrel was mentioned because the trap used was too large for this species. Remarkable is the large number of Northern (Hen) Harriers captured by F. HAMERSTROM during migration. Merlins and Peregrine Falcons should be added to the long list of raptors captured with bal-chatri traps.

Fig. 329 A Snowy Owl captured with a raptor bal-chatri trap. A lure dove is in the cage of the trap. Photo: T. MEICKLEJOHN.

Fig. 330. Of 69 Snowy Owls observed in the State of Wisconsin (USA) in the winter of 1960/61 54 were captured and banded. With rich booty from left to right: F. HAMERSTROM, N. MUELLER, D. D. BERGER, C. SCHACHTER, E. HAMERSTROM, H. C. MUELLER. Photo: O. PETERSEN.

Most species of owls can probably be captured. F. HAMERSTROM mentions Barn Owls, Hawk Owls, Long- and Short-eared Owls, and Barred Owls as well as Great-horned Owls. BERGER and MUELLER usually catch Poreal Owls in mist nets. Even eagles can be captured, but this is sometimes difficult as the traps cannot be left alone for even an instant, thus often requiring much time.

The crowning glory of their work for the authors mentioned came in the winter of 1960/1961. Of 69 Snowy Owls observed in the state of Wisconsin they captured 54, albeit after much exertion and persistence (Fig. 329 and 330).

The usefulness of this trapping method needs no further proof.

In the USA catching with bal-chatri traps is widespread and the literature mentions time and again various experiences (see WHITMAN, 1962; CLARK 1969, 1970). In the fall of 1968 in the late afternoon more than 50 American Kestrels were sighted in a field near Cape May, N. J. Quickly 6 traps with mice were set. The success was phenomenal. In the 1 ½ hours until darkness, 22 birds were captured, and three traps caught two kestrels simultaneously.

Meanwhile, more experience has been gathered on other continents as well. Australian banders (LANE, 1966) adopted this method after raptor trapping with mist nets proved unsatisfactory. They fitted the trap bottom with metal strips to make it heavier and to prevent the trap from turning over. Manufacture followed American models. The Australian banders used white mice as lures, but these will probably not meet all trapping requirements.

SIEGFRIED and BROEKHUYSEN (1971) used this method in South Africa to capture the palearctic Common Buzzard. The banders constructed the trap in the form of a conical hat as birds get caught equally well no matter whether they try to strike from the top or from the side. Agouti-colored house mice were used as bait. These banders investigated in exemplary manner the behavior of the Common Buzzard towards the baited traps—probably a first for this type of trap.

16.3 Catching other birds with nooses

16.3.1 Catching of shrikes

The American ornithologist CLARK (1967) developed a combination of bal-chatri and noose carpet for the capture of Loggerhead Shrikes. A regular bal-chatri trap was not quite right as the shrikes grab and kill their prey with the bill and only then carry it away. A shrike therefore usually does not land on the top of a trap where the nooses are, but rather beside it. The shrike then walks around the trap trying to get at the lure bird. As this is not possible, most shrikes fly away unbanded from a conventional bal-chatri trap.

The 15 cm long and 8-10 cm high cage for the lure animal is fastened onto a flat sheet of narrow-meshed hardware cloth aproximately 30×30 cm. Clark gives a height of only 5-7 cm. Under the cage, a door 6×8 cm is cut into the sheet: the edges of this opening must be filed smooth. The base door is also made of hardware cloth but must be somewhat bigger than the opening itself. The hinges and closure are made of soft steel wire. Clark uses two 120 g weights similar to those used to balance tires and fastens them to the inside of the cage with wire. This makes the trap too heavy for the shrike to lift. The entire trap gets a flat black coat of paint. Once this coat has dried, parts of it are painted dull green and red giving it excellent camouflage.

Clark uses nylon angler's line of 4 kg weight for the nooses. The end of a strand pulled from a roll is knotted into a small loop around the end of a crochet hook. Then the strand is cut 20 cm further on. The free end is now threaded through the small loop to make the noose that catches. The noose must stand up and be open. Next the nooses are fastened to the trap with triple knots. Nooses must stand upright, as lying flat they cannot catch. Clark secures the knots against opening with a drop of glue. The diameter of the nooses is about 3.5 to 5 cm. 20 nooses on the top and 20 on the base sheet are enough for this type of trap.

Wild house mice are the best bait according to Clark, but tame brown or black mice tire less quickly. White mice are not particularly suitable as they are often rejected as bait.

K. Hahne of Bremen (Germany) baited with mealworms and beetles in summer and with newly caught European Siskins. Shrikes attacked every trap occupied by a siskin but mostly ignored House Sparrows and European Tree Sparrows.

16.3.2 Catching of Prairie Chickens

F. Hamerstrom also catches *gallinaceous* birds, i.e. Prairie Chickens, with noose carpets which should be at least 30×30 cm. It is useful to put down two carpets with a small path between them and to fasten them down on all 4 corners with wire hooks. These hooks should not be tightly in the ground as the captured bird is supposed to tear them out when it attempts to leave after capture. Both carpets are tied to a strong string with a weight attached to the other end so that the captured bird cannot escape. When catching Prairie Chickens, F. Hamerstrom sets a stuffed hen between the carpets before the cocks appear (Fig. 331).

Fig. 331. Double noose "carpet" for catching prairie chickens. After F. Hamerstrom.

Matschie (1897) reported on the capture of the Red Junglefowl on Malaka (Southeast Asia). Chickens and doves are captured with a snare which has been fastened to a long bamboo pole or switch. The concealed hunter quickly slips this snare over the bird attracted by bait. The Malakans are very adept at this catching technique. A boy made a simple snare from cotton string, attached it to a long pole, picked some berries, put them under the tree and hid behind it. He held the pole over the berries as if it were an angler's rod. As soon as the doves alighted to feed, the boy skillfully slipped the noose over the head of one dove and pulled it from the midst of the flock that the other birds were not in the least disturbed. He was able to catch several birds in this manner before the flock flew away.

This method is used often to catch the Red Junglefowl. A tame cock is tied to a pole with jesses and string. The bird is used to this and starts calling as soon as it is on location. When a wild bird answers, the 'Sakai' hides and waits

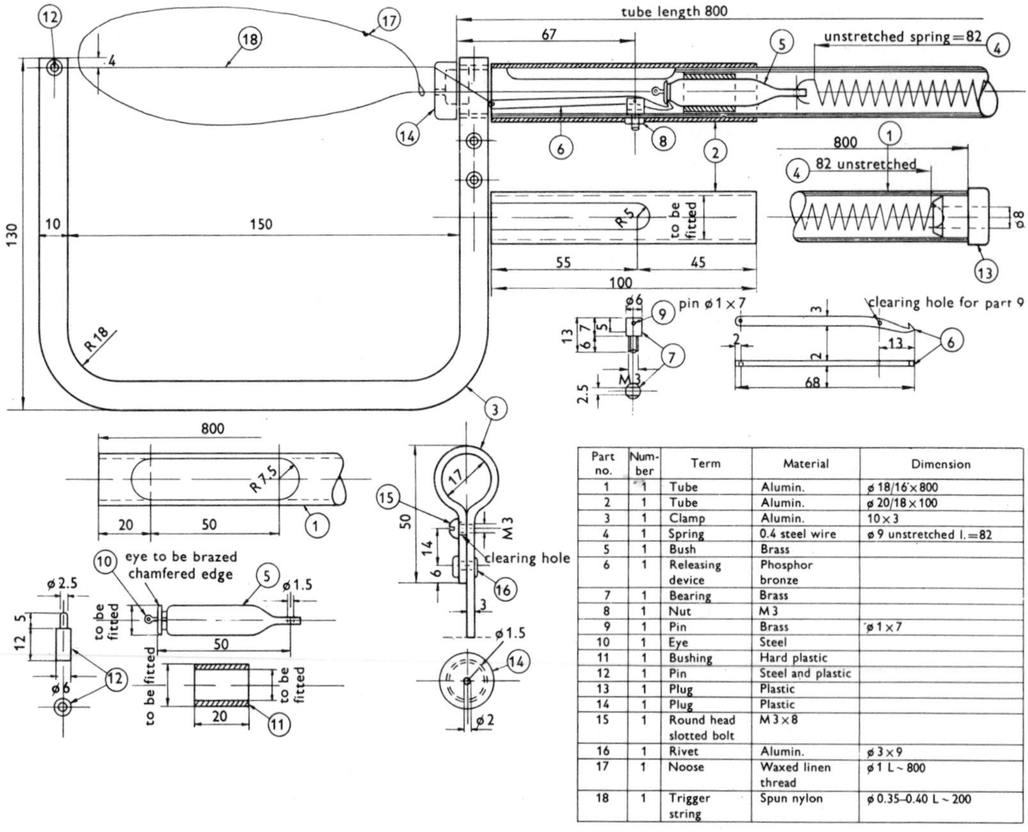

Part no.	Number	Term	Material	Dimension
1	1	Tube	Alumin.	⌀ 18/16×800
2	1	Tube	Alumin.	⌀ 20/18×100
3	1	Clamp	Alumin.	10×3
4	1	Spring	0.4 steel wire	⌀ 9 unstretched l.=82
5	1	Bush	Brass	
6	1	Releasing device	Phosphor bronze	
7	1	Bearing	Brass	
8	1	Nut	M 3	
9	1	Pin	Brass	⌀ 1×7
10	1	Eye	Steel	
11	1	Bushing	Hard plastic	
12	1	Pin	Steel and plastic	
13	1	Plug	Plastic	
14	1	Plug	Plastic	
15	1	Round head slotted bolt	M 3×8	
16	1	Rivet	Alumin.	⌀ 3×9
17	1	Noose	Waxed linen thread	⌀ 1 L~800
18	1	Trigger string	Spun nylon	⌀ 0.35–0.40 L~200

Fig. 332. Construction details for a noose trap for Willow Grouse (Ptarmigan). Measurements are in millimeters. From N. H. HÖGLUND, 1968.

until the jungle cock attacks the supposed rival to drive him from his territory. As soon as the birds are fighting, the hunter catches the unwary fighting bird in his snare.

To catch ground doves and gallinaceous birds as well as the Great Argus Pheasant, a lure bird is tied with one leg to a peg on a smooth lawn. A long rope made from bark is laid around the bird in spiral form. Thin nooses, made from fibers of pineapple leaves or other material, are fastened to this rope at short intervals. Much corn is strewn between these nooses. The birds get caught by stepping into a noose which tightens around the leg as soon as the birds try to walk on. The Great Argus Pheasant is sometimes caught in this manner without lure birds.

HÖGLUND (1968) of Sweden describes a method to catch ptarmigan (*Lagopus* sp.) in winter. Marking of unfledged birds was not enough because of the high mortality rate among young birds. Therefore it became necessary to develop a trapping method for older birds capable of flight. After a few years of trial and error a trap was developed which makes it possible to catch ptarmigan in winter. The trap consists of a noose attached to a spring concealed in a tube. The noose is tightened around the bird's leg as soon as it touches the trigger (Fig. 332-334).

This noose trap is set up in a space between bushes, which are often artificially arranged (Fig. 335). The twigs in the snow should be fresh and have a lot of buds to tempt the ptarmigan to eat. It is useful to cut down a few small birches and to place 2 or 3 traps in the vicinity. The traps should be set in a way that blowing snow does not cover them. During periods of sunshine water sometimes condenses in the tube so that the trigger mechanism may be frozen in ice during a cold night.

The traps must be patrolled at least 3-4 times

Fig. 333. The trap is set in a "door" in a hedge made of birch twigs. The noose lies over the trigger string but is covered lightly with snow. Photo: N. H. HÖGLUND.

daily, the last time after darkness has set in. Catching with these traps was quite successful. From January 24 to March 25, 1967, 69 ptarmigan were captured in 64 traps set in 32 'fences'. Thirteen were recaptures.

HÖGLUND puts the captured birds into a 20×60 cm bag of dark cloth and ties it with a string. Near the end of the bag is a zippered opening. The birds are quiet in this bag. The birds' wings can be pulled through the opening for measuring and marking (Fig. 336).

The ptarmigan are marked with wing markers and, in addition, receive a color pattern on their primaries (Fig. 337). Color and pattern must be varied carefully for each bird. These markings are easily recognizable in the flying bird, but only partially or not at all in the sitting bird.

The later recovery of molted primaries is important—the primary molt takes place between July and September. They indicate summering and molting grounds. As the molt progresses in sequence, at times the exact date can be pinpointed. The paint is so durable that it can often be determined even after 2 or 3 years.

17. Catching by hand and with dip nets on land and in water

17.1 Catching by hand

Such catches are by no means always accidental. But they do happen accidentally when birds get into houses or otherwise into trouble. BREHM (1855) already talks about Black Redstarts which are sometimes caught inside churches.

Fig. 334. Captured Willow Grouse (Ptarmigan): behind the bird is the trap. Photo: N. H. HÖGLUND.

Fig. 335. Catching area with two traps (left and right). Photo: N. H. Höglund.

Shorebirds and Gulls. H. Behmann captured a phalarope by hand. He needed one hour to accomplish this. K. Kliebe managed three times to catch a Jack Snipe by hand. It is a well known fact that Common Dotterels can also be captured by hand. We are not talking about behavior on the nest as observed by Bengt Berg (1925) and others but behavior while on migration. Prof. Drost told of a bird that he alone encircled and then covered with his hat. In such as case it is important to crowd the bird more and more by walking around it in continuously smaller circles. P. Becker approached a Common Dotterel with a mealworm in his left hand and caught the bird in his right hand.

Better known is the method of catching wintering Black-headed Gulls by hand. The birds are fed on ledges, window sills, buildings or even the ground. After a while they will lose their caution and come close so that one can quickly grab them. Some gulls take bread offered from the hand while flying by and can thus be captured at times by a quick grab. In Prague (CSR), in one winter alone, 500 Black-

Fig. 336. Measuring the wing of a Willow Grouse (Ptarmigan). The method of measurement and the support for the wing are worthy of attention. Photo: N. H. Höglund.

Fig. 337. Color coding on the inner vanes of the primaries of a Willow Grouse (Ptarmigan). Photo: N. H. Höglund.

headed Gulls were captured by putting pieces of meat on bridge piers and making a quick grab.

It is not usually known that some European species (i.e. Common Redshank, Common Oystercatcher, Black-tailed Godwit, Herring Gull) may be petted while sitting on the nest. But they should not be taken off the nest as the brood would be endangered.

Rails. Near Braunschweig (Germany) R. Berndt tried out the following method when catching Common Coots. When a lake thickly populated by coots suddenly froze over the birds sought shelter in the adjoining reeds when disturbed by humans. There, if discovered at all, they let themselves be captured easily by hand (7 birds in 15 minutes). The others disappeared during the following night.

Rails often behave curiously. Many a Water Rail and Corncrake has been captured by hand between the rocks or in doorways on the island of Helgoland (Germany).

An alert observer will be able to catch several of these birds—including the Spotted Crake—by hand in the field during excursions. C. L. Brehm (1855) knew this already: "Many times one can catch Corncrakes and Water Rails by hand. The former often hide under a patch of grass or corn and can be captured by hand. This becomes even easier during the harvest as they lose all primaries at once. This applies to all rails."

"Rails of the genus *Porzana* are captured similarly to the Corncrake with nets or sometimes by hand during hay-harvest."

"When Water Rails have left a secure place, they lose all common sense. I possess a stuffed bird which, flushed from a marsh by a shot, fell down in a field and was captured by hand—despite the fact that is was completely uninjured."

"Common Gallinules are also often hunted with dogs which catch them in their hiding places. I had a trained dog which caught not only young but also old birds with great skill. Gallinules are also often captured by hand after being driven from ponds onto meadows or into brush. I have obtained several this way. One was captured by hand in a wooden shed, the other in a well house, the third at the wall of a barn, the fourth in a well, and so on."

According to Schifferli (1936/37) Common Coots were caught by hand when being fed at a dock in Zürich (Switzerland).

W. Prünte describes catching Common Gallinules in Westphalia (Germany). One person walks alon each side of a stream keeping a close watch on the gallinules. As soon as a bird hides in the high grass or other vegetation because it feels threatened, the companion is given detailed instructions on where it is. Prünte and friends have captured up to 30 gallinules in one day—a considerable number—and have caught a total of about 600 to-date along with a few Common Coots, Water Rails and Spotted Crakes. A short dip net comes in handy.

Raptors. Tschudi (1860) writes that Golden Eagles become less alert after a good meal. Macpherson (1897) had reports from Scotland that shepherds occasionally captured White-tailed Sea Eagles which had stuffed themselves to the very limit with mutton. Once one of two fighting Golden Eagles was caught with the aid of a dog.

A report from Siberia given to Macpherson must probably be listed as a fairy tale. The traveler heard from local hunters that they were easily able to capture Bearded Vultures (Lammergeier) with stones of a size easily swallowed by the bird. A number of such stones were coated with blood and thrown in a heap where the vultures occured. The clotted blood on the stones induced the vultures to swallow the stones. But the heavy meal gave the birds such indigestion that they were unwilling to fly.

SCHENK (1952) reports from Hungary that a farmer hit two buteos with a whip so well that neither could fly away. The flight feathers of two others froze together after a rain.

Other species. According to VÖMEL (1938) a migrating Common Crane flew against a high mining company building whose lights had blinded the bird. It slid down along a wall and was captured. The catcher, having grabbed the crane by the neck, ran along with the bird until he had entirely overpowered him. The bird was banded and released.

MACPHERSON (1897) reports on the hunt for the Common (Brown) Kiwi with dogs wearing muzzles. The kiwi is a night bird and spends the daylight hours in his hiding places. W. BULLER and his hunters were able to catch kiwis of all ages in a one-week expedition. After being 'run down' by the dogs, the birds were caught by hand, but not without defending themselves against capture with their sharp claws.

SCHÜZ (1957) acquaints us with a rather barbaric catching method from Persia. In winter Caspian Snowcocks are hunted in the Taligan valley of the Elbrus Mountains. These birds like places swept free of snow by the wind. In a large area a hunter stands at each one of these places on a typical hunting day. He chases the birds away as soon as they arrive. They finally get so very tired that they hide under the snow and can then easily be grabbed by hand.

SUNKEL (1938) reports that a pupil caught a Common (River) Kingfisher because his wing feathers had frozen together. A number of Common Starlings crash-landed after a winter bath and were captured by hand—although birds bathing in winter usually do not lose their ability to fly. GROSS (1924) writes that another Common Kingfisher landed on the fishing rod of an angler sitting quietly and was caught. The perfectly healthy bird may have been a youngster.

Generally it is impossible to catch wild songbirds by hand. However there are instances of birds which are migrating from unpopulated areas and have little or no fear of man. Dozens of Goldcrests have been captured by hand on one day on Helgoland (Germany). Their trust is so great that they will sit on the head or shoulders of people.—Knowledgeable inhabitants of Innsbruck (Austria) catch Yellow-billed Choughs at the windows of their apartments.

17.2 Catching with a hand net

Aided by a dip net there are other specific methods of bird catching. Most dip nets have oval or round frames. The diameter varies but should be at least 30–40 cm for small and medium-sized birds. The light-weight pole is between one and several meters long. The mesh size and strength of the net depends upon the birds to be captured. The depth of the net bag is about 50 cm or more. When tilted, the net must have enough room for the bird, in other words, the net bag must hang over the frame in sufficient length. In Southern Hessen (Germany) K. HILLERICH captured more than 30 Tawny Owls living in common white willows with an oversized 'butterfly net'. About 10 birds escaped out of the dip net or before being caught at all.

During high snows in winter Canadian ornithologists catch Northern Hawk Owls in this manner. They put a mouse on the snow close to a bird sitting in a tree. The hungry owl often comes right away and can be captured with the long-handled net.

Even the capture of trusting shorebirds is possible with a hand net. One British bander caught a Sanderling with the type of hand net illustrated in Fig. 338. Ancient Egyptians used hand nets to catch quail (ROSER, 1960).

According to PFANNENSCHMID (1882), on March 4, 1882 he succeeded in catching six Snow Buntings out of a large flock with a hand net. The capture of Goldcrests and Firecrests, *Phylloscopus* warblers, tits and others is of course possible if they are particularly tame.

The dip net is very useful in the capture of just fledged birds which leave the nest early.

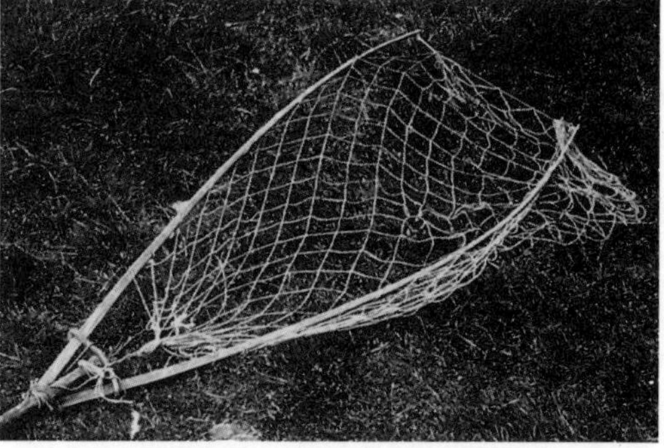

Fig. 338. This type of hand net has only a shallow net. Photo: S. MATHIASSON.

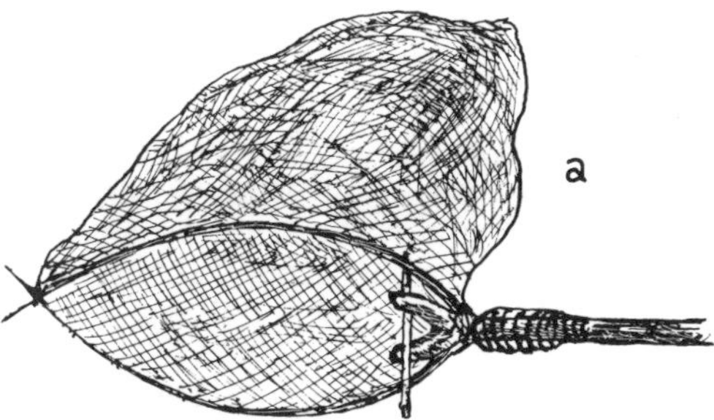

Fig. 339. Catching the Pacific Fruit Dove on Samoa with lure dove and dip net (a). After KRÄMER, 1903; from ANELL, 1960.

Such birds are not capable of sustained flight. In overgrown areas the catcher needs to watch carefully where the bird lands.

An interesting method of capturing Pacific Pigeons was practiced until the beginning of this century on the islands of Micronesia such as Samoa (KRÄMER 1903, 1906). This catching tradition was slow to die then and may have been revived since. Catching was done with a net held by two spreading switches attached to a stick (Fig. 339). A cross piece like that used with fish nets held the net in place. The net pole often was 3-4 m long.

Generally catching was done from the ground (Fig. 339), but on Apo also from trees. The trapping season began in June and lasted until October. Catching was done in specific locations which were cleared of trees, on the borders of which hunters erected their leaf blinds. These blinds were a shelter against sun and rains as well as hiding places for the catchers, who in addition had covered themselves thoroughly with vines in order to resemble a bush as much as possible. Each catcher had a tame pigeon which was attached to a thin string by one leg. Upon a signal from the leader of the hunting expedition, all hunters let their pigeons fly, higher than the tops of the surrounding trees. Noticing this flock, the wild birds suspected a feeding place and approached. This caused a conflict between the tame and wild pigeons; the hunters now quickly pulled their tame birds down. The wild ones followed in the heat of the fight and were captured with the dip nets. This required great skill. But we would like to remark that we do not believe in the theory of a fight between the pigeons tied to the strings and the wild ones. Ethologically as well as from the point of view of animal psychology, it is likely that the tame and wild pigeons formed one flock that was slowly brought close to the ground by pulling down the strings tied to the tame pigeons. A parallel is the behavior and reaction to live lure birds at a double clap net.

In former times this sport of pigeon catching, seuga lupe, was the most noble occupation of the Samoans, and the tribal chiefs were especially devoted to it. Often they spent many weeks without interruption in the bush and could not be persuaded by anything to restrain their enthusiasm. The story of Tupuivao proves the point. He was excluded from succession to the crown because of his excessive passion for this sport.

The pigeons were not caught for food. Mostly they were trained as new lure birds. A great host of words applies solely to this dove and its capture.

Krämer (1903) writes: "Unfortunately this beautiful sport has almost entirely disappeared, through the introduction of firearms which have decimated the doves, and also through the influence of the missionaries who considered this innocent game damnable because it kept the newly converted from church too much."

17.3 Catching by hand and dip nets in water

In the area around Braunschweig (Germany) H. Wehfer and several friends tried to catch Little Grebes on small streams with good success. The creek or stream should not be wider than 4-6 m. If the stream is wider it becomes impossible to catch the Little Grebes. Narrow tributaries into which the birds can be driven are perfect. Wehfer and his friends constructed large triangular dip nets with sides of 40-60 cm long. The pole was 1-2 m long, the net bag about 40 cm deep. The net must be light so that it can be handled easily and quickly.

The actual catching proceeds as follows: at least two, preferably four people slowly walk along the banks of the stream, two on each side. Never go downstream as you may soon lose the birds from view as they move faster when swimming with the current. Of the four persons, two walk somewhat behind the others in order to spot birds that are "breaking through". If a bird is sighted it is necessary to determine in which direction it dives. This can be inferred from a "bow wave", or sometimes from air bubbles.

Fig. 340. Duck catching in Hungary. After Anell, 1960.

Fig. 341. Duck catching in China. After ANELL, 1960; THIERSANT, 1872.

The bird soon approaches the shore as it is exhausted after several dives. Often careful observation is needed in order to find it again. The catchers must communicate with each other. The diving bird must quickly be covered with a dip net, or if it tries to dive away, the net must be held quickly in its diving path.

An experienced group of catchers can, with practice, expect to catch 60% of the Little Grebes in residence. The catching season is from fall until spring as long as migrating or winter birds are in the area. Catching is said to be best before cold spells. MACPHERSON reports that the bird catchers of the Wagri (Africa) run after larger rails in the reeds. The birds generally dive and are grabbed as soon as they rise. As already mentioned rails can also be sniffed out by dogs.

Grabbing ducks and other swimming birds by the feet is a method that was used widely on all five continents and may still be found occasionally today (THIERSANT 1872, QUINET 1904, ANELL 1960 etc.) The hunters sometimes use hollow pumpkins or similar fruits into which they cut openings for the eyes (Fig. 340). ANELL (1960) shows a hunter with a "hat" in Fig. 341.

SCHÜZ (1957) reports that, in the region of the Caspian Sea, hunters approach the ducks in the water after the cotton harvest and before the first cold spell, pulling the birds into the depth by the feet. A pumpkin with holes for the eyes serves as camouflage. This same hunting method is said to have been used in Mexico as well. Six or more ducks were reportedly captured during an expedition. Keeping the ducks alive for banding purposes is difficult to imagine without having to give up further catches after the first one. Perhaps they could be put into a wooden box which the catcher pulls after himself and which is fairly inconspicuous at dusk. Nonetheless it seems a difficult enterprise.

MACPHERSON (1897) describes the clever method of Australian aborigines for catching cormorants. The men set up poles in the water where no natural resting places were available for the birds. When the birds alighted on these poles, the catchers swam to the unsuspecting birds and grabbed them.

17.4 Catching sea birds from ships

17.4.1 Catching by hand, with dip net or net

ELIOT (1933) described a successful catching expedition along the North American coast by W. E. D. SCOTT, a prominent ornithologist of the last century, in 1881. SCOTT left port early one morning in August with one fisherman. After two hours sailing, land was out of sight and they had reached the catching location. There the fisherman caught several codfish, cut out the livers, and threw them into the water. Immediately the water was covered with an oil layer which the current carried away, thereby attracting sea birds. Storm Petrels, Shearwaters and Skuas circled the boat in great numbers. Now the fisherman took a piece of string, fastened a large piece of cod liver to it and dragged it behind the boat. As soon as the birds had discovered the liver they hovered over it like flies. Slowly, the string was pulled in until it was no more than 1 m from the boat. Now the fisherman took an old crab net on a long handle and caught birds as other people catch butterflies. He put the captured birds on the deck of the boat where they waddled around helplessly, not being able to take off from level ground. With one sweep of his net the fisherman caught nine of the smaller birds at once. Eliot (1933) reports:

"'Well, what could be done in 1881 can be done in 1932. I doubt if any one has ever banded these sea-birds. What fun to be the first!' So early one August morning eight of us, all more or less seasoned ornithologists, set out in a fisherman's launch and chugged steadily out to sea. Wilson's Storm Petrels *(Oceanites oceanicus)* flitted by us more and more commonly, and as the land sank behind us to a mere sandbar, a Cory's Shearwater was seen. Several came within our purview, but their usually more numerous cousins, the (smaller) Greater Shearwaters, made no appearances, and the Sooty Shearwater only an unsatisfactory one, at a great distance. Eager for cod-livers wherewith to tempt them nearer, most of us started fishing, and by-and-by I was crumbling the oily stuff over the side with mounting excitement: for the little black chippering petrels found it at once and began flying fearlessly within easy reach of the fisherman's long-handled net. The weather, however, was not the flat calm that must have favored W. E. D. Scott, for our launch was drifting before a breeze up which all the petrels flew. It was necessary to drop the liver-chum over the lee side, well aft; then, as the boat drifted, the chum passed under her stern, and rapidly away to windward; and only the boldest, hungriest petrels pounced on it before it was several yards away from us. We did get a constant stream of them flying under our stern, and I began valiantly wielding the net.

"It was no butterfly net. Designed for scraping whelks off the sea-floor for bait, it had a heavy iron hoop, a short, coarse-meshed bag, and a ponderous wooden handle. A dozen lunges at the dodging birds 'winded' me entirely. And my skeptical comrades would do nothing but jeer and boo at my panting efforts. For perhaps half an hour I doggedly struggled, crumbling liver over, seizing the great net, and swooping at the elusive flutterers who thronged towards it. At last I caught one, and the tune changed. Others asked for turns at the net. One man would bang it down over a petrel, completely submerging the poor bird. Others failed to twist the net inboard fast enough, and the prey escaped from the shallow bag. But they began to catch birds faster than I could band them, and I remembered that Scott's captives had been unable to rise from the cockpit. Alas, on this breezy day, petrels turned loose in the cockpit had little difficulty in getting a puff of air that wafted them neatly out of our reach. Each one had to be held, and soon some of my companions had a petrel in either hand. As rapidly as possible I wrote down each number, pinched the band on above the little yellow-webbed foot, and watched the released bird fly straight and fast as an arrow down wind away from that boat! They never paused nor swerved, and were indifferent to the hordes of their fellows pressing up wind past them toward the feast of bait.

Twelve I banded, before accident befell. His hands all slippery with cod-liver oil, one of the party grabbed the net for his turn, made a mighty lunge, and away went the greasy pole through his palms, entirely out of reach and promptly dragged to the bottom by its iron hoop. We continued to pull in cod and strew liver, and had the petrels in jabbering, squeaking hordes almost in hand's reach, but the catching and banding were over. Attracted by the throng of birds, several Parasitic Jaegers investigated us, and just as we turned about for the return trip a splendid Pomarine Jaeger came. So despite the mischance, the trip was voted a complete success, and I had proved my point that petrels, if not shearwaters, could be caught wholesale by a properly prepared banding expedition.

"Such an expedition should be equipped with several light weight, deep-bagged nets at the end of long, light poles, perhaps of bamboo. It should carry some bait along, in order not to be dependent for it on fishermen's luck. Cod-livers are probably purchasable, and I have heard that cheap, veterinarians' castor oil is equally alluring to the birds. To prevent too rapid consumption or loss of bait, especially on a breezy day, a cod liver might be made fast to a shingle or bit of cork, so as to float naturally and yet be tied to the boat. It was amply clear that petrels would hover in clouds over such a piece of bait as that: one might catch several together. In the cockpit there should be a coop or deep barrel, into which to stow the captives that cannot at once be banded. Hundreds, literally, could be tagged in a few hours by such a trip, and were the practice an annual one, recoveries would pretty certainly be made. Then sooner or later the far-southern islands where Wilson's Storm Petrels breed would be visited, and birds that had been banded off Massachusetts recovered in their home burrows."

In European waters a similar attempt was made by Danish bander BRIAN ZOBBE in 1959 on board a Danish trawler off the coast of Norway. He caught and banded 213 Northern Fulmars in 1.5 days. The dip net, fastened on a 3-4 m long bamboo pole, was 1 m in diameter and 0.75 to 1 m deep. These dimensions worked very well. The birds were attracted with the innards of the netted fish.

BOLAU (1903) reports on the catching of Black-legged Kittiwakes by dip net on the North Sea coast. Many flew around between the trawlers. They were so tame that they often alighted on a boat to pick up a fish. Many were thus caught by hand, others with a hand net. "One bird does not learn from the misfortune of another; one may catch one bird after the other without its companions being disturbed by it or flying away."

During a trip on the research ship "Anton Dohrn" in the summer of 1971, R. MOHR saw Black-legged Kittiwakes use the bow as a resting place. If it was occupied, a new arrival would sometimes push a less aggressive companion to the 'back', where it would sit behind the bulwark. The bird could be picked up from there as he was unable to fly off.

Fig. 342. Australian ornithologist J. D. GIBSON catching albatrosses with a ring net. Photo: V. WOOD.

Catching albatrosses. The large numbers of albatrosses found near the coast of South East Australia, especially between the middle of June and the middle of September, led to the formation of two groups of banders from Malabar and Bellambi near Sydney together in a joint effort.

At first GIBSON and SEFTON (1959) used triangular metal devices baited with octopus. They gave up on this method soon as being too slow and unsure. Trial and error with various net designs led to a loop of cane (approximate diameter 1.4 m) with a net of loose mesh (20-25 cm) (Fig. 342). Several meters of strong rope fastened to the loop on one end and the boat on the other permit net and bird to be hauled in after a throw. The success of this method depends primarily on whether one can get close enough to the birds. The Wandering Albatross is particularly easy to approach. Two circumstances may make it impossible for the birds to fly, both of which are useful for the bander to know: one is the absence of wind and the second an overabundance of food. Lack of wind requires the albatross to expend considerable effort to become airborne; with several pounds of food in his crop it becomes impossible for it to take

Fig. 343. The net is thrown toward the albatross. Photo: V. WOOD.

Fig. 344. The albatross has been caught and is pulled on board. Photo: V. Wood.

wing. In practical application one approaches the bird with the wind. When it turns to swim away the catcher in the bow throws the loop with the net over the bird (Fig. 343). While the catcher holds the line taut the boat heaves to and a helper takes the bird on board. With some practice birds 6 m and farther away can be captured this way, even when they turn to flee. Usually it is only necessary to hold the bird back with the net covering it (Fig. 344), or one can even pull a flapping albatross to the boat by its wings alone. The boat does not stop for the catching. While one bird is being banded the boat heads for the next. If circumstances are favorable 15 or more birds can be banded in one hour. The method described is used by the Bellambi group whose 14 foot boat—a sturdy fishing boat with a 4 hp motor—is manned by 3-4 people. The boats of the Malabar group are occupied by only two people. They are much lighter (fiberglass and plywood) and are equipped with outboard motors. Catching methods are essentially the same. Sudden movement, though, must be avoided in such light boats. The birds are marked with Monel bands which last longer. By 1966 the Australian ornithologists had banded 1,700 Wandering Albatrosses.

Gill, Slader and Huntington (1970) report on catching petrels and shearwaters in the North Atlantic. They used the same net-throwing technique so successful for Gibson and Sefton in catching albatrosses.

"The major components of the nets are nylon netting and a frame of plastic hose (outside diameter 30 mm, walls 3 mm thick) held in a circular shape by a dowel fitted tightly into the ends. A circular piece of mist-netting (36 mm mesh), cut large enough to make a shallow, but slack, bag when threaded onto the frame, is most effective for storm petrels. For shearwaters, however, nylon gill-netting of 60 mm mesh is as effective and much more durable than mist-netting. A hoop about 1.3 m in diameter is best under normal conditions, but when birds can be lured to within a few meters of the boat, we can catch two or more at a throw with a loop as large as 1.8 m in diameter. The hoop is retrieved with a length of cod line.

During slack tide on calm, clear days tubenoses are easily attracted by a stream of chum, squeezed from the livers of freshly caught fish. A thin trail of oil keeps Wilson's Storm Petrels in a relatively compact group; however, if large pieces of bait are doled out, the petrels tend to be dispersed by the aggressiveness of Greater Shearwaters. The latter are surprisingly bold, often approaching within a meter or two of the boat. The hoop is tossed from the fantail of the boat over a bird feeding nearby. They seem to come closer when the thrower is crouched and the hoop is held inconspicuously low. With practice even flying birds can be netted. Since petrels are frequently pinned to the water surface rather than actually entangled, they often roll out if the hoop is retrieved too quickly or escape when the hoop is lifted aboard. Such losses can be reduced by scooping the trapped birds up with a longhandled dip net. Because the plumage of petrels usually becomes wet during the process, the birds were dried out in draw-string cloth bags for one-half hour before release."

From 1965 to 1968, 250 birds of 4 species were banded: 119 Greater Shearwaters, 1 Sooty Shearwater, 117 Wilson's Storm Petrels, and 13 Leach's Storm Petrels. Fifty-seven Wilson's Storm Petrels were caught on a single trip in early August 1965; on another occasion 12 Leach's Storm Petrels and only 2 Wilson's Storm Petrels were netted.

Greater Shearwaters, whose noisy calls attract others of their kind, are especially easy to catch at times. In the afternoon of August 18, 1966 the American ornithologists banded 72 birds within 3 hours. Sooty Shearwaters often accompany Greater Shearwaters, but are less common and bold. But where abundant, they and other species could probably be captured in greater numbers.

GILL, SLADEN and HUNTINGTON continue: "We hope that the effectiveness of this method will encourage others to band birds at sea. The colonial habits of sea birds make it possible to band large numbers at the breeding grounds. Wilson's Storm Petrels have been banded at their nests in the Antarctic by the British Graham Land Expedition (ROBERTS, 1940), the British Antarctic Survey (Sladen and Tickell, 1958; Hudson, 1963 and 1967), and the United States Antarctic Research Program (Sladen et al., 1968), but in relatively small numbers. They would be easy to catch in large numbers by our throw-net technique in Antarctica, where they concentrate in sheltered harbors such as Deception Island (62 57'S, 60 38'W). Several thousand Leach's Storm Petrels have been banded at Kent Island, New Brunswick (GROSS, 1947), in Britain (SPENCER, 1959), and elsewhere on their breeding grounds, as have Greater Shearwaters on Tristan da Cunha (HAGEN, 1952; HUDSON, 1967). If this work is supplemented by more banding at sea, we can hope to extend our meagre knowledge of the movement of these birds." During a trip to Antarctica (BIERMANN and VOOUS, 1950) VAN DER LEE and SAMUELSEN caught a considerable number of Snow Petrels and Antarctic Petrels while standing on the forecastle. A dip net with netting of 100×70 cm on a 3 m handle served to capture the birds. The net was made according to the instructions of the fishermen of the Faeroe Islands.

S. WELLERSHAUS criss-crossed the Atlantic on board the research vessel "Gauss" from July 29 to September 25, 1958 without being particularly interested in bird catching. Nonetheless, 33 Leach's Storm Petrels were banded which came—or better, fell—on deck at night when the ship was anchored (at sea) and the lighting on deck was switched on. Other storm petrels and seabirds did not come on board.

17.4.2 Catching with fish hooks

Catching larger seabirds with fish hooks whose barbs have been carefully filed off is common practice on research ships and among ornithologists aboard trawlers.

This centuries old catching method (NAUMANN, 1840; H. P. v. FIRDENHEIM/LINDNER, 1959) today can only be justified under circumstances as described here. With strict observance of all guidelines, this catching method has proven to be without danger for seabirds.

During a voyage on the Atlantic H. W. NEHLS gathered much experience "fishing" for Northern Fulmars. He used a 20 m long black nylon line, attached a medium-sized fish hook without barbs with cod liver as bait on the hook. The hook then got caught in the bill or maw. Under favorable circumstances the line could be thrown repeatedly and pulled in with a bird. The birds did not take alarm upon capture of one of their companions; they did not seem to comprehend the capture. Few Black-legged Kittiwakes were caught as they mostly tore the bait off the hook.

During a voyage on the South Atlantic, MAHNKE (1971) banded a number of birds of the open sea: 82 Black-browed Albatrosses, 6 Giant Petrels, 133 Cape (Pintado) Petrels, and 2 Southern Fulmars. All birds were captured according to the aforementioned method from a slowly steaming ship. Bait was mostly hake liver. There were no injuries.

Further experiences from 1971 and 1972 on the North Sea and in the area around Ireland are shared by R. MOHR. After the disappointment of an earlier trip he used fresh bacon from the ship's larder. Bacon has the advantage of floating—very important—as opposed to fish which is easily dragged under by the fish hook (nobody else has reported on this so far). A bit of the rind was left on the bacon pieces. They were basically cut in a pyramidal shape with a

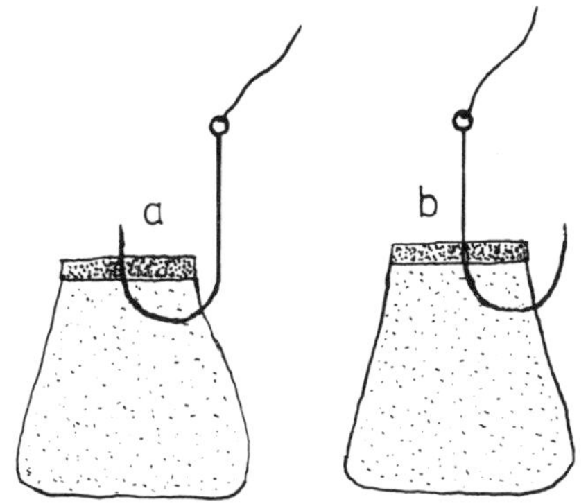

Fig. 345. Catching seabirds with anglers' hooks from ships. (a) wrong, (b) correct. After R. MOHR.

2×2 cm base area (or a bit larger) and the height of the bacon slab. A piece of rind 0.6 to 0.7 cm square was left on top. The hook was then guided through this bit of rind. The hooks were either bought without barbs as "trout hooks" or else the barbs were carefully filed off and a subsequent finger test was done to check the smoothness.

In catching birds it was essential that the hook be fastened as shown in Fig. 345. A 10 lbs. nylon line was fastened to two neighboring bitts. Mohr says specifically: "If a bird is hooked, there is no time to put the line neatly around the bitts. If possible, make sure that the line lies loosely and with some room on deck. Then it can quickly be reused and at day's end brought into the 'starting position'."

The Northern Fulmar was captured only on those occasions when more than one was flying with the ship and they were watching each other. If fulmars were around, Mohr (who apparently used mainly bacon) then would throw additional bait into the water to draw them to the ship. Of course, he threw the bait near the hook. The ship was kept in position so that the port side was to windward in order to avoid going over the lines for plankton net, water scoop, etc. which were all on the port side. Through wind pressure the ship always floated away towards starboard; consequently the bait floated away towards port. Often Black-legged Kittiwakes were useful in transporting the bait. It was not always possible to throw it into the group of fulmars sitting apart. The kittiwakes came more readily to the ship to pick up the baited hook and carried it away before letting it fall again, sometimes not until all 100 yards of the line were unrolled. While nibbling on the bait, they apparently noticed something not quite right. Consequently their kind was hardly ever captured.

Once the fulmars became interested in the bait, they forgot all caution. If only a few birds were present, they would try to tear off the bacon, which was impossible due to the rind. When more birds arrived, the morsel had to be swallowed quickly. Often the hook got caught in the tip of the bill. As long as the bacon was visible in the beak, any pulling on the line was in vain; the birds were always able to free themselves. The hook had caught, usually in the gape, only when the bacon could no longer be seen. During the scrabble for the bait it would happen that the hook would catch a foot or a leg. Only once was there a greater blood loss when the hook punctured a leg. Otherwise there was hardly a drop of it. Once or twice the line got wrapped around a wing and the bird was pulled aboard. But generally fulmars which are only entangled in lines are able to free themselves again. It is possible to hold fulmars and loosen the hook with bare hands, but a quick and sure grip is necessary. Because fulmars are virus-carriers, disinfectants are necessary for bites and must be carried on all trips. Injuries from other seabirds should also be treated.

On this trip Mohr captured only one skua, using sand eel bait. When it was thrown through the air, one of several skuas immediately dashed after it and swallowed it. The hook got caught directly behind the bill.

On his 4th sea voyage bacon as bait again proved best. The main advantage was that there were no problems obtaining it. Mohr also had an opportunity to experiment catching with young fish of up to 20 or 25 cm in length and fresh fish liver. The filed-down fish hook must be pulled through the backbone near the tail fin; otherwise the fish will not stay on. Fish are swallowed head first so there is no problem with deeply swallowed hooks. Mohr took young ocean perch only up to a length of 15 cm. They are relatively big and could not be swallowed at once otherwise.

Catching with fish liver proved to be difficult as there is no way to attach it securely to a fish hook. Mohr improvised by drawing the hook several times through each piece so that the string kept it together somewhat. As always, the prerequisite for successful catching is the gathering of large crowds of birds. The birds stimulate each other and are less careful. Catching with liver bait went surprisingly well. The Northern Fulmars sat at the stern and watched the liver being thrown overboard. As soon as it hit the water it disappeared in a thick knot of birds. Happily, quite often a bird got hooked by the bill, foot or wing. On this trip 117 Northern Fulmars, 18 Great Skuas and 3 Black-legged Kittiwakes were caught. When Mohr used liver, no species other than Northern Fulmars had a chance to get near the bait.

Mohr experimented with various hook sizes. Size 1 hooks seem to work best. Size 3 is too small, but may possibly be used for catching Black-legged Kittiwakes.

Mohr made some interesting observations on the behavior of Black-legged Kittiwakes after capture. Upon the first pull on the line they fly up and follow it so readily that they almost glide into the catcher's hand. They are probably more aware of the hook and avoid all pulling. Fulmars, on the other hand, have to be dragged first over the water and then on up onto the ship.

Lesser Black-backed Gulls never touched the baited hook, even when several were fighting for the other pieces of bait. Even when diving for the bait they must notice something and therefore avoid the hook.

Mohr also shows that fish-catching is not absolutely necessary to banding success. It is enough if the ship stops or proceeds slowly.

It is probable that fishermen and seamen have long captured albatrosses with fish hooks and bait, but only to supplement their dinners. Macpherson (1897) heard from a seaman who caught 150 albatrosses on a voyage to Batavia (Djakarta). This exceptional zeal apparently even led to a ship wreck.

17.4.3 Catching with strings

Macpherson (1897) and Sunkel (1927) reported on ways to capture birds with strings. Sunkel talks about catching shearwaters and Black-legged Kittiwakes: "Several pieces of twine are knotted to a piece of wood the length of a pencil. The catcher on board the ship holds the twine in his hand. When the wood floats on the water, the twine and the wood form an angle. The catcher throws his bait into this angle. The birds dive so eagerly for the morsels that the wood jumps from the water on impact and the twine winds around the wings securely." The use of several separate pieces of twine has not been tried recently.

W. Mahnke gathered new experiences in the North Atlantic when, on January 28, 1963 within a limited area, he encountered about 50,000 Black-legged Kittiwakes. The birds were attracted by offal from fish and approached the ship to within a few meters. An attempt at capture was made only when at least 300 birds were close by. During dead-slow speed or drifting he threw a black-braided nylon line 20 to 25 m long overboard (dia. 1 mm, weight 10 kg). A thin wire hook was fastened to the end of the twine. The birds flying to and fro got tangled in the fluttering line, and the weight of the hook assured that the line would wrap around their bodies. At the same time the hook prevented the twine from slipping off the feathers. Often the hook was not needed because the birds did not get tangled at the end of the line but higher up. During periods of drifting, a small piece of cork was fastened to the line 1 m before its end to make it float. The line remained above the water surface and permitted the continued capture of birds. One throw of the line captured at most 3 Black-legged Kittiwakes or 2 Northern Fulmars.

Mahnke used yet another method. He fastened a 1.5 m long nylon line (see above) to a broomstick of the same length. Again a thin wire hook was attached to the end of the line. The line was then thrown after birds flying close by and the captured bird was hauled in. Both methods captured about 100 Black-legged Kittiwakes and 100 Northern Fulmars in approximately 30 trapping hours. There were no injuries to the birds.

Lambert (1971) reports on the results of two trips in 1966 and 1967 in the Eastern Atlantic and off the coasts of Southwest Africa where the following species (with the exception of 1 British Storm Petrel, 1 Leach's Storm Petrel and 1 Common Tern) were captured and banded: Shy Albatross, 3; Black-browed Albatross, 19; Yellow-nosed Albatross, 52; Giant Petrel, 1; White-chinned Petrel, 321; Sooty Shearwater, 24; Cape (Pintado) Petrel, 114; British Storm Petrel, 5; Wilson's Storm Petrel, 3; Leach's Storm Petrel, 3; Cape Gannet, 33; Great Skua, 5; Pomarine Jaeger, 2; Parasitic Jaeger, 6; Common Tern, 2. Lambert caught some of these birds with baited fish hooks, but most with lines cast overboard and weighted with small metal pieces such as a screw. Particularly well suited as weights were unbaited fish hooks (treble hooks) which Lambert had filed down. The birds were captured in the lines. Success with this method is only possible at dead-slow speed and the number of birds following the ship must be sufficiently great.

Roberts (1967) talks about Wilson's experiences in Antarctica in the years 1904 and 1910. This is an extract from his diary: "A bent nail was knotted to a line the other end of which was fastened to the rigging at the stern of the ship. The rigging was 10–13 m above the deck so that the line was high above the water for a long distance. The other end with the nail was dragging in the wake of the ship where the birds mostly stay. Flying to and fro it happens that they touch the line with their wings. It seems as if the bird sensing the touch of the line turns around in the air thereby wrapping a loop around its wing. The bird can then be hauled on board." ROBERTS had some success with this method. During such trapping the speed of the ship did not exceed 3 nautical miles. This kind of catching does not seem terribly efficient to BUB, but it may depend upon conditions such as the number of birds present.

In 1968 C. F. KINSKY told BUB details of capturing albatrosses off the coast of New Zealand. This method is an extension of the Australian catching method of GIBSON and SEFTON (1959), but generally only usable during simultaneous fishing operations. Small fishing boats are often used to clean the fish after a drag net catch and before the start of a new catching operation. The offal is thrown overboard, and if the weather and location are right, many birds—especially albatrosses—are attracted close to the boats. They fight for the remains and lift their wings in the process. This behavior makes it easy to catch them with wire hooks on a line (Fig. 346). These relatively large hooks must be fastened securely but not rigidly on the line. The birds are not supposed to swallow the hooks, but their ends must nonetheless be blunt or bent so that no bird will be injured. If the birds are near the boats fighting for food with lifted wings the catcher throws the line over the birds' backs and manoeuvers it in such a way that one hook or another will catch the "upper arm" of a bird. The bird is then carefully hauled

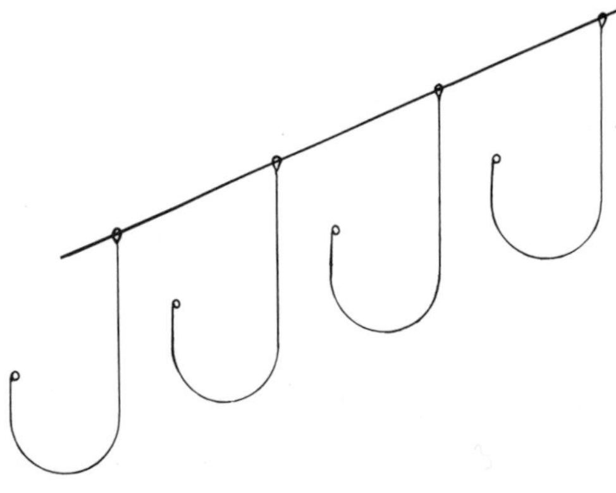

Fig. 346. Catching albatrosses with wire hooks. After C. F. KINSKY.

on deck. In such a manner a number of birds (all albatrosses) were captured and banded. There were no injuries. After a night of catching fish KINSKY captured 25 birds in the early morning hours in a very short time. In one instance he even got two birds with one throw.

An interesting way of capturing Great Frigatebirds as observed by KRÄMER (1906) on the Gilbert Islands and on Nauru (South Pacific) should be reported here. Catching Great Frigatebirds is a beloved sport of the natives. "Many houses on the beach are truly full of the tamed birds and the natives feed them with flying fish and other sea creatures disregarding their own stomachs. The Nauruan prefers to go hungry himself rather than denying food to his darling bird. Within the villages one often stumbles every few steps on small racks on which a Great Frigatebird sits. At certain times, when strange frigatebirds arrive, the young men of a village will prepare for the catch. A patch of ground near the beach is cleared and fenced. Now each hunter lets his bird—which he recognizes on certain marks under the primaries—fly to attract the strange animals."

For the actual catching they used a long string on which they fastened a walnut-sized stone of coral limestone or something similar. The natives first swing this 'bullet' in short circles over their heads and then fling it over the frigatebird so that the string falls over the large spread wings and in which it gets tangled. Once the bird is on the ground he cannot get airborne again without wind and is easily captured.

17.4.4 Catching with triangles

Even ancient seafarers already knew how to catch albatrosses and large petrels with triangles. ROBERTS (1967) describes the observations and experiences of the English zoologist EDWARD WILSON in Antarctica shortly after the turn of the century. Cut a triangle from sheet metal, brass or some other metal and fasten it to a piece of wood so that it will float on water (Fig. 347). A piece of meat is tied inside the triangle and, thus baited, it is towed on a long line in the wake of the ship. As soon as the albatross snaps at the meat, the upper mandible hooks into the opening. The line now has to be hauled in tightly because the bill will 'unhook' if the line is slack.

ROBERTS (1967) found catching good when

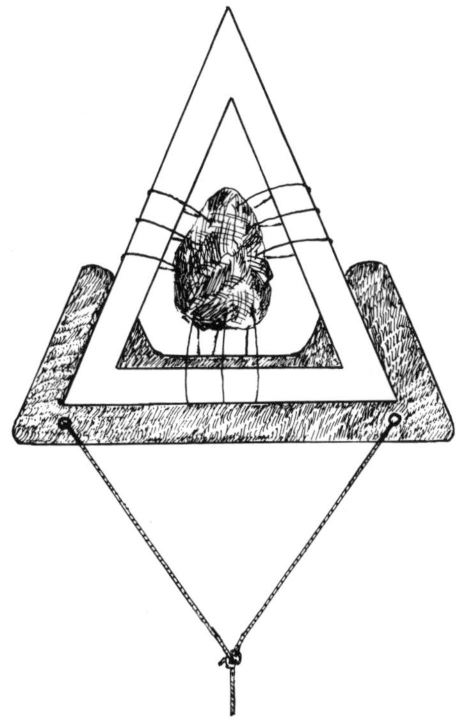

Fig. 347. Triangle to catch albatrosses with. After E. WILSON, from ROBERTS, 1967.

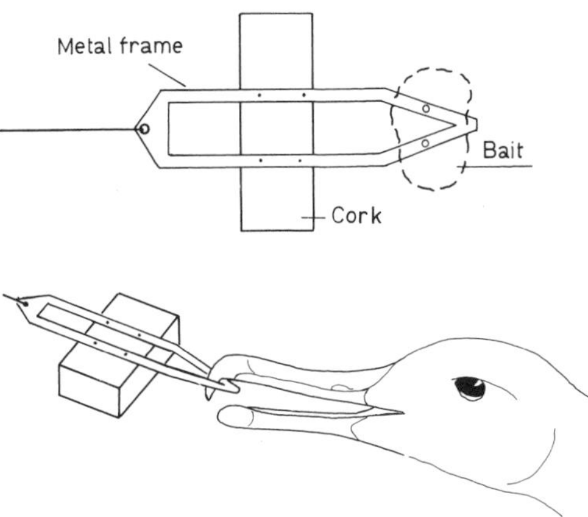

Fig. 349. Triangle to catch albatrosses. These triangles are used by Australian banders. After GIBSON & SEFTON, 1959.

the ship was lying still or was at dead-slow speed. Once it sailed faster than 3 nautical miles he caught no albatross.

E. WILSON also mentioned and drew an elaborate "Albatross Catcher" without stating how successful catching was (Fig. 348). Near the Southeast coast of Australia, GIBSON and SEFTON (1959) used another version of the triangle (Fig. 349) which seems more practical than the simple triangle. They soon abandoned it though in favor of a method better suited to the coastal waters and their small crafts (Fig. 342).

17.5 Catching northern sea birds on land

17.5.1 Catching with dip nets or by hand

In the northern countries bird catching was traditionally done near 'bird mountains' or in their vicinity, as along the coasts of Iceland, on the Faeroe and Shetland Islands, on St. Kilda (Iceland), on the Norwegian coast and other similar places. For reasons of conservation, capture of the species which were sought is now generally severely restricted or totally prohibited.

The Common Puffin was and still is especially desired as it seemed to be very abundant in many areas. But today its numbers have diminished as compared to a few decades ago.

Many authors and researchers have reported on catching methods: NAUMANN (1844, 1903), HARTWIG (1870), BACHMANN (1902), LOCKLEY and RUSSELL (1953), PEDERSEN (1954), MYRBER-

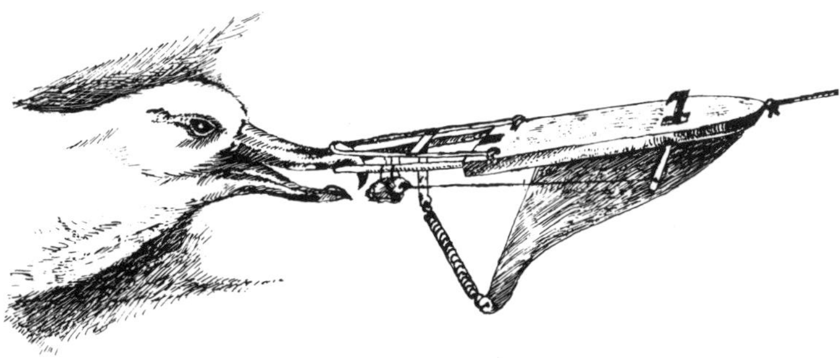

Fig. 348. A device for capturing albatrosses. After E. WILSON, from ROBERTS, 1967.

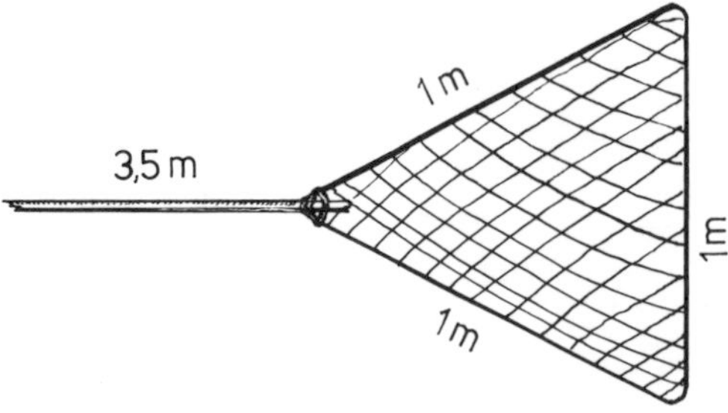

Fig. 350. Dip net of Norwegian bird catchers. After MYRBERGET, 1960.

GET (1960), MATHIASSON (1962), to mention only a few. When catching at bluffs the catcher positions himself at a place where the birds are carried with the wind close to the rock and then turn outward again. The long-handled dip net is turned quickly after the catch so that the bird cannot escape. This dip net (Fig. 350) is a sturdy device. MYRBERGET shows such a net in its typical triangular form from the Norwegian island of Lovunden. Its dimensions are considerable. Other dip nets often have shorter handles (3 m) and the net itself is often oblong. The shallow net bag is typical of all dip nets.

MATHIASSON (1962) and PEDERSEN (1954) write in detail about puffin catching by the inhabitants of Mykines, an island of the Faeroes. PEDERSEN has an interesting description (see Fig. 351 and 352). After the catchers already have been trapping birds for weeks with auk hooks at their breeding places, the time arrives when the puffins flock. Let's listen to PEDERSEN himself: "Puffin catching had lasted only a few weeks, when I became a witness to an impressive phenomenon one evening during a warm southwest breeze. The army of over half a million young puffins floating on the water rose and began to swarm. Almost in one stroke the air was buzzing with the beating of innumerable wings. The swarm circling continuously over the island was so thick that it literally darkened the sun. This magnificent sight continued for hours and only dusk brought a change. A few birds separated from the rest and landed with stiff wings on the island. Some immediately rose into the air again, but those that stayed enticed others down. Their numbers grew with every instant and suddenly, as if on a secret sign, the entire swarm alighted on the island. There they sat, perfectly quiet, row upon row with their white bellies turned towards the sea, listening almost expectantly to the uniform muffled thunder of the surf, the only sound in the evening's calm. But their new surroundings did not keep them long: hardly had an hour passed and they were airborne again, the dull thunder of the surf again drowned out by the continuous beating of innumerable small wings, until it too was ended by a renewed visit to the island. The

Fig. 351. Puffin catcher at his catching location on one of the Faeroe Islands. Sometimes catchers set up stuffed birds near the catching location, but this is a practice rarely followed. Photo: S. MATHIASSON.

Fig. 352. Puffin catcher on one of the Faeroe Islands during catching operations. Photo: S. MATHIASSON.

spectacle continued in this manner throughout the bright night and into the late morning hours. Only when the sun was high in the sky the entire flock sat again on the sea, diving for small fish and sleeping, rocked by the rhythmic waves, to awake again to new deeds when dusk fell on the remote island.

"The bird catchers had waited for this event. They called this time of puffin flocking "Landkommedage" (land-coming days) and they were the beginning of the greatest bird catching of the year. When the events of the preceding evening were repeated the next, all men skilled with dip nets had left the village. Every single projecting niche on the steep rocks was occupied. Here the daring men would stand on tiny ledges directly above the precipice and swing the immense dip nets into the swarm of birds with almost superhuman strength. Catching itself, i. e. swinging the monstrous dip net which is almost 5 m long requires not only very good physical condition, especially well developed back and arm muscles, but also a good share of practice. The dip net was not held so that the bird would fly into it, but rather was swung from the rear over the bird which naturally requires considerably greater dexterity."

Up to 1,000 birds a day can be captured by one catcher. Often a catcher will pick such a dangerous place that he will plunge to his death or that he is unable to extricate himself again from among the rocks and must die miserably.

Pedersen also describes the catching of Northern Gannets: After the gannets have started permanently occupying their galleries on the rocky cliffs in the spring, the time for catching has come. A dozen men left the village around midnight in a dense fog and the blackest darkness. The equipment consisted solely of a strong rope 80 m long. The spots where the men were to be lowered down were marked. One man now tied the rope securely around his chest and the upper part of one leg. While three other men held the rope, he leaned backwards over the precipice and glided down. A strip of wood fastened to the basalt kept the rope from breaking through abrasion. About 50 m below the rim the man untied himself from the rope. The party above continued on to the next site where another man descended. Only the 3 men operating the rope lift remained on top. The gannets sleep so deeply and feel so secure on the cliffs that they only wake up when danger is literally touching them. This is the reason for catching them at night.

"In order to have as many birds as possible within reach, the men moved along the very edge of the ledges, using the sturdy nests of the gannets as supports, as they would otherwise not be able to find a hold. Here they lay down full length upon as many sleeping birds as possible. Then they had to kill the gannets one after the other as quickly as possible before the birds realized the danger they were in. If it happened that a bird woke too early and noticed the human near by, the catcher had to retire quickly into the darkness. With no room for the catcher to maneuver, the strong bird could easily be dangerous to life and limb or give deep bloody wounds with its sturdy beak. These daredevils were fully aware of the danger into which they ventured. But they themselves claimed that the catching provides so much excitement that they hardly have time to give a thought to the possibility of losing their grip and falling into the abyss. Indeed, so far the catchers only have had one death to mourn: about 10 years ago a young man disappeared so totally that despite weeks of search no trace of his remains was ever found."

This night's catch was 420 birds. The method of catching did not seem to result in any decimation of the bird population, as the empty spots were immediately occupied by roaming Northern Gannets. From the reports of others we know that some banding was also done.

Northern Gannets who have gorged themselves can supposedly be captured by grabbing them by hand. But surely such opportunities must be rare.

Catching with a dip net is not restricted to Common Puffins, but is also possible with fulmars and shearwaters, Black-legged Kittiwakes and other birds. For example, according to Lockley and Russell (1953) up to 200 shearwaters were captured within a few hours on the Faeroe and Vestmannaeyjar Islands (Iceland).

During an expedition in 1962 to Spitzbergen, Norwegian ornithologists wanted to band as many Dovekies as possible because they had hardly ever been banded. They used a variation of the dip net. The iron hoop was fastened to a 4 m long pole. The net was coarse and had a small pouch. According to Norderhaug (1963) 5 such dip nets captured 400 birds in 5.5 hours. Near the end of July catching was suspended after 2100 bandings: the Dovekie bands had been exhausted. Between 1962 and 1965 11,000 Dovekies were banded at Spitzbergen; in 1965 alone 3 000 adult birds were banded (NORDERHAUG, 1966).

The Norwegian ornithologists used the same method for Common Murres, Parasitic Jaegers, Black-legged Kittiwakes and Northern Fulmars with varying results.

American ornithologists were successful in catching Great Skuas with dip nets (pole length approx. 1.5 m) in Antarctica. Breeding birds flew barely above the head of the intruders near their nests, as is common with other gulls and terns.

Fig. 354. Short alcid hook used to catch from holes dug into the ground. Photo: S. MATHIASSON.

17.5.2 Catching with hooks

An old bird catching tool is the 'auk hook' (Fig. 353 and 354) which was as important to people in northern regions as the dip net. Of course, the hook, particularly as a tool for fishing and agriculture, is widely known all over the world in many variations. LAGERCRANTZ (1940) reported a number of different hooks from East Africa alone, as used by fishermen there then and probably still today (Fig. 355).

KOHN and ANDREE (1876) mention the auk hook as the catching device for the puffin colonies of Kamchatka (USSR): On the sheer ocean cliffs of the bird island the Kamchatkans often dared "horrifying" descents as the cliffs were only accessible from above, because of the surf. "The safety of each climber is only in his surefootedness. They generally climb barefoot, but cannot use their hands because they carry baskets or buckets in which they collect eggs. They only leave a cliff when their containers will hold nothing more. They pull the birds from their holes with iron hooks. An experienced climber will catch 70 or 80 birds." The authors apparently refer to the Rhinoceros Auklet.

Fig. 353. An alcid hook. After LOCKLEY & RUSSELL, 1953.

Auk hooks have long since become a tool of scientific bird catching. LOCKLEY and RUSSELL (1953) describe experiences in Great Britain. In general, an auk hook consists of a thick bamboo pole 1.5 to 4.5 m long and a piece of sturdy galvanized wire 60 to 90 cm long. The latter is fastened to the thin end of the pole. The last 5 cm of the wire form the hook. The width of that hook can be adjusted according to the birds to be captured. The wire opening for Common Puffins and Razorbills is 12 mm, for Northern Gannets 18 mm. The point of the hook should be bent slightly to the outside.

Auk hooks are considerably less conspicuous than nets. If the catcher is patient and moves cautiously he can catch one bird after another right out of the colony. It is amazing how birds such as Northern Gannets, Common Puffins, Razorbills, Common Murres, Northern Fulmars, Black-legged Kittiwakes and other gulls will permit the bander to push the pole carefully so far along the ground or the rocks that he can guide the hook around the leg of a bird. Once the hook is around the bird's leg the pole must be pulled in quickly and evenly, otherwise the bird will pull its foot out of the hook. While banding the wings have to be held, as the bird can otherwise disturb the colony with its flutterings. Near Grasholm (Great Britain) up to 60 adult Northern Gannets were banded in one afternoon, and up to 50 Common Puffins were banded one summer evening near Skokholm (Great Britain), both using long hooks. If catching in nest holes, shorter hooks are better suited (Fig. 354).

Banders in Lundy (Great Britain) developed a shorter but more easily transported version of

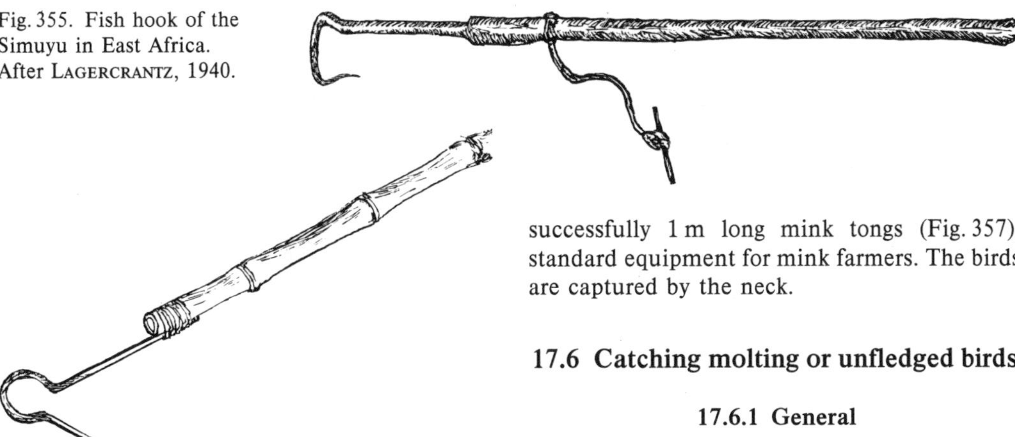

Fig. 355. Fish hook of the Simuyu in East Africa. After LAGERCRANTZ, 1940.

Fig. 356. Swan hook. After WALKE, 1965.

the auk hook. Those hooks are made of copper telegraph wires bent into shapes similar to shepherds' staffs. They are attached to telescoping radio antennas which extend from 38 cm to a length of 3.3 m. Usually the birds will let one approach that closely. The hook is then guided around the upper part of a bird's leg. If the bird flies or tries to fly, it thereby facilitates its capture. The hook can also be used to catch large nestlings which will not let one approach, and to return them to the nest.

Auk hooks are also useful for banding unfledged shearwaters, alcids and other hole nesting species which otherwise can be reached only with great difficulty or not at all.

A hook for catching swans (Fig. 356) is shown by WALKE (1965).

For catching adult and juvenile Common Cormorants, some Razorbills, Common Murres and Common Puffins MYRBERGET (1968) used successfully 1 m long mink tongs (Fig. 357), standard equipment for mink farmers. The birds are captured by the neck.

17.6 Catching molting or unfledged birds

17.6.1 General

Molting ducks, geese, swans and other large and small waterbirds are incapable of flight. Many people around the world have hunted these birds with diligence and passion. This hunt often easily netted the large amount of game necessary for the survival of some Eskimos as well as the nomadic tribes of Northern Asia. For all it was certainly a welcome addition to their diets. Ducks and geese, being the most numerous, were the natural targets.

This type of hunting is practiced around the entire northern hemisphere of the globe. Of course, people elsewhere have also hunted flightless birds of these and other species, and still do.

BIRKET-SMITH and DE-LEGUNA (1938) report on the hunting methods of the Eyak Indians in Alaska, to take one example from North America. Ducks, geese and swans were killed mostly in August while molting. The entire population of a village would drive the birds along marshy river banks to a shallow spot where they could be herded ashore. The best timing for this kind of hunt was the early morning or the evening.

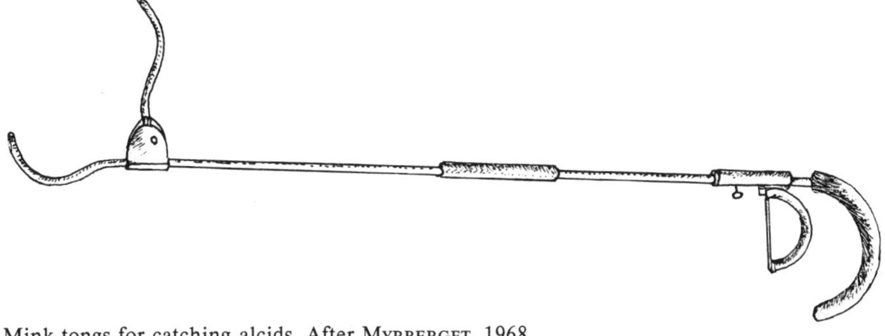

Fig. 357. Mink tongs for catching alcids. After MYRBERGET, 1968.

The people made lots of noise while herding the birds.

Abercrombie observed American Indians while hunting molting geese in the marshy lowlands near Alaganik, also in Alaska. The entire village population—men, women and children—ran after the birds. In order to catch the geese, which were faster than people, the Indians made a large circle which they tightened slowly. The birds tried to hide behind driftwood and some succeeded in breaking out.

Molting loons were not hunted because a folk tale tells of an Indian boy who once was changed into a loon.

In Northern Eurasia the hunt for molting ducks and particularly geese was widespread. According to Spörer (1867) 15,000 geese were killed in two hunts on the island of Kolguyev (USSR). While many tribes used large numbers of people to encircle the flightless birds, the Samoyeds for example also used net walls.

Macpherson (1897) tells us how the Samoyeds, according to a carefully laid plan, drove a large flock of molting geese in front of seven boats along a river. They drove reindeer herds along the banks to prevent any escapes by the geese and then guided the birds into a large net enclosure which they had prepared beforehand.

Kohn and Andree (1876) report on the hunting methods of the Yukaghirs in Eastern Siberia. These people even employ trained dogs in the hunts.

It seems that both authors encountered molting geese on a river island in the tundra around Dauriya (USSR). Kohn and Andree explain: "Here the geese seem to be secure from persecution by the Mongols, especially as religious beliefs forbid them to kill an animal on the water. When you kill a bird on the water, it says, its blood mixes with the clean water and the herds drinking from it later will surely die."

From Europe there are early records regarding this catching method. According to Gessner (1600) a certain day in August in Zürich (Switzerland) was reserved for catching molting Great Crested Grebes. The birds were driven into nets. H. P. v. Firdenheim, whose 17th century work was first unearthed by Lindner (1959), tells of catching young or molting flightless ducks in Southern Germany. The birds were driven into the rope bags through leads on either side. Döbel (1746/1910) and Winckell (1878) describe catching unfledged young and molting adult ducks and geese with a sturdy trammel net. Döbel describes these nets: "This is a triple net. The outside nets must be knotted with strong twine at 12″ (approx. 28 cm) intervals. Net height is 90 cm. The inner net has 28 rows of mesh, but the mesh is only 3″ (7–8 cm). The outer nets are about 90 m long, but the inner net is 150 m so that it hangs very loose and the bird gets fully entlangled."

Macpherson (1897) describes a method of catching geese on the German island of Rügen (Baltic Sea). The number of molting Greylag Geese must have been considerable there. The birds ate on certain meadows near shore at night or during the twilight hours. The island's inhabitants laid nets on the beach along this area. The nets were first covered carefully with sand. As soon as the geese had come ashore the catchers erected the nets. The men then drove

Fig. 358. Net for divers (loons), cormorants, and other species sleeping on the beach. The birds are surprised during their rest and do not have time to fly up because of the net quickly pulled into place. The banders must have dip nets. The net should be closer to shore and longer than shown here.

Fig. 359. Eskimo drive molting Canada Geese into a pen. From E. W. NELSON, 1928.

the flightless birds toward the water where many got caught in the nets. This seems to be a useful method even for today's scientific banding of not only molting ducks and geese but also for water birds which sleep or rest near shore (such as loons or cormorants) but which glide back into the water as soon as people approach. Such catching methods, adapted to local conditions, might net several of these birds (Fig. 358).

This leads us back to scientific bird catching.

17.6.2 Catching molting and unfledged geese

NELSON (1928) reports the activities of some North American ornithologists, who were certainly among the first to band flightless geese in large numbers (Fig. 359). The Eskimos, experienced goose hunters since long ago, showed them how.

PETER SCOTT (1950), during his Perry River Expedition (Northwest Territories, Canada) in 1949, captured a number of geese for his research: Canada Goose, White-fronted Goose, Snow Goose, and Ross's Goose. Catching was done with stationary nets arranged to form a funnel into which the geese were driven from the water.

An unexpectedly large development in the banding of molting and unfledged geese in the Arctic started in 1951. When, in the fall of 1950, the first Bean Geese were captured in the Scottish migration and wintering grounds with the recently developed rocket net, PETER SCOTT and JAMES FISHER soon developed a plan to follow these banded birds to their nesting places in Iceland. There they intended to catch as many birds as possible in the summer when the adults molt their flight feathers and the young cannot yet fly.

The first expedition into the interior of Iceland started from Reykjavik on June 22, 1951 (SCOTT, FISHER and GUDMUNDSON 1953, SCOTT and FISHER 1953, 1957). SCOTT and FISHER had their ideas on how to catch the birds, but their effectiveness was as yet untested. The researchers had set themselves a goal of banding 500 Bean Geese. They carried several long nets, the longest of which measured 80 m. They also had 2 "side wings" of 32 m length and a "net cage". Waterproof pants and boots were taken along for river crossings.

After one month, on July 24, 1951, 323 birds had been banded (Fig. 360). This figure was the result of many small catches. In general, this result was not very satisfactory or encouraging. The key to catching large congregations of geese had not yet been found. The researchers were still groping in the dark, although they often found the remains of the goose pens which undoubtedly served the ancient Icelanders to catch the very same geese (Fig. 361).

The first major success was 247 bandings on July 25, 1951 (Fig. 362). The following description shows how difficult it really is to direct large numbers of geese with only 4 people:

"We set off about 12.30, crossing the Blautakvisl and dividing up immediately. (Dr.) Finnur (GUDMUNDSSON) and I (PETER SCOTT) forked left across the marsh while Phil (Miss PHILIPPA TALBOT-PONSONBY) and Valli (VALENTINUS JÒNSSON) rode up the Djòrsà bank. We had agreed to emerge upon the marsh (now known as Falcon Marsh) at 14.10. Finnur and I found a party of geese at the exact spot where we made our best

Fig. 360. Catching molting Bean Geese on Iceland. The birds are already in the pen. From SCOTT & FISHER, 1957.

Fig. 361. An Icelandic stone enclosure for geese, now fallen into disrepair. Photo: PHILIPPA SCOTT.

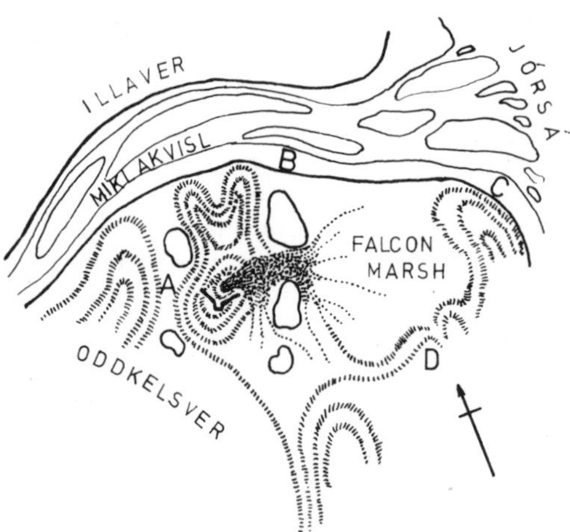

Fig. 362. The goose drive of July 25, 1951. A–D are the positions of the participants. The geese are walking towards the catching pen. After SCOTT & FISHER, 1957.

pony-hunt. They moved on to the hill ahead, and it seemed we might get them to the nets, so just before the Fox Earth we divided, Finnur with the pack-pony going up over the ridge while I rode round the foot of the hill. Near the Fox Earth the geese and many more came into view, and began heading for the North Crest (where the net was set). But they slipped out to the left, streaming down in flocks and families towards and across the braided channels of the Miklakvisl ... When the last of the goose families—200 strong in all—had crossed the stream into Illaver, I rode over and joined Finnur. Then I went off to take my place, feeling already slightly despondent and frustrated. All our trust must now be placed in the marsh on the other side of the hill, where we could hear at least one goose calling.

"I dismounted when I came to the point at Miklakvisl (B) and peeped over. Not a goose was to be seen on the green marsh. When there were still two minutes to go I saw Valli ride out from his corner (C), so I mounted and emerged. At first I could see nothing. Then I saw three or four adults and a few young making their way up to the top of the hill. It looked as though we should not draw completely blank. I moved up to the ridge to make sure that they did not leak out. I could see Phil riding up and down far across the marsh. Evidently there were geese in the dead ground between us. As I breasted the ridge an amazing sight was disclosed. Coming towards me was a flock of some 200 adult geese, and further behind another flock of not less than 300 mixed goslings and adults. The first lot saw me and turned uphill, then by working the ridge carefully I could keep them in view while keeping out of sight of the second batch. Certainly many of the second lot went the wrong side of the wing of the net, but a great and awe-inspiring crowd of geese was already at the top. I rode over to Finnur and together we moved up on to the summit, with Valli on the left. There in the middle of the V of the net they all stood, tightly packed, adults and goslings together. From what we subsequently learned there must have been at least 500 birds. For one thrilling moment we saw them move down towards the cage. They came to the wing at one side of it. It held them up for perhaps five seconds, then, under the weight of birds, it fell down and they all ran out down the hill. It was a bitter blow. But Valli and Phil were galloping round them and soon many were on their way up again. I had put up the net again, but in a few minutes the geese were against it on the wrong side. Again in a matter of seconds it was down and the stream of geese had run through to the top of the hill; now they were back on the right side of the net but heading for the Miklakvisl. Finnur and Valli came in and cut off the stream before many had gone, and once more a phalanx of geese was standing on top of the hill. Now the four of us were on all four sides of the geese and a state of near-equilibrium had been reached. This was the critical moment. Could we keep them there long enough to repair and strengthen the net? How long would they be ready to stand in a bunch on the crest of the hill without making a determined effort to break out? Already some individuals were making minor sallies to test our siege. But the goose crowd seemed to offer some kind of refuge to them, for as soon as they were headed they ran back into the crowd and were, it seemed, relieved to be in the thick of the flock again."

Now the question of the old Icelandic goose pens also had been solved (Fig. 361). Neither side wings of bushes nor nets had been necessary, but only the stone pen. A few more riders were there certainly, but 20 were surely enough. The birds were surrounded on top of the hill and then slowly pushed towards the pen.

The following day the expedition suddenly came upon 4 goose families, but only one adult and 3 goslings were captured. The participants conclude that their inability to bring the group together was due to their sudden closeness in the beginning. When the surprised birds were set upon so suddenly, their instinctive reaction was to flee singly and to duck down alone. When they are driven from some distance, the herding instinct is much stronger. According to Scott and Fisher there are two influences: (1) the security in the flock, and (2) panic, dissolution of the flock, each goose for itself. It is therefore important to maintain (1) and to avoid (2). Once (1) is lost, it is difficult to reassemble the herd; but the larger the flock the easier it becomes.

The last large catching day of this first Icelandic Bean Goose expedition was July 31 with 200 additional geese. The total bandings amounted to 1151 geese (382 adults, 769 goslings). On August 2 the expedition members started on their return journey.

In October 1951 Scott and Fisher caught 530 Bean Geese with rocket nets in Scotland. Among the birds were 9 geese they had banded 3 months earlier in Iceland.

In 1953 a second expedition took place. Scott, Boyd and Sladen (1955) give a detailed account. They reject the possibility of the report being misused; on the contrary, they point to the importance of publishing such experiences in order to have the best results with the least mishaps during future expeditions, particularly as Scott needed two trips to find the best catching methods!

Only during the last weeks of the 1951 expedition did the participants conceive of the idea of surrounding the geese as a herd—as the ancient Icelanders did—and to drive them into a net pen. In order to exploit this method of catching to the fullest, the number of expedition members was increased to eight in 1953. Even more participants would have been ever better, but that was impossible at that time.

In 1951 two important observations were made in connection with the new catching method: geese run uphill when pursued and, having reached the top, they stop to look around. In 1951 and during the first catches in 1953 the tactic followed was to position "backstops" (people who prevent the geese from running too far) behind a suitable hill near a marshy feeding area. The geese were driven up the hill and when they had reached the top their escape was cut off.

The birds are distributed irregularly over the region but there are areas where food is so abundant that geese are regularly present. This area may be only a few hundred meters distant from the next hilltop. Such drives can net several hundred geese depending also upon the surrounding terrain. Some hills provide better cover for the approaching "backstops" than others.

After beginning to try larger drives, new techniques were developed. In the middle of a large feeding area a hill or wall—no matter how small—was chosen. Expedition members were posted fairly equally around it in a large circle. They then moved forward simultaneously and according to a set time table towards the center giving the birds enough time to walk ahead. It is essential to the success of this catching method that the birds not be rushed.

Requirements for this simple plan are the natural features of the land: the glacier tongues with their various moraines and the river Djòrsà. On the other hand neither the moraines nor the river presented any obstacle to the geese, on the contrary, they provided them with ideal possibilities for escape. For man and horse the walls of soft grey gravel were practically impassable and crossing the Djòrsà was an additional obstacle. Between these two obstructions the tributaries flowing from the glaciers to the Djòrsà were, in the majority, only to be crossed by horse and at selected fords. The big drives were thus limited to a certain extent by the landscape.

"Nonetheless, the big drives which netted 1000 to 3000 geese were the most successful and did not seem to split up the congregations any more than smaller parties. With today's knowledge and experience it would have been possible to mark the same number of geese in maybe four big drives rather than 36, plus maybe four or five smaller ones at more remote sites. Such a program could have been accomplished in 2 weeks between July 18 and 31. Had a crossing of the Djòrsà been possible, an additional four drives would have been required." (Scott, Boyd and Sladen 1955). Lay-over days between drives are necessary to give horses and people a rest. Iceland ponies can work two days and then need a day's rest to eat in peace.

As soon as a flock of geese assembles (on a hill), a feeling of togetherness develops (provided there are more than 50 birds) which each bird experiences as part of the flock; as a member of that flock the bird feels safe.

This is true of most herd animals, and the herding of geese resembles in many ways that of sheep. Thus a flock of 3 000 geese can be kept quiet by 4 or 5 people. In such a situation goslings and sometimes even adult geese begin to eat, preen, sit down and even sleep. The lack of all panic and even all evidence of fear on such occasions is amazing. The fact that even those birds that could easily fly away stay in the flock until they are driven into a pen illustrates the particular feeling of security they experience within the mass of other geese.

In 1951 the nets were normally set up before the start of the drives (sometimes several days ahead); but on August 1, 1951, the geese were rounded up before and the nets set up later while the geese were kept together by the other "hunters". This was the method followed in 1953, and it proved highly successful.

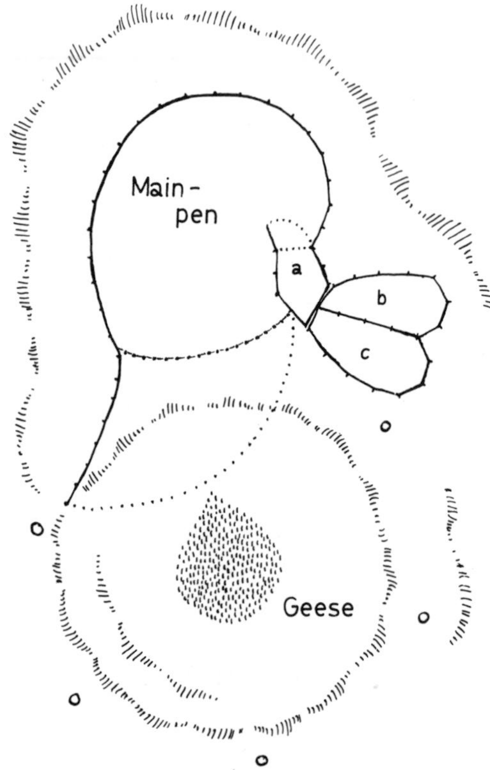

Fig. 363. Iceland-expedition 1953: A large flock of Bean Geese on its way into the pen. At some distance five participants are posted, slowly driving the geese. (a) pen; (b) pen for adults; (c) pen for fledglings. After Scott, Boyd & Sladen, 1955.

It is very interesting what Scott, Boyd and Sladen (1955) have to say about the last stage of catching and banding as well as the equipment used. In 1951 a considerable length of net was carried along, which also permitted the setting up of long wings. Now only nets for the corral and holding pens were necessary. The catching equipment consisted of three nets for a total length of 120 m, about 0.9 m height and 13 mm mesh. The net was attached to a 1.2 m high pole at 1.8 m intervals. Unfortunately the net was not strong enough. The geese damaged the net and a small number escaped. After that the net was doubled thereby reducing the length by half. The netting must be rip resistant. In addition it proved necessary to decrease the distance between poles to 1 m maximum. A sufficient number of sturdy but—due to transportation reasons—light poles must be carried. The poles must also be smooth so that the nets won't tear as happened to the English ornithologists several times. The nets remain attached to the poles. This facilitates and speeds the set-up.

After the geese were surrounded, 5 men evenly distributed around the flock kept them together (Fig. 363). The other three unloaded the nets (off the ponies) and picked a spot 40 or 50 m distant for the holding pen. This spot should be largely free of stones as geese easily damage their toenails. Two men unfurled the net while the third drove the poles into the ground with a camping hammer. The main holding pen was shaped in the form of a horseshoe. One side was lengthened 4-5 m. On the other side the net was doubled appropriately but pulled towards the inside to form a small catching pen approx. 4 m long and 1.5 m wide. It is advisable to set up two additional holding pens, one for adults and one for young birds. If kept separately the young cannot be stepped upon by the adults and families find each other more easily after release.

After the flock of geese has been driven into the large enclosure (Fig. 363), the long net side should be carried around to close the entrance. The actual size of the pen varies with the number of geese—a number usually estimated too low without at least some counting. A circle of 10-12 m diameter proved to be sufficient for 600 geese. Both of the big catches of over 3 000 geese required all 60 m of the double net, providing a diameter of 20 m.

Then banding began. The bands were unpacked and the strings of 100 bands laid out sequentially. The ponies are turned lose to forage, corn hauled along in bags is fetched to feed the geese while they are waiting to be banded. The successful catchers too must think about eating as many hours of labor are ahead of them.

The captured geese remained unexpectedly calm; many young birds slept, while others ate. It took about 45 minutes before banding was started. Groups of 30-50 animals were driven into the small holding cages (Fig. 363 a). During this process it was possible to separate adults and young. Three members of the expedition were in the cage with the birds. Each goose could be grabbed easily, which helped speed banding. The men grabbed two or three geese at once and held them so that three people outside the cage could attach the bands. Each bander had two consecutive strings of 100 bands each, one over his shoulder, the other hanging from a pole for adults and young respectively. As the wing markers which had been used on a number of young birds in 1951 had not been very successful, they were no longer used in 1953. Banders managed to band 120 birds per hour, including the time needed to drive new groups of birds into the holding cages. The 7th member recorded recaptures and noted when a bander started a new string. The 8th person patrolled the pen to see whether the geese were pressing anywhere against the nets thereby possibly trampling other geese.

After banding the geese were transferred to separate pens (Fig. 363 b, c) and were released in groups of about 200 birds. There were not always enough nets for these pens and indeed it was a bit dubious if they were really needed. After release the goslings generally stayed outside the main pen, tried again and again to enter and then wandered off to eat down to the meadows just below. After the adults had also been released a general reunion of the former groups began—a happening which was later confirmed through recaptures in Iceland and Great Britain.

A large percentage of families reestablish the old bond again after capture and banding. Those goslings which did not find their parents again afterwards formed a new group. At that time they were no longer dependent on parental care.

From 10 July to 6 August, 1953, a total of

12,310 Bean Geese were captured during this Icelandic expedition:

New birds:		
Adults 3,884	Young birds 4,861	Total 8,745
Recaptures during this expedition:		
Adults 1,647	Young birds 1,659	Total 3,306

The large catches are distributed as follows: 21 July—3,167 (including one Brant Goose); 29 July—3,115; 1 August—1,892 Bean Geese.

These two Icelandic expeditions of the British Wildfowl Trust in 1951 and 1953 begot a host of similar enterprises. In 1954 West Spitsbergen (Norway) was the destination of GOODHART, WEBBE and WRIGHT (1955) after an exploratory visit in 1952 to the archipelago during which 42 goslings had been banded. In 1954 in the area of Gipsdalen 568 Bean Geese, 23 Barnacle Geese and 80 Brant Geese were banded.

During a 1962 expedition to West Spitsbergen Norwegians banded 685 adult Barnacle Geese within 4 days (LARSEN and NORDERHAUG 1963, NORDERHAUG 1963). They used a 90 m long and 0.9 m high nylon net (40 mm mesh) as well as aluminum poles. Contrary to catching methods on Iceland, the travelers to Spitsbergen erected the net out of view of the geese. Catching Barnacle Geese was different insofar as these birds were mostly encountered on the water and had to be driven off it. The men therefore carried a raft along—otherwise any success would have been doubtful.

Other observation and catching expeditions of Norwegian ornithologists seemingly met with little banding success (for example NORDERHAUG 1964).

DENNIS (1964) captured 39 of 153 molting Canada Geese on the Scottish Coast, aided by two boats, 8-10 helpers and the usual nets. It was a difficult enterprise.

A 6-member expedition force went to Greenland in 1956 to determine the numbers of Bean and Barnacle Geese and to band as many as possible (GOODHART and WRIGHT 1958). Only a few Barnacle Geese were actually banded. Banding figures worth mentioning were achieved by an expedition in 1961 (MARRIS and OGILVIE 1962)—450 adult and 119 young Barnacle Geese as well 6 Bean Geese. The English ornithologists carried all equipment themselves: an inflatable raft, a one-man kayak and 2 nets (approx. 60 and 30 m long). During catching the nets were fastened to aluminum poles every 4.5 m. In between were bamboo poles. Nets and poles were rolled up ready for catching. Actual catching was usually done right next to water. Soviet ornithologists have banded many of the numerous geese in the arctic of their country.

During the three summers from 1952 through 1954, COOCH (1957, 1953, 1955) banded 4,450 Blue Geese and 10,550 Snow Geese in Canada. In general, the catching method used was that applied by SCOTT in Iceland, which in turn had been known in its original form to the Eskimos for centuries. COOCH (1957) describes the simple but effective catching technique of the Eskimos. His own experiences, too, are informative and of special interest.

At the start of an Eskimo-hunt a group of men walks almost parallel to the direction taken by the geese. In the warmth rising from the earth the birds see the figures appear only hazily, and when these figures do not approach them very directly they only slowly walk away from the intruders. If approached directly, many will flee. The technique consists in approaching them indirectly and leaving a man behind from time to time in order to cut off their retreat. The drivers who continue try to have the geese move in concentric circles slowly and companionably. Otherwise they would get frightened and run apart. This procedure is followed until one man is left who continues on his way. Often one driver must run very quickly ahead of the herd in order to close the circle. The men left behind from time to time stay in hiding until they see that one driver has walked around the herd, and that the geese now begin to wander back to their original feeding area. No matter in what direction they now turn their path is cut off by figures which now appear on the horizon. Confused they look for the largest gap between the men. These gaps are closed by running across the path of the geese. This is fairly easy as the men running on the horizon appear very fast. Finally the geese are totally confused and give up. They form a large herd milling about aimlessly and call so loud that oral communication between the drivers becomes all but impossible. Now the birds are under control. Except for one person, the men now begin to walk slowly towards the herd. The former now walks—without turning once—towards a pen. The geese run from the drivers surrounding them and follow the one who apparently retreats.

When the man leading the herd reaches the pen, he walks through the entrance and climbs out at the other end while the geese throng into the pen. The Eskimos undertook long drives. Even today stories are told on Southampton and Baffin Islands (Canada) of drives lasting several days. These drives were interrupted when the geese got tired and then continued until the goal was achieved. In this way it was possible to lead a small flock of geese into a tent. It truly happened in 1952 when a small group was needed for studies.

The stone pens in which the Eskimos caught their birds are not particularly impressive. SCOTT and FISHER describe and show pens they found in Iceland and which are not substantially different from those Cooch found. According to reports from MAUNDER (1852), the inhabitants of certain parts of Siberia used a similar method for catching geese.

Many researchers have found stone pens in locations far from today's breeding areas. Some were of the opinion that this meant these areas were used as breeding places. This is not necessarily so. According to information obtained by Cooch from Eskimos on Southampton Island and confirmed by his own experiences, the pens were often constructed close to camp and the geese were driven there from miles away. The procedure followed during the bandings of 1952 and 1953 (COOCH 1953) made use of the extensive mud flats of the Bay of God's Mercy (Southampton Island, Canada). At low tide two or three people went out on the flats until they were no longer in view of the camp. Then they walked parallel to the water until they believed to have cut off a large part of the flock. The intent was aided greatly by a small island just beyond the water line which permitted them to approach the geese unseen and to guess at the number and location of the birds. Two other men remained in camp until they were sure that the mud flat group was far enough away. Then they went to a predetermined location about 3 km inland where they separated. One man walked parallel to the route taken by the group, the other walked slowly towards the island. Finally the two groups came into view of each other or could at least see how the geese moved away from the mud flat group on the horizon.

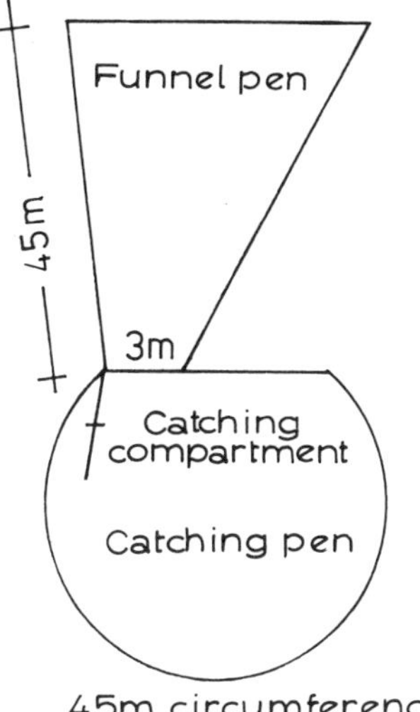

Fig. 364. Goose catching pen with a large funnel pen as used for catching Snow Geese in the Canadian Arctic. The entrance width of this pen is not mentioned, but is possibly around 20-30 m. The funnel pen is closed with wire mesh as soon as all geese have entered it. After COOCH, 1957.

Thus the geese were already caught between the two groups even before they had really noticed their presence. The only time the men had to hurry was during the first few minutes on land. Once the geese were forced aside, it became easy to drive them into the pen.

Almost unbelievable results can be achieved with this method. In 1953 15,000 birds were surrounded in two drives. Such large catches are not desirable: in both cases they happened accidentally because of inaccurate timing.

Erecting an appropriate enclosure was simple (Fig. 364). The pen of about 50 m circumference and the approx. 100 m long wings were originally made of heavy 1.2 m high nylon net with 25 mm mesh. The net was attached to 1.5 m high bamboo poles and held down with rocks and clay sod. This nylon netting was totally inadequate. The geese constantly got entangled and several injured themselves in escape attempts. In addition, the bills and tongues of Blue and Snow Geese are sharply serrated which enables them to cut nylon without too much difficulty. A large number of birds was thus able to escape. In order to cut such losses and to prevent injuries to captured birds, the central pen was reinforced with 1 m high chicken wire of 25 mm mesh diameter.

The catching pen now in use consists of chicken wire. The wings have been shortened to 15 m and the bamboo poles have been replaced with aluminum poles. Scott and Fisher pointed out correctly that such wings are not necessary, but one can use them as additional or provisional pens.

Unfortunately, Cooch had the sad experience that a number of birds in the pen died—it seems that Blue and Snow Geese are more delicate than other sturdier geese. The heaviest loss occurred on 1 August 1952 when 78 young and 19 adults died out of approximately 4,000 birds. Young birds are pushed down and stepped on and die of overexertion or shock. The mortality rate seems to increase with every passing hour that the birds have to stay in the pen.

In 1953 therefore Cooch and T. W. Barry changed their existing technique so that mass catches could be done with little or no losses (Cooch 1955 a). This method mainly considers the behavior of a large flock of geese and consists of the way in which the birds are driven into the pen. Goslings like to remain at the end of a flock when it is being driven. About half a mile before reaching the pen, a person divides the flock by crossing its path at a right angle. The front part which consists mainly of adults is kept back, the rear part of which 90% are goslings is led into the holding pen and the gate is closed. The flock of adults is then led between the wings which are then closed and thus forms a second pen.

Through this procedure the goslings are protected from the adults. They can be banded in the main pen in half the time needed otherwise and then released. As soon as the main pen is free, additional goslings from the second flock between the guides are driven into the main pen. This is possible because the goslings like to gather in one spot, usually close to the flock in the main pen. Additional goslings are driven into the main pen several times more until there are only adults left for banding. When the goslings have been released, they usually remain close to the pen. Each time an adult female appears among them after release, one or two young will follow her. Thus the number of birds without parents to look after them is substantially reduced.

When the cages were held separate this way 1,694, 1,722 and 3,719 birds were captured and banded in 1953 without one casualty. At the beginning of the season no less than 5% of the goslings had lost their lives during small catches with goslings and adults penned together.

In concluding, let us do some comparing between catching Bean Geese in the hilly terrain of Iceland and Blue and Snow Geese in the tundra of the Eastern Canadian Arctic. The catching methods employed by Scott in Iceland were specific for the Bean Goose and the landscape and are only generally applicable to tundra and mud flats. It also seems that the smaller Blue and Snow Geese are more delicate and therefore require somewhat different treatment. Scott always drove old and young birds into the holding pen together.

The goose banding expeditions in Iceland as well as in Canada and the other areas have shown that astounding successes are possible with determination and selfless efforts.

17.6.3 Catching swans during molts, as fledglings and in the hunger period

In August 1962 Kinlen (1961/62) and three companions were in Western Iceland in order to band adult Whooper Swans during their molt and unfledged young birds. Most of the methods that are designed to capture many birds at once proved futile because of the shyness of the birds, their speed when fleeing, and their disinclination to leave the water when danger threatened. By driving them and wading into the often shallow lakes, the banders nonetheless managed with swan hooks to catch 49 cygnets and 6 adults out of the many hundred birds observed. Some of the young were caught on land when they and their parents were on their way from one lake to another. Despite these poor catching results it does seem possible to Bub that larger catches should be possible if properly planned.

Catching Mute Swans in winter on the icy coasts of the Baltic in Germany does not involve flightless molting birds, but the circumstances of catching these hungry and therefore trusting birds resembles a bit the goose catching of Iceland and Spitsbergen (Norway). During the winter of 1969/70, according to H. W. Nehls, about 800 Mute Swans and about 10 Whooper Swans were banded on the Baltic Sea coasts of Germany. In February 1970 Nehls and several companions captured 152 birds, a maximum of 70 during one operation, in one day at four different locations. The swans were fed, usually next to or close to a wall of some sort (house, pier, etc.) which later facili-

Fig. 365. A method to catch unfledged or flightless swans. After Grant, 1968.

tated the catch. A 15 to 20 m long coarse meshed net stood ready at the feeding location. When the birds were busy eating the grain on the ground, the catchers surrounded the entire assembly with the net. There was always considerable commotion among the swans as soon as they noticed their capture which then required great skill and labor to keep the swans inside the netting.

Single birds were also captured during feeding, but this proved hardly worthwhile mainly because the swans quickly become shy. When large numbers of swans are present, catching should be planned on a large scale from the beginning.

Another method for catching swans singly on smallish streams is illustrated by Grant (1968). The net does not have to reach from bank to bank. Each person has enough rope on his side so that he can pull the net quickly to his side if the need arises (Fig. 365). Three people are involved in the catch: two follow the banks, the third awaits the swan.

If the driven swan happens past the end of the net, the sight of the rope is often enough to discourage it from turning around. It is much more desirable, if each catcher has enough rope so that the net can be drawn close to shore. When the end of the water has been reached, two methods can be used. Either the swan climbs out of the water on to the shore where it is captured by the third person, or, and this is the better solution, the net is dropped over the bird as it reaches the shore.

There is no need to shout during catching. That would only panic the swan and cause it to squeeze past the net forcing its escape route.

Bub considers it possible to drive whole groups of swans into a prepared net pen with this method paddling in boats.

In Sweden Mathiasson (1973) captured molting non-breeding Mute Swans in a small boat with a powerful motor (8-24 hp). A small part of the large molting flock (which cannot be driven onto land) is guided from the shallows into the deeper water before the coastline, where there are no rocks which make using a boat difficult. The swans remain in formation close together and can be kept together by a circling boat. One by one the swans are separated from the flock and captured either with a net or a neck loop which is fastened to a 3-5 m long bamboo pole. As soon as swans are put into a boat they become totally docile, and one can deal with up to 8 birds at once and without any difficulties. If more than 8 birds have to be dealt with at once, the banders put small black hoods over their heads. The birds then remain quiet. In general, the crew of a boat consists of 3 people: one operates the boat, one catches, and one bands and records the data, which includes age and sex. Sometimes weight and measurements are also taken. Before release the Swedish ornithologists mark the swans' heads with color dots for two reasons: one, to avoid catching the same birds over and over again, and second, to gain data on movements after the molt when the large flocks break up. There are different color-dot codes for the various areas where swans congregate to molt.

From 1964-1967 645 Mute Swans were banded through the capture method described above. There were 231 recoveries and recaptures.

17.6.4 Catching molting Common Shelducks incapable of flight

North of the Weser delta (Germany) lies the "Knechtsand"—a large sandbank with a small island core. There, tens of thousands of Common Shelducks molt each year. No systematic catching and banding of these birds has been carried out. The vastness of the landscape and the many escape possibilities for the ducks makes any catching difficult from the beginning (Fig. 366).

FREEMANN (1956) worked with several methods as the opportunities arose. One was to send to the large sandbank several fast runners who were to drive the shelducks residing there into the water, where a number of helpers waited for them. Most birds moved to the left or right, but nonetheless a substantial number of them could be captured. Another time the fishing boat made for the sandbank full speed ahead and turned 90° when it still had about 10 cm of water under its keel. Now, while the boat was moving along, one catcher after the other jumped out. Thus a line of people was placed in front of the sandbank, cutting off escape by the ducks to a tidal channel. This method yielded the most birds. Occasionally good results were obtained by largescale operations to surround the birds on the sandbank. Good and fast runners are needed for this. The number of helpers in all instances can hardly be large enough. Nonetheless, disappointments happen. The maximum for one year was 400 bandings (Fig. 367).

Great care is necessary while carrying and banding the birds. The plumage on their bellies must not be disturbed, as this would render the birds no longer truly waterproof and they would die miserably.

E. RADDATZ made an interesting observation on the behavior of Common Shelducks. When he surprised the ducks at close range on the sandbank, for example by suddenly appearing

Fig. 366. Molting Shelducks on the Knechtsand (Germany). Aerial photo: Dr. F. GOETHE. Released by Reg.-Präs. Düsseldorf Nr. BN 21.

Fig. 367. Two molting flightless Shelducks were caught by hand in water. Photo: Dr. F. GOETHE.

from behind a dune, the birds would lie flat on the ground with stretched out necks. If the observer changed his location, the birds would follow his movements by turning their heads and necks. They allow a person to approach as close as 3 m, and then they flee suddenly.

Since 1969 H. OELKE and coworkers have been catching and banding during darkness, preferably with a cloudy sky and rain. On the other hand one should not be on mud flats during thunderstorms. During low tide the large mud flats present good catching opportunities. The Common Sheldrucks are blinded with powerful flashlights. Thus it is possible even to surprise large aggregations of geese. Under good circumstances one catcher can obtain a maximum of 40 birds during one tide.

Holding the captured geese always presents the greatest challenge. OELKE and coworkers each carry a woven basket on their backs, 50 to 60 cm high. However, more than ten birds cannot be kept in it without endandering them. In addition, the weight of more than 10 birds is a burden greater than the average bander can carry.

17.6.5 Catching molting or unfledged mergansers and other ducks

We have already mentioned catching molting ducks. One or two centuries ago the opportunities for catching many ducks during their molts must still have been very great in many parts of Europe. Today this may only be true for eastern Europe where nature and the environment are less subject to interference from man. Catching ducks was done mostly with funnel traps (Fig. 368) and stationary nets which were set at some distance from each other similar to the arrangement used for lark catching. BIRKNER (1693) shows us just such a hunt. WINCKELL (1878) mentions a method in which small funnel traps and leads are set up in ditches leading through overgrown wet meadows or marshes. When molting ducks and other flightless birds

Fig. 368. Catching of molting and young ducks. After PAYNE-GALLWEY, 1886.

are pursued, they usually run towards the ditches as the best route to reach the close-by rivers, ponds or lakes. Thus they get into the funnel traps and are caught.

GAWRIN (1964) describes a catching method for diving ducks in Kazakh S.S.R. (U.S.S.R.), where the natives have always captured molting ducks and geese with great skill. The hunters prepare light-weight (stationary) nets with a mesh of 50-60 mm, 50-60 cm high and 20-25 m long. These nets are fastened to sturdy and smooth ropes. The hunting party consists of 4 people who, on a windy day, row in light boats to a previously determined location where the ducks molt. It is unwise to try to hunt the ducks on windless days as they—and especially the geese—are too attentive and cautious. It is almost impossible to get close to them on calm days.

The most experienced hunter among them approaches the noisy flock and determines the place where the nets should be placed. Now the hunters—two to a boat—carefully begin to set the necessary number of nets. To do this they make a half circle around the reeds in which the ducks are hiding. The hunter in the stern carefully pulls the boat along the reeds, the second in the bow sets the nets, fastening the top line to the reeds in a way so that the lower edge of the nets is 5-10 cm submerged in the water. Experienced hunters sometimes set their nets only 10-20 m away from the molting flock without causing panic. After the nets have been set in a semicircle—an operation taking about 20 to 30 minutes—the hunters start the second and most important part of their task: driving the ducks into the nets. The difficulty consists of shooing all birds at once in the right direction.

The number of captured ducks is often great: It can happen that 10 nets with a total length of 200-250 m catch 200 to 300 ducks with one drive because they are totally entangled with head, wings and feet. Four experienced hunters can thus bag 400 to 500 ducks in one day.

KARTASHEV (1962) made a remarkable observation on the flight-worthiness of molting ducks. On a small lake in the tundra of the Kola Peninsula (U.S.S.R.) about 1 km from the sea in August of 1947 he saw a flock of 6-7 Oldsquaws in flight of which he shot three. They were adult females whose primaries and secondaries were missing. The wing surface was made up solely of coverts, none of which were missing. All retrices were also present. KARTASHEV had the impression that the frightened birds were afraid to fly over solid ground and therefore always circled down to the lake again. They repeated this performance several times and did not leave the lake despite the fact that they were shot at several times.

Catching goslings hatched at several Bavarian lakes (Germany) proceeded as follows according to E. BEZZEL. In mid-summer when the young are ¾ grown a suitable river is closed with nets. Water is low at this time in the small Alpine rivers and a narrow spot can easily be closed in a U-shape. However, it is important that a large crew is ready which will slowly drive the young 2 to 3 km upstream toward the net. This must be done very carefully so that the adult in attendance does not fly away but rather stays with its young. Otherwise the goslings will disperse and hide and refuse to be driven any further. Shortly before reaching the U-shaped net they must be driven somewhat faster, and then it becomes important to grab the birds fast without hesitating. Capture of all young is not certain with this method as the goslings are able to tear through the thin net and therefore a funnel-type trap leading to shore should be tried out.

17.6.6 Capture of molting pelicans, cormorants, grebes, cranes and bustards

These birds also were hunted with greater or lesser success at the time of their wing molts. According to BREHM (1855) the Arabs captured many cormorants and pelicans. They barred with nets the narrow spots into which these birds drove the fish and then drove the birds into the nets with boats.

Catching molting cranes is extremely difficult as one can imagine with these cautious and shy birds. GUNDA (1968/69) gathered valuable information on hunting cranes by Hungarians. During the second half of the 19th century cranes still often nested in the Great Hungarian Plain where the birds molted at the end of the summer. In the lowland Hortobagy plain the running bird was pursued by horse herders on horses. The herders often fastened a lead ball to the end of a 2-3 m long whip. The whip (similar to the Bola) wrapped itself around the bird's neck thus catching it. During a chase cranes often would lie on their backs defending themselves with legs and bills. The herders caught such birds by throwing a large piece of clothing over them. If the bird was desired alive its legs would be bound together. Molting cranes have also been caught with a lasso or in fall during the times of freezing rain when the birds' wings would freeze together.

Sometimes the hunter or herder would use primitive camouflage to come close to a group of cranes. He would bind a bunch of reed or rushes on his back and try to approach the group crawling slowing on the ground. It was also common for crane hunters to bind a bunch of rushes to their front and, thus camouflaged, to advance upon the birds. Then a lasso was thrown over the bird closest to the hunter.

In some of Hungary's marshes it was customary to stalk young, still-flightless cranes. The hunter would slip into a loosely tied bundle of reed, and thus hidden from their view, would try to catch the cranes by hand.

According to Gunda (1968/69) the Hungarian horse-herders, at the end of the last century, still captured bustards on horseback and with a lasso when the birds' wings were frozen together during fall's freezing rains.

17.6.7 Catching unfledged precocial birds

General. A few hints will be given here which are useful for the catching and banding of unfledged precocial birds, particularly as far as marsh and water birds are concerned. Catching flightless geese and swans has already been described.

Banding shorebirds, gulls and terns. Bechstein (1821) specifically mentions catching semi-fledged young Northern (Common) Lapwings with trained dogs which captured the young birds and carried them unharmed to the hunter or herdsman. Today patient observation with binoculars—often from a car—will reveal not only young lapwings. Every year E. Raddatz catches up to 50 birds near Bremerhaven in this manner. Lindau (1935) used a novel method by having a herd of sheep driven over a meadow. The young lapwings got up, ran ahead of the sheep and thus betrayed their hiding places.

R. Heldt catches young plovers and other shorebirds on the shores of Schleswig-Holstein (Germany) from his car. The birds run around freely even in the immediate vicinity of the car.

P. Gloe, in the same coastal area, waits for the spring tides. The rising water makes the young birds, which love to duck into low spots, get up and thus become visible.

For catching American Woodcock Mendall (1938) suggest setting up a low wire fence around the nest. The parents will jump over it while the young will remain inside until they can be banded, at which time the fence is removed. Young shorebirds may already be banded on their first day. Tieing the young to the nest site causes the parents to abandon their offspring. Search with a trained dog proved very successful. Between 9 May and 15 June the dog still found 83 unfledged young. The dog also discovered several nests with incubating birds but with one exception all were abandoned afterwards. Sandpipers *(Scolopacidae)* seem to be more frightened by dogs than by humans. While searching for young one has to be careful that the dog is very well-trained and well-chosen as he will catch the birds otherwise. The dog should drag a long leash behind him with which one can tie him as soon as he points so that he will not run further and be seen by the birds. The young will certainly scatter upon the alarm call of the adults but generally can be found again easily.

Catching unfledged gulls often has its own problems as Pepper (1965) describes. One has to try to prevent the scattering of the almost fledged young which often gather in "kindergarten" flocks. For this reason Lyon (1926) surrounded a larger number of Caspian Terns with netting so that he could band them in peace.

Pepper's description which he calls "Problems of a Gull Bander or To Run or Not to Run" may serve to show the kinds of "difficulties" that can arise during banding young gulls: "Once having arrived at the nesting grounds one has to decide which method to use: Plan A, 'The scientific method', where one divides the colony into exact areas and covers each square methodically, foot by foot; or Plan B, the aimless wander back and forth throughout the area.

"Having chosen one of these there still remain other questions. Does one frantically run after each and every young miler one sees disappearing into the distance, disregarding the dangers of stepping on a youngster hidden in the ever-present poison ivy? Does one peer into every bush and under every clump of grass in one's path, or just walk until one sees a victim—or part of him—sticking out from under the ivy? From a few test runs of these choices I think the latter brings the best results. But I'm not sure yet.

"Then there is the question of what to do when one comes upon two or more half-grown young ones and is able to catch them all at once. Just try holding even two almost full grown young black-backs by yourself, and try to band them without being well-bitten and scratched in the attempt. I have even more or less sat on one while working on the other.

"One should also try to time one's banding to periods preceding feeding times of the gulls—whenever that may be? There is nothing I know of that makes one more attractive, on returning to one's boarding house, than to have received a young gull's last meal down one's entire front.

"A few more random bits of bitter wisdom: don't kid yourself into believing that one of these teenagers that can just fly will soon tire and you can catch up to him. I've found I usually tire much faster. There is also the lesson to be learned from trying to catch a youngster that has run across the beach and into shallow water. It's no use following—unless this effort is really an excuse to enjoy a refreshing swim. These escapees can swim faster than you, and seem to enjoy just drawing you on. The best plan here is to leave the area for a short time, and then return along the beach to where the swimmers may have, as they often do, returned to shore for a siesta in the grass just above the beach.

"Having decided the above problems and learned your lessons, odorous and otherwise, all you have to do now is to tell the difference between a young herring gull and a young black-back, at all stages of their childhood, wet or dry. The one you mistake will be the one that is recovered by the expert—(he even collected it)—I think just to prove me wrong.

"Having decided which gull it is, be sure and place the band on a leg with the numbers upright when the bird is standing, so those reading

Fig. 369. Catching flightless flamingos in the Camargue. Photo: Dr. H. HAEFELFINGER.

bands by sight, i.e. binocular experts, won't have to circle the gull on their heads through the garbage dumps."

So says PEPPER. Of course, it won't be so exciting all the time. Mostly the young gulls duck into the high vegetation. But careful as one may be, equipped with dip net and sacks for covering the young, excitement will spread in varying degrees throughout the breeding colony. The dip net not only serves to catch fleeing youngsters but one can also keep it close at hand during banding to defend oneself against attacking adults.

We endured many difficulties during the years 1952-1959 while banding about 650 semi-fledged Herring Gulls. The nests were on the ruins of concrete harbor structures in Wilhelmshaven (Germany). Doubled over, BUB sneaked from one nest to the next, covering the young with sacking and if necessary leaving them beneath the cover after banding. Even so he could not always prevent one young gull or another from jumping off the concrete and into the water several meters below. It would then later try to scramble back up at another spot. But many were covered with a dip net just in time before they jumped (cf. DROST, FOCKE and FREYTAG 1961).

Banding flamingos. Many unfledged Eurasian Flamingos are banded each summer in the Camargue (Southern France). Fifty to sixty people surround the birds, at first in a large circle which slowly but steadily draws closer together. With such a drive the birds are pushed into a pen (Fig. 369). 1,800 birds have been captured within 3 hours. The extensive experience of L. HOFFMANN indicates that the wire mesh pen for the unfledged young must be located on absolutely dry ground. If necessary the ground must be covered with straw as the birds otherwise get absolutely filthy. More than 1,000 birds should not be in one pen as they might suffocate each other if too many are together. Furthermore, one must make sure that the youngest chicks are older than 1.5 months. Younger birds are too delicate for this particular catching method. The birds are banded above the heel. This does not harm the birds and the bands are protected from corrosion. WILLIAMS (1963) reports on similar catching in East Africa.

Catching ducks. A group of Dutch banders (F. J. KONING, H. VADER, P. VAN SPANJE) developed a method for catching unfledged Tufted Ducks (KONING 1970), which can also be used for catching other ducks and water birds. The net which is described is the result of years of testing with various sizes of horizontal and vertical nets in water. It was used successfully in an area with many ditches and small lakes which supply Amsterdam with water. Whole duck families can be caught in this manner.

The net is 6 m×4 m and is simple to set up. It is attached to eight bamboo poles 1.5 m long. The connection in the middle is made of metal, the connections between the outer poles are pliable plastic (water hose). The outer poles are connected with strong nylon ropes which lead to the shore of the ditch or small lake (Fig. 370). If the ropes on both sides of ditch or lake are pulled at the same time the outer poles lift out of the water and fall on the inner poles.

The best results were obtained with four people: two to operate the net and two to drive the ducklings to the net. The latter must be done very slowly as the ducks will dive and scatter otherwise. If the ropes are held directly above the water surface, the ducks will swim on wide paths over the middle of the net. This is possible if the catcher has the opportunity to sit a few feet above the water level. As soon as the net has closed it can be pulled to shore and the ducks can be taken out. During strong winds the net cannot be used. If it is to be used in flowing water it is advisable to tie it to shore with a small line.

If more than one family of ducks is around,

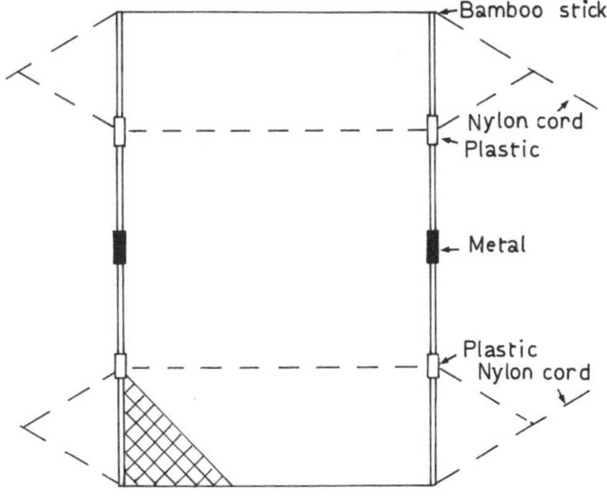

Fig. 370. A clap net to catch flightless duck families. After KONING, 1970.

they should be caught one after the other, preferably so that they do not notice the capture of the other families as they are quick to recognize danger. During 7 years of banding the Dutch ornithologists captured approx. 1,200 Tufted Ducklings. Many interesting recoveries were reported.

17.7 Catching in snow holes and tunnel traps

According to MÈRITE (1942) French catchers in the Alps catch ptarmigan by poking holes of the right size in the snow with champagne bottles (Fig. 371). These holes must freeze and ice over if catching is to be successful. In mild weather the ptarmigan free themselves easily as they can work free of the soft snow.

According to HÖHN (1969) ptarmigan reach their sleeping quarters not by walking but by flying there in order to avoid leaving a "scent trail" which might endanger them. The sleeping quarters themselves consist of protective holes dug into the snow which they will leave in the morning by flying out.

CHAIGNEAU (1961) describes a tunnel trap or better, a "hole" trap. The narrow funnel is 3 m long and 10-13 cm high, depending upon the size of the bird. This funnel slants downward about 40 cm to the other end. The catcher spreads bait at the entrance (A) and to the end of the funnel. Once the bird reaches the end of this trap it should not be able to turn around.

17.8 Catching by hand from a blind

17.8.1 Catching raptors

American ornithologists have adopted this catching method from American Indians. The bander digs himself into sand or earth at a location with a good view to all sides. Head and chest are covered with a straw hood or bushel basket which will permit some visibility. In one hand the catcher holds a live dove, the other remains free to catch the raptor with. Of course, this method requires much patience and endurance; on the other hand, it is supposed to have been very well worth the effort. Still, it can only be recommended for those areas where raptors are quite common.

SCHÜNEMANN (1957) reports in a travel diary from northern Norway that catching raptors by hand was also common in Europe and even today may occur in isolated instances. Catching places are simple holes in the ground, into which two men fit and which are covered and camouflaged. The oldest 'traps' are lined with stone and have wooden roofs most of which are

Fig. 371. Catching ptarmigan (*Lagopus* sp.) in holes in the ground. After MÉRITE, 1942.

decayed today; the newer ones are mostly made of wood only. The catcher arrives at the "eagle's blind" before dawn. Meat is put out near the trap and tied to two strings which lead to an opening through which one can reach with two arms. During the weeks preceeding the catching the blind's location has been baited with meat (such as a dead sheep) regularly, in order to accustom the eagles (mainly White-tailed Sea Eagles).

Speaking of the catching itself, SCHUNEMANN writes: After entering his blind the eagle catcher must wait for dawn. The hungry eagles flying over the island then see the bait, and generally one will land to have a closer look. Others will follow, mostly young birds, as no old eagle will touch a piece of meat that has suddenly appeared from nowhere. They are much too suspicious. It would have to lie there several days before they would get used to it. But young birds can be tricked. There is always one among them which is daring and will begin to tear a few pieces from the bait. Then the others will join in and soon it will be a free for all fight for the meat. This is the moment the catcher has waited for. Slowly he pulls the meat closer with the two strings. The greedy fighting young eagles do not notice that they are next to the dark opening of the 'eagle's blind'. Now the catcher quickly grabs an eagle with both hands and pulls him into the blind (Fig. 372). Of course the bird will flap its wings and scream, but the others are doing so as well in their voracity and do not notice that one of them disappears.

The catcher ties the bird and puts it beside himself on the bench. He now repeats the procedure catching one eagle after the other until the rest notice something and promptly fly. Now the eagle catcher must wait until dark before leaving his blind. Should an eagle see him come or go at the catching location, all hope of further catches would be in vain there. In former times the captured eagles were killed because they were considered a danger to grazing sheep.

Those were terrible times for the eagles! As late as the beginning of our century, it is said, one catcher alone without help caught 32 eagles and on another occasion 22 with a helper. In that same fall two eagle catchers once caught 40 eagles within 2 weeks. One man caught 10 eagles, another 16 in one day. When Schünemann visited the coast of northern Norway about 20 years ago, only one eagle catcher on

Fig. 372. A White-tailed Eagle is caught by hand from an eagle blind. After SCHÜNEMANN, 1957.

the island of Vaery still plied his trade. The slow reproduction rate of eagles makes such decimation of their numbers intolerable.

17.8.2 Catching gulls

Whoever might believe that there are no more possibilities to improvise in bird catching has not yet heard of Danish ornithologist EDDIE FRITZE. Inspired by the method American Indians used to catch eagles, he dug himself into the refuse of the city of Copenhagen (Denmark). From May 23, 1970 until March 1, 1971, he captured—mainly on weekends and during his vacation—the following birds: 3,582 Herring Gulls (2,384 adults and 798 juveniles were banded), 77 Greater Black-backed Gulls (63 adults and 1 juvenile bird) and 1 Lesser Black-backed Gull (Fig. 373-Fig. 375). This is a great record! Recaptures from among the Herring Gulls came from the following countries: Sweden, 34; Finland, 25; Soviet Union, 1; Germany (East), 67; Germany (West), 5; Belgium, 1.

How exactly did he do it? First, sturdy and appropriate dress is needed. Even if plastic bags and other packaging materials form a large part of any dump, one must protect oneself against wet refuse. FRITZE by no means digs himself in entirely, but rather covers himself half lying

Fig. 373. A small section of the dump of the city of Copenhagen. In the middle E. Fritze surrounded by Herring Gulls. Photo: P. Fritze.

down up to his chest. His most important piece of clothing is a 'magic hood' made of a plastic bag with openings for eyes, nose and mouth. Similarly equipped, he waits for the gulls. The length of the wait depends upon various circumstances. During weekends, when nobody is working around the dumps, other people looking for usable items or even illegally hunting

Fig. 374. The Danish bander Fritze with captured Herring Gulls. Photo: P. Fritze.

Fig. 375. The camouflage hood has been taken off. Photo: P. Fritze.

the gulls can disturb the catching; dogs running loose do not help either. Preferably, all disturbances have disappeared preceding the catching—something that cannot always be achieved. The gulls are used to the people and machines working there during the week. The catcher is, however, helpless when he has already installed himself in his catching location and people picking over the dump appear. Nevertheless, one time two men fled precipitously when one of them attempted to lift the 'magic hood' and noticed two arms which suddenly held it fast and the other even saw a hand appearing out of the glove.

First the Black-headed Gulls come close to feed off the refuse, but they are not captured. Fritze himself often has two sacks of old bread next to him so that the gulls have an incentive to approach, especially when few edibles are around. The Black-headed Gulls seem to be the most unconcerned and boldest. Where they are, the Herring Gulls soon appear and then immediately dominate the feeding place.

Fritze does not immediately begin to catch, and often he would not even be able to if he wanted. First he must support the weight of 20 to 30 Herring Gulls, who perch on his head, shoulders, arms and chest. But within a few minutes the gulls disperse and now catching begins in earnest. Fritze grabs the birds with his gloved hands, avoiding sudden and fast movement. Normal arm movement is not noticed by the gulls as by now they are busily searching and feeding. With a firm grip around the back or wing close to the body the bird can be pulled close for banding. Certainly the bird screams, but because many other gulls are doing the same, it is ignored although one would imagine that those screams differ somehow from the others. After banding Fritze immediately frees the gull. Under favorable circumstances, during calm periods in the surroundings, with much foodstuff and when the gulls were hungry, he caught and banded up to 50 birds in one hour, i. e. almost one bird per minute. During August and September it is easiest to catch the young birds who are least shy. On 23 August 1970 Fritze banded 210 birds.

Sometimes the gulls all fly away together. Either a disturbance is approaching or the birds are full. But generally they reappear in about half an hour.

Catching is possible in all types of weather. During times of heavy rain or storm the birds seem to be somewhat more cautious. Greater Black-backed Gulls are considerably more suspicious than Herring Gulls, and catching them requires greater caution. On 27 December 1970 Fritze managed to catch 20 adults in a single day, but every catch of a Greater Black-back is an event. If a Greater Black-backed Gull was around, Fritze had to forego catching Herring Gulls in order to avoid scaring the bird away.

Other species react differently: when the gulls were absent, House Sparrows sometimes came, but they were difficult to catch by hand. Common Starlings came rarely, but they too reacted quickly to attempts at grabbing. The one Common Starling that was caught wore a band from Helsinki (Finland).

Common Jackdaws and Carrion Crows were surprisingly cautious. Fritze had the impression that the birds saw his eyes under the 'magic hood' and they kept their distance at 10 to 15 m. One Common Jackdaw was captured anyway, but no crows.

Banding captured birds is not without its problems, as the bander can only use his pliers and the bands which are fastened to his arms with strings. Notebook and pencil usually cannot be used. Fritze therefore bands the birds with different strings of bands in order to be able to distinguish adults and young birds afterwards.

Finally a word about the numbers of Herring Gulls occuring here. From January through the middle of August 1970 there were 11,000 to 15,000, in September the number fell to 2,000, then rose steadily until year's end.

17.9 Catching by 'calling down'

Macpherson (1897) and Timmermann (1938) report on a strange and often smiled upon (from ignorance) hunting method used in Iceland for Whooper Swans. Timmermann describes it in detail: "Several of the swan-hunting farmers took up their places in spots where they knew from experience that the post-molt swans and their young would fly close by on their way to the sea. Once the birds were sighted and had come within reasonable distance, the farmers started, upon a prearranged signal, an ear-splitting noise by yelling and screaming, barking of dogs and banging on all sorts of instruments.

Fig. 376. Hunter with dip net, after the Livre du Roy-Modus. Henry de Ferrieres, Chambery, 1486; from Lindner, 1940.

The young swans, scared to death, would fall to earth where they could then be grabbed while the adults seemed to pay no attention whatsoever to the incident and continued on their journey." Timmermann explains this surprising behavior of the just-fledged young swans: "At first flying, just as pedaling a bicycle or dancing, seems to have to be learned consciously and can only become a reflex after sufficient practice. Therefore, the young bird cannot immediately fly and pay attention to other things as well. But if his attention is forced through some unusually strong stimulus, the bird loses control over its flying motions and falls to the ground."

18. Catching in the evening and at night

18.1 General

The hunting day nowadays ends at dusk in all developed countries. In almost all European countries, and many others as well, hunting at night is prohibited. On the other hand hunting at dusk and at night was still generally practiced until the last century (Fig. 376). People used the opportunity to catch animals that were active and wary during the day in their sleeping quarters at night.

Nocturnal bird catching of sleeping flocks of starlings, Common (Gray) Partridges, European Quail, larks and others with pull nets, dip nets, spotlights or other lights was well known and widely used. These hunting methods are a thing of the past.

The only exception is scientific bird banding as there are basically no objections to catching birds in the evening or at night outside the breeding season. Of course, one has to be circumspect about it so that the birds do not come to harm. Often the birds will only be released at dawn except for nocturnal species (owls, nightjars) or where freedom will not endanger the birds (swamp and water birds in coastal areas and on lakes). Songbirds may occasionally be released on moonlit nights under very favorable circumstances as will be discussed later on.

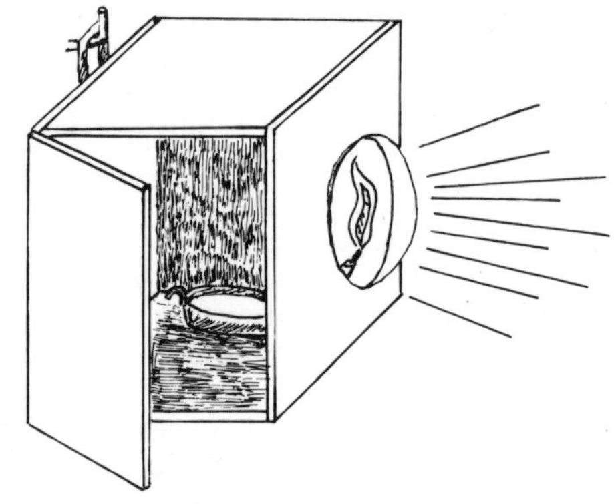

Fig. 377. Blinding lantern with lens after Leonardo da Vinci, around 1500. After Feldhaus, 1914 and 1970.

One thing should be pointed out particularly which may influence comfort and well-being of both man and bird considerably: wind and rain already make things miserable during the day and at night completely spoil all possibilities of good catches. Nonetheless, there are certain bird species which are easier to catch under just such circumstances. We should therefore make our own observations and listen to weather forecasts before each and every nocturnal catching operation to find out whether heavy rains are predicted which might spoil everything. In addition we should watch the wind, its direction and speed. The English ornithologist WATSON (1966) advises good relations with a weather station and regular listening to weather forecasts.

18.2 Light sources

Today we possess good and durable light sources. In earlier times people had to rely on torches of twigs or wax-soaked rags, oil and kerosene lamps, and similar contraptions. But FELDHAUS (1914 and 1970) shows us already a spotlight dated around 1500 designed by LEONARDO DA VINCI (Fig. 377), which apparently was fed by wax or fish oil.

18.3 Catching migrating birds at lighthouses

The attractiveness of lighthouses along coastlines and on islands to night-migrating birds is well known (Fig. 378), and we are aware that many thousands are killed and injured at such places when they are blinded by the light and fly against the structures. Many lighthouses—as for example the one on Helgoland (Germany)—have taken protective measures as a result.

It is obvious that one can catch birds for banding in such locations. The flight during migration is usually particularly heavy on dark nights. Foggy weather and light rain favor the attractiveness of the light sources. Basically, we can expect to catch all species of birds that migrate at night.

We need large dip nets, approximately 100×50 cm, for catching. The pole can be up to 3 m long. It is best to keep several dip nets with poles of varying lengths handy. During nights with large masses of birds many can even be captured by hand.

It is important to have enough space available to keep the birds. Starlings, thrushes, etc.

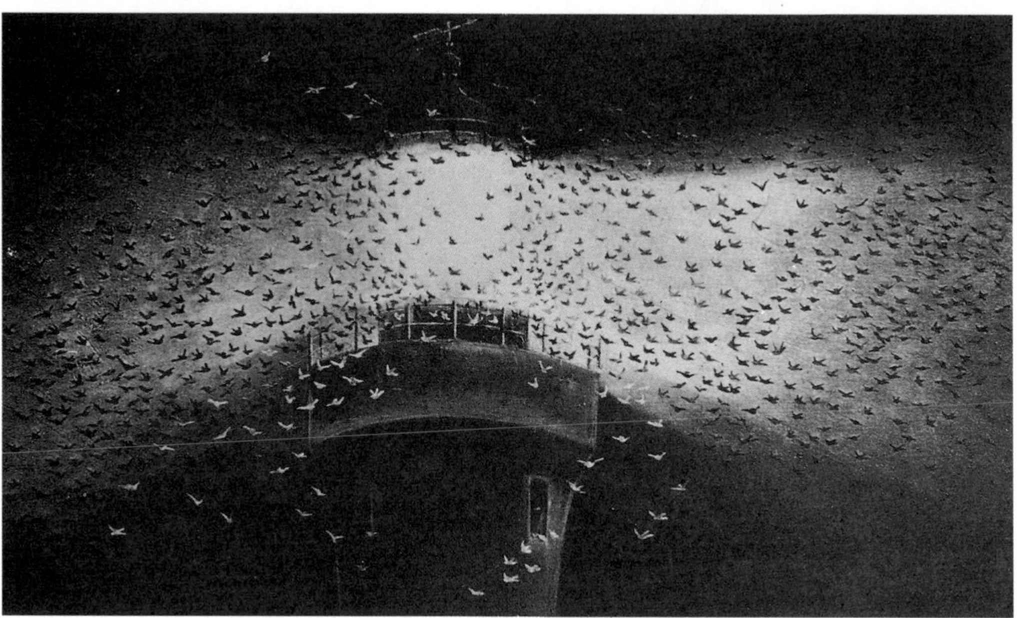

Fig. 378. Impressive rendering of the attraction of a beacon on nightly bird migration. From CLARK, 1912.

are kept best in light airy bags and then laid out on the floor in cool rooms. Banding itself starts in the morning so that the birds are not again attracted by the lighthouse and possibly injure themselves and die. In addition it is easier to determine the species in daylight and to see whether all birds are flightworthy. The catch at night varies depending on many circumstances. For example on Helgoland (Germany) in the night of 24/25 October 1927, 1,558 Common Starlings were captured and banded.

Drost (1928, 1952) describes this migration and catching night so vividly that we will listen in: "Around 8 pm the first migrating birds appear; here and there birds are seen in the beam of the lighthouse. The sky is quite cloudy and a light wind is blowing. Half an hour later a large flock of starlings—many hundreds—circle around the lighthouse. Then, for many minutes at at time, the sky seems to clear and occasionally a star is seen twinkling through the clouds. Impatience and tension are high as stars mean light for orientation and they diminish or obliterate the attraction of the lighthouse light. Several times it becomes light and dark alternately, and several times the flock of starlings disappears and reappears. Finally the weather decides in favor of darkness.

Around 11 pm a light rain starts and from then on it becomes almost totally dark. The wind is stronger and later becomes stronger still. Now the birds have come to the light in almost unbelievable numbers. The one large flock of starlings has become a veritable army. Besides them there are innumerable larks and thrushes. The three beams of the lighthouse, during each of its turns, are filled with dots, as far as the eye can see. It snows birds. Snow most easily conveys the impression to the viewer and gives an idea of the number of birds involved. Around the lighthouse is a teeming mass. At the windows where the beams originate it is a swarming as if with insects. On the gallery below untold numbers sit and run around. There's work for us! Lying, crawling and standing we grab birds, often several at once with one hand, and put them into bags.

The bags are soon full, as full as one dares make them with living birds. Now what? Where do we get empty bags, what do we do with the full ones? A brilliant thought: dump the birds in the cellar of the ornithological research station's house! Off we go, down the tower. Each of us has a heavy load to carry, so heavy that we get quite hot. Who has ever sighed beneath a load of living starlings? It is indeed quite a weight we had to carry several times that night down from the height of the tower and then on to the station, a total of 107.36 kg of living starlings! This sum is based on an average weight per starling of 74.3 g. The number of captured starlings can be calculated from this, but I will tell the number myself: 1,445!—

The last captives which are carried down the tower around 6 am, remain in the bags until banding time. We stop about 6 am, even though we could catch more, but we've had it. Everything has its limits! Even bird catching can get too much, especially if it's done all night at great height and during wet and windy weather. We also have to think about the captives which need to be banded and released. He who has ever marked birds knows how much time the banding of a single bird requires. But now we have 1,558 birds—besides the starlings a few larks, thrushes and other species—to band!

After a short rest we start banding around 8 a.m. We don't know how long it will take as we don't even know the number of birds we have captured. In addition the starlings have to be captured a second time around in the cellar. The room where the birds are is indescribable. Imagine a room about 6×5×2 m large in which 1,000 starlings fly around and babble. Our neighbors, we later found, were unable to sleep all night because of the racket and at first believed our water main had broken, which is quite descriptive of the noise. It is difficult to answer questions as to the total number of migrating birds that night. The 1,558 birds captured for banding most certainly did not even make a dent in the swirling mass. These 1,558 were surely less than a hundredth of the total. We can probably say hundreds of thousands, possibly the million mark was passed."

The bandings of this migration night in 1927 resulted in 30 returns (2.8 %) from England, Ireland, Belgium, France, Germany, Denmark, Sweden and Norway.

During nights of heavy migration, catching with lamp and dip net is also possible at the foot of a lighthouse (Dennis, 1965). From observations on Helgoland we know that many species then flutter down on house walls and fly around streets more or less helpless. At the end of September 1967 ornithologists meeting on Helgoland were able to observe this for themselves in a quite impressive display. At such times dozens or more birds can be captured with a dip net.

18.4 Catching marsh and water birds

18.4.1 Catching at night with dip net and spotlight

This is an ancient hunting method. Glasgow (1958) reports that birds were already being captured in the 15th century in Europe, Asia and Northern Africa using artifical light (torches). English literature reports that in 1528 Eurasian Woodcocks and other birds which sleep on the ground or hide there were captured this way for the English markets. Catching with artificial light was one of the most productive methods. The torch of European bird catchers consisted of a bowl filled with straw, wood or rags soaked in wax which was fastened to the top of a sturdy pole. It was held high above one's head. The catcher found the birds by the gleam of their eyes or their light-colored feathers. Besides a long-handled dip net he usually also carried a bell which he rang continuously.

Glasgow (1958) also describes the great successes he had during his nightly catching expeditions in North America. In 8 years he caught and banded 7,869 American Woodcocks plus several other birds. Catching equipment consisted of one 6-volt head lamp and a dip net with a 3.5 m long bamboo pole. The net bag had a diameter of 60-80 cm and a depth of 30 cm.

In their winter quarters in Louisiana the woodcocks spend the day in the thicket. Only at dusk do they fly to feed in the wet pastures and the corn, sugar cane, cotton or fallow fields where they are captured and banded. A seasoned catcher can catch about 80 % of the birds.

Taapken and Mooijman (1960) of the Netherlands have much experience in this catching method. They consider dark moonless nights the best, especially those with strong wind and heavy rain. They encountered many birds on the coast during high water. They also searched for birds on polders several centimeters under water. Here it was possible to pick up the birds in the light beam at greater distances. On meadows and overgrown fields the birds are difficult to discover. Thick fog is unfavorable as the birds are very difficult to find and catchers often lose their bearings.

A bird caught in a light beam has to be approached slowly and calmly. Then it is covered with a dip net. If all this is done carefully the birds close by do not notice and will remain were they are.

Taapken and Mooijman use flashlights with 3 batteries (4.5 volts) which they replace every night. Trial and error showed Mooijman and Dome that it is best to search for the birds with a weak light, for example with a carbide lamp of a bicycle, and once found then to blind them with a bright flashlight. It is important that both banders work well together. The beam of a flashlight can be made smaller with a ring of black paper pasted on the glass. Watertight flashlights were not useful as they were hard to switch on, but regular flashlights were full of water after half an hour in a rain storm. They were therefore put in a plastic bag which left only the glass uncovered. In order to protect the lamp and have both hands free it should be carried on a string around one's neck.

Taapken and Mooijman approach the birds as quietly as possible. The birds seem to frighten particularly at irregular noise in or above the water. Pulling one's boots from the mud often produces loud noises which scare the birds away. Putting one foot flat on water splashes noisily. If there is enough water on the ground it is preferable to leave the feet under water and walk as if on skis. Catcher and net must never get into the light beam. (Schmidl walked only barefoot during the warm months. On mud flats without extensive shell beds it is probably the best choice.)

All species of birds must be kept within the light beam as long as possible. This requires a lot of patience and perseverance. If the light beam wanders quickly from one bird to another the flock becomes agitated and flies off. The beam therefore must be moved slowly during the search. Generally speaking approaching a flock is more difficult than single birds, but it is possible to get close to large flocks of oystercatchers, Black-headed Gulls, ducks and sandpipers. The most favorable catching weather is usually cold, windy and rainy and the catchers' clothing must be appropriate. A southwester and rubber boots are highly recommended. Plastic and oilcloth coats are unsuitable as they are noisy during walking. A leather jacket with-

Fig. 379. (a) and (b) Catching in North Sea mudflats at night. Photo: D. SCHMIDL & J. PILASKI.

out sleeves and a watertight windbreaker keep the back warm and protect the front from wind and rain. We recommend that one keep dry clothing in the car. A compass is a must at all times.

TAAPKEN and MOOIJMAN use dip nets (diameter 60 cm) on 2.5 m long bamboo poles. A heavier net was used for Common Eiders. During rain captured birds must be kept in protected cages if they are not banded immediately.

On the English coast near Skokholm, LOCKLEY and RUSSELL (1953) regularly captured many birds (Manx Shearwaters, gulls, oystercatchers, Northern Wheatears, etc.) in calm, foggy nights with a hand-held spotlight or a flashlight and a dip net; sometimes even without the last. A thick wet fog seemed to confuse the birds. They sat quietly or flew up reluctantly in the lightbeam and could be captured by hand out of the air. The net poles were 1.5 m long and the net diameter was 40 cm.

HOLLOM (1950) informs us further: Once the gull is caught in the light beam, net holder and lamp holder walk closely together and very quietly towards the gull. Under no circumstances let yourself be tempted to rush forward in the last few meters. Once the catch has been made, the lamp is put down in front of the net so that its light hits the next birds and they cannot see the bander and get excited. Banding is done outside the circle of light in the dark with the help of a weak or screened flashlight. Arranging bands in sequential numbers saves unnecessary writing and use of the flashlight.

The time between the first and last quarter of the moon is generally too bright, especially when the sky is clear. During windy and stormy weather it is best to approach the birds with the wind. They then sit facing the bander and are often easier to recognize due to their lighter underparts. Most favorable are those dark foggy nights in which the fog lies low on the ground and everything is quiet and still, especially if it is somewhat chilly. Under truly favorable circumstances it is possible to walk around among the gulls, to cover 20 or more birds with the net and to grab the others—standing quietly on one leg—by hand. Black-headed Gulls, Mew (Common) Gulls, Herring Gulls, Lesser Black-backed Gulls, and Greater Black-backed Gulls were thus captured. The greatest number of birds banded in one evening was 159 (only because all bands were used up). The average was 20 birds. H. M. SALMON (according to HOLLOM, 1950) captured some gulls as a test in spring near Skokholm, when the gulls were already at their nests, but did not yet have eggs. He used a flashlight, approached the gulls with the wind and caught Herring Gulls and Lesser and Greater Black-backed Gulls.

As bird warden on the North Sea island of Scharhörn (Germany) H. WEHFER devoted much time to nighttime bird catching. He described the nights during new moon with heavy cloud cover and wind as most favorable. During one rainy and stormy night in August 1963 WEHFER caught 174 birds. He specialized in recognizing banded birds and capturing them. He caught a Little (Least) Tern from Poland and one Common Tern each from Finland and Sweden. His list of captured birds is quite impressive: Red-breasted Merganser, Common Oystercatcher, Northern (Common) Lapwing, Ringed Plover, Snowy (Kentish) Plover, Black-bellied (Gray) Plover, Eurasian Curlew, Bar-tailed Godwit, Common Redshank, Common Greenshank, Red Knot, Little Stint, Dunlin, Sanderling, Greater Black-backed Gull, Lesser Black-backed Gull, Herring Gull, Mew (Common) Gull, Black-headed Gull, Little Gull, Sandwich Tern, Common Tern, Arctic Tern, and Little (Least) Tern. During 1965 and 1966 SCHMIDL further captured these species: Mallard, Common Shelduck, Eurasian Golden Plover, Ruddy Turnstone, Black Tern, Northern Wheatear, and Meadow Pipit. Catching was done with a flashlight containing 5 batteries (Fig. 379a and b).

Fig. 380. Dunlins in Mudflats at night during rain.

In stormy nights with heavy rain and hail showers, catching can also be successful during full moon. In the fall of 1965 on Scharhörn (Germany) SCHMIDL was able to achieve good catches during full moon with short heavy rainfall and strong wind. As soon as it starts to rain the Dunlins crouch low on the mudflats turning their heads into the wind and closing their eyes (Fig. 380). The bill points upwards as can also be observed with breeding birds during a hailstorm.

SCHÜZ (1957) describes another method of catching ducks at night: "The fresh water areas along the South coast of the Caspian Sea are the winter quarters for immense flocks of ducks from Siberia (the same applies to the saline bay of Bandar-e Gaz on the Southeast coast). Numbering these flocks into the millions is certainly not an exaggeration. This abundance has led to a peculiar kind of bird catching at one of these sites, the lagoon Murd-Ab at whose outlet to the Caspian Sea the Iranian seaport Bandar-e Pahlavi (Enzeli) is located. Mountain rivers feed the lagoon which is slowly silting in and is therefore very shallow; open water has receeded to about 20×3-4 km. During the course of this silting a row of long, very shallow lakes such as the "Wet Sea" formed in the west on the spit and some other ponds are now completely cut off from the lagoon. During February and March these lakes are at times completely covered by a blackish mass of ducks of various kinds. It is quite an event to be able to hear the calls of the birds (especially those of the Common Teal) melt together in one vast and steady sound.

"The hunt in question takes place between August and February/March and only on dark nights. For about 6 days before the moon is full and about 6 days afterwards the hunt rests entirely. Hunting is carried out only in shallow waters, especially where burr reed *(Sparganium erectum)* grows. The ducks love its seeds and prefer the areas where it grows. Special paths are cut for this hunt so that the boats may pass without hindrance. Two boats with two men each work together. In the bow of the first boat on a non-flammable surface stands a vessel with a bunch of oil-soaked reed tassels which can be renewed at need. Behind it stands a small (less than 1 m high) screen which conceals the catcher from view. The catcher has a 3 m long dip net in his hand. The pole is not quite half its length, the rest consisting of two sticks in the form of a fork. At the very end a string holds them at an approximate distance of 30 cm.

"Between this fork hangs a baggy net with fairly small mesh. In the stern of the first boat stands a man with a stake. In the second boat a man is in the bow and in the stern hangs a gong of 40 cm diameter on a gallows-like support. The man at the gong has an important job because it depends upon his quick beat and the uniformity of the sound—which may increase and decrease in loudness—whether the ducks will remain. Under these circumstances the ducks will let the boat approach quite closely, some even seem to coming swimming from their hiding places towards the boats. Once they are right next to the boat they are captured with one sweep of the dip net. The wings of the captured duck are then tied in a knot and the bird is thrown into the boat.

"In his "Wild Chorus" PETER SCOTT describes this strange hunt vividly and includes an artistic drawing. He also reports that, according to his experiences, it is most difficult to judge exactly when a duck has come into the reach of the dip net and when the catch must be made. He writes that the ducks often sit so tight that one can grab them by hand, or else the catcher whistles a certain note which causes the ducks to get up. SCOTT witnessed himself how a small crack in the cloud cover diminished the chances for catching, how on the other hand rain and especially snow made the ducks sit tight. He also confirms the importance of the gong. (I assume that the ducks equate this sound with the rushing of the wind in the reeds and therefore are not disturbed, because the ringing camouflages the sounds of the catchers.) As soon as the catchers interrupt the gong—for example if they need to bail out their boats together—all the ducks around them immediately take to the air so that soon there are no ducks around them for 50 m. PETER SCOTT was present when one duck catcher netted 65 ducks in one hour. He was told that in special cases in the fall soon after the arrival of the ducks 600 or even over 1,000 ducks could be captured by one team of catchers."

DENNIS (1966a, b) reports on catching at night with artificial light and dip net around Fair Isle Bird Observatory (south of the Shetland Islands, Great Britain). The British ornithologists were especially interested in catching those species which cannot be trapped by any other means. They captured 650 birds of 51 species from 1959 to 1965.

DENNIS describes their equipment: "The most important item of equipment is a powerful lamp. We now use, after trying a variety of electric torches, a converted Tilley lamp. Originally, this lamp was a Tilley radiator heater, which has a heating mantle mounted on a vaporizer and a dish reflector. The base of the heater is a pressurized tank for holding paraffin; a pump is fitted in the side of the tank for pumping the paraffin oil up the vaporizer and into the heating mantle. We had the reflector silvered and replaced the heating mantle with a 500 candle power lighting mantle shielded by a glass dome for outside work. Another lamp we converted has a 300 c. p. mantle and we find this nearly as efficient and easier to keep alight.

"This equipment gives a very strong wide beam. A powerful torch with a narrow beam is not as efficient, because one has to spend so much time sweeping with the torch beam to find a bird, whereas the Tilley lamp gives off a beam of, say, 150 arc and illuminates all the birds in front of the operator. Also, on Fair Isle, we find it cheaper and handier to run a lamp on paraffin, rather than buying batteries for a large torch, especially as the batteries are only of use for dazzling when they are new; once they are slightly run down they lose most of their effectiveness for dazzling."

"The other piece of essential equipment is a good hand-net: really one needs several handnets suitable for different weather conditions and species. We make our nets from a length of stout fencing wire, which we shape into a circle, from one-and-a-half to three feet diameter. A net is fixed on to the wire; for smaller birds we use a small mesh and do not have much 'bag' on the net and for larger birds we use a larger mesh and have more 'bag'. If the mesh is too big, the birds tend to get entangled and time is wasted in extracting them. The ends of the wire are twisted together and bound tightly to the end of a bamboo cane or long stick. For all purposes we prefer the longest and lightest pole and the largest diameter of net. Our best handnet has a 12ft. bamboo handle, but often the wind is too strong and the large handnets become unwieldy. It is important that the handle is firm and does not whip in a wind and that the wire frame is tightly bound to the handle so that the net will not swing in the wind or rattle against the handle when in use. We use binding wire and string to join the net to the handle and finish it off with adhesive tape."

"When dazzle-netting we always carry a small rucksack containing bird-bags and sacks for holding the catch, an electric torch, note book for recording retraps and matches for re-lighting the Tilley if it blows out. We wear long rubber boots so that we can wade into wet areas. We find the oilskins make too much noise as we walk and scare the birds, so we wear anoraks."

Now some details on the catching methods themselves. On Fair Isle the summer months are very light. The first suitable nights are in August. The catching team consists usually of two people, the lamp carrier and the catcher. Both must be well practiced. The catcher must always walk closely behind the lamp carrier and must not switch on his flashlight to find his way over bumpy ground. During a suitable night the ornithologists of Fair Isle usually make a circuit of the island visiting various small lakes and meadows and walking along the streams. Almost all marsh and water birds sleep at night on or near water. The choice of these roosts is often determined by the direction of the wind and its strength.

Ducks tend to swim either slowly away or in circles when dazzled by light. As quietly as possible the catchers approach the bird through the water, the beam pointing at the bird. In particularly dark nights some birds will swim towards the light. If the bird is swimming quietly in front of the light man the dip net should be lowered to head height so that the distance may be judged correctly and then lowered quickly over the bird. The catcher does not leave the shadow of his companion until the bird has been caught.

If no other birds are around the catcher quickly extracts the bird in the light. If it is one of a flock, the light carrier continues to walk forward shining his light on the other birds. The catcher then extracts the bird from the net in the dark, puts it in a bag and passes it on to his companion.

Generally it is easier to catch single birds. The bird can be approached from any direction, but it seemed to be most successful if the catchers approached it into the wind. Geese and shorebirds tend to walk away from the light at first, then to stop and finally turn towards it. The catchers follow the birds quickly and quietly until they have come within easy catching distance.

Diving ducks on shallow pools tend to be blinded easily but they dive as soon as they become worried.

When catching Whooper Swans it is best to use swan hooks. Swans are easy to dazzle but it is no easy task to keep them under control in the dark. When a swan flies away, it is worthwhile to call like a swan. Usually they answer and land close by.

Most of the captured birds are examined at the station and banded there. They are held until the morning in a dry tea chest. If a bird is banded immediately after capture it has to be held away from the light for a few minutes so that it can adjust to the darkness.

The aborigines of the Philippines use a hand net and throw net with weighted rim when catching shorebirds—including rails and small ducks—at night. Most of the catch consists of shorebirds from the Northern Hemisphere, but they also net local species. The birds generally sleep deeply from exhaustion after their long flights so that they are easy to approach. Sometimes an entire flock is captured with one throw. Three catchers told Murphy (1955) that each of them had captured an average of 50 birds per night during migration.

Macpherson (1897) reports that the fishermen of the Canary Islands (Spain) occasionally caught White-faced Storm Petrels at night when they were out catching fish with torches.

The following report indicates the extent to which birds were persecuted in former times (Macpherson 1897). During fall migration the bird hunters in the state of Virginia (U. S. A.) hunted the Sora Rail which migrated many thousands strong. In a dark night the hunters took a light canoe which had a lighted fire on top of a thick pole in its middle. About one hour before high tide the canoeists, using light 3 m long paddles, pushed the boat through the broken reeds floating on top of the water. The fire lighted the surroundings far and wide. The birds were so dazzled that they could be killed by a blow on the head with a paddle. Similarly the Clapper Rail was hunted with torch light in the South of the United States. How easy it would have been to catch the birds alive with a dip net!

18.4.2 Catching rails by hand

Another method of catching birds at night is briefly described below by PETER BECKER (1979):

"With this method, which can be used throughout the pairing and breeding season, the ringer needs practically no equipment but a torch, since trapping is carried out at night. Being too large and resistant, moorhen and coot are difficult to catch by hand. In this case a hand-net will be more effective, all the more because these species occur in a more open vegetation permitting an unhindered use of the net. As the voices of water-rails are difficult to imitate by humans, it is advisable to use a tape recorder, whereas—with some experience—the voices of corncrake and the three *Porzana* species can be imitated quite well, so that a recorder is seldom needed as far as these three species are concerned.

The calling rail is approached as far as possible. A few meters from the bird the vegetation is quietly and carefully trodden down in order to make a small open space which the bird will have to pass when approaching the recorder. This open space is important because the bird can only be caught if the ringer's hand does not get entangled in the vegetation at the critical moment. The ringer himself should stand on solid ground permitting him to turn round if necessary, as some rails tend to run at their supposed "competitor". Care must also be taken not to make any noise when bending down, to prevent the approaching rail from being scared away. The bird is now attracted (by recorder or sound imitation), while the torch is held in the direction of the calling rail. The beam should be as narrow as possible so that the surroundings remain in the dark and the rail will not see the ringer. Most rails creep silently, though sometimes rather fast, towards the recorder. Some birds, however, also call according to the degree of their excitement. The torch, still directed towards the bird, should be brought as close to the bird's head as possible so that the other hand, which is going to catch the bird, remains in the dark. Shortly before reaching the recorder or the imitator the bird is caught with a sudden grip (Fig. 381).

Dark and calm nights are most suitable for catching. On windless nights the creeping rails are better to be seen and heard. On moonlit nights it is more advantageous to keep the bird

Fig. 381. Trapping of a Water Rail at night in April 1981 by PETER BECKER. Photo: D. TAYLOR.

between the moon and the ringer so that the bird will not notice the ringer against the bright moon.

The author has successfully used this method with all three *Porzana* species, the water-rail and the corncrake. He has not only approached calling males, but also attracted and caught females. If a territory is known where the birds are no longer calling, they can still be attracted and caught provided the worker is inside the territory. In this case the rails mostly approach the recorder silently so that the ringer must permanently search for them with the aid of the torch. The author, for instance, did not notice an approaching rail until it fluttered up his leg towards the recorder. This method of catching birds by hand was also used with other night-calling birds, e.g. Grasshopper Warbler, Savi's Warbler and River Warbler. The birds often approach the recorder rather fast and violently, and can be easily caught in the light of the torch when they continue singing by or even on top of the recorder.

Various interesting ways of behaviour have been observed on these occasions, e.g. two male rails approached the recorder at the same time and carried out territorial fights in the light of the torch, uttering various intimate sounds which would otherwise have never been heard. Also in this direction the use of a recorder will undoubtedly provide further valuable information."

18.5 Catching gallinaceous birds

Catching gallinaceous birds with nets at night is very rare in Western European countries. In earlier centuries partridges and quail were captured with drop nets, or with circular nets or hand nets, almost always with artificial light.

According to Winckel-Tschudi (1878) the hunters used drop nets for catching Black Grouse at night during snowy weather. The men went with flaming torches through the sleeping quarters of these birds which then flew towards the light and dropped to the ground again close to the hunter. They then could be covered with a net.

Labisky (1968) reports on catching pheasants, Sharp-tailed Grouse and quail in North America. Morris (1970) describes nighttime catching in New South Wales (Australia) of 651 quails (573 Stubble Quail, 1 Brown Quail, 44 Little Quail, 33 Red-chested Quail) from 15 January 1969 to 22 March 1970. Also captured were larks, pipits, ducks and rabbits. Trapping was done from an automobile. The catching party ought to consist of three people: one to operate the lamp, the second to do the catching with hand and dip nets, the third to drive the car. If only two people are available, the driver also has to do the catching.

One has to drive around a pasture or field until a quail flies up. Now the headlights are switched off because one has to keep the flying bird in the light beam of the lamp. The car follows the bird, and stops close to the spot where the bird landed. With the lamp still on, the bird is covered with a net. The captured birds are banded and are released only when the catching party move on to the next field.

The lamp is a 'Lucas Long Range', 12 volt, 48 watts, type SLR 7005. The dip net (diameter 80 cm) has a 1.5 m long bamboo pole which does not have to be absolutely stiff.

The results? Morris started on 88 trips of two hours each and captured 6.8 quails per expedition. The best result for one night was 55 quails. From July to November only a few trips were undertaken because of the crops on the fields and the rainy season.

Wind and a bright moon diminish the success of this method. The best results were achieved in cloudy nights, or when the moon was only a quarter full. Approximately 70% of the flushed birds, and more males than females, were captured. According to Morris the males fly shorter distances and are therefore more easily found again.

Macpherson (1897) reports on an unusual hunting method of the Maoris in their quest for megapodes. The Maoris light a fire in the woods which draws the birds. Then they bait the birds with a red rag tied to a pole. In their belligerence the birds attack the rag and are beaten to death by the Maoris. The birds could probably be captured with a dip net in the same manner.

18.6 Catching songbirds

18.6.1 Catching songbirds in reeds in the evening

Earlier centuries also saw this method of bird catching. Rordorf (1836) (see Lindner 1964) reports on his observations in Northern Italy. In dark nights on reedy shores the Italians went after the sleeping birds with stationary nets and torches. In one night 180 starlings, over 50 swallows and 5 kingfishers were netted.

For decades now ornithologists have used this catching method for scientific bird banding. Hundreds of thousands of swallows, wagtails, buntings, starlings, etc. have been captured and banded using this method. Between 1950 and 1960 during fall migration the Radolfzell Ornithological Station (Germany) captured tens of thousands of Barn Swallows and Bank Swallows (Sand Martins) on Lake Constance. The trammel nets were set up in long lanes which had been cut in the reeds beforehand.

The lower edge of the nets should be at least 50 cm away from the water's surface (as a precaution against drowning of catches and molesting by Common Coots). As soon as it is sufficiently dark the swallows fly into the nets without further human action being required. Similarly White (Pied) Wagtails and Yellow Wagtails can be netted. Catching Common Starlings in trammel nets should be avoided as taking out large numbers of them is rather difficult and time-consuming.

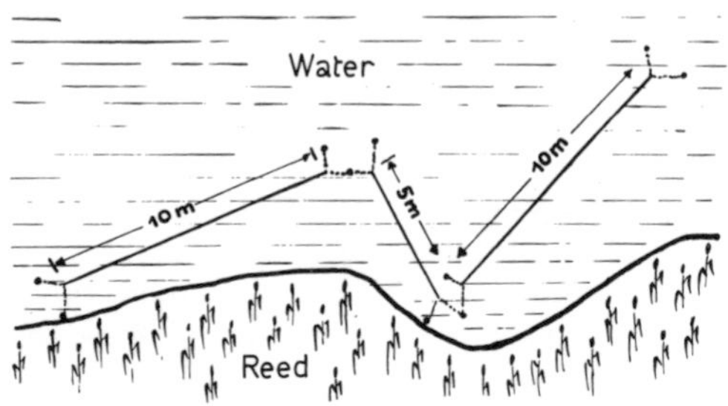

Fig. 382. Arrangement of nets for catching swallows at dusk.

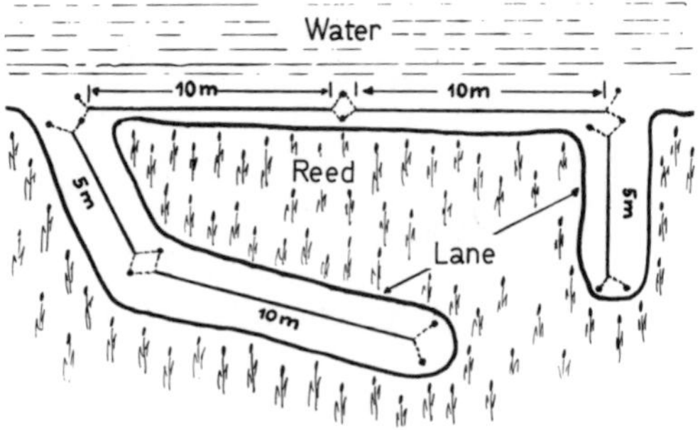

Fig. 383. Arrangement of nets for catching wagtails at dusk.

Near smaller bodies of water, too, we can capture many swallows, wagtails and pipits as well as Common Reed Buntings and Corn Buntings. Preparations for catching must be completed before the birds arrive at their roosts. The nets are set up in narrow lanes or even along the edge of the reeds. Part of the flock gets caught when approaching or while searching for a place to sleep. The rest of the swallows can be scared up several times. After dark they do not readily take to the air and if so, fly low and only a few meters. In order to take the birds out of the nets flashlights may be necessary. In the fall of 1960 near Wilhelmshaven (Germany) Bub captured over 200 Barn Swallows and Bank Swallows (Sand Martins). Most of those birds were caught in nets set up on dry land next to a 10 m wide reed bed.

K. Greve set his nets at angles when catching around Braunschweig (Germany) (Figs. 382 and 383). One August evening 65 swallows and 25 wagtails were caught within a few minutes. It is often advantageous if the birds fly to and fro before finally settling down. When catching wagtails, two-thirds of the nets should be above the reeds. Then many birds will already get caught at dusk (Fig. 384).

After banding the swallows at their roost the Ornithological Research Station Radolfzell released them a little to one side of the light source. Only in pitch-black nights were the swallows held till the next morning.

Besides trammel nets, mist nets and other similar nets are fully serviceable, although they are somewhat inferior to trammel nets in their effectiveness.

M. Dornbusch set up his nets in the following manner:

1. Perpendicular to the shore across the reeds 6 to 30 m distant from land to open water in natural and cut lanes.
2. Parallel to the shore in small open water patches within the reeds.
3. To block off narrow gaps in reed patches leading from one area of open water to another.

During bright nights swallows were released immediately after banding (they landed right away again among the reeds). In dark nights they as well as all other birds were kept until dawn—separated according to species—in thin airy bags (40×20×25 cm) or in boxes with air holes. Several birds—but not too many—of the same kind can be put into the bags at once. According to DORNBUSCH it was better to release *Acrocephalus* (reed) warblers, wagtails and Common Reed Buntings one by one into the bushes near shore with a flashlight—particularly during bright nights or at times of large catches. Wagtails and Common Reed Buntings remain there until dawn, while *Acrocephalus* (reed) warblers sometimes hop further along in the bushes.

Water Pipits sat quietly on fingers and could be grabbed again with the other hand. This fear of darkness seems to indicate that Water Pipits probably do not migrate at night. It should be mentioned here that many birds are night-blind.

Catching at the roost in the reeds should only be done once every week or, during one season, once two nights in a row. Otherwise catching should only be done at the edge of the roost and without otherwise greatly disturbing the birds—something we recommend in principle anyway. Catching should be terminated before midnight.

The mortality rate is relatively small. During eight expeditions with a total catch of many hundreds of birds one bander lost the following birds (hanged, drowned, killed by a Water Rail, weasel or cat): 0%, 1%, 1%, 2%, 3%, 1.5%, 2.5%, 2.5%. BUB himself has rarely heard of any mortalities occurring during evening catches. Of course, catching over water requires special watchfulness as the birds can drown. The set nets should not exceed the number that can be handled comfortably with the help available.

In the last few years J. SADLIK captured mostly Yellow Wagtails at two roosts near Merseburg/Saale (Germany). Roost 1 (water depth 30 cm) consists of a reed bed 100 m long and 50 m wide. Roost 2 (water depth 60-70 cm) is only 12 m wide, but 250 m long. SADLIK avoided setting the mist nets in the middle of the reed bed or directly beside the roost in order to avoid disturbing or scaring the birds away permanently when taking down the nets in the darkness.

The birds got caught either while approaching the roost or by a slow drive of two or three people. A drive was conducted three or four times in one evening. SADLIK found that the birds avoided the nets when they were visibly higher than the reeds. It proved to be advantageous to widen the net lanes to 2 m. The reeds were not cut but only stomped down. The wagtails liked to sit on these broken reeds before bedtime. A wagtail struggling in the net and calling from time to time seemed to excite the curiosity of

Fig. 384. Large catches at nocturnal roosting places. Photo: H. BEHRENS.

others who had landed to the sides. These in turn got caught while approaching to investigate the affair.

Several lay-over days were observed between two catching days. Results for Yellow Wagtails: 1966-316; 1967-418; 1968-571.

Catching Barn Swallows at their roosts with a tape recorder can be very worthwhile. SPEEK (1971) had made recordings at a roost when he noticed by chance that the swallows dove at the recorder while he replayed their evening twitterings. On 24 July 1971 at 4 a.m. SPEEK retaped the twitterings approaching the birds to within 4 m. He used a 'Uher 4,000 Reporter 1' tape recorder. The microphone was an AKG D 202CS and was used in connection with a 24" parabolic screen. The best part of the tape was duplicated continuously, so that the tape would play for one hour at a speed of 3 ¾" per second. The result sounded to human ears like the normal song of a Barn Swallow.

On 25 July, without nets, SPEEK staged a dress rehearsal at a heavily frequented roost. The roost consisted of a reed bed 500 m long and 50 m wide and had 8 net lanes. Generally the birds arrived at their roost around 8:45 p.m. The tape recorder was therefore set up one quarter hour earlier in the middle of one of the net lanes. Approximately 500 swallows flew 1-3 m high over the recorder evidently searching for the swallow singing there. After 5 minutes SPEEK relocated the tape recorder in the next lane, approximately 80 m distant. Within one minute the swallows gathered around the new spot. Speek let the tape recorder play until darkness. When he switched it off, the swallows were sleeping in the reeds around this spot. After these observations the following results were obtained during two nights of catching on 26 and 27 July with a 45 m net in lane 3 and a 36 m net in lane 4:

lane	3	lane 4
26 July	with tape recorder 107	without tape recorder 9
27 July	without tape recorder 1	with tape recorder 130

As the swallows generally did not settle at the roost before 8 p. m., Speek and his coworkers took the birds out of the nets at 8:10 p. m., 8:25 p. m. and 8:45 p. m.

After such success catching was repeated over the next week. The results mounted to 1,200 birds. During later excursions the Dutch banders used only one net lane and set the tape recorder with the swallows' song at one end of the net 1 hour before roosting time. During this time they shifted the recorder periodically from one end of the net to the other. With a net length of 39 m it became possible to catch birds at one end while earlier captures were being taken out at the other. Speek also did solo catching, but then he was able to process only 150 birds per evening at the most.

South African ornithologists have banded many tens of thousands of swallows, particularly Barn Swallows, at their roosts within the last 20 years. During the winter months of 1965/66 the Witwatersrand Bird Club alone banded 18,000 birds. Hewitt (1966) reports of catches near Rosherville and Vrischgewaard. 14,000 birds were caught here in mist nets, among them a number of already banded birds: 1 from Italy, 1 from Belgium, 8 from Great Britain. Of the birds banded in South Africa some were found during the following year: 13 in the Soviet Union, 4 in Great Britain, 2 in Poland, one each in Denmark, Finland and Germany. Especially remarkable are birds found east of the Urals.

Lawson (1966) completes this report from Natal and Macleod (1966) from the area west of the Cape. Catching there was generally done with mist nets.

On 30 December 1965 N. Myburgh captured 125 Barn Swallows with a hand net. Three banders managed to catch 1813 swallows with hand net and "head lamp".

18.6.2 Catching with drop nets

For hundreds of years many European peoples caught birds in the evening and at night with drop nets. This method is also useful for bird banding. Of course, it may only be used under strict observation of applicable laws and regulations covering such activities. The birds in question are almost exclusively those species of songbirds which live in sparsely vegetated areas, rest there during migration and spend the night as well. Catching with drop nets during the day has already been described earlier. For birds active during the day as well as at night such as sandpipers, good catching results may perhaps be achieved at all times. Kliebe (see above) had good results catching Jack Snipes near Marburg/Lahn (Germany).

The netting of a drop net must be particularly sturdy so that it will not tear quickly. If nothing better is available, mist nets may be used in meadows or where no major vegetation can damage the net.

C. L. Brehm (1855) mentions dimensions of 20-25 m length and 8-10 m width for a drop net to catch larks at night. A thin string is threaded around the net. In order to stretch it nicely, thin strings are threaded through the net at 1 m intervals. The ends of the strings are tied to the main string at the edges, leaving a piece for tieing to the carrying poles (Fig. 385).

According to Brehm, the lark catchers observed carefully where the larks would sit down to sleep. Today too we have to find out equally carefully if our catching is to be successful. By the way, the drop nets do not have to be quite as big as mentioned by Brehm. But nonetheless they must not be too small either, otherwise the catches are going to be small too.

K. Hillerich used mist nets to catch Meadow Pipits. After the birds have settled at their roost at the onset of darkness, a 12 m net, carried on poles, is laid down over those spots in the meadow where birds have settled. One October

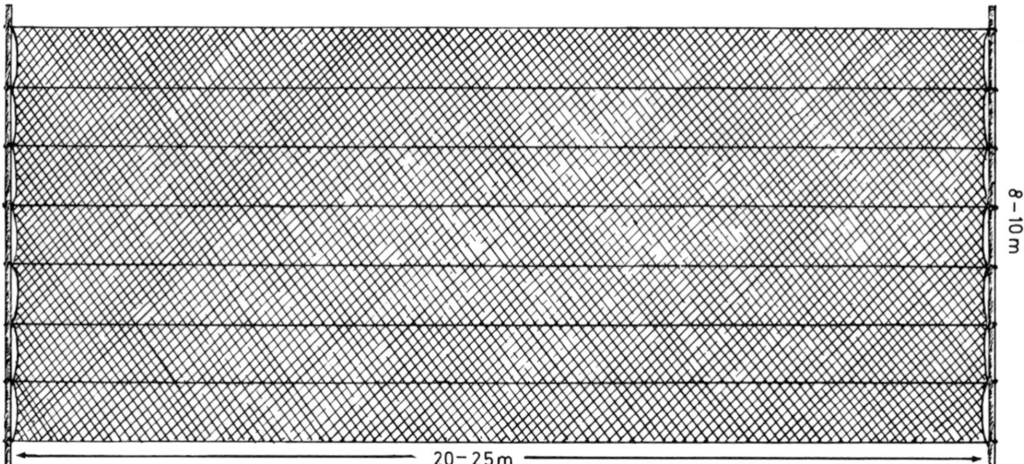

Fig. 385. A draw net for catching larks at night. After BREHM, 1855.

evening 29 birds were netted. Similar methods can be used for Common Skylarks and Wood Larks. Larger nets yield better results.

GRUNER (1972) describes one catching expedition on the island of Helgoland (Germany). A large mesh net of about 3×5 m hung on 2 bamboo poles is carried horizontally by two people closely over the ground at night, as quietly as possible. A third person walks in the center behind the net with a screened lantern and makes moderate noise. The birds (Meadow Pipits, Common Skylark, various thrushes, etc.) sleeping in the dense vegetation during migration wake up from the noise and light and flutter up against the net which is promptly laid on the ground. 50 to 100 and more Meadow Pipits during one night were not uncommon. Catching at night with a drop net is especially important on Helgoland because Meadow Pipits, Common Skylarks and other field and meadow birds generally do not visit the catching garden of the Ornithological Research Station and therefore hardly ever get caught and banded there.

It should be mentioned here that nowadays the people living on Helgoland do not catch birds on their own, although they certainly did a lot in earlier centuries. The ire of German bird enthusiasts during the last century was especially provoked by catching at night with lantern and hand-net. ROHWEDER (1900) describes this catching: "During dark nights in October and November when easterly winds and overcast skys promise good catching the catchers walk in twos over the upper island and search the grass and potato fields systematically. One carries the lantern and lights the ground around his feet. Next to him walks the man with the hand-net. Each bird sleeping in the brush—mostly larks, but also pipits, Bramblings and Chaffinches, Eurasian Linnets, etc.—on which the light shines flutters up in a fright and, before it can decide to fly off into the blackness around it, is captured on the ground by the long-handled net smashing down over it. An experienced catcher can also practice this by himself."

As already mentioned earlier, catching larks was a common practice in all European countries in earlier times and many thousands of larks were captured. Let's examine some of these catching methods.

BREHM (1858) reports on the catching of Calandra Larks in the province of Murcia (Spain) in which he participated himself. The equipment consisted of a shiny lantern with mirrors, a bell such as used for goats and cows, a hand net and a bag for the captured and killed birds. (Such a bag is unsuitable for keeping live birds, a number of small airy bags are better.) Having checked on the sleeping places in the early evening, the catchers started after dark. They wore 'alpargatas', sandals made of hemp, unsurpassed for this kind of hunting. Around the sleeping places strict silence was observed. One man (a shepherd) carried the lantern in front remaining in its shadow. He opened the mirrors and with the other hand began ringing the bell. The birds, used to cattle moving around in great numbers in many of these places, ignored the sound. Three steps behind the shepherd came the catcher with the net holding it closely above the ground and next to him walked another shepherd with the bag. 30 larks were captured that particular evening, some even by hand.

SALVATOR (1897) describes a very similar catching operation for several species of larks on the Baleares (Spain). MACPHERSON (1897) deals in detail with this catching method and its variants in different countries.

"There was a time in 'Merrie England' when the right of catching Larks by such means was so highly valued as to be restricted in practice to the owners of land. The title of 'Low-belling' was employed to distinguish this variety of fowling from other methods. An old indenture of lease between LAWRENCE ROGERS, citizen and clothworker of London and FRANCIS AUNGER of East Clandon in County Surrey, Esquire, dated 24th Eliz. Nov. 20 (1582), expressly reserves 'to and for the said EDWARD CARLETON and Marie, their heirs and assignes, all views of Frankpledge, felons goods, wayfes and estraies within the said mannor, ... together with liberty to go to a batt fowling, liberty to go with lowbell, liberty of hawking, and liberty of hunting the hare, fox, and other beasts of warren.' ... The compound term of Low-belling requires an explanation, because the word 'Lowe' has become nearly obsolete. Yet it survives in the homely dialect of the north of England. As an instance of this, I may mention that one of the Esk poachers described to me in the vernacular how a light is often employed in gaffing salmon that are lying alongside of a weir. He told me that the reflection of the lanterns or candles, cast upon the fish, is still known to the members of his fraternity as the 'Lowe' or 'Low'. Hence it will be readily understood that 'Low-belling' is the combined use of a light and bell. ... MARKHAM also gives an interesting account of the Low-Bell. 'After the night', he says, 'hath covered the face of the earth (which commonly 'tis about eight of the clock at night), the Ayre being mild, and the Moon not shining, you shall take your Low-Bell, which is a Bell of such a reasonable size as a man may well carry it in one hand, and having a deep, hollow, and sad sound, for the more quick and shrill it is the worse it is, and the more sad and solemne the better: and with this Bell you shall also have a net (of a small mesh) at least twenty yards deep, and so broad that it may cover five or six ordinary Lands or more, according as you have company yo carry it (for the more ground it covers, the more is your sport, and the richer the prey that is taken); with these instruments you shall go into some stubble field, either Wheat, Rye, or Barley, but the Wheat is the best, and he which carrieth the Bell shall goe the formost, and toll the Bell as he goeth along, so solemnly as may be, letting it but now and then knock on both sides; then shall follow the net, being borne up at each corner and one each side by sundry persons; then another man shall carry an old yron Cresset, or some other vessell of stone or yron in which you shall have good store of sinders or burning coales (but not blazing), and at these you shall light bundles of dry Straw, Hay, Stubble, Linkes, Torches, or any other substance that will blaze, and then having spread and pitcht your Nette where you thinke any Game is (having all your lights blazing), with noyses and poles beat up all that are under the Net, and then presently, as they flicker up, you shall see them intangled in the Net, so as you may take them at your pleasure: as Partriges, Rayles, Larkes, Quailes, or any other small Birdes of what kind soever which lodge upon the ground, which done, you shall suddenly extinguish your lights, and then proceede forward and lay your net in another Place." A rather inappropriate method for today.

In conclusion two further references. The well-known lark catching which lasts at most until late dusk—no light was used—was already described earlier in the chapter on stationary nets. NAUMANN (1824/1900) also describes catching larks at night with drop nets. ROHRBACH (LINDNER, 1964) reports that larks were also driven into stationary nets set across fields at night and with artificial light (straw torches).

According to SCHMIDL suddenly awakened Common Skylarks generally rise only 10 to 20 m into the air during windy and rainy weather. If the bird can be kept in the light beam, it can easily be captured after landing.

18.6.3 Large trapping station for Common Starlings and icterids

The Patuxent Wildlife Research Center in Laurel, Maryland (USA) built a large trapping sta-

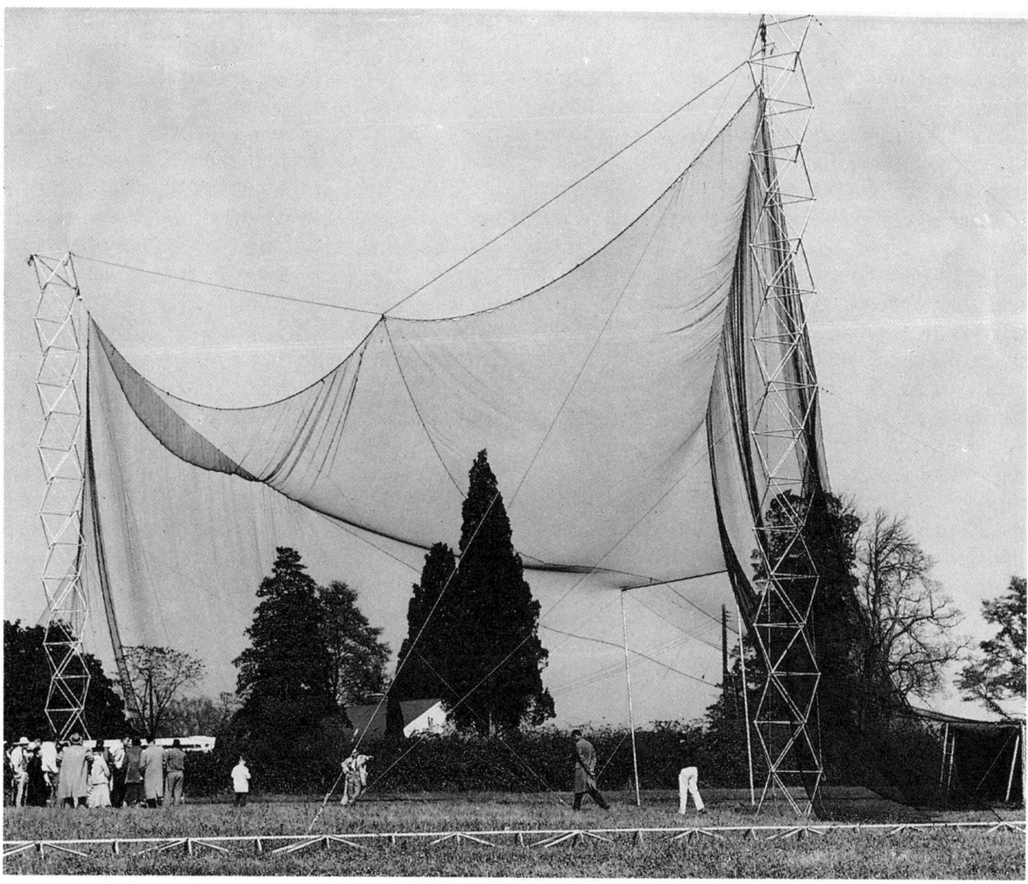

Fig. 386. Giant funnel trap to catch starlings and icterids at night. From SEUBERT, 1963.

tion for icterids in order to reduce the numbers of several species which had become agricultural pests (SEUBERT 1963). Some of the birds captured at their roosts in fall and winter, using bright lights, were banded. From 1957 to 1962 the following birds were captured: 414,162 Common Starlings, 135,492 Common Grackles, 89,880 Brown-headed Cowbirds, 30,816 Red-winged Blackbirds, 1,351 Rusty Blackbirds, for a total of 671,701 birds.

This gigantic trap was built mainly for the agricultural reasons mentioned. For bird banding purposes it is much too large as the number of birds captured would be impossible to process adequately. The results of three nights of catching show this without question, overshadowing all previous catching records: respectively 80,000, 102,000 and 120,000 birds in one night. Only at roosts of 10 to 15 million birds are such enormous catches possible. These numbers clearly show us the limits of bird banding because the simple banding of 1,000 birds already requires several banders and additional helpers. Imagine the preparations required to band 5,000 birds in one sitting maintaining some semblance of order!

Nonetheless we shall deal with this trap here to show its possibilities for bird catching. Constructing such giant funnel traps can possibly be of value even for European purposes as Soviet ornithologists have shown at Rybachiy/Rossitten (Baltic Sea) as several tens of thousands of bandings during one migration period can increase recoveries considerably. This particular structure is a giant funnel trap (Fig. 386) made of netting which is supported by two aluminum towers approximately 15 m high and other frames. The towers are octahedrons and are mounted in 3 foot sections.

The funnel ends in a long sail cloth enclosure

about 9×3.5×2.5 m. Two to four powerful 1,000 Watt flood lights beam their light from this funnel out of the trap and onto the grounds in front of the structure. The sail cloth enclosure can be zipped closed in front and back and the birds can be killed (with gas) in it after capture. For use in banding operations, however, this funnel end does not have to be made of sail cloth.

The two aluminum towers are about 30 m distant from the end of the funnel. SEUBERT does not mention anything about the distance between the two towers, but it is probably between 40 and 50 m. The American catchers erected the trap close to the roosts so that the birds could be driven directly into it. Catching was done in dark cloudy nights by an appropriately large number of drivers—in other words, under the usual favorable conditions. The flood lights illuminated the area in front of the funnel, but not the funnel itself.

18.6.4 Mass catching of Common Starlings and icterids at night, but without spotlight

In the marshy areas of Colorado (U.S.A.) SPENCER and DE GRAZIO (1962) captured many Common Starlings and icterids with hand nets at night, without any light. When catching in this manner, four conditions must be met: (1) a dark night; (2) low plant growth in order to be able to sweep the hand net freely; (3) solid footing which in winter a good frost provides by covering marsh and water with ice; and (4) dense concentration of birds at their roost. If the birds are disturbed under these conditions they fly up reluctantly and land again after only a few meters. The catcher can usually get in among the birds with a quick sprint of 10 to 30 m where he can catch up to 40 birds with one sweep of his net. The first time SPENCER and DE GRAZIO tried this method, they captured 550 Common Starlings and Red-winged Blackbirds in three hours. For three catching expeditions on dark nights in the marshy area around Denver the total approached 1,200 birds. During a fourth expedition at new moon six men with two hand nets easily captured 600 birds in three hours.

In bright nights the starlings awoke earlier when approached from a distance and flew further than in dark nights. They did not leave the marshy area, though, and could be driven from one spot to another. Several catchers were stationed near larger gatherings while one person slowly drove the birds in front of him until they settled within reach of the catchers. In this manner catching did become a success, but not more than 25 birds were caught with one sweep of the hand net. On the other hand this method also permitted catching during mild weather and in those areas where the growth was so high and dense that the catchers could not move quickly. The water level in marshes also prevented quick sprints by the catchers.

The hand net had a diameter of 70 to 80 cm. The small mesh net bag was 1.2 to 1.8 m long. The telescoping poles of the hand net could be lengthened to 3 to 6 m. They were made of lightweight sturdy piping material. The birds get easily entangled in fine mesh—a somewhat heavier material is preferable. The net bag must be long enough so that enough room remains for the birds after the net opening has been twisted closed. Portable holding cages for 100 birds each kept the birds safely until banding. Less than 1% of the birds died.

Today catching of starlings is less important. Nevertheless, reports on new catching methods appear in various journals as the species constantly presents new problems. ISSEL (1937) developed a simple method in Bonn (Germany). Starlings sleeping in the English ivy on a house wall were to be captured. Catching with a trammel net was abandoned after one try (it takes too long to take a flock of starlings out of the net). ISSEL made a 1×1×0.5 m cage of thin slats, covering it with narrow wire mesh. Only the front remained open. ISSEL was able to close this opening. 1×1 m in size, by narrow mesh netting which he pulled along a rod. The cage was screwed to a long pole. ISSEL and coworkers captured 50 starlings the first evening, and during the next few weeks the numbers mounted to 400 (maximum for one evening was 146). Then the box was lightened. Only the front sported a strong wooden frame, the rest of the 'box' was only wire mesh. The net was replaced by a light weight dark green cloth which now was pulled up from the bottom.

The starlings were not very sensitive to the disturbances. They flew up and landed again in the ivy a few meters away. One bird was even captured a second time a few days later.

18.7 Catching sleeping Common Flickers at dawn

Wilcox (1963) reports on some interesting catching of these woodpeckers sleeping on telephone poles. His banding station is located on Tiana Beach (Shinnecock Bay, Long Island, New York, U.S.A.). A stretch of about 15 km along a barrier island has hardly any trees, and Common Flickers therefore often sleep on telephone poles. During September and October 1962 Wilcox and Terry caught a total of 176 Common Flickers—44 of them shortly before sunrise on such 'sleeping poles'. Most of them were sitting about one to two meters above ground, some only half a meter high, others up to three meters up the pole. Besides the Common Flickers they captured 1 Brown (Common) Creeper, 1 Hairy Woodpecker, 2 Downy Woodpeckers, and 5 Common Sapsuckers. All slept with their heads under their wings and did not fly away when a car drove by at 6 m.

When Wilcox noticed the shape of a woodpecker on a pole, he stopped the car and approached on foot as quietly as possible (Fig. 387). Nonetheless several birds heard him, but did not fly away immediately although they did not have their heads under their wings any longer. Wilcox approached such birds from the back rather than from the side. In this manner he managed to catch even some of these birds which were already awake. During September Wilcox captured 37% of the birds, during October—after perfecting the technique—he netted 65% of the woodpeckers sighted on telephone poles. Most of the birds were grabbed by hand. If the birds sat too high to reach on the poles he used a hand net with a diameter of 30 cm and a pole 1.8 m long. Six birds were the maximum for one morning.

18.8 Catching birds sleeping in the open by hand

Let us take a quick look at this possibility. We refer here to birds that sleep in trees, bushes, under roofs, near or in buildings or other open places, not those sleeping in nest boxes or holes.

During the winter of 1964/65 J. Bernd-Bärtl captured by hand 700 Twites in the windows of a factory in Merseburg (Germany). He was able to reach the birds from the inside.

Similarly House Sparrows can be captured by hand under roofs. Prerequisite is total darkness and somewhat turbulent weather. The birds are grabbed after flashing one's flashlight briefly to note the exact sleeping place. During migration sleeping Common Swifts can be captured in the same manner; the banded birds are released at dawn.

Fig. 387. Catching Common Flickers and other woodpeckers in treeless areas of the USA. After Wilcox, 1963.

Brehm (1855) describes a French method of catching crows at night. Catchers dressed entirely in black climb roost trees on dark evenings just before the crows arrive. According to the report the birds are not frightened away by the immobile figures and let themselves be grabbed.—Today we use crow traps, pull nets, double clap nets or even cannon nets.

In Haiti Kaempfer (1924) saw how natives captured Ruddy Quail Doves by hand with torches. The birds generally slept on low branches.

Kale (1966) reports on catching by hand of sleeping birds on the islands off Florida's coastline (U.S.A.). The birds in question are generally exhausted by long migration flights.

18.9 Systematic catching of migrating birds

The reports of several Swiss ornithologists show that evening and nighttime bird catching provides new opportunities for banding (Godel and de Crousaz 1958, Vuilleumier 1959, Dorka 1966, among others). Planned catching at night is important and necessary for migration research, considering just how little of this phenomenon is known and has been observed.

Catching was done at the Alpine passes of Col de Bretolet and Col de Cou. According to Godel and de Crousaz (1958) the Col de Bretolet proved to be the better catching place after some trials in 1952 and 1953. The Col de Bretolet is less windy and the nets can be set up openly without losing their effectiveness. In addition, the alders distributed around the pass entice some birds and at the same time provide some camouflage to the nets during the day. This is of importance especially for migrants stopping for a rest in the morning.

At Bretolet the nets are arranged in a 100 m long and 2.5 m high "barrier" which closes almost the entire pass. Several additional nets are set on the Swiss side below the pass, hidden among some bushes. Those could be replaced by some small Helgoland-type funnel traps as catching with funnel traps requires less time than with nets and may have other advantages. At locations such as Bretolet the funnel traps should be portable, being exposed to more severe weather conditions.

The "high nets" fastened to 5 m high bamboo poles proved especially effective for nighttime catches during calm weather. During the day most birds got caught in the nets hidden by bushes, at night in the nets out in the open. Four banders were just barely able to handle all these nets (see also Fig 258). All birds whether captured during the day or at night were released as soon as possible.

Of course, all labor, toil and selfless efforts do not suffice to catch a significant number of nighttime migrants of most species; nonetheless these bandings provide many valuable insights into nighttime migrations.

Dorka (1966) in particular presents the results of research so far and we would like to refer those interested to his publication. Worthwhile results can be obtained from continuous day and night catches, as shown by the migration pattern of Northern Wheatears (Fig. 388). Northern Wheatear migration regularly reaches its peak at Cou and Bretolet around the middle of September (in four years on 16, 14, 10, and 15 September). The first birds appear in the middle of August, the last during the middle of October. The species is a typical nighttime migrant with greatest flights in the hours around midnight. The sparse catches from sunrise to afternoon are almost exclusively resting birds.

19. Catching at watering places

19.1 General

As for all living things, water is indispensable for birds. Water is not only necessary to birds for drinking, but also as an integral part of caring for their feathers in order to maintain their ability to fly. Therefore birds are constantly on the lookout for suitable watering places.

In areas with reasonable amounts of rain birds can easily find spots just right for their drinking and bathing needs. Except for dry summer months there is almost everywhere enough water. Things are considerably different in steppes, semi-deserts and full-fledged deserts. It often becomes necessary to travel considerable distances as shown through the example of North African sandgrouse.

The dependence of birds upon water for drinking and bathing has been exploited

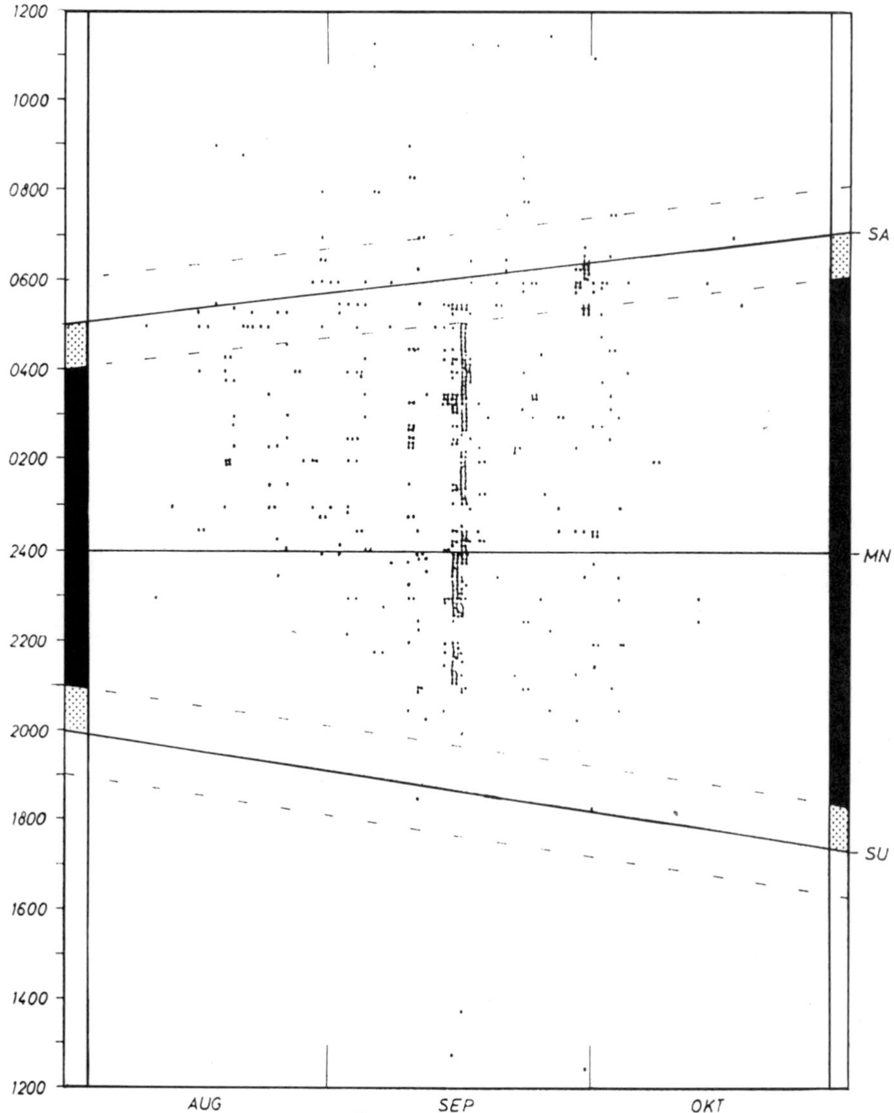

Fig. 388. Northern Wheatear *(Oenanthe oenanthe)*: density distribution of catches (n = 520) according to time of day and season from 1962 to 1965 at the Alpine passes Cou and Bretolet (Wallis, Switzerland). From Dorka, 1966

through the ages. Vast amounts of literature deal with hunting and catching birds at watering places. Ornithologists have long since adopted catching at watering places for scientific bird banding.

Watering places offer many possibilities for bird catching. Lure birds are not absolutely necessary, but they can enhance catching. In general one should not expect large catches, but at a well-planned watering place, always full of fresh water, many of the birds in that area will eventually stop by. Therefore this method lends itself well to population studies. Some species such as Eurasian Nuthatches, Common Bullfinches and Great Spotted Woodpeckers (only rarely captured otherwise) come often. Doves, Eurasian Jays, Creepers, European Robins and Coal Tits come too, while others such as Tree

Pipits and Common Redstarts do not come as often. But maybe this differs from locality to locality depending on how numerous the species is.

Besides using natural watering places, artificial ones can be very promising indeed, especially in dry wooded areas or parks where birds may not have any natural water sources but are restricted to artificial structures for their needs. These are good possibilities for artificial watering places where one then has good chances of catching birds.

C. L. BREHM (1855) describes in detail the "watering spot" and describes the ancient practices of the bird catchers in some of the wooded mountains of Germany, which are not only confined to these areas. FEUERSTEIN (1939) and JÄCKEL (1941) among others also describe bird catching at watering spots.

Elsewhere generally, natural creekbeds or other water courses were used for bird catching, but on arid Helgoland (Germany) DROST in the twenties built a small artificial pool in the catching garden of the Ornithological Research Station. After 1960 VAUK built further small watering places which greatly increased the attractiveness of the catching garden (Fig. 389). On dry Helgoland where the large masses of migrant birds can only expect a rain water puddle, DROST (1933) developed the 'water lure' (small bird funnel trap) which contained a dish of water as its sole attraction (Fig. 52, 60).

The inclusion of water as an attractant in catching activities at Helgoland indicates the important role which artificial watering places can play.

The season most favorable for water as a lure is summer, when most of the available puddles and many streams have dried up and the birds must of necessity use an artificial watering place or a particular section of a creek which has been set up for catching. But water can also be scarce during spring and fall, and areas generally short of drinking water (such as Helgoland) are different again.

Naturally occurring watering places must be sought and explored. The bander who knows his area well will soon know the preferred spots. But sometimes we find such a watering place by chance in a spot we would not have otherwise suspected. Watering places seem to appear almost suddenly, and soon individuals of various species will frequent them.

SUNKEL (1961/62) employed the following method in order to find watering places in the woods in Hessen (Germany): he takes a map

Fig. 389. One of the pond funnel traps in the catching garden of the Ornithological Research Station on Helgoland, Germany. Photo: GISELA VAUK.

(scale 1:25,000) and follows brooks until they become mere trickles from a wooded valley. Here he finds spots with a great variety of plants where many birds of different species will come to the water to drink and bathe.

19.2 Building artificial watering places or bird baths

There are various possibilities for building artificial watering places. On the ground we excavate a hollow the depth of which does not have to exceed 15 or 20 cm and which tapers gradually towards the rim. A small branch put across the watering place and extending beyond it simplifies access for the birds. The hollow is then covered with a cement layer and thus made impermeable to water (the ratio of cement and sand is 1:1). For catching, smaller watering places of about 40 by 50 cm are preferable. Larger areas are hard to cover by the nets used for catching. Preparing the bird bath becomes even easier if one uses roofing paper to waterproof it. With a bit of effort it can be formed into a hollow. Larger dishes as used in developing pictures are also usable if the rims are tapered towards the middle with some gravel and coarse sand.

MÖHRING (1955) covers a hollow—10 to 15 cm deep in the middle and tapering on all sides, with a diameter of approximately 70 cm—with a thin sheet of plastic. Around the outer perimeter he digs a 15 to 20 cm deep ditch in which the ends of the plastic sheet are weighted down by earth and sod (Fig. 390). The entire sheet is then covered with gravel and flat stones to prevent the birds from slipping. This bird bath is not as permanent as a cement construction would be. It is advisable to remove the plastic during the winter months.

MÖHRING also describes a pole bird bath (Fig. 391) which is more secure against ground predators. Vegetation for about 3 m around a watering place on the ground should be well mowed at all times for the protection of the birds, but this is not as critical for a pole bird bath. If such a bath is to be used for bird catching it should not be higher than 60 to 80 cm.

Fig. 390. Bird bath built with a plastic sheet. After MÖHRING, 1955.

Fig. 391. A bird bath on poles. After MÖHRING, 1955.

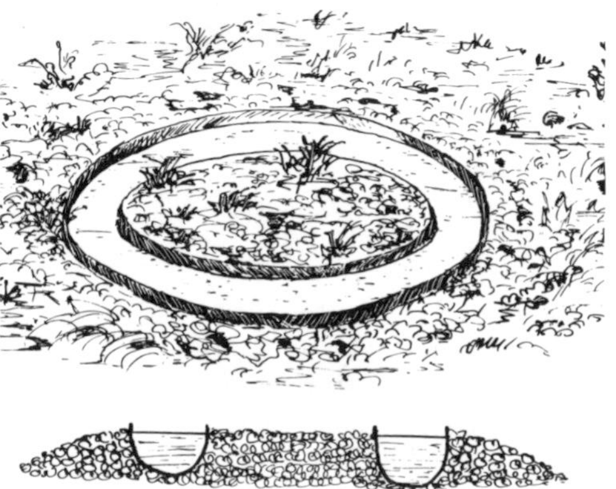

Fig. 392. Bird bath made of an old car tire sliced in half. After KEIL, 1957.

Two poles and a few old boards form the basis for a 'table' about 100 cm long and 75 cm wide. We nail together a wooden frame 50 cm by 70 cm by 5 cm high and tack the platic sheet on top with staples and thin wooden slats. The sheet must form a shallow pool. This frame and sheet are set in the middle of our 'table'. The table in its turn gets a rim about 8 cm high. The space between the rim and the basin is filled with gravel, earth or sod; but it can also be planted with sedum, saxifrage or similar plants if the bird bath is to be attractive as well. The plastic sheet itself has to be covered with gravel

Fig. 393. Bird bath with water-drip. The continuously dripping water attracts birds especially well. Photo: H. Bub.

and sand to give the birds a foothold. Möhring brings the basin inside in the winter.

Keil (1957) suggests to use halved tires as watering places. Making a clean cut along these tires may be difficult (Fig. 392). See also Fig. 393.

19.3 Catching at watering places in Southwest Africa

In semi-deserts and deserts water is a particularly effective lure during droughts. Hoesch (1959) talks about his experiences in such areas in Africa. First, one picks those biotopes which contain the birds one intends to catch. This is particularly important because some of the native seed eaters may specialize in certain grasses. A large flat bowl is dug into the ground—see the discussion on establishing artificial watering places and bird baths above—and is patrolled and refilled daily for one week as mammals find the water too and drink it up.

In addition the habits of the species must be catered to when establishing a catching station. The Black-faced Waxbill, for example, likes best to go deep into a well under the earth even if the water is slimy. Common Serins and sandgrouse prefer to drink on flat spots where for example the water spills from a leaky watering place for cattle; doves like the rim of the watering place. Pale Rock Buntings and Red-billed Queleas like flat watering places with stony bottoms. Yellow Canaries, Golden-breasted Buntings and Black-eared Finch-Larks again prefer spots with sandy bottom. Peach-faced Lovebirds and Yellow-bellied Greenbuls regularly sit on the ends of water pipes from which the ground water—pumped up by diesel or wind engines—flows into an open basin. They often sit 'in line' on the horizontal pipes and wait patiently until 'their turn' comes, as Hoesch says, even if there are plenty of puddles all around from which they could drink easily. These difference in drinking behavior must be observed if certain species are to be captured.

Birds which do not frequent such watering places—most of the insect eaters do not—are difficult to catch even if lure birds are used. At watering places Hoesch preferred the drop trap for catching; he pulled the string from a distance of about 30 m. "There is nothing for it but to sit, often for hours on end in the blazing sun, trigger in hand, until the desired birds are under the trap. As long as it is cool and a pleasure to sit in the open country only few birds will come to the water. The attempt to catch them at night with a flashlight is soon doomed to failure as the majority of our trees and bushes have thorns which do not let go willingly. That too is one reason why catching with nets is difficult in our corner of the world."

Bow nets, funnel and other repeating traps were not used as mostly undesirable birds got caught, in addition to mice, ichneumon and other animals. Another disadvantage is that such traps cannot be left untended even for short periods. Gabar Goshawks and Red Meerkats are immediately at hand and the new catch is either killed or chased to death. Whoever does not wish to catch specific species should certainly be able to use stationary nets, vegetation permitting.

E. M. Arnold (1964) also describes observations at watering places in Southwest Africa but without adding to Hoesch's remarks.

19.4 Catching in water

Catching in or near water—without relationship to catching at watering places—will be mentioned only in passing. This technique has been mentioned in chapters on double clap nets, pull nets, stationary nets, drop traps and funnel traps.

RADDE (1884) describes catching Common Coots near Lenkoran (U.S.S.R.) on the Caspian Sea. The birds can be driven by the boat and usually do not fly up if the lane has been marked by reeds which have been broken off no less than about a foot above the water level. The multitudes dive and swim slowly in front of the boat until the drive ends in a small bay. One net is already in the water and the catcher holds a second net while hiding in the reeds. Once the birds are close together in a bunch he throws the net over the birds which can no longer escape.

SCHÜZ (1957) reports further details on the catching of swans on the southern shores of the Caspian Sea referring first to HABLIZL (1783). A pull net about 20 m long and 3.5 m wide is set in shallow water. The wild swans were lured close to the net by two birds, either tame or with clipped wings. RADDE (1884) also witnessed catching Whooper Swans in one bay on the Caspian Sea. "Reed huts are built in shallow water. The birds get used to them and the hunter can wait there unseen. Usually he has a few tame lure birds and the wild swans alight trustingly. At these spots diagonally set nets are located under water. As soon as the swans are close together they are pulled together and the birds are caught." On 22 January 1880 over 80 Whooper Swans were apparently captured with one pull of the nets. Catching swans today is possible and can be productive as many Whooper and Bewick's Swans can often be found on flooded meadows. In addition, a greater variety of techniques exist today, including pull nets, double clap nets and rocket and cannon nets.

SCHÜZ (1954) reports on the time when a large number of Arctic Loons (Black-throated Divers) got caught in salmon nets and most of them could be extricated alive.

At the beginning of this century BLUSSIUS (1956) tried an original method for catching flamingos standing in water and well able to fly on the shores of the Black Sea. He used a net about 50 m long and 30 m wide with a mesh of about 20 cm. He spread the net on the bottom in very shallow water where flamingos had been standing and fastened it with strings to stakes on shore. The flamingos were carefully driven toward the net. When all or at least a good part of them were standing on the net, fishermen in boats pulled the strings of the net up. Many birds got caught in the net with their long legs. People from the boats as well as from shore helped take the birds out of the nets. On the first try only 7 flamingos out of 60 were captured, but later the catches were more successful. Once they had 20 birds plus two pelicans within a few hours.

The flamingos were not afraid of the boats as they had never been bothered by the fishermen before.

19.5 Catching at feeding stations

It is hardly necessary to talk about catching at feeding stations. The possibilities are generally well known and can be applied year round.

Details have been mentioned while discussing various catching methods, including the combination of watering places with feeding stations.

Artificial feeding stations can be established by planting fruiting bushes and by simply throwing out food as bait. Which method is chosen depends largely on local circumstances and the chances for catching birds.

20. The clap nets

20.1 General

The clap net is mainly thought of as a bird catching technique. Clap net designs can be traced as far back as ancient Egypt 4,000 or 5,000 years ago (MACPHERSON 1897, SCHALOW 1905, SCHÜZ 1966). In those times ducks and geese were the main catches. Romans in pre-Christian times—possibly through the Egyptians—were aware of catching possibilities with clap nets. MACPHERSON emphasizes that Italian clap nets are very similar to those of the ancient Egyptians. LINDNER (1940) speculates that the

Fig. 394. Double clap net in the Tacuinum-sanitatis manuscript (Cod 4182) in the Bibliotheca Casanatense, Rome (Italy). Work from Verona, last quarter of the 14th century; from LINDNER 1940.

use of clap nets was introduced to Germany via the Sabian Franks living under Roman overlordship.

Catching with a double clap net requires time and the setting-up is not quick. That is presumably the reason why catching with a couble clap net is practiced only regionally in Central Europe. Alterations in landscape and increases in human population density have certainly also contributed to the decline. Nonetheless, catching with double clap nets and pull nets (the "one-winged" clap nets) is still necessary and rewarding for scientific bird banding.

The double clap net should be set up in locations where birds appear regularly, even if not in great numbers (Fig. 394, 395) Double clap nets also have to be in places where human intrusion is at an absolute minimum. A hut is always necessary for the catcher to watch the clap nets.

Fig. 395. Trapping place with a double clap net on the Isle of Malta. The place is situated directly on the coast in order to induce arriving migrants to invade. Lure birds are placed on the stone-wall in the middle of the trapping-place. The catcher, whose head is visible, is sitting in a hiding-place built of stones. The top of the hiding-place is open. This is the general practice which obviously is not detrimental to trapping. In the background the Isle of Gozo. Photo: H. BUB.

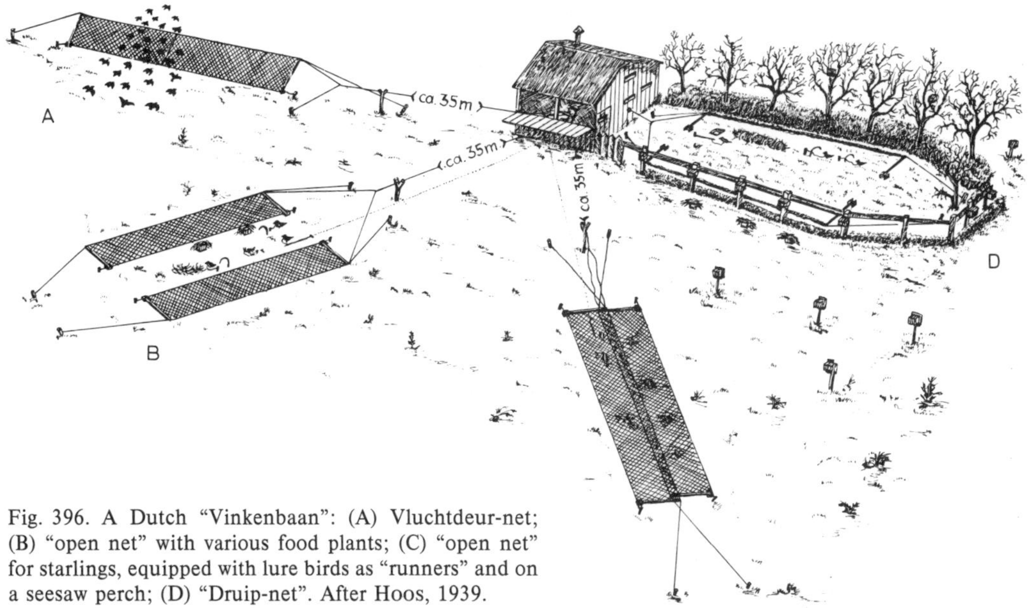

Fig. 396. A Dutch "Vinkenbaan": (A) Vluchtdeur-net; (B) "open net" with various food plants; (C) "open net" for starlings, equipped with lure birds as "runners" and on a seesaw perch; (D) "Druip-net". After Hoos, 1939.

20.2 Dutch double clap nets

Double clap nets have been known for centuries in the Netherlands. In the 16th and 17th centuries almost every farm had such devices set up. Universities even had special holidays for their students during migration times! Since 1936, however, the trapping of birds for reasons other than scientific banding has not been permitted.

What the Dutch designate a 'Vinkenbaan' is not only a certain type of double clap net, but a variety of trapping methods with double clap nets and pull nets (Fig. 396). They can all be erected on a 'Vinkenbaan', thus greatly increasing the effectiveness of the station. The set-up has been simplified compared to earlier ages. The Dutch do not plant berries which, even today, are considered obligatory in some other locations. One explanation for the missing berries may be the lack of berry-bearing bushes in this country of few forests. According to B. J. Speek, twigs of sea bucktorn, rowan and similar berry-bearing trees are stuck in the ground if thrushes are to be trapped with an "open net".

Hoos (1937 and 1939) and Pelkwyk (1941) describe Dutch double clap nets in great detail.

To set up a 'Vinkenbaan' and its various nets the bander picks a location along a regular fall flyway that make the time, effort and cost worthwhile. The construction of the hut comes first. A small stove is installed to prepare meals as well as to give warmth on cool days. Minimum size according to Hoos (1939) is 3×2.5 m plus a small anteroom of 1.5 m. The appropriate "peep holes" and string holes for double clap nets and pull nets must also be considered.

Fig. 397. "Garden" at a Druip net location. The food plants often hide several lure bird cages. In the foreground a bird in a harness and a water dish. Photo: P. Nijhoff.

The "Druipnet" is a regular double clap net. Dutch banders usually set it up close to the hut (Fig. 395, D) towards the northwest to avoid looking into the sun. On the clap nets shown here, a 1 m high earth barrier was laid out to the left of the hut and was covered with sod. Dead trees 5 to 7 m high and branches were set up here. They will help attract the birds. The branches should be leafless so that one can see the birds. On the northeast is another lower earth barrier.

The entire double clap net is often 10 m long and the nets are 2 m wide. The poles—1 m long and 3 to 5 cm thick—are often made of wood. In former times the nets were up to 20 m long. The pull strings add another 10 m to the space required. At the net, various lure birds (in harnesses) and several food plants are grouped together in one spot (Fig. 397). This creates the impression of a garden. For Eurasian Siskins, European Goldfinches and Common Redpolls alder branches make a nice "garden". The branches should be 60 cm long and all leaves must be removed to make the fruit visible for the birds. Other food plants can be used in such a "garden"—for example thistles which goldfinches like very much. The hoops (Fig. 397) of sturdy willow twigs or firm iron wire protect lure birds and food plants.

The "open net" (Fig. 396, B, and 398) is another version of the double clap net. Here the nets are not pleated back as with the "Druipnet" and other double clap nets and pull nets, but are left lying open upon the ground. The poles are 1.6 m long. This particular net serves to catch birds that like to stay in flocks such as Common Starlings, larks, and buntings. "Open nets" are never set near trees the way "Druipnets" for catching thrushes and finches often are.

Even birds sitting on the netting are captured. The catcher often pulls the string—made of steel wire 3 to 4 mm thick—when the birds are just coming in. This is similar to catching methods for Eurasian Golden Plovers and Northern (Common) Lapwings.

Placing the nets so they overlap after closing, as shown in Fig. 396, prevents the escape of captured birds. It is a good idea to fasten two or three short pieces of wood at those spots where the net strings hit the ground to prevent the strings from injuring or killing birds accidentally.

Today the nets are mostly made of nylon. The mesh size measures 15 to 18 mm on a side (25 mm for thrushes and starlings). The captured birds should be able to stick their heads through the mesh. The difficulty in removing birds increases if they are able to flutter to and fro under the net. Unfortunately, this cannot be avoided altogether with double clap nets and pull nets that have a bow for the protection of lure birds.

Nylon nets as well as pull strings made of steel wire or thin wire cables have the advantage of not suffering from wetness and humidity and therefore being always fully usable.

Third, the "Vluchtdeur" net must be mentioned. It consists of only one net, 20 m long, and is basically a pull net (Fig. 395, a). The pole is 1.6 m long. The net is set up across the path of the wind. It is only used when a strong breeze blows from the west or southwest. Under such

Fig. 398. An "open net" at the well-known Wassenaar banding station on the Dutch coast. Only a few lure starlings wearing harnesses are on location. Photo: P. NIJHOFF.

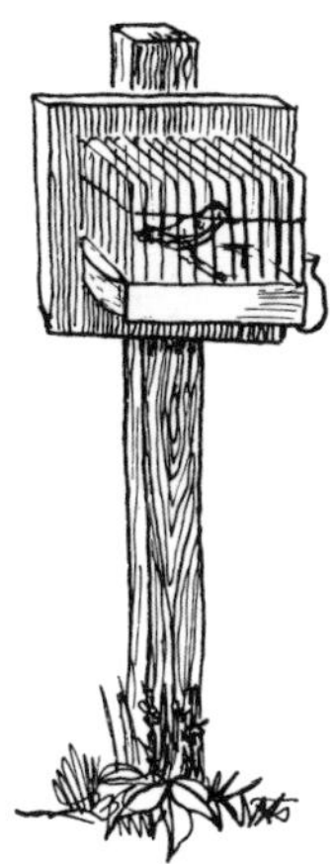

Fig. 399. Lure bird cage on a pole. After PELKWYK, 1941.

conditions the birds (finches, larks, thrushes, starlings, etc.) fly low over the ground and can be captured with the net closing in the opposite direction to their path of travel. When the catcher pulls the string and the net clears the ground, the wind catches the net and speeds its closing. Catching is done without lure birds as the migrating birds, travelling close to the ground, do not react to them. Up to 158 birds have been captured with one closing of the net. Hoos has more than once captured over 50 birds using this method.

Lure birds: Without good lure birds—birds that call a lot—successful catching is much harder most of the time. Hoos (1939) thinks it necessary to keep some birds in dark quarters before fall migration to suppress their singing. When they are brought out into full light again at the onset of migration, they begin singing again. Song can be stimulated by certain additions to the food of the birds without having to put them into darkened cages. In addition Hoos also recommends that the lure birds should be caged for about a year prior to their use. The following birds are good lure birds according to Hoos: Eurasian Siskin, Chaffinch, Brambling, Eurasian Linnet, Twite, European Goldfinch, European Greenfinch, Yellowhammer, Common Reed Bunting, Snow Bunting and the Eurasian Tree Sparrow. The Common Redpoll is not suited for a darkened cage.

The cages with the lure birds are set up in varying ways. Some are sunk halfway into the ground so that the wild birds will not be put off by the hopping and possible flutterings of the lure birds. Others are camouflaged with food plants. Still others are hung in rows—visually screened from each other—on the fences near the banding hut.

Those lure bird cages attached to poles often face towards the clap nets in Dutch stations (Fig. 399), while OTTO (1910) and other German authors point them away from the nets.

Large banding stations should have a very diverse collection of lure birds, as many birds only react to their own species. The lure birds should be protected against cold and strong winds wherever possible, if catching is done at such times at all, so that they will remain active. For protection against rain, catchers put window glass on top of the cages. Some banders cover each cage with a piece of sod, especially if they are placed close to the catching site.

As already mentioned by BREHM (1855), lure birds are brought out onto the catching site after the nets have been set up. Dutch banders follow

Fig. 400. Starling on a seesaw perch. After PELKWYK, 1941.

Fig. 401. European Siskin on a seesaw perch; a harness. After Hoos, 1939.

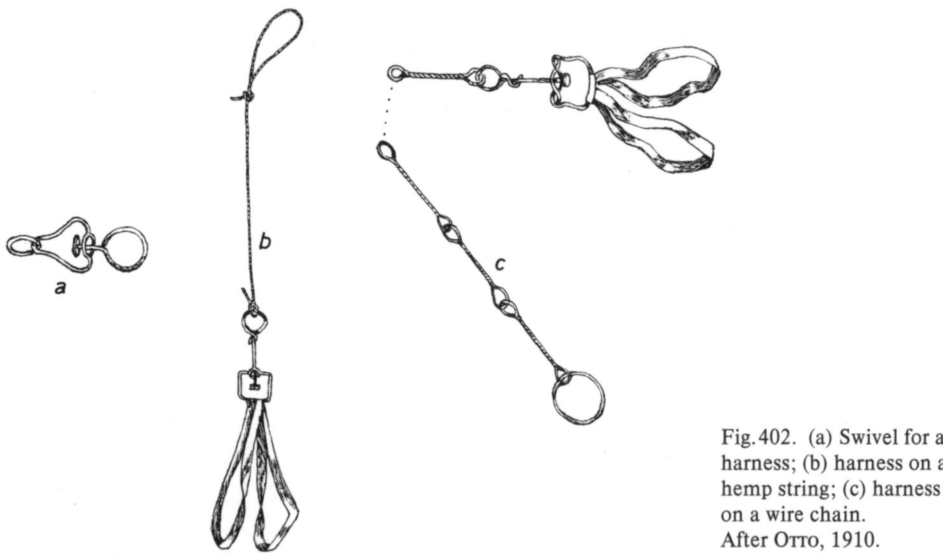

Fig. 402. (a) Swivel for a harness; (b) harness on a hemp string; (c) harness on a wire chain. After OTTO, 1910.

this advice. One thing, however, is unique to their set-ups. After the clap nets have been set, the catchers erect the "bordjes". These are short poles with a board nailed to them so that a lure bird cage can be hung up. Often they will put freshly caught birds into these cages while the "experienced" lure birds are located very close to the nets.

The lure bird or "runner" on the seesaw perch is of great importance (Fig. 400, 401). This device consists of a small pole which holds a stick which in turn is moved by a string from the hut. The bird, wearing a harness with a sufficiently long string, sits at the end of the stick (see also Fig. 402). The harness must be fitted exactly, for the bird will otherwise either get open sores or escape. If the catcher pulls the string to the see-saw the stick rises. If he releases the string the stick goes down causing the bird to flutter. The wild birds are supposed to see this and it often causes them to come and settle down. Hoos prefers young European Greenfinches on the see-saw, but certainly other species can be used as well. However, Hoos does not use Eurasian

Fig. 403. Tape recorders for transmitting bird songs to the catching location. Photo: E. WANDERS.

Fig. 404. Switchboard for controlling volume and for transmitting the song to each of the nets. Photo: E. WANDERS.

Linnets as they often get overly excited. In summer, if he did not have a Common Starling for his seesaw, he would use a black rag. Sometimes the starlings would respond and come flocking to the station.

The "runners" with a harness made of thin leather and tethered to a black nylon string are especially important to Dutch stations, as they employ few caged birds at the nets themselves. When the Dutch do use cages—such as with double clap nets—the cages are dug halfway into the ground and are camouflaged. Hoos often uses new catches as "runners". The birds generally get used to this new state of affairs in a few days—those which don't must be released.

Fig. 405. Loudspeaker hidden near the net. Photo: E. WANDERS.

Since 1969 Dutch ornithologists have used tape recorders (Fig. 403) to attract birds to their banding stations near Wassenaar. They set a loudspeaker between the two sides of a clap net (Fig. 405). Then they play back the song of those birds which are not represented among the lure birds (i. e. Meadow Pipit, Horned (Shore) Lark, Snow Bunting, Lapland Longspur (Bunting)). With the help of a switchboard (Fig. 404) the volume can be controlled and different songs can be played back at each of the nets. Amazing results have been obtained with this method.

20.3 Belgian double clap nets

Besides hedge and tent nets Belgian bird catchers use double clap nets of various sizes at many locations. Fig. 406 shows a simple station at Halle, near Brussels, with many lure bird cages. Numerous twigs of food plants have been planted in the ground and seeds have been scattered. All net and pull strings are made of hemp. Tension is achieved through strong rubber rings or rubber bands. Towards the nets the rubber bands lie on wooden pins as illustrated for the hedge net (Fig. 412). The blind is basically a dugout, covered and camouflaged, and is about 10 to 20 m distant from the nets. The nets themselves close very quickly as soon as the catcher pulls the strings. The catcher braces himself against the earthen walls while pulling the strings.

In Belgium J. SPAEPEN caught 191,493 birds of 165 species from 1960 to 1979. He used a double clap net, each half of which was 25 to 35 m long and 2 m wide. At times he used clap nets up to 45 m long. Among others he captured 48,952 Meadow Pipits, 40,340 Eurasian Linnets, 34,349 Common Skylarks, 14,642 Redwings, 9,216 Eurasian Tree Sparrows, 8,058 Chaffinches, 7,923 European Greenfinches, and 7,608 White Wagtails. From 1947 to 1954 he netted 3,600 Tawny Pipits, 3,100 Blue Headed Wagtails and 550 Ortolan Buntings.

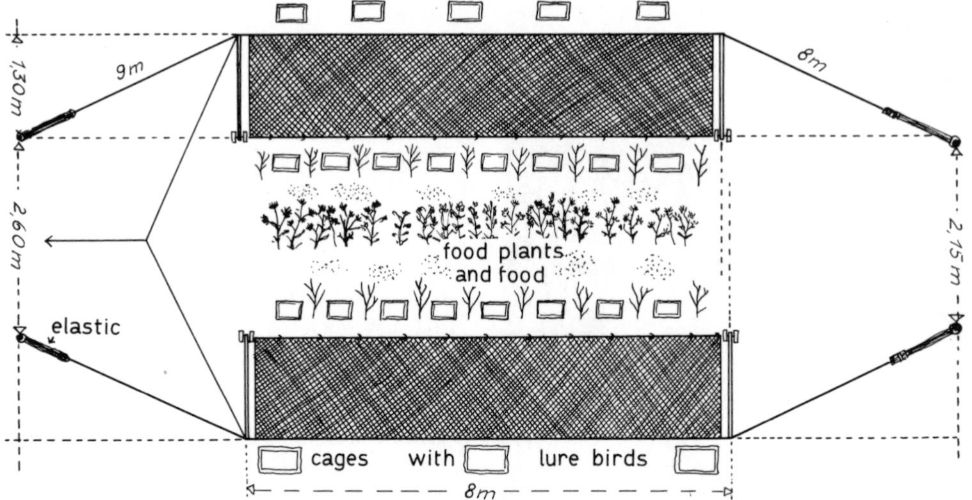

Fig. 406. Double clap net of a catcher near Halle in Belgium.

20.4 Italian double clap nets

Double clap nets were already used in ancient Rome. Even today there are many places where the main catching methods are double clap nets and pull nets, although for Italian bird catchers the trammel net has probably replaced these methods in importance. But the days seem numbered in which the catching of song birds for gastronomic purposes may still be practiced. There is a strong movement under way in Italy to end this practice of mass catching of song-birds for the table.

Toschi reported in 1938 that the zoological department of the University of Bologna had put a double clap net into use for ornithological research and established additional stations for this purpose. He stresses the value of this catching method to the science of ornithology.

Sunkel (1956) mentions the "larga" as a device much used with the double clap net. A counterweight causes the poles to flip over. The nets are up to 30 m long and 1 to 2 m wide. The catchers often operate several nets from a blind.

D. W. Stürmer reports on a catching station at the Lago di Garda (Italy) where bird catchers have set up six double clap nets on a meadow 2,500 m² in size (Fig. 407). The meadow—without stones or rodents—is mowed shortly before catching starts. Rodents are so effectively controlled that the nets can stay set even at night. A permanent blind is situated at the east side of the station, where the catchers have an excellent view and can release the nets. The nets—40 m long and 2 m wide—are fastened to strong wires. They are kept in position with large springs. It goes without saying that the pivot stakes for the net frames must be anchored solidly in the ground; it is often done with concrete blocks. The nets overlap about 50 cm and are slack. So that the two sides of the double clap net overlap properly, either the right or the left side may be released first. The choice depends upon where most birds sit. The force of the set springs is so great that the nets flip over at a very great speed. It requires extraordinary exertion on the catcher's part to reset the springs by folding the nets back again. The number of injured birds is surprisingly small, presumably because of the vegetation covering the ground and the fact that the nets are slack. About 10 m² of the ground between the nets of the center clap net are dug up, and a bare tree stands in the middle of the entire catching set-up. This tree is removed once no more pipits can be expected. All trees around the catching station are limbless except for the crown, and the bushes are low and without undergrowth. Lure bird cages are not hung in the trees but rather on poles 6 m high. Two lure birds of different species are put in double cages that are raised and lowered by means of a pulley.

As at all double clap net sites, here too we

find a bush clap net (Fig. 408). The bush clap net is 8 m long and 3 m wide. It is at right angles to the other nets and has been raised about 75 cm by a berm. In the center is a row of bushes 6 m long and 1 m wide. These bushes are cut so short in spring that in fall there are only soft pliable branches 1 m to 1.5 m high. The bush clap net sits on flat ground on top of the berm which slopes down gently; very gently towards the east. One of the two nets in this clap net can cover the entire row of bushes. The second more narrow net is triggered by the release of the first larger net and catches those birds escaping from the other side. About 3 m distant, rows of cages with lure birds have been set up: these contain about 30 House Sparrows and Eurasian Field Sparrows, some Dunnocks, buntings, Chaffinches and European Greenfinches. Within the bushes themselves, several species of *Sylvia* warblers in a wire cage act as lure birds. This cage is fashioned of bicycle wheel rims and hardware cloth. It is bell-shaped with several perches and a water bowl. Food for the birds consists of figs squeezed through the wires. The "inmates" are new captures; *Sylvia* warblers are not kept through the winter.

The cages with the lure birds are almost always on the east side of the clap nets (only rarely on the west side), and mostly they are on the ground or hung from wooden poles ½ m high (two cages to a pole). Only a few cages are on the ground between two clap nets. Tethered birds are the only ones actually within the clap nets themselves. No cages are under closed clap nets because that would interfere with the removal of the captured birds.

Tethered birds are protected from the closing nets by wire hoops. For starling, finch and bush clap nets many birds are tethered with a string instead of a seesaw perch, and for starling catching many individual birds are tethered to sticks to give the impression of a flock to migrating

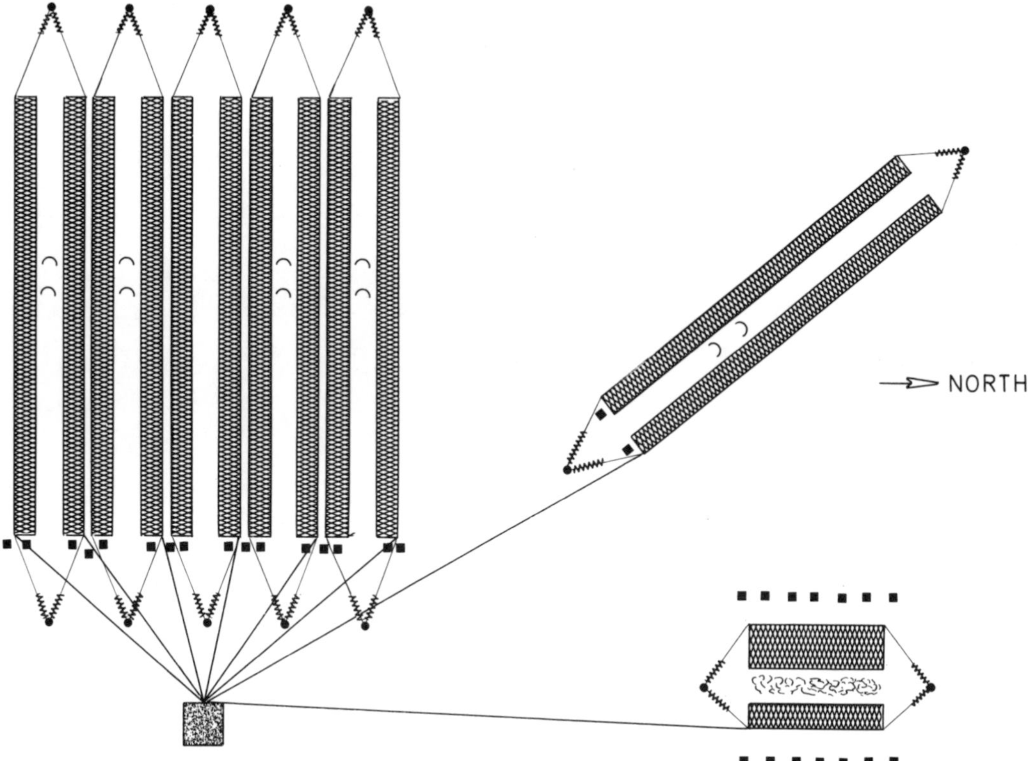

Fig. 407. Italian trapping location on the Lago di Garda (Italy). The blind is about 12 m in front of the nets. The double clap net for starlings diagonally to the right and the bush clap net at right are each 40 m distant from the blind. Lure bird cages in front of the nets have been indicated. In the middle of the nets are some lure birds. After Dr. W. STÜRMER.

Fig. 408. Bush clap net at an Italian catching location.
Photo: Dr. W. Stürmer.

birds. Starlings—live or stuffed—are also set on bare branches and twigs as well as on high poles all around the catching area. Before dawn each morning all lure birds are taken out to the catching area from their nightly quarters (a solid house nearby). By sun-up lively twittering and singing is under way. Catching is best during early morning hours, gradually tapers off, and picks up again around 4 p. m. During very hot days the lure birds in their cages are all taken back to their quarters at noon.

The catching of songbirds commences on 15 September and ends by 31 December. First are the Ortolan Buntings, Tawny Pipits, Tree Pipits, Yellow Wagtails, Garden Warblers, Greater Whitethroats, Black-cap Warblers, Lesser Whitethroats, sparrows, and Dunnocks. The lure birds comprise these species. At the beginning of October, Meadow Pipits, the first White Wagtails, and Common Starlings appear. The following birds are captured without lure birds: Stonechats, Whinchats, shrikes and Northern Wheatears. Last come Water Pipits, Wood Larks, Common Skylarks, and a few other larks. Each year rare species are also captured in the nets.

Stürmer gives additional descriptions of the handling of the lure birds. After capture they are kept in individual cages and fed a mixture

Fig. 409. Westphalian (Germany) double clap net shortly after 1920. Photo: Dr. H. Reichling.

Fig. 410. A Belgian hedge net. Photo: Firm of PRIEM-VERLINDE.

of millet and grits, meat meal and silkworm meal, or simply a grain mixture as used for laying hens. Insectivorous birds are first accustomed to their new life by some cut-up grasshoppers and mealworms, thrushes also with figs and berries. Those birds designated as lure birds for the following year are put into darkened rooms come next May. Only about two weeks before catching starts the birds are reaccustomed to daylight. They immediately start singing. Such treatment of lure birds is to be condemned for humanitarian reasons and is actually forbidden by law in some European countries. An appropriate diet, with some vitamins and hormones added, will also produce song in the fall.

In addition, the catchers use calls so skilfully that even caged birds will answer. The "falcetta" is an especially interesting call: it imitates the whistle produced by raptors while stooping for prey. Flocks of birds, still hesitating to settle on the ground or in the bushes, are thereby literally forced to take cover.

Besides numerous lure birds in cages, tethered birds are equally important on clap net sites. They sit on a seesaw perch and can be made to fly up and settle down at will. They wear a soft harness (see p. 20) under their feathers, which will hinder them neither while feeding nor while preening nor while flying up. A harness for these birds can also easily be made by hand with a double string of cotton. Seesaws are operated with a string from the banding hut. It is very important that tethered birds on a seesaw fly up at the right time. If the birds one wishes to catch are already sitting on the ground, or if they are just about to come down, the flutterings of the tethered bird will either cause them to fly up or even leave altogether.

Italian clap nets have achieved a high degree of perfection. It is well worth some attention for bird banders. The bush clap net with its great potential for catching can be recommended in particular.

21. The hedge net

This is a net whose net walls do not touch the ground as with double clap nets, but merely turn 90° to cover a hedge (Fig. 410). This method is used extensively in Belgium to catch finches, thrushes and other birds which like bushes. To set up a permanent catching place the bander needs a hut about 20 to 40 m distant from the nets. Size and selection of bushes depends upon the species to be captured. To catch thrushes one needs sturdier branches than for finches. This is easily observed. The preferably bare branches should be sturdily sunk into the

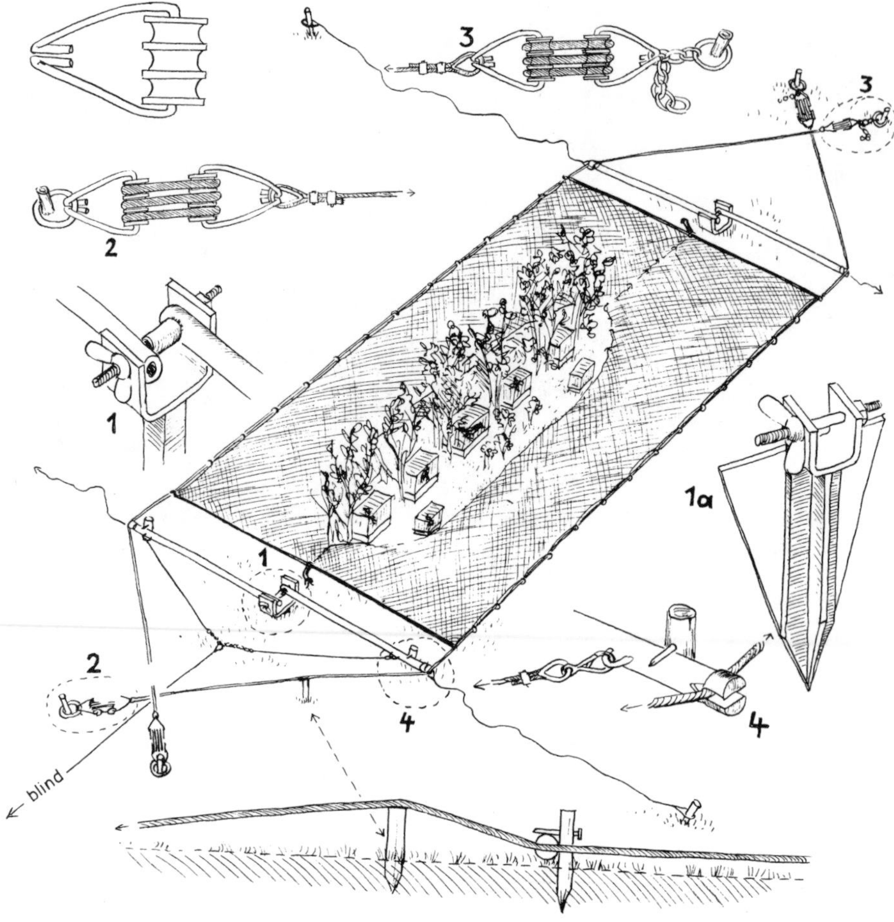

Fig. 411. A Belgian hedge net. For further explanations see text.

ground and provide the birds with a quiet resting place. Lure birds, food plants and berries are indispensible. The bare area behind the bushes provides space for taking the birds out. The long sides of the net must be fastened to the ground with stakes. The mesh of the nylon nets is 18 mm. If there is a possibility of catching smaller birds such as *Phylloscopus* and *Sylvia* warblers a smaller mesh is necessary. These birds are generally not captured with hedge nets. The lure bird cages among the bushes are well camouflaged or are set in holes dug into the ground or are outside the net area altogether. Tethered birds are useful for catching but may be omitted.

The size of the nets can vary. One Belgian store sells nets 5 to 20 m long and 2.25 to 3.50 m high. Following are some technical data: Size of the wires which run along the outside of the net walls is 3 mm. The trigger wire may be thinner. The net poles, made of galvanized iron, are 20 mm in diameter with 1.5 mm wall-thickness. Each pole receives a very hard wood top with a deep cut (Fig. 411, 4) through which the wire runs. The metal of the pole should be cut slightly too. The net poles should be fastened at the base to the pivot stakes. Fig. 411 shows two possibilities (1 and 1 a), although (1) seems more practical. The two poles must be separated by an intermediate piece so that they do not touch after release. The wide "wings" on the sticks (1 a) prevent their turning in the ground when the nets are being set and released. This must be observed with all nets of this type. After release the net poles are restrained by strings attached to their tips. These strings must be se-

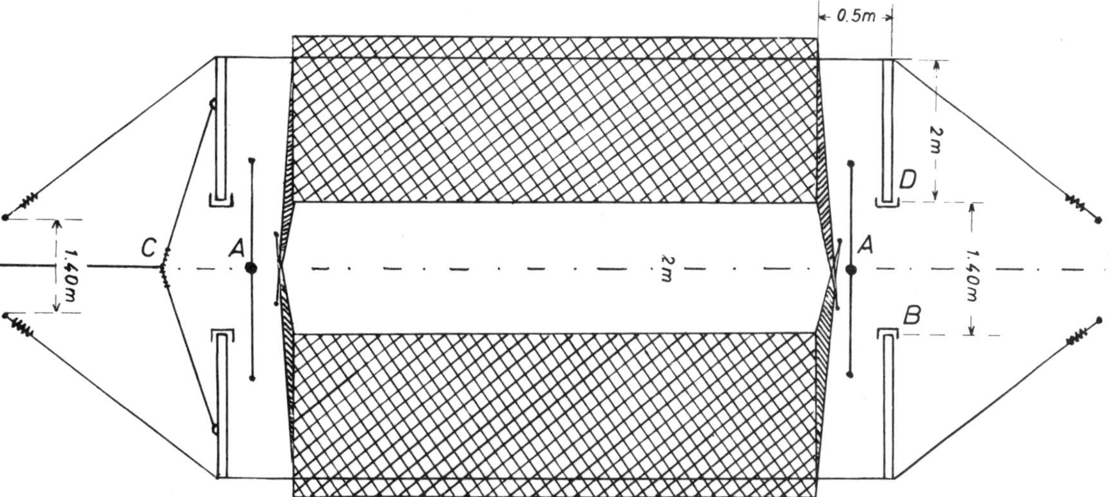

Fig. 412. A set Belgian tent net. After PRIEM-VERLINDE, 1967.

curely attached a few meters away and of course, must be long enough to accomodate the poles. It is very important to attach the nets to cross poles at 30 to 50 cm intervals (different intervals for each side). The nets should overlap about 20 cm so that no birds can escape.

The inside diameter of the four iron rings—two in front and two in back (Fig. 411, 2 and 3)—is 11 cm, and their thickness is 10 mm. The thickness of the chain links on the iron rings at (3) is 6 mm. The tension on the net strings is regulated here with the help of the chain links. Prerequisite of any net tension are the rubber rings (thickness 12 mm, inside diameter 12.5 cm) at the four anchor points. As shown at (2 and 3) there are three to each point, i.e. a total of 12. The rubber rings are held in place by iron hooks, 6 mm thick and 14.5 cm long, together with wooden spools for the rings (Fig. 411, above 2).

In order to increase the tension of the nets to a maximum two pegs are set up (Fig. 411, below). The nets are so taut now that they have to be prevented from clapping shut at the top. A nail in the two little stakes near the top of the net poles (see 4) keeps the poles in place. When the net strings are pulled the poles are freed and can close in the blink of an eye. The two retainer stakes are only in front.

We can recommend this hedge net highly to all bird banders.

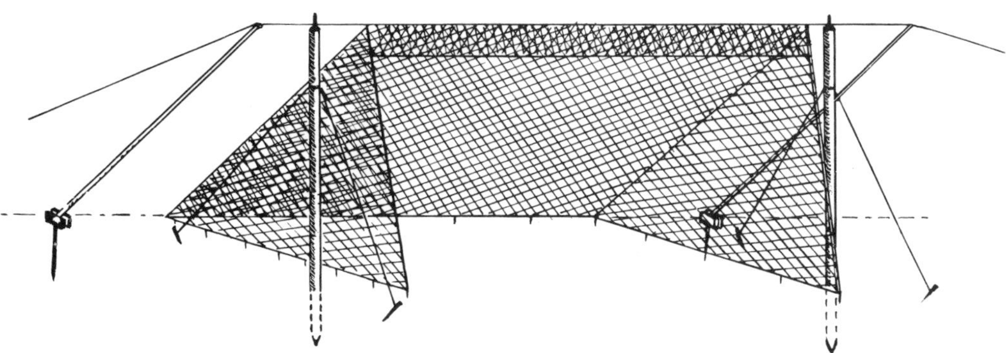

Fig. 413. A Belgian tent net, released. After PRIEM-VERLINDE, 1967.

22. The tent net

The tent net is similar to the hedge net. After release the nets take the form of a tent because the pivot stakes (Fig. 411, 1 a) are not close together but rather 1.4 to 2 m distant from each other. This distance increases the space available for bushes and branches. These are stuck in along the entire length (Fig. 412), which is rarely done with a hedge net (Fig. 410). Catching capabilities are therefore increased as compared to

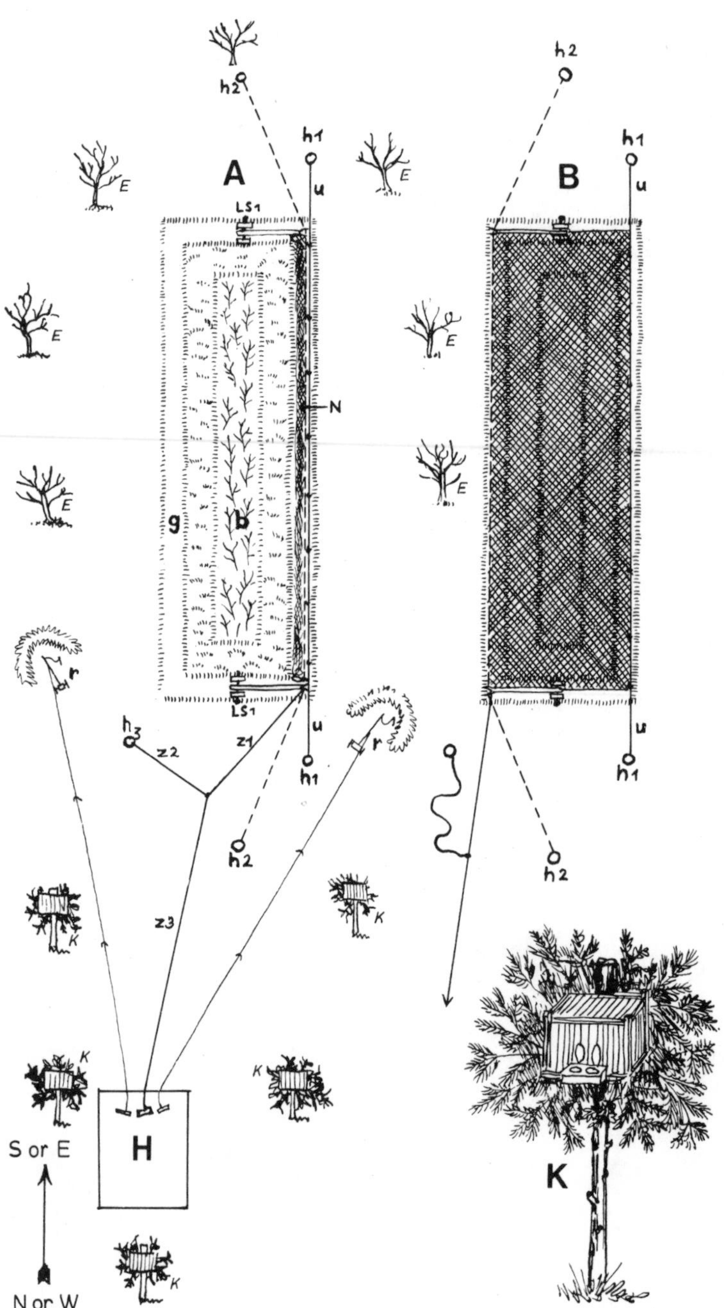

Fig. 414. Pull net after H. Otto, 1910. (A) pull net, open; (B) pull net, closed; (b) berry bushes; (E) landing trees; (g) ditch; (N) folded net in the ditch. (h1) pin for the lower string. (h2) pin for the upper string. (h3) pin for the pull string. (H) blind. (Z 1-3) pull and lower strings. (K) cages with lure birds, turned away from the catching location. (r) mounds for harnessed lure birds on seesaw perches.

Fig. 415. Closed pull net in Westphalia (Germany) around 1920. Observe the places for the lure birds in the small bushes. Photo: Dr. H. Reichling.

the hedge net. On the other hand sometimes a hedge net might take higher branches. It is also easier to construct. Bub has no preferences here, especially as length may vary considerably (from 5 to 20 m).

The tension lines of the nets, fastened to 2 m long poles, meet the two strong steel-capped poles (1.70 m high) when the net closes. Other catchers string a rope between the two poles (Fig. 413). The two net walls must overlap 20 cm so that captured birds cannot escape. The side nets must also overlap sufficiently to close any escape routes. In addition they must have an elastic band along their edges.

Fig. 416. Closed pull net in a wooded area in Westphalia (Germany) around 1920. The lure birds are covered with pine branches at left next to the net. Note in the foreground the tension holder for the leading edge of the net. Photo: Dr. H. Reichling.

23. Catching with pull nets

23.1 General

Pull nets are essentially single clap nets. They are especially useful where it would be impossible or impracticable to set up a double clap net. BREHM (1855) did not like pull nets because he considered any restrictions on the use of double clap nets inappropriate. Today banders rather like them because they largely suit their needs. The number of variations invented by yesterday's bird catchers and today's bird banders are truly amazing. Setting up a pull net is not always done quickly, although it may be somewhat easier than a double clap net. Besides the usual release of the net by pull string, today it is sometimes done electromagnetically or even via radiotelephone. Therefore it is necessary to consider several different methods.

23.2 Catching songbirds

Closely related to the double clap net is HUGO OTTO's (1910) pull net with which he successfully captured thrushes and finches (Fig. 414). The net is 8 m long and 2.5 m wide. For thrushes the mesh is 30 mm. The bed of berry bushes is 5 to 6 m long and 1 m wide. The bed is banked by grass and on one side is a 20 cm deep "ditch" which receives the net. The net poles are likewise recessed into the ground. The berry-bearing branches and twigs of rowan and juniper are pushed deeply into the ground (so they will stay fresh) with their tops pointing towards the hut. Branches gathered in September can be kept fresh in wet sand in a cool cellar.

Fig. 415 to 417 illustrate pull nets of Westphalian (Germany) bird catchers earlier in this century. They clearly show the application of lessons learned with double clap nets. To keep the net lines taut a tension holder is present, but only at the far end of the net.

F. GRAH of Solingen (Germany) has yet another method (Fig. 418). The net is 5 m long and 2 m wide. The tension ropes on the sides also measure 2 m, the pull string is 40 to 50 m long.

a) The net is laid flat on the ground and the ends of the tension ropes are secured with wooden pegs (the size of the pegs depends upon the ground). The bander drives in the pivot stakes 15 cm behind the upper edge of the net.
b) The tension rope is loosened on the right side, looped around the right net pole and reattached to the right peg. The end of the tension rope should have a sturdy rubber band attached so that proper tension can be

Fig. 417. Set pull net as seen from the blind. The photo was taken in Westphalia (Germany) around 1920. Observe the berry bushes and the lure bird on the seesaw perch in front of the bushes. The net lies in a ditch in the ground, to the right of it are the camouflaged lure birds. Photo: Dr. H. REICHLING.

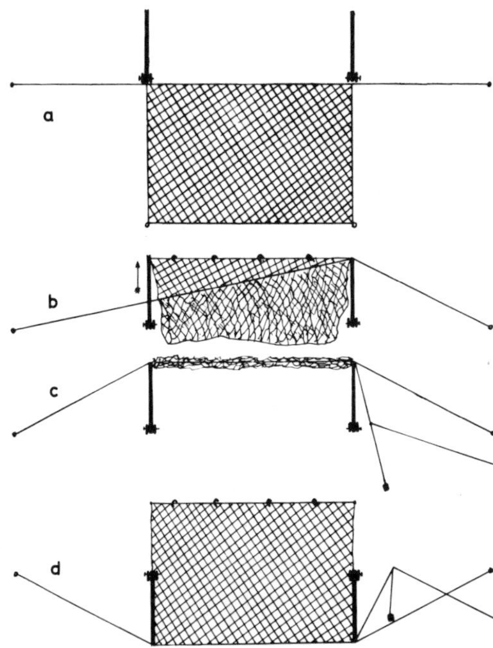

Fig. 418. Simple pull net. After F. GRAH. (a) before being set; (b) while being set; (c) set; (d) The net has been released. (c) The catcher ties another string of about 3 m to the top of the right net pole and to that string the actual pull string.

achieved. Now the tension rope is guided in a small furrow along the left net pole. The front end of the net is folded back (under the rope) and staked to the ground.

c) The catcher attaches another tension rope—about 3 m long—to the right net pole and to it the actual pull string.

d) The net has been released.

H. MÜLLER used a simple 3×1.5 m pull net bound with a 6 to 7 mm thick hemp rope (Fig. 419 and 420). The net should lie loosely and netting of at least 3.8×2.2 m is needed. The net poles are 85 cm long—they should not be longer for this kind of pull net—and have a screw eye at the bottom. This screw eye attaches to a U-shaped iron wire. The tension ropes are the same length as the net sides. The pull string is attached near the top of the net pole and runs over a forked stick approximately 15 to 20 cm high. This forked stick facilitates the release of the net. Near the net the pull string is attached to a sturdy wire or cable so that the net immediately flips over when the bander pulls the string.

Banders H. MÜLLER and H. KONOFSKY of Radolfzell (Germany) banded 136 European Goldfinches in one afternoon in 1953 with a 2.5×1.4 m net at a watering spot. MÜLLER captured 1,300 Eurasian Siskins with a 3.5×1.8 m pull net from January to May 1966.

HOLLOM & BROWNLOW (1955) show a pull net, which is at least 3×1.20 m and which may even be 3 times as big (Fig. 421). Noteworthy here is the simple pivot stake, which consists of an iron hook whose size depends upon the firmness of the ground. This is truly a simple version of the pull net. Fig. 421 a shows a device made of steel rod (diameter 5 mm) for use in sandy and muddy ground. The cross piece in the middle is welded only at the back. This "pivot stake" can be anchored more securely by a wooden peg set in the ground on the right.

The Dutch "Vluchtdeur" net (Fig. 396) mentioned earlier under Dutch double clap nets catches mainly starlings, thrushes and finches in flight. The particular net under discussion here is set only when a medium to strong wind blows from the southwest or west. At that time many birds fly very low over the ground.

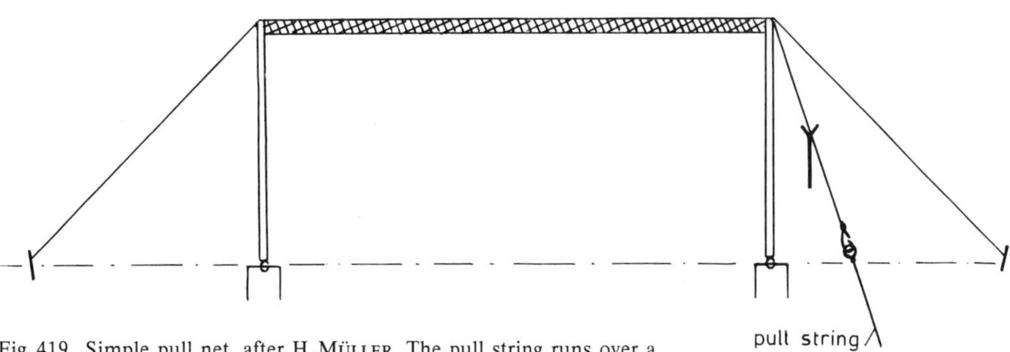

Fig. 419. Simple pull net, after H. MÜLLER. The pull string runs over a short forked stick to facilitate the closing of the net.

Fig. 420. Pull net after a catch. In November 1965 H. MÜLLER caught 46 Bohemian Waxwings with one pull. During the winter 1932/1933 K. WARGA captured and banded 1371 waxwings within 12 days with one 5 m pull net at two locations in Budapest (Hungary) (Aquila 1939, p. 491). Photo: H. MÜLLER.

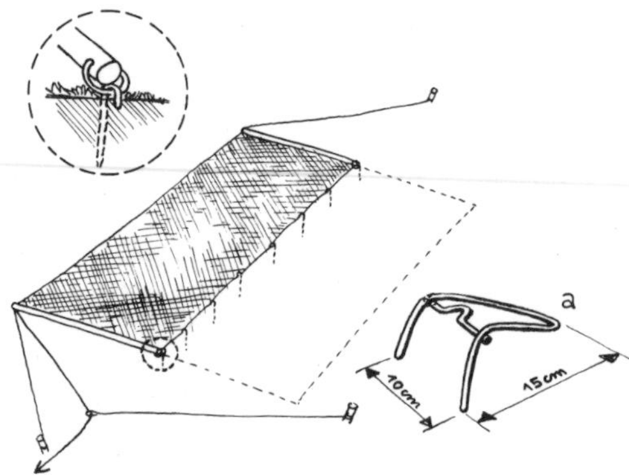

Fig. 421. English pull net with anchoring for the net poles; (a) Anchoring for the net poles in sandy or wet ground (right). After HOLLOM & BROWNLOW, 1955.

23.3 Catching crows

Traps for crows were already described in detail earlier. But the catching of crows with pull nets is much older. It was very well known at Rybachiy (Rossitten) (U. S. S. R.). THIENEMANN (1927) was influenced to take up banding here in part because of the crow catching practiced. The pull net (Fig. 423) is about 6×2 to 2.5 m. The picture shows the simple net, the tension rope at the rear end of the net and the net pole with which the net was made taut. The pull string would not stay taut if the net pole had not been forced under a small hook fashioned out of a tree branch by the catcher. Pulling on this string frees the net pole and the net slams shut (Fig. 422). A second pole can be used on the end toward the trapper. The correct tension must be found by trial and error. Closing of the net—usually from a hiding place up to 30 m away—requires not just pulling but also flipping the string by grabbing it just behind the peg at the end (see Fig. 423).

23.4 Catching Eurasian Golden Plovers and Northern (Common) Lapwings in the Netherlands

Northern Holland has a long tradition of catching these two species. In earlier times the birds were needed for food or else—especially Eura-

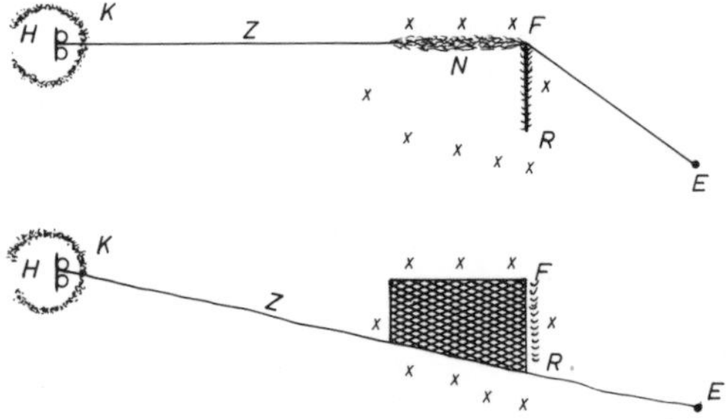

Fig. 422. Pull net for catching crows. Above: The net is set when the net pole lies in the ditch (R). After closing of the net is has been thrown away and the ditch in the lower drawing is empty. (H) blind; (K) locking handle on the pull string tied behind two posts; (Z) pull string; (N) net; (R) ditch for the net pole; (F) hook to hold the net pole; (E) end post. The x's represent lure crows. After SCHÜZ, 1932.

Fig. 423. Crow catcher on the Baltic coast near Rybachiy (USSR) with a pull net. Several new catches are used as temporary lure birds, besides the regular long-term lure birds. Photo: E. Schüz.

Fig. 424. A large pull net for catching Northern Lapwings and Eurasian Golden Plovers, shown while closing. Photo: F. Haverschmidt, 1943.

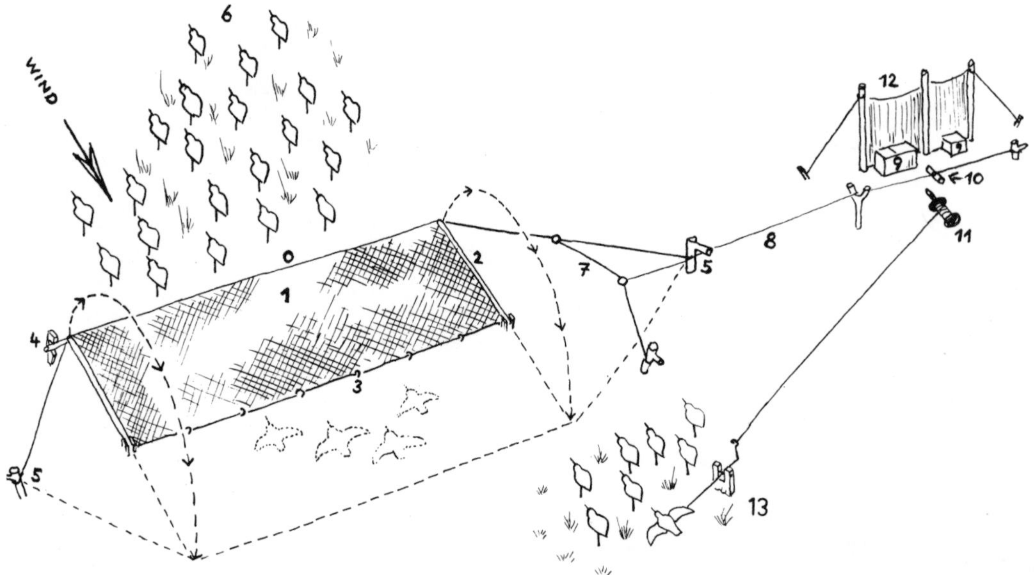

Fig. 425. Pull net for catching Lapwings and Golden Plovers. (1) net; (2) net poles; (3) net sides staked to the ground; (4) small stick which holds the net to the ground, pulling on the pull string releases the net; (5) end post for tieing the net string; (6) wooden decoys in front of and behind the net. In the foreground a bird on a seesaw perch; (7) a cross string; (8) pull string; (9) boxes for wooden decoys; (10) locking handle for pulling the pull string; (11) device attached to the string of the seesaw perch; (12) wall for protection and camouflage; (13) the seesaw perch with decoy which has been fastened to a rubber plate here; (14) lure calls (see Fig. 427). After Eenshuistra, 1956.

sian Golden Plovers—were exported. Today there are only about thirty licensed Golden Plover trappers, people who grew up practicing this trade. After their retirement this catching for profit will cease. In the meantime, though, banders have begun to use their methods for scientific research.

According to law, nets in the Netherlands may not be larger than 25×3.5 m at the most (Fig. 424). Generally they are about 20 m long and 2.5 m wide and have a mesh of 6×6 cm. Banders should use a mesh of 4×4 cm. Nothing further need be said about net poles and pivot stakes. The net rope is a wire guided between points 5 and 5 (Fig. 425). The catcher puts a sturdy stick through a loop at each end of this line. The net is kept as flat and as close to the ground as possible (see also under hedge net). Both net poles lie in grooves in the ground. The next step is the "scissors", a piece of wire (Fig. 425-7) which is fastened to the 30 m long pull string. At the end of the pull string is the blind which is designed primarily to shelter the catcher against the wind.

The boxes near the blind serve as storage facilities for stuffed and wooden decoys as well as other bird catching paraphernalia, and they serve also as seats when no birds are around. From here too runs a string to a seesaw perch (Fig. 426) where a live Eurasian Golden Plover or Northern Lapwing sits. The bird has its feet in a woolen sock. As soon as the catcher pulls on the string the seesaw goes up about 30 cm and falls back when the string is slack. This causes the lure bird to flap its wings and thus draws the attention of the wild birds. In addition about 30 to 35 wooden decoys (Fig. 425 and 427) are set at the catching site, about 25 of them on the windward side. Their bills all point into the wind.

It is surprising that newly arriving Eurasian Golden Plovers do not seem to be bothered by the relatively short distance of 30 m from the catcher. As soon as he has seen flying Golden Plovers, he imitates the two-note call of these birds on his bird call, and immediately afterwards the "murmuring" of a flock of Golden Plovers ready to land. At the same time the catcher pulls the seesaw string and the lure bird flaps its wings. As this species likes to settle down where there are conspecific birds already around, they fly low over the ground towards the lures. At the right moment the catcher pulls the pull string and quite often catchers a large part of the flock. The net always closes with the wind and can thus cover the Golden Plovers flying low and into the wind.

HAVERSCHMIDT (1943) quotes many numbers. One man caught 5,232 Eurasian Golden Plovers in 9 years. That is an average of 581 birds per season. In all, several tens of thousands of birds were captured each year. During a 48-year career of bird catching one trapper managed to capture a record 114 birds in one day.

KLUIJVER and VAN DER STARRE (1943) report on a different method of catching Northern Lapwings, one that VAN DER STARRE practiced for years near Reeuwijk. He preferred to set up his pull nets—12 m long and 3 to 4 m wide—near water, especially on small peninsulas in order to direct the birds more easily towards the

Fig. 426. An Northern Lapwing on a seesaw perch. Photo: F. HAVERSCHMIDT, 1943.

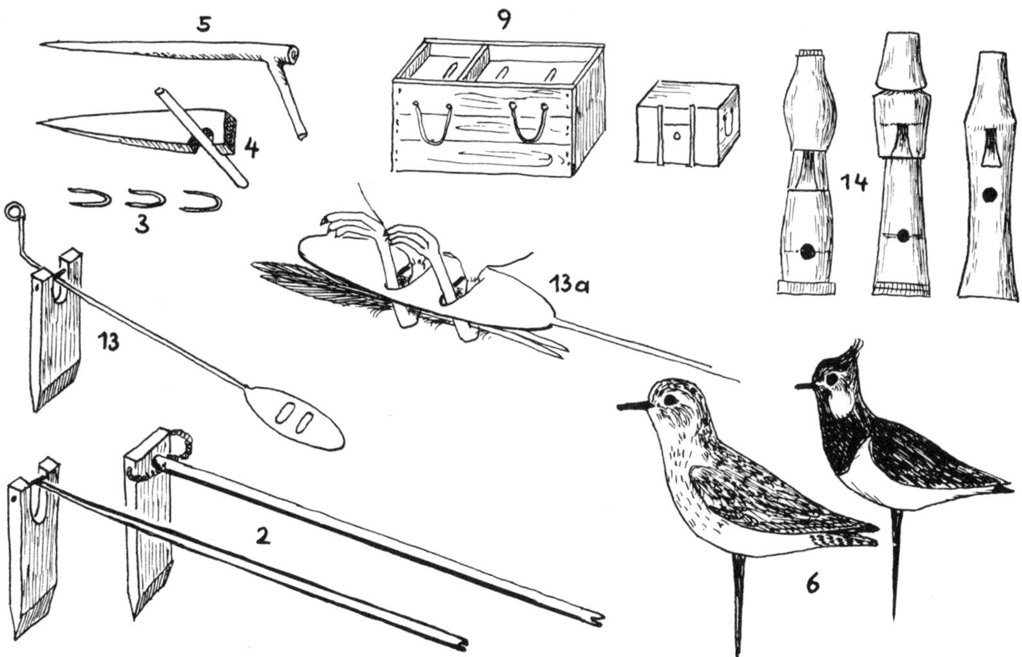

Fig. 427. Aids of various kinds for catching Northern Lapwings and Eurasian Golden Plovers in the Netherlands. The numbers correspond to those in Fig. 425. After EENSHUISTRA, 1956.

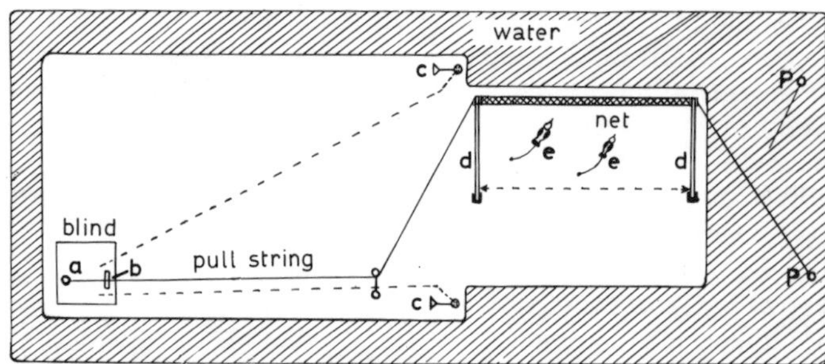

Fig. 428. Pull net for catching Northern Lapwings in the Netherlands. The drawing indicates an island catching location. Even better would be peninsulas or banks of rivers and lakes. (a) pole for tieing the pull string; (b) locking handle for pulling the pull string; (c) birds on seesaw perches and their pull strings; (d) net poles; (e) catching area; (P) post for tieing the net string. The upper post is used when the net closes in the other direction. After KLUIJVER & VAN DER STARRE, 1943.

limited catching site (Fig. 428). VAN DER STARRE had several catching places so that he was not dependent on the wind direction. As already mentioned, the nets must close with the wind. The catching site is mowed very short. Two birds with harnesses serve as lure birds. Two additional birds on seesaw perches also lure wild birds. Females are preferred because they are calmer. The blind of the catcher can be made of reeds but must have view slits on all sides. Often the catcher uses a call on which he can imitate lapwing vocalizations.

Nets 20 m long have also been used, but they should incorporate stronger springs for the release. Catching results during fall and spring are variable. VAN DER STARRE captured 680, 100, 581, 429 and 454 Northern Lapwings in the fall from 1938 to 1942, and in the spring 900, 161, 478 and 81. He did not do any catching in the spring of 1938. On good migration days catches of 100 birds per day are not uncommon. On 12 November 1946 VAN DER STARRE captured 345 birds, of which 9 % were recovered. The total of Northern Lapwings banded from 1938 to 1959 was over 11,000.

23.5 Catching gulls and terns

Pull nets are very useful catching these species. H.-W. NEHLS and W. MAHNKE captured about 3,500 Herring Gulls, Black-headed Gulls, Mew (Common) Gulls and Greater Black-backed Gulls in four winters around 1960 in the fishing harbor of Rostock (Germany) (Fig. 429 and 430).

The nets are tied securely only to rope b (Fig. 431). The other three sides of the net are

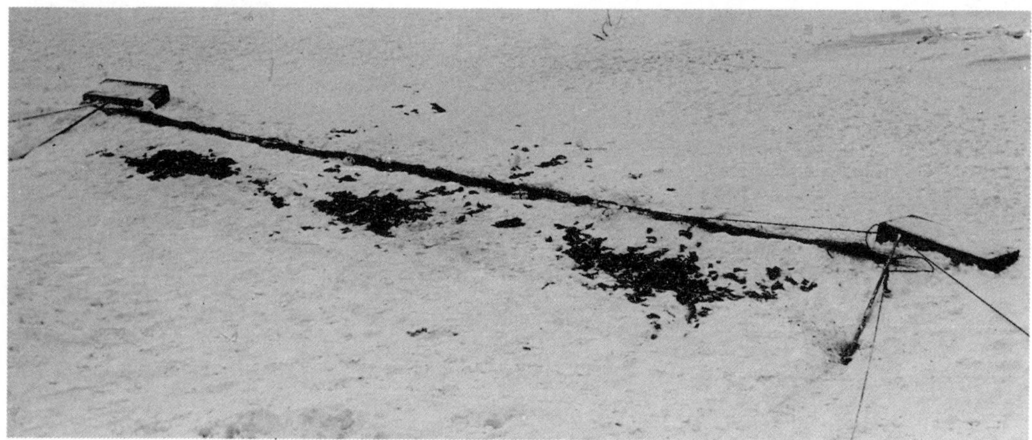

Fig. 429. Set pull net for catching gulls in the fishing harbor of Rostock (GDR). Photo: H.-W. NEHLS.

Fig. 430. Captured Herring and Great Black-backed Gulls. See also Fig. 429. Photo: H.-W. NEHLS.

also threaded with rope, the side ropes having double the length of the poles e. When setting, the net rope a, to which the back of the net is tied, must be stretched between two pegs in a way that the ropes at the sides are firmly on the ground when the net has been released. NEHLS and MAHNKE tied the pull string to the pole one third of the way down from its end, thereby increasing the closing speed and shortening the pulling distance. See also Fig. 432.

It is important to weigh down rope b with lead or other weights. This increases the speed and prevents it from springing back. The net (30 to 40 mm mesh) itself must be sufficiently large so that the captured gulls can move. The 8×2.60 m net has also been used at the beach to capture shorebirds, and can even be used to catch small birds provided the mesh is small enough.

NEHLS and MAHNKE captured a maximum of 73 Herring Gulls at once using plenty of fish offal. Catches of 50 to 60 birds were not uncommon. Prerequisites were careful setting and tension of the net as well as camouflage, especially when catching often at the same location.

D. AMMERMANN caught gulls in the fishing harbor of Cuxhaven (Germany). He specialized mainly in Greater Black-backed Gulls, banding about 80 in the process. Catching with the usual pull net was not possible as the ground was paved. AMMERMANN developed his own method (Fig. 433): First he tied the basic ropes (dotted line) and fastened them to hooks at the fish market door and on the wooden beam at the edge of the harbor basin. On these tightly stretched ropes he fastened the pull net. The pivot stake must be particularly well tied to the base rope so that it does not pull to the side when closing. The ropes must also be tight as the net will sag otherwise when flying up. Setting this net (mesh 30 to 40 mm) is somewhat time consuming, but on the other hand it can be used in many places.

AMMERMANN particularly mentions that the gulls are very suspicious after catching and only after a longer rest period can another attempt at catching be made. Perhaps one reason may be that this net could not be camouflaged.

The gull trapping by the Dutchman ENGELEN (1959) should be mentioned (Fig. 434). He used a 13 m long double clap net. Each net was

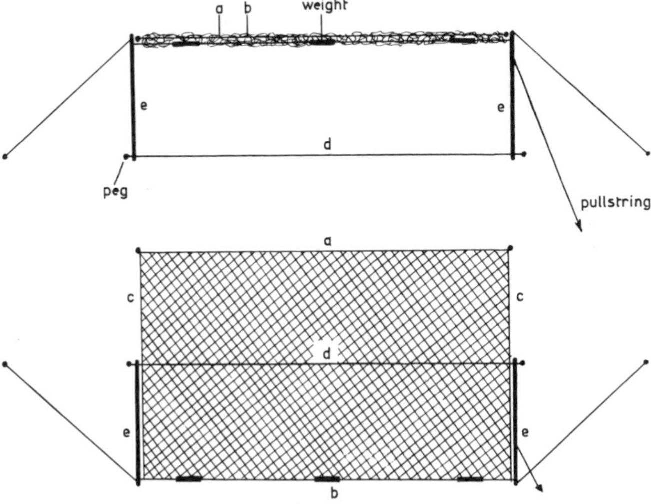

Fig. 431. Pull net for catching gulls. Above: set; below: after closing. See text. After H.-W. NEHLS.

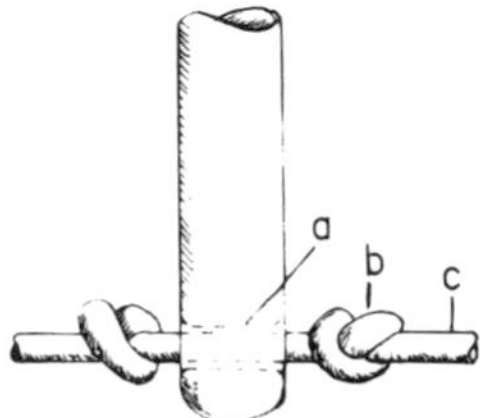

Fig. 432. Method for tieing the net string to the end of the net poles on the gull pull net. After H.-W. NEHLS.

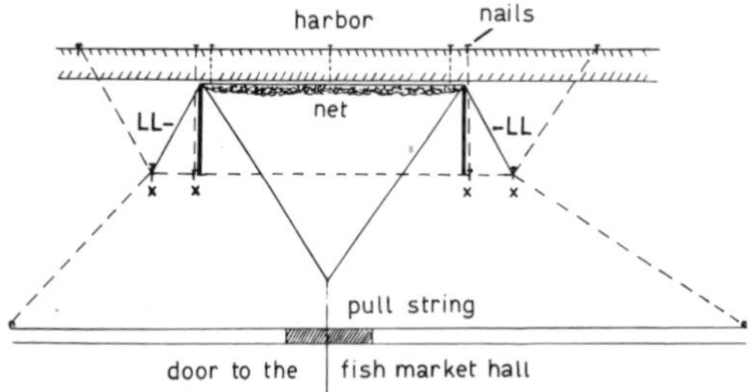

Fig. 433. Pull net for catching gulls. At the places marked with an x the device is tied down with rings and spring hook. After D. AMMERMANN.

1.90 m wide. The net poles (a)—notice their position relative to each other—are 1.25 m long and overlap so that no birds escape. ENGELEN uses old fish nets. The ropes are thin steel cables. Fig. 434 shows the arrangement of the ropes. The outer net rope is held in place by a notch in the top of the net poles (see also under hedge net) (Fig. 411).

The trapper's blind is at an angle of 90° to the net (which is probably not always necessary) and about 25 m from it. ENGELEN uses three or four lure gulls and puts two between the open nets. Not every day is well suited to gull catching. Sub-freezing temperatures are particularly favorable in cities.

Resting and sleeping terns can also be captured near the perimeter of cities. Simple pull nets can be used, or those whose net poles flip over with the help of springs. J. PILASKI had good success at the coast of Southwest Africa (Fig. 435).

24. Cannon netting

by P. L. IRELAND, C. M. LESSELLS, J. M. MCMEEKING and C. D. T. MINTON

WARNING—Cannon netting equipment is potentially dangerous to both birds and people. Only suitably competent people should be allowed to use the equipment and great care must be taken with the explosives and firing devices.

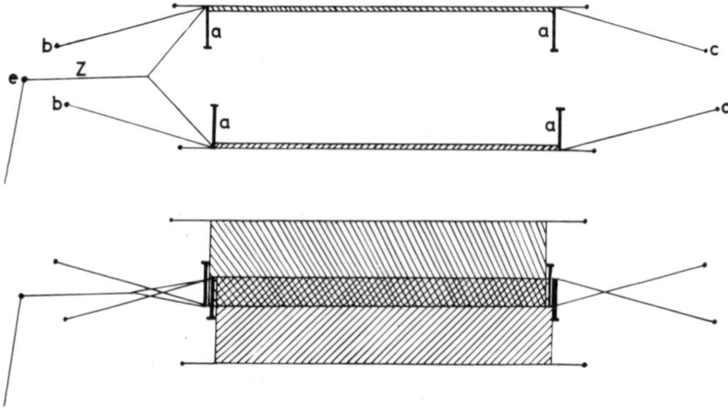

Fig. 434. Pull net for catching gulls. Above: set; below: after release. The pull string is guided through a hole drilled at point e in a wooden post. See text. After ENGELEN, 1959.

Fig. 435. Catching terns with a gull net in Southwest Africa. Photo: Dr. J. PILASKI.

24.1 Introduction

Cannon netting has been developed primarily for catching roosting waders. It has also been used to catch gulls, ducks, geese and other species which assemble on the ground including corvids, starlings and finches.

At its most basic the technique involves laying a net in a line on the ground to which is attached two or more projectiles. These projectiles are then inserted into cannons. When the birds are in the desired position the cannons are fired, the projectile pulls the net in an arc over the birds and the netting then gently settles on the birds thus entrapping them. Experienced handlers can then remove the birds from the net and process them as for birds caught by other means.

Cannon netting is an adaptation of rocket netting, the principle difference being that the propelling device is a static cannon whereas in rocket netting the rocket which is attached to the net is powered.

Cannon netting is essentially a team operation. Although very small catches could be made by an individual, significant sized catches will require a number of people, dependent on the number and species of birds expected.

Successful cannon netting depends on good advance information about where the birds will be and about conditions that are likely to occur at the catch site on the day of the catch.

The team leader will select the actual catching site after considering the advance information available. The net or nets must be set according to the leader's instructions in good time before the arrival of the birds.

The success of the attempt to catch requires the birds to be caught to go within the area on the ground that the net will cover when fired, but not so close to the net that it will kill or injure the birds when it is fired. The skill in setting the cannon net lies in placing it where the birds will gather although gentle persuasion may on occasion encourage birds outside the catching area to move to the required place.

When the required number and species of birds are safely in the catching area the net will be fired over the birds. Care must be taken that no more will be caught then the people available can handle and that no bird is in the anticipated trajectory of the net when it is fired.

Following capture most species of birds will need to be covered in the net immediately to prevent them injuring themselves by flapping. They will then be removed from the net and put in some form of temporary holding cage awaiting the normal banding process.

24.2 The equipment, setting and catching procedure

These details are included here to give a complete picture of what is involved in the manufacture and operation of cannon nets. It cannot be stressed too strongly that anyone contemplating cannon netting should first get in contact with someone who has already gained experience in the technique so that all the safety aspects can be fully considered. This is not only something which needs to be done from the bander's point of view but also because if something went wrong with cannon netting in one instance somewhere within Europe it could affect the possibility of cannon netting for all other banders throughout the continent.

24.2.1 Equipment/Glossary of terms

For Setting. *Net:* The standard size for a cannon net has evolved as being approximately 26 × 13 meters, although half size nets are fairly common and various other sizes are in use. A suitable mesh size is 3 cm square for waders and gulls although larger meshes would be adequate for larger birds. Smaller mesh sizes are not recommended. Twine size number 18 is commonly used and gives a fairly light yet strong net. The twine itself needs to be fairly soft and the net will work best if dyed a color to blend with the places that it will be set. Details of how to make a net start on page 305.

Pegs: 30 cm long, thin pieces of wood or metal used to secure the back of the net to the ground.

Cannon: A substantial piece of ironwork, normally made from steel seamless tube, 50 mm in diameter and 60–80 cm long with a chamber at the end to hold the cartridge which contains the explosive. The standard sized net described above usually has four cannons to fire it.

Cartridge: Reloadable brass cartridge of approximately 25 mm diameter and 80 cm length into which is loaded the gunpowder and fuse. One is required per cannon.

Gunpowder: Cannon nets are fired with a slow burning, charcoal based gunpowder. Choice of powder will depend on what is available locally but it cannot be stressed too strongly that a coarse grained, slow burning powder is required and experiments to find the correct quantity in each cartridge should start with very small amounts. 15-25 g per cartridge is typically what is required.

Detonators/Fuses: A small electrical fuse is required to go in the cartridge to ignite the gunpowder. As with powder, local availability will govern the choice of fuse, but any which contain explosive as well as the fusehead should be avoided.

Projectile: A heavy (about 3 kg) steel object to attach to the net's front edge and to be fired by the cannon. Modern designs have a solid mass of a diameter just less than the diameter of the cannon to go down the barrel and a thinner rod (15 cm) which comes out of the top of the barrel and onto which the ropes of the net are attached by means of a shackle.

Dropper Cable: Electrical cable with five connecting points, one for each of the four cannons and another to connect to the main cable.

Main Cable: Long (up to 400 m) cable used to link the blind or observation point with the dropper cable. Ex. Army field telephone cable is very suitable for this purpose.

Jiggler: Cord with small pieces of cloth or feather at approx 50 cm intervals, laid 20 cm in front of net and extended to observation point. For use if any bird is on or close to the net in an attempt to get the bird to move and hence allow the safe firing of the net.

Blind: Standard bird watching blind, to be set as an observation point when attempting to catch.

Circuit tester: Avometer for testing that the electrical circuitry is connected correctly. This item of equipment must be designed or modified so that under no circumstances will it set off the net being tested.

Ancilliary equipment: Mallet, Spade, Spanner.

For Catching. *Firing Box:* Specially made device for instantaneously delivering a high voltage when the "firing button" is pressed. Depending on the fuses and main cable in use, 200 volts or more is required to ensure the cannons fire immediately the firing button is pressed. Firing boxes should be regarded as a potentially lethal piece of equipment.

Decoys: In some circumstances stuffed birds may be placed in the catching area as an encouragement for others to come and join them, thus increasing the chance of a catch. Poor quality decoys will hinder catching!

Radio or telephone: Useful or essential means of communication between members of a team.

Covering Material: Hessian cloth or other lightweight yet dark material for covering birds caught in the net. Essential item of equipment to prevent birds flapping and damaging their wings.

Holding Cages/Boxes: Adequate provision for keeping large numbers of birds between removal from net and banding is essential. Details of a design for waders are given on page 307.

Bands and processing equipment: As required for project being carried out.

24.2.2 Cannon Net Setting

Instructions for setting a standard, four cannon, net are given below:

1. The team leader decides where to put the net(s) and marks this with pegs. Basically an area, clear of obstructions and with a viewing point such that an observer can see along the line of the nets is required. As a loose projectile can fly several hundred yards, nets must not be set firing towards people, property or roads.
2. If not already present (e.g. plough on field) make a shallow grove (5 cm deep × 15 cm wide) to put net in.
3. Lay out net in position marked.
   - 3.1 Start by finding the right hand back corner (marked red) of the net and peg this to ground at back of grove (net side of elastic).
   - 3.2 Lay net in line on ground. Find left hand back corner and peg to ground so that the back line is in the position required, straight but not taut or stretched.
4. Untwist net.
   Work from one end ensuring the net is not twisted, pulling the back line free from the

rest of the netting. The less the rest of the net is spread at this stage the easier later stages will be—try to avoid tide wrack, crop debris etc. getting caught up in the net.

5. Square net.
   With a person at each end of net, find front corners. Pull net so that each corner is level with corresponding back corner. (It may be necessary to remove snags in the netting to do this—pulling too hard could result in a torn net.)
6. Furl net.
   6.1 Each person on the two front corners should gather together net by working along side rope and putting resulting netting on back line.
   6.2 Work from one end of net picking up leading edge rope and then furling net by gathering up over arms. At same time check no debris is left in the net. Furl until back line is reached but this should not leave the ground. Place furled net on top of back line in as small a bundle as possible.
   6.3 Continue furling until the other front corner is reached ensuring it is done at sufficiently frequent intervals that all debris is removed and that the leading edge is on top of the resulting pile of netting. When furling one bit of net,

Fig. 436. Cannon. Photo: Phil Ireland.

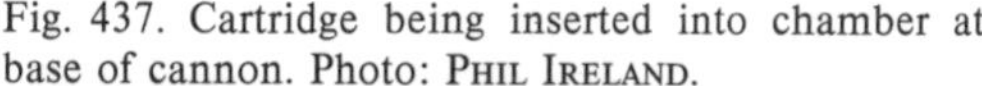

Fig. 437. Cartridge being inserted into chamber at base of cannon. Photo: Phil Ireland.

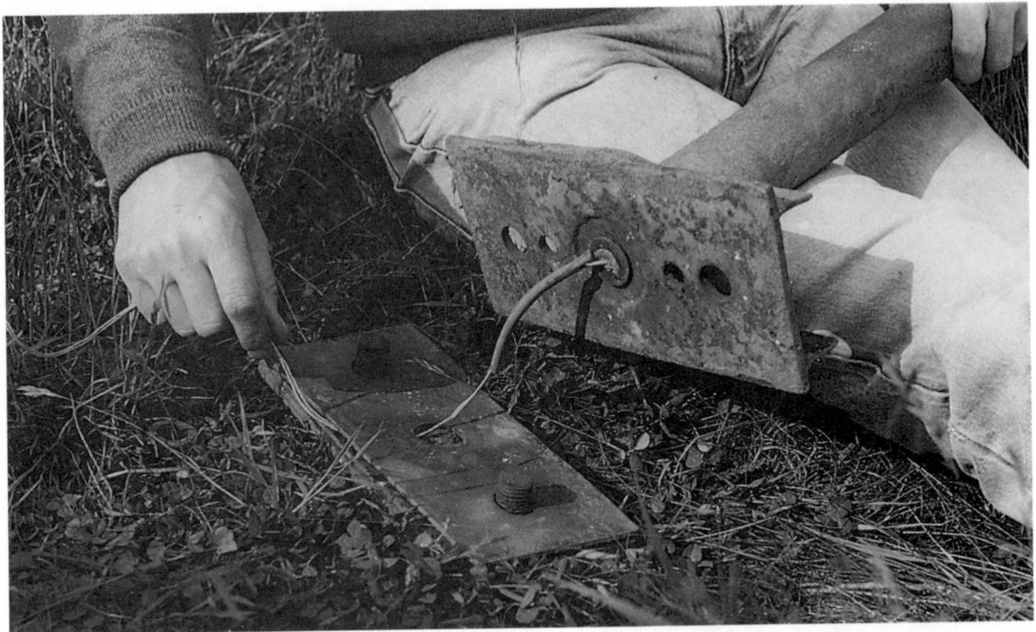

Fig. 438. Cartridge in cannon and fuse wires being fed through shock absorbing rubber on cannon base plate. Photo: PHIL IRELAND.

Fig. 439. Base Plate being bolted to cannon. Photo: PHIL IRELAND.

ensure previously furled net is not disturbed.

During furling the three intermediate jump ropes should be thrown out behind the net to avoid tangling.

7. Peg the three intermediate jump ropes about 10 cm behind the back line. This can be done using the loop at the end of the jump rope and fastening so the rope will not pull off the peg. (i.e. push doubled rope through loop and then put over peg—pull tight.)
8. Fasten projectile to trace ropes, i.e. pass all trace ropes for that projectile (4 ropes on central ones, 3 at corners) through shackle and tighten pin completely.
9. Dig holes for cannons.
   9.1 At each end, 3 meters from end of net, angled so that the set cannon will point at the front corner of the net when fired.
   9.2 For center cannons, immediately behind projectile and so cannon will fire at right angles to net.

In general, holes should be dug so that the muzzle of the set cannon is about 15 cm behind net. The hole must be dug deep enough that the whole cannon except the muzzle is below ground. The back of the

Fig. 440. Net being furled (See Cannon Net Setting). Photo: PHIL IRELAND.

hole should be parallel with the back of the set cannon to resist recoil in firing.

10. Set cannons.
    10.1 Place cannon in hole, with fuse wires uppermost.
    10.2 Set cannon at correct angle for firing. For small birds this is four or five paces out and then kneel upright. Barrel should then be set so that it is firing straight between the eyes! This will give an angle of approximately 15 degrees to the ground. For larger birds a higher angle is required. 30 degrees (i. e. two paces and kneeling) is probably about right for geese and species such as gulls which hover.
    10.3 Cover all except muzzle of cannon, restoring ground to original condition as far as possible.
11. Place projectile in cannon, ensuring set

Fig. 441. Projectile attached to trace ropes. Photo: PHIL IRELAND.

Fig. 442. Hole dug for cannon. Photo: PHIL IRELAND.

Fig. 443. Placing projectile down cannon. (This is given as stage 11. in the notes on Cannon Net Setting but can be done before or after setting angle of cannons which is stage 10.2). Photo: PHIL IRELAND.

Fig. 444. Placing cannon in hole. Photo: PHIL IRELAND.

Fig. 445. Setting cannon at correct angle for firing. Photo: PHIL IRELAND.

angle is not disturbed. Arrange projectile ropes on top of netting and ensure projectile when fired will clear the net. Also ensure shackle is arranged so it will not catch net on firing.

12. Lay dropper cable just behind cannons and connect in fuse wires. Ensure no bare wire is outside the connecting block and put blocks so dampness cannot cause a short circuit.
13. Lay firing cable from net to firing position. Particularly when set on a field, avoid laying cable in a direct line between the two points (normally lay at right angles to net for 10 metres and then turn and take to firing position). Person laying cable must be responsible for knowing which cable is connected to which net and leave the cables geographically arranged at the firing position.
14. Set jiggler in front of nets
    14.1 Peg elasticized end about 3 meters beyond end of net line and about 50 cm in front of net line.
    14.2 Run jiggler cord in front of net and then continue to hide. Just on blind side of net line put in two pegs to form an upturned "V" to guide the jiggler cord. Also do this in front of

Fig. 446. Cover cannon and tidy net putting ropes on top of the netting. Photo: PHIL IRELAND.

Fig. 447. Finished net with team still working on a second net. The net has been marked with debris found on the beach. Photo: PHIL IRELAND.

net if (because net line is not straight) when pulled taut the jiggler will be ineffective or foul the net.

14.3 Return to the net and pull some slack in the jiggler at the pegged end. Also arrange jiggler in front of net so that it lies limply on the ground (so that on first using this the slight movement before it becomes taut is enough to gently disturb the birds.)

15. If necessary camouflage the net using any available vegetation which will not snag the net when fired. The camouflage must be as light as possible.

16. Person who will be responsible for firing the net marks as clearly but inconspicuously as possible.

16.1 Ends of net (and joins if more than one net set) on net line.

16.2 For each net the distance out that it is expected birds will be caught. (About 10 meters as birds further out will escape before net reaches them.)

16.3 The width of the "danger area" (i. e. the two meters close to the nets in which birds would be in danger of being struck by the net: The net must not be fired if any bird is closer to the net than this)

17. Test that there is an electrical circuit through all cannons from the end of the dropper where the firing cable will be connected. It is ESSENTIAL during this operation to ensure NOBODY is in front of the net. The circuit tester used should not be capable of setting off a net.
Also test that there is no circuit through the firing cable. If both test satisfactorily join dropper and firing cables.

18. At each end of net take net off peg (see 3 above) and reattach to peg using the loop on the end of the jump rope.

19. Team leader makes final check of nets, particularly ensuring:

19.1 Net on jump ropes correctly and that spare rope from these is between peg and net.

19.2 Shackles on projectiles are properly done up.

19.3 Cannons correctly in position and that the projectiles will clear the net when fired.

19.4 No projections in the catching area will snag the net and the net is not caught on any of the pegs used in setting.

19.5 All ropes are laid along line of net.

Fig. 448. Testing electrical circuit. Photo: PHIL IRELAND.

20. Finally when the whole team is clear of catching area, check circuit from the firing position. Also check the markers (see 16 above) can be clearly seen from the firing position.
21. If the firing position and blind are not the same place it is essential to arrange telephone or radio links. In any case, unless the whole team is to be at the hide, communications of some sort should be set up to keep everyone informed of events in and near the catching area.

24.2.3 Making Catch

Firing. Once the nets are set, the team leader's main responsibility is to catch the required sample of birds of the desired species, whilst ensuring that they can be safely caught/extracted/banded/processed by the team.

1. To ensure the safety of the team/public, check:
   a That firing boxes are not connected or armed until the team is clear of the nets.
   b That if anyone returns to the nets the firing box is disconnected.
   c That "TWINKLERS" (team members with/without vehicles who are moving birds towards the nets ARE NOT IN THE LINE OF FIRE—or are instructed to move out of it.
   d That if the public can approach the catching area (e. g. along a beach), "sentries" are posted to ASK them to keep clear, and to warn the firing party if they do not do so.
2. To ensure the safety of birds:
   a Ensure that you have a clear view of the catching area.
   b Ensure that all of those in the observation blind(s) know the position of all the markers and the extent of the danger area.
   c Ensure that the numbering of nets is clearly understood and that main cables are clearly identified.
   d Ensure that jigglers work properly, if set.
   e Ensure that phones/radios work properly, especially if the nets are to be fired from another position, and that any hand signals (e. g. given to twinklers) will be clearly understood.
   f Ensure that there is ample covering material/ holding cages in the correct place.
   g Designate one experienced observer to keep a running total of the number of birds in the catching area, and advise the team leader.
   h Designate another experienced observer to watch the danger area and keep the team leader advised. (If there are two blinds for parallel lines of nets, pick observers for both lines).

Fig. 449. Net being fired over a flock of Oystercatchers. Note that birds wings are not yet raised even though the net is well out. Photo: GRAHAM COUCHMAN.

Fig. 450. Catch of gulls coming into bread on a snow covered beach. Note higher angle of the net compared with waders. Photo: PAUL DODD.

i Consider the size/experience of the team in deciding the size of catch to attempt. Remember mixed catches of large and small birds are more difficult to extract. Also consider the weather conditions, the state of the ground, and the likely condition of the birds (e.g. birds may be in poorer condition during hard weather, and birds in molt should be extracted as quickly as possible to avoid damage to growing feathers).

In general, do not attempt to catch over 1,000 birds as a maximum for 10 experienced people, and, for a beach catch, 500 is a maximum for 10, except in ideal conditions on a falling tide. In special circumstances—dry site, good weather, large experienced team—catches at least twice as large are practical (if adequate holding cages are available)

j Do not ARM firing boxes until birds are in the area. Do not SWITCH-IN nets until an opportunity to fire seems imminent. Only SWITCH-IN nets that you expect to fire.

k Warn the team that a catch is imminent.

l Immediately before firing:

- i Check that there is no danger to people/property. (In particular that no vehicles have moved into the line of fire.)
- ii Decide which nets are most likely to give the desired size and species composition of catch; SWITCH-IN those nets.
- iii Check that the nets and danger area are completely clear of birds—use jigglers if necessary. Remember that any change in wind will affect the trajectory of the nets.
- iv Check that no birds are in flight over the catching area—never fire if they are—and be extra careful if the flock is alert and likely to take off.

Immediately After Firing. The procedures below must be completed as rapidly as possible to make safe unfired nets and to ensure the well-being of the birds which have been caught.

1. Disconnecting nets
   If some nets are unfired, the team must, whenever possible, avoid going in front of unfired nets.
   The person who fires the nets must therefore also stay behind and:

   a Turn all firing box switches to OFF.
   b Disconnect ALL firing cables IMMEDIATELY. (Cables to unfired nets first). Firing boxes should be designed to assist quick disconnection.
   c Place the firing boxes in a safe position,

well away from cables and protected from the rain.

d Check that this procedure is complete before leaving the firing position. In particular BOTH strands of wire from each net must be disconnected; it is still possible to fire a net accidentally with only one strand connected.

2. Covering birds
Unless the catch is VERY small, cover the birds in the net immediately using lightweight opaque material; they struggle much less in darkness and therefore do not graze their wings on the ground, or become entangled. Various types of Hessian and synthetic sacking and light tarpaulins have been used successfully. Continuous light material 1.5 to 2 meters wide, in lengths of up to 10 meters, is ideal for covering large areas, and easily handled by two people. Remember VERY light covering material may not be opaque, and will blow away easily in wind. Take great care not to tread on birds under the net; insist that NO member of the team EVER treads on covering material. Covering material must be close to the nets for quick transport on foot, or in vehicles to drive to the nets immediately after firing. Team members not in the blind(s) or "twinkling" must be concealed where they can reach the nets quickly—speed is vital to prevent birds injuring themselves by prolonged struggling.

3. Beach Catches
Part of the net may fall in the sea, with birds under it. Should this happen, it is essential that:

a All members of the team (except the person disconnecting the nets) get into the sea in front of the nets.

b They then lift the net, and the birds in it, up the beach with maximum care and speed. Work together so that no tension is put on any part of the net with birds in it; never pull or drag the net up the beach as this would damage the birds. Whenever possible move the projectiles first, or carry them so that they do not cause tension on the net.
In deciding how far to take the net, consider how much further the tide will reach. The further the net is taken, the more difficult it is to spread the front edge in order to be able to extract birds, but the net should be taken far enough to avoid having to lift it again.

24.2.4 Extraction/Use of Holding Cages

The team leader's next responsibility is to get the birds removed from the nets and into holding cages.

1. Extraction should commence when covering is almost complete Extraction is generally done by the members of the team with most experience of extracting from cannon-nets, with the remainder of the team carrying birds to holding cages "Learners"—even qualified mist-netters—should be taught to extract under supervision on small catches, or with the last few of bigger catches if conditions are good.
2. Extracting birds should be as swift as possible, consistent with maximum care. Too much haste may harm birds, and may be no quicker.
3. Persons extracting should not crowd too closely, but give each other room to work. Avoid tugging/pulling on the net as this may harm birds, especially ones that your neighbors are extracting.
4. Be especially careful when there is a tight mesh over the carpal joint; it is usually best to draw the whole of the primaries and secondaries through the mesh, after which the wing can be drawn back undamaged in the spread position.
5. Any person having difficulty extracting should ask a more experienced person to take over; this is nothing to be ashamed of; a slightly different approach will often solve the problem. If anyone else's extracting/handling is unsatisfactory, draw his attention to it and advise the team leader (later, or at once if necessary).
The most experienced extractors should space themselves among the rest of the team to advise them when necessary. The team leader should watch the performance of all team members.
6. When extraction is complete the team leader should detail a responsible person to check that all the nets are completely emptied of birds. This is especially important when the net has been gathered up, for instance on a beach catch.

7. Holding Cages
Holding cages are needed for all except tiny catches. They should be concealed as close as possible to the nets before firing; erect them as necessary during covering (one for the odd birds coming to hand) and extracting. Normally, cages should be as close as possible to the main mass of birds, but they may need to be further away to avoid the rising tide, to use natural shelter (for birds and the banding team), to avoid an unsuitable substrate (e. g. mud that may become sticky when trampled by birds), or because another catch is sought immediately on the same site. If several nets catch, erect holding cages for each; banders can move to them. Erect cages far enough apart for people to walk easily between them. Never step OVER cages.
   a Appoint a mature person (who need not be a qualified extractor) to allocate birds to cages. Separate birds of different species for convenience when banding/processing. In any case, only birds of approximately the same size should be placed in the same compartment. Control the number per compartment to avoid trampling or over-heating.
   If long-legged birds (e. g. curlew/godwit/ Spotted Redshank) may be caught, provide taller cages to minimize the possibilitiy of "leg-cramp". Such cages are essential for Curlew, advisable for others. Old birds of these species should be banded and released at once, or put in improvised tall cages (e. g. a large box, a blind, or a trailer covered with a tarpaulin) and processed quickly. (See Bainbridge, 1975. B. T. O. Wader Study Group Bull. no. 16 for a design for tall holding cages).
   b Ensure cages are secure; e. g. pile sand/ earth against the bottom of the walls, "guy" both ends to keep the material taut, cover entry slits to prevent escape. Problems will vary with the species.
   c In continuous hot sun, provide additional shade for cages to prevent overheating. In these conditions, or if birds are crowded, take a few birds from each compartment when processing starts; otherwise empty one compartment at a time. Check carefully when a cage appears to be empty; lift compartments one-by-one; this gives a chance to prevent the loss of odd birds, and ensures that none are gathered up in the cage.
   d One advantage of holding cages is the help given to cold/wet birds. Waders get wet naturally, but not with their feathers artificially ruffled. Time to preen in the warmth/shelter of a holding cage is invaluable. Ensure that damp birds are distributed evenly between compartments of dry ones; their warmth/activity will help to dry birds. Wet (as opposed to damp) birds should go in a separate compartment to prevent trampling.
   e If birds become muddy (at sites where this is likely only small catches should be attempted and with great caution, the mud should be washed off in a bowl of clean water, and the birds placed in cages to preen/dry. Do not crowd the cages; mix with dry birds if available.
   f Do not process/release wet birds until thoroughly warm/dry. In extreme cases make use of a car heater to dry birds.
   g People carrying birds to cages should never carry more than two at a time. Portable holding boxes can be very useful and can also replace holding cages for small catches. Ensure that they are cleaned out regularly. (See Sheldon and Williams, 1977, B. T. O. W. S. G. Bull. no. 20 for the design of a portable holding box).
   h The team leader should allocate people to duties as extractors, carriers, and cage supervisors in advance, duties may have to be varied after firing due to circumstances.

24.2.5 Banding/Processing/Releasing

1. Depending on the size and species composition of the catch, the team leader must decide whether all are to be processed, or only a small/large sample. The decision must depend on the time of day, the time already taken to complete extraction, and the recent/present/forecast weather. Do not impose excessive strain on the birds by keeping them too long. Process long-legged birds first. Generally aim to release all birds within a maximum of four hours after capture. This time can be exceeded if the birds are in good condition and feeding is easy (not

hard weather), provided that later release will still allow time to feed before high tide/dusk. Avoid very long periods in conditions so trying for the team that handling standards suffer (e. g. extremes of heat/cold/wind/rain).

As with setting nets, except when the team is small, the team leader should not be tied to a particular task, but be free to oversee the operation and deal with any problems that arise.

2. Set up one or more banding teams. Mature non-banders can be used to issue bands/record band details; at least one "A" or "B" bander experienced with the species being handled is needed for each banding team to check banding standard and ageing, instruct trainees properly, and deal with overlapped bands, queries about re-banding etc. After they have been banded some/all of the birds will be processed; for large catches the use of a holding cage as a holding site between the banding and processing teams allows the banding and processing teams to work independently. Instructions as to the classes of birds to be held for processing can be varied as the work progresses.
3. One or more processing teams will be needed. These should include some of the most experienced members to ensure correct ageing and consistent measurements. Do not put all your best workers in processing teams—some will be needed to supervise banding, but remember that the data collected is worthless unless it is both accurately measured and correctly recorded.
4. When releasing birds, do not launch waders into the air. Put them down to walk away and fly when ready, or let them fly from your hand near the ground. Remember that birds usually take off into the wind. At night, release birds away from bright lights and give them time to get used to the dark before flying. Ensure that the release area is free of obstacles, and that the team do not need to cross it; and at night rope off the release area to prevent access.

   In daylight ensure that all birds fly before they are lost from sight. After night releases the area should be searched in daylight if possible so that, should there be any non-flyers, they can be located and dealt with suitably.

24.2.6 After The Catch

The cause of any net or cartridge failing to fire correctly should be investigated prior to removing equipment from the catch site.

The fired net should be refurled from the leading edge to the back line, removing any vegetation or other debris tangled in it.

Fired cannons should have the fuse wires broken off and unfired cannons should have the fuse wire wrapped round the barrel to distinguish which are which.

All equipment should be collected together and counted as it is loaded into vehicles for transport back to the team's base. Cannons in particular can disappear in loose shingle or soft mud.

24.3 Other Considerations

Public Relations. Cannon netting is more conspicuous than other techniques—you cannot hide a big team and vehicles, still less a big "bang"—and an emotive subject for some bird-watchers. You must have a landowner's full consent and communicate with other people/bodies with genuine interest in the area-tenants, wildfowlers, local bird-watchers, local banders, etc. At a new site, advise local police in advance, and don't upset the locals by a first firing in the early hours. Avoid places where the public are numerous—they may be endangered, can hamper operations, and will need several team members detailed to answer questions. Always be ready to delegate a mature/experienced member to enlighten interested members of the public who appear; they may be valuable as "friendly locals", but could be a real handicap if allowed to become hostile through inconsiderate behavior by the team. Similarly, always be extremely polite to farm-workers etc. on private land, or to public who have to be asked to keep off a right of way until you have fired. If you have written permission for a site, carry a copy with you; not all farm-workers, local police constables etc. may have been informed of your right to be there. Ensure that the team observes the "country code"; remove litter, close all gates, and drive considerately on farm roads, giving way to farm operations.

Any major public relations problem or opportunity (a hostile visitor or the local press are examples) should always be referred to the team leader personally. Cannon netting is an invaluable technique for wader, wildfowl and gull banders. There will certainly be other applications for it. Its continued acceptability and legal approval could be jeopardized by a single serious accident to a bander or the public, by heavy casualties of birds which could have been avoided by applying known rules and techniques, or by overworking a favored site to the point at which its desertion really could be due to cannon netting. By following this code, in the spirit as well as the letter, you should avoid these dangers and help to ensure that cannon netting ceases to be regarded as an "emotive" or "controversial" technique and is acceptable to all.

24.4 Other Species

24.4.1 Cannon Netting Gulls at Rubbish Tips

This section only contains relevant modifications or additions to points already mentioned.

Setting nets

1. Catching Site—Nets are usually set to catch gulls dropping down to feed on fresh rubbish, or alternatively on nearby loafing sites, but the chance of catching at the latter is low.

   It is important that the ground in front of the net should be reasonably flat, and all large pieces of refuse such as planks, boards jagged sheets of metal and pieces of wire which may become caught in the net should, if possible, be cleared away from the catching area.

   The areas onto which the edge of the net will fall should be inspected to ensure that there are no projections which might hold up the edge of the net, or hollows into which the net cannot fall, for these may allow the birds under the net to make a quick exit!

   Rubbish tips are often in built-up areas, so special care must be taken to avoid damage to people or property; cannons should not point in the direction of buildings, huts, structures, vehicles, railways lines or any areas where people or livestock are likely to be moving about, or any areas that are unobservable. Projectile ropes must be strong and in good repair. Take no chances.

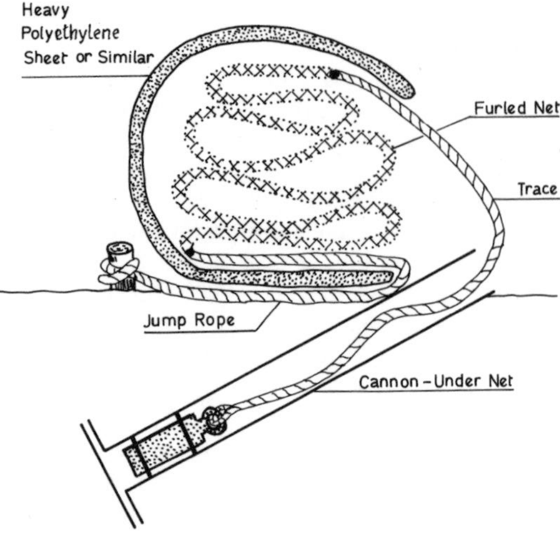

Fig. 451. An alternative method of setting the net when catching gulls at rubbish tips. (This method can also be useful for wader catching when the space available to set the net is limited.).

2. Net Setting—Rubbish tips contain vast numbers of small and large plastic and metal objects which readily tangle in the net. They can damage the net or prevent it from spreading properly if fired in a "dirty" condition. To avoid as much tangling as possible it is advisable to put the net down on a strip of polyethylene or other material one to two meters wide, and to furl and set the net on this. To prevent the birds standing directly on the furled net, which prevents firing, the net can be covered with cloth or polyethylene sheet. On firing the net is pulled out from beneath the material.

   An alternative to the usual method of setting nets may be followed (see Fig. 451):

   a Choose the baseline site and by pacing along it locate the cannon positions. Dig the cannon holes, which need not be deep but check that the cannon base is well bedded in. Place the cannon in the hole and set the cannon at a high angle of 30-35° (see "Cannon Angles" below). Do not insert the projectile yet.

   b Place a sheet of covering material or heavy gauge polyethylene sheet along the proposed baseline and over the cannons leaving only the muzzle uncovered. (In the final set the cannon is under the net and the muzzle just in front of the net.)

   c Furl and set the net on this material taking care to prevent the net contacting surrounding rubbish. Peg the baseline on jump ropes. These are led forward under the net and then backwards under the material on the ground before pegging if one piece of material is used both as a "groundsheet" and covering. Obviously, leading the jump rope forward over the net and then backwards under the "groundsheet" for pegging will lead to disaster on firing!

d Cover the net from the rear either by folding forward the material on which the net was set, or with polyethylene sheeting weighted or pegged at the rear. These precautions help to prevent birds standing on the furled net, and as it seems likely that birds will be least willing to stand on slippery polyethylene, this kind of material may be best. On firing, the net is pulled out from under the covering material which usually remains where it is, thus making it possible to fire the net safely when birds are standing on top of the polyethylene sheet. However, it is best if the net is free of birds, and it should not be deliberately fired with many birds on the covering material.

The covering keeps the net dry if it rains, which is an additional advantage, but windy conditions obviously make the use of this technique difficult.

e Push the projectiles down the barrels. By leaving this until this stage there is no chance of getting the traces and projectile ropes looped round the net. (With some cannon nets this will be impossible because an air lock is created in the barrel. Care should be taken if the projectiles are inserted at an earlier stage because of this).

f Connect up the dropper cable last. This avoids electrical connections being kicked out by the net-setters.

It is not necessary to camouflage the net; indeed birds must be discouraged from standing on it. It may be advisable to have a strip of covering material or earth covering the rubbish immediately in front of the net to discourage birds from standing there.

3. Cannon Angles—When wader catching it is usual practice to wait before firing the net until all the birds in the neighborhood are settled on the ground. This is often impossible when catching gulls on rubbish; birds are continually in the air, and often hovering over the feeding area. The net must be fired when there is minimal aerial activity and when any gulls flying over the area are within a few feet of the ground and not in the trajectory of the net.

   The net must pass over these birds and it is therefore fired steeply at an angle of 30-35° above the horizontal. If the net is fired too steeply, most gulls will have flown out from beneath the net before it falls to the ground. Obviously the person firing the net must have a very clear view and be free from distraction.

4. Marking nets—Care must be taken that net markers are clearly distinguishable from the firing position.

Blinds

A vehicle can usually be driven close to the net site and makes an excellent blind from which to fire, as gulls usually ignore vehicles. Alternatively it is often only necessary for the team to move 50 to 100 meters from the net for the birds to come down. Main cables from the nets to the firing position should not cross vehicle routes—tracked vehicles and heavy lorries break cables.

Safety

Particular care should be taken when cannon netting on rubbish tips because of the proximity of people and vehicles. Keep a lookout for either moving into the area, and before arming firing boxes check that there are no people or vehicles in or approaching the line of fire. (See also "Liaison with Refuse Men" below).

Extraction

Small gulls may entangle their carpal joints in nets of the mesh size used to catch waders. A smaller mesh is desirable; 38 mm (1.5 inch) stretched diagonal is satisfactory.

In general it is best not to use covering material when catching gulls; they do not injure themselves by struggling for a brief period, and immediate extraction gives them a shorter time to become entangled, resulting in easier and quicker extraction. In addition, on wet ground the weight of the covering material may make the birds muddy. However, waterproof covering material may be useful if it rains (see "Wet Conditions", below).

Extraction should be completed as quickly as possible, and the site cleared to allow refuse men to return quickly to work.

Holding

Large gulls (Herring and Black backed) must be kept in separate sacks or suitable keeping cages to prevent them injuring each other. Large sacks can be fitted with ties half way along so that they can be divided into compartments, each holding a single gull. Up to five small (Black-headed and Common) gulls may be put in one sack. Large and small gulls should never be mixed. Soft but strong wire can be used as a sack closure that is quicker both to use and untie than string. Species should be segregated for ease of processing, and the sacks arranged neatly in rows on dry ground. Sacks should never be placed on top of each other.

Handling

Large gulls have dangerous beaks which can strike out long distances in all directions, and can inflict painful injuries on the handler. Always keep them well away from people's faces. The following procedure should minimize injuries:

1. Extracting. It is useful to place a sack over the bird's foreparts so that it cannot see what is happening. Then roll back the net to expose the rear parts, and gently draw it backwards from the net holding onto the legs, tail, and wingtips which can be grasped in one hand. The bird is unable to peck the hand which holds it, and can safely be transferred into a sack. The bird can later be removed from the sack using the same procedure.
2. Processing—During processing gulls can be restrained in a plastic or metal cone as described by Seel, D. C. (1975) "A tube for holding birds during examination of molt in the flight feathers", Bird Banding 46: 74-75. This tube has a slit in one side that allows a wing to be withdrawn for measurement and examination of molt. The bird can also be weighed in it. Use of such a cone speeds up processing immensely, and several can be used at once.
3. Hygiene—People who handle gulls are advised to take the precaution of having anti-tetanus injections. Conditions on rubbish tips and other places where gulls feed are generally unhygienic; have a supply of clean water, soap or detergent, disposable paper towels, and first aid kit available.

Wet conditions

Gulls and starlings (which may be caught together) quickly become soaked if they struggle beneath the net on wet or muddy ground. Large catches should not be attempted under wet ground conditions, and small catches only if the ground is not too wet and muddy, and the birds can be extracted in less than ten (preferably five) minutes. Circumspection should be exercised when considering catches in rain or snow.

Starlings become soaked quickly; it is best to release them by lifting the net and letting them fly out—taking care not to let the gulls go! If this is done quickly after firing only a few will have become entangled and need handling.

To prevent birds getting wet unnecessarily, puddles in the catching area can often be drained quite quickly when the nets are set by making one or two channels with a spade, or be filled in or covered over with any available material.

Wet gulls that are unwilling to fly may be kept in a shed or cardboard boxes (not cramped in bags) and preferably in a warm place until they are dry; this situation should not normally arise.

Release

Gulls may fly up to 50 km to their roost sites and all birds should be released in sufficient time for them to complete this flight in daylight. When unforeseen problems have arisen gulls have occasionally been kept in huts overnight and successfully relased the following morning.

Liaison with refuse men

Bulldozer drivers at tips are usually interested and helpful partners in gull catching operations, and are often instrumental in making successful catches. An obliging driver of an earth moving vehicle may assist considerably in levelling the site and tipping earth where required, and bulldozer drivers can usually be persuaded to spread rubbish in the right place in front of the net. During such operations the firing box must not be connected however great the temptation. Be prepared to connect the box quickly as gulls will often pour down when the bulldozer is backed off. Always be sure that the machine and driver are out of line of fire before arming the firing box. Drivers usually show caution after seeing the nets fired once! It is extremely difficult to work successfully without the cooperation of the men on the tip site; care must be taken to avoid hindering their work and good "public relations" are essential.

24.4.2 Cannon Netting Duck

Relatively little duck catching has been attempted with cannon nets, but at present cannon netting wildfowl appears to present few extra problems. The advice given previously in the code should be followed, together with the following additional points.

1. In general a smaller team is needed because of the smaller expected catch. An experienced person can deal with 30 ducks even if the net is fired over water. Nets should not be fired over water deeper than 0.3 meter unless the net is to be lifted ashore immediately after catching (as for wader beach catches; see after firing, 3).
2. Duck are quiet in the net and do not tangle apart from putting their heads through the mesh. As with waders, it is best to have a person to carry the birds once they have been extracted, but they must not walk across the net, because of the danger of treading on a duck lying flat on the water under the net.
3. The precautions used are the same as those for catching waders on a flat field; the catching area must be clearly marked; on moonlit nights peg markers stand out in the water.
4. Duck on The Wash (England) have been baited with tail-end corn spread to within one meter of the line of the furled net. Shelduck can be similarly baited on saltings or the higher parts of mudflats.

How to make up a net

The rope positions for the full net (13×26 m) and for the half net (13 × 13 m) are shown in Figure 452 and 453 respectively.

To attach ropes to a full-sized cannon net (13×26 m) the procedure is as follows:

Thread 4 mm rope along the sides (A-D and B-C) and the back (C-D). Thread 8 mm rope

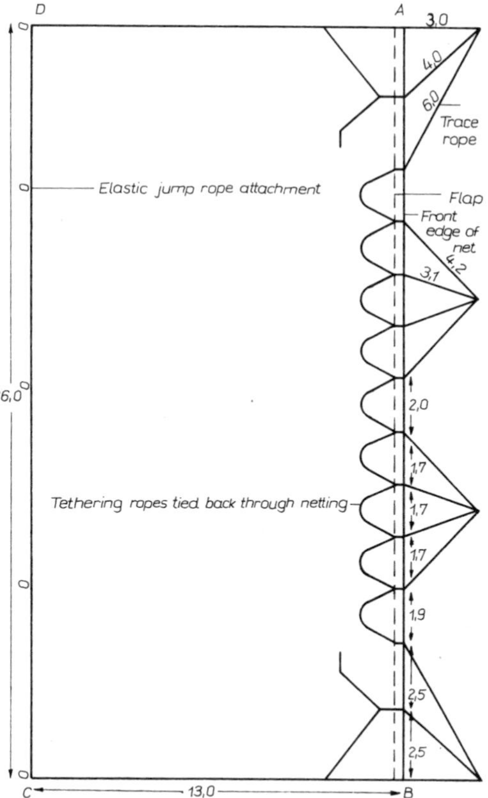

Fig. 452. The design of a full-sized net. All measurements are in meters.

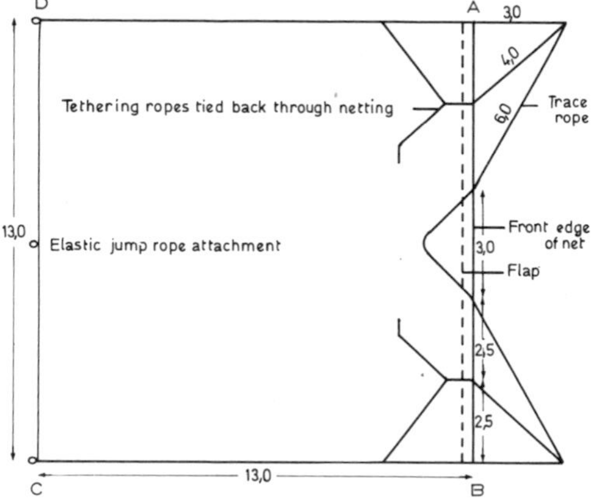

Fig. 453. The design of a half-sized net. All measurements are in meters.

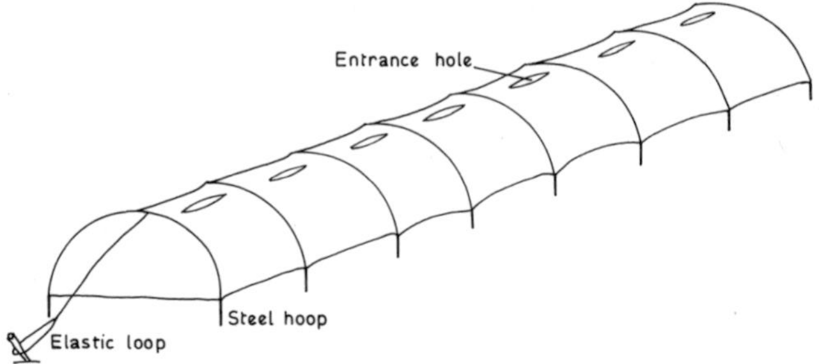

Fig. 454. A holding cage set up ready for use.

through the front of the net (A–B), 0.5 m from the front edge, leaving three meters spare at B (this will form an end trace rope) and 10 m at A (this leaves slack on the rope which will be used when tying in the 12 central trace ropes, and will form the other end trace rope).

Tie the center trace ropes to the front edge of the net, starting near B (leave two meters spare to form a tethering rope which will later be tied into the netting). To attach the center trace ropes to the front edge rope, loosely tie a clove hitch in the edge rope around a small stick. Then carefully remove the stick and tie another clove hitch with the trace rope through the clove hitch in the front rope. Gently tighten all four ends, applying more pressure as the knot tightens. When all the trace ropes are attached there should be some slack netting along the front edge. Cut the long end of the front edge rope at A to a length of 3 m, to form an end trace rope. Fold the 0.5 m of loose netting under the front edge to form a flap. Thread 0.5 m of each tethering rope back through both the net and this flap. Untwist the remaining 1.5 m of each tethering rope so that one strand can go to one side and two strands to the other. Thread the strands through the netting (remember to leave slack netting) so that each meets the tethering rope from the next trace rope. Each trace rope has a small loop spliced on the end so that it can be attached to the projectile.

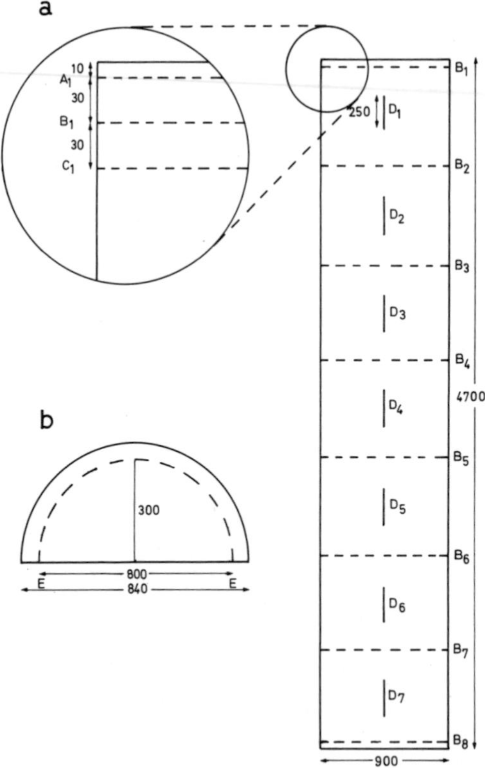

Fig. 455. The material for a holding cage marked out for assembly. Points A_2–A_8 and C_2–C_8 are not shown but are in the same positions relative to B_2–B_8.
Inset a shows the positions of A and C relative to B. All measurements are in mm.

Using thread of the same denier (thickness) as the netting, sew all ropes (ropes around the edge of the net first, then tethering ropes) to the net using clove hitches or camel hitches (Ashley 1944) at regular intervals. When the roping is finished there should be no points of stress on the netting, since these will cause the net to tear when it is fired.

Ropes are attached to a half-sized net (13×13 m) (Figure 453) in a similar way to a full-sized net, except that the middle eight trace ropes on a half-sized net are omitted.

To finish the net, three meter long jump ropes (ropes to anchor the back of the net to stop it going too far when fired) must be attached by means of elastic to the back of the net.

Fig. 456. A holding cage with some boxes for larger birds, used by the Dutch ornithologists. Photo: B. J. Speek and A. J. de Zwart.

Holding-cages

The design is shown in Figure 454. The best material has proved to be hop sacks made from 50/50 Hessian and polypropylene. This is strong, and dark enough to deter the birds from escaping. Each cage consists of a 4.2 m long semicircular tunnel which is divided into seven compartments, each with a hole in the top for inserting and removing birds. The cage is kept rigid by eight 1.4 m long wire hoops, the ends of which are pushed into the ground. A guy rope at each end, with an elastic loop, keeps the structure taut and prevents birds from escaping through the slits.

A holding-cage is constructed as follows:

Mark out a 4.7 m × 0.9 m piece of sacking as shown in Figure 455 a. The sacking is now folded along B_1 and A_2 is sewn to C_2 to form a narrow tube. Repeat through A_8-B_8, making sure that all the tubes are on the same side.

Now cut along D to make the slits for the entrance holes and on each side sew a flap that extends about 20 mm over the slit. This overlap will help prevent birds escaping.

Cut eight semi-circular pieces as shown in Figure 455 b, fold along E-E and sew this flap down. Attach each of the semi-circles to the strip by sewing E-E to B. This is best done with thick twine and a sacking needle.

When all the semi-circles are sewn in, thread the wire hoops through the tubes. The best hoops are made from 5 mm galvanized prestressed steel wire. Ordinary fencing wire is not strong enough to be pushed into the ground without it bending.

Finally, tie a guy rope around the top of the hoop of each end wire. On one end guy rope tie a 300 mm loop of thick elastic. This not only tensions the cage, but when the cage is folded up (like a concertina), the loop can be placed around all the hoops and sacking to keep the cage closed during transit.

When birds are caught in winter or the cages have to be erected on damp ground, they should be placed on top of Hessian covering material or, better still, on strips of old carpet. This prevents the birds getting cold and wet whilst in the cages. A complete holding cage will normally hold about 35 Oystercatchers, 100 Knot or 140 Dunlin.

Special, taller cages should be used for curlew and other long legged species.

References

Clark, N. A. (1986): A Method of Roping a Cannon-Net. - Wader Study Group Bulletin No. 46: 31-32

Clark, N. A. (1986): Keeping-Cages and Keeping-Boxes. - Wader Study Group Bulletin No. 46: 32-33

Lessells, C.M.: McMeeking, J.M. & Minton, C.D.T. (1978): Cannon-Netting Code of Practice. - BTO.

24.5. Bibliography of Cannon Netting

ARNOLD, K.A. & COON, D.W. (1972): Modifications of the cannon net for use with cowbird studies. - J. Wildl. Manage. 36: 153-155

CLARK, N.A. (1980): A simple rapid method of moving a set cannon net with minimal disruption. - Wader Study Group No. 28: 32

DILL, H.H. & THORNSBERRY, W.H. (1950): A cannon-projected net trap for capturing waterfowl. - J. Wildl. Manage. 14: 132-137

ELLIOTT, C.C.H. (1974): Cannon-net success. - Safring 3: 5-6

GIESEN, K. M., SCHOENBERG, Th. J. & BRAUN, Cl. E. (1982): Methods for trapping Sage Grouse in Colorado. - Wildl. Soc. Bull 10: 224-23

GRIEB, J. R. & SHELDON, M. G. (1956): Radio-controlled firing device for the cannon-net trap. - J. Wildl. Manage. 20: 203-205

LACHER, J.R. & LACHER, D.D. (1964): A mobile cannon net trap. - J. Wildl. Manage. 28: 595-597

MALTITZ, I. v. (1982): Exploder box for detonator-launched cannon nets. - Safring 11: 49–57

MARQUARDT, R.E. (1960): Smokeless powder cannon with lightweight netting for trapping geese. - J. Wildl. Manage. 24: 425-427

— (1961): Investigations into high-intensity projectile equipmet for net trapping geese. - Proc. Okla. Acad. Sci. Vol. 41: 218-223

MINTON, C.D.T (1980): A glossary of cannon netting terms. - Wader Study Group No. 28: 46–47

NASS, R.D. (1964): Sex- and age-ratio bias of cannon-netted geese. - J. Wildl. Manage. 28: 522-527

O SAKI JOMEI (1982): Report of the Bird Migration Research Center (Februar 1. 1981-January 31., 1982). Shinhama Station: 167-187. - Bird Migration Research Center Yamashina Institute of Ornithology Shibuya Tokyo, Japan

PARRIS, R. W. (1977): A method for capturing adult Great Blue Herons. - Proc. Conf. Colonial Waterbird Group: 163-165

PIENKOWSKI, M.W. & DICK, W.J.A. (1976): Some Biases in Cannon- and Mist-Netted Samples of Wader Populations. - Ringing & Migration 1: 105-107

RAVELING, D.G. (1966): Factors affecting age ratios of samples of Canada Geese caught with cannon-nets. - J. Wildl. Manage. 30: 682-691

ROWLEY, I. & CHAPMAN, G. S. (1981): The „Nonnac“ Method of Netting Wild Birds. - Corella 5: 77-84

SCOTT, P. (1952): Marking of Wild Geese. - Severn Wildfowl Trust. Fourth Ann. Report 1950-1951: 12-13

— (1953): Rocket-nets.—Severn Wildfowl Trust. Fifth Ann. Report 1951-1952: 22-24

— (1955): Rocket-netting in 1953 and 1954. - The Wildfowl Trust. Seventh Ann. Report 1953-1954: 56-57

SHARP, D.E. & LOKEMOEN, J.T. (1980): A remote-controlled firing device for cannon net traps. - J. Wildl. Manage. 44: 896-898

SILVY, N.J. & ROBEL, R.J. (1968): Mist nets and cannon nets compared for capturing Prairie Chickens on booming grounds. - J. Wildl. Manage. 32: 175-178

SOUTHERN, W. E. (1972): Use of the cannon-net in Ring-billed Gull colonies. - IBBA News 44: 83-93

SOUTHERN, L.K. & SOUTHERN, W.E. (1983): Responses of Ring-billed Gulls to cannon-netting and wing-tagging. - J. Wildl. Manage. 47: 234-237

THOMPSON, M.C. & DELONG, R.L. (1967): The use of cannon and rocket-projected nets for trapping shorebirds. - Bird-Banding 38: 214-218

— (1973): Migratory patterns of ruddy turnstones in the central pacific region.—The Living Bird, 12. Ann. 1973 Cornell Labor. of Ornithology: 5-23

TREE, A.J. (1979): Bird ringing in Rhodesia. - Safring 8: 12-14

TURNER, L. B. (1956): Improved technique in goose trapping with cannon-type net traps. - J. Wildl. Manage. 20: 201-203

UNDERHILL, L.G. & D.G. UNDERHILL (1987): The Zap Net: An elastic-propelled variation of the cannon net. - Safring News 16: 21-24

25. Other books and manuals for Bird Banding

BALDWIN, S.P. (1931): Bird Banding in America from 1912-1929. A collections of 41 papers by several authors

BARBER, J. (1954): Wild Fowl Decoys. New York

BARDI, A., L. BENDINI, E. COPPOLA, M. FASOLA, F. SPINA (1983): Manuale per L'Inanellamento Degli Ucceli a Scopo di Studio. - Instituto Nazionale di Biologica della Selvaggina. Bologna

BATEMAN, J. A. (1971): Animal Traps and Trapping. Newton Abbott London

BATEMAN, J. A. (1979): Trapping: A Practical Guide. Newton Abbott London

BTO (1984): The Ringer's Manual. 3rd Edition. Beech Grove. Tring

CABALLOS, P., J. MOLINA, A. FRANCO, B. PALACIOS (1984): Manual del Anillador. Madrid

HOLLOM, P.A.D. a. H.G. BROWNLOW (1955): Trapping Methods for Bird Ringers. BTO Field Guide Number One. Revised Edition

LINCOLN, F.C. (1928): A Bibliography of Bird Banding in America. Suppl. to The Auk XLV, p. 1-73

LOCKLEY, R. M. a. R. RUSSELL (1953): Bird Ringing. London

MACKEY, W.J. Jr. (1965): American Bird Decoys. New York

MACPHERSON, H. A. (1897): A History of Fowling. Edingburgh

McClure, E. (1966): An Asian Bird-Banders Manual. Hong Kong
McClure, E. (1984): Bird Banding. Pacific Grove
Mead, C. (1974): Bird Ringing. BTO Guide Number 16
North American Bird Banding. Vol. I and II. U.S. Fish and Wildlife Service, Canadian Wildlife Service. 1984 and 1987
Payne-Gallwey, R. (1986): The Book of Duck Decoys. London
Perdeck, A. C. a. B. J. Speek (1976): Ringersboek. Vogeltrekstation Heteren, Boterhoeksestraat 22, NL—666GA Heteren
Schemnitz, S. D. (1980): Wildlife Management Techniques Manual. The Wildlife Society. Washington. D. C. 4rd Edition. - The most comprehensive textbook for people working on wildlife and related natural resources
Stonehouse, B. (1978): Animal Marking. Baltimore
The Studia Ethnographica Upsalliensia can be obtained from Institutionen för Allmän och Jämförande Etnografi vid Uppsala Universitet, Kungsängsgatan 19, Uppsala or from our agents, viz.: Kegan, Paul, Trench, Trubner & Co., 43 Great Russell Street, London W. C. 1, and Klaus Renner Verlag, Tangastrasse 22, 8000 München 59.
So far over 40 papers have been published. Besides the series „Occasional Papers". Editors are S. Lagercrantz, A. Lööv and L. Sundström.

26. Periodicals for Bird Banding

Journal of Field Ornithology (formerly Bird Banding). AFO, P. O. Box 368, Lawrence, KS 66044. USA
North American Bird Bander (formerly EBBA-News and IBBA-News). (Eastern Bird Banding Association, Inland Bird Banding Association, Western Bird Banding Association). USA
Journal of Wildlife Management.—The Wildlife Society. 5410 Grosvenor Lane, Bethesda, MD 20814. USA
Wildlife Society Bulletin. The Wildlife Society. 5410 Grosvenor Lane, Bethesda, MD 20814. USA
See also in "North American Bird Bander". Appendix B (B-1-B-6): Reference to books and periodicals useful to banders.
Ringers' Bulletin. B. T. O., Beech Grove, Tring, Hertfordshire HP 23 5 NR. England
Wildfowl. Published by the Wildfowl Trust, Slimbridge, Gloucester, GL2 7BT. England
Wader Study Group. Bulletin. N. & J. Clark, Department of Zoology, University of Edinburgh, West Mains Road, Edinburgh EH9 3JT. U. K.
Corella (Journal of the Australian Bird Study Association). Vol. I 1977. Formerly "The Australian Bird Bander". Postal address of the Association is: P. O. Box A 313, Sydney South, N. S. W. 2000. Australia
The Ring, International Ornithological Bulletin. Exchange and business address: Natural History Museum of University, Sienkiewicza 21, 50-335 Wroclaw, Poland
Safring News, published by the South African Bird Banding Unit, University of Cape Town, Private Bag, Rondebosch, Cape, 7700. R. S. A.
Op het Vinketouw. Contactblad voor Ringers. Uitgave Vogeltrekstation Heteren, Boterhoeksestraat 22, NL—6666 GA Heteren. Netherlands

27. Bird Trapping in World Literature

E. Stresemann collated a short bibliography of older books on bird trapping for W. Sunkel's book of 1926. A selection from this bibliography is presented here so that the reader can gain a picture of the historical trapping literature.

As can be seen from this survey, the oldest book on the subject dates from 1492, printed in Flemish. It is true that almost 1500 years before that the Roman writer Gratius Faliscus composed a didactic poem on bird trapping techniques, but this survives only in fragments and appeared for the first time in print in 1538. A particularly excellent work in English is Macpherson's book of 1897. We trust that the reader will appreciate that the titles are best left in their original languages.

Latin

1538

Gratius Faliscus (a contemporary of Ovid): Gratii Falisci Cynegeticon et de venatione Liber I atque eiusdem de aucupio fragmenta duo. Venetiis 1538.
A poem on bird trapping from the 1st. Century A. D.

1566

Angelio, Pietro da Barga: De Aucupio liber primus ad Franciscum Medicem (etc.). Florentiae 1566. 4°.
(Bayr. Staatsbibliothek München P. o. lat. 64)
Later editions: Florentiae 1568, Romae 1585. 4°.
A didactic poem.

1571

Heresbach, Conrad: Rei rusticae libri IV universam rusticam disciplinam complectentes ... Item de venatione, aucupio atque piscatione compendium. Coloniae Agrippinae 1571. 8°.
2nd. edition 1573.

1582

Lonicerus, Johann Adam: Venatus et aucupium, iconibus expressa et versibus illustrata. Frankfurt 1582. 4°.

1588

MEDICES, SEBASTIANUS: Tractatus de Venatione, Piscatione, et Aucupio. Coloniae 1588. 8°.
Later editions: Helmstädt 1592, Cöln 1598, Marburg 1698, all 8°.

1676

WILLUGHBY, FRANCIS: Ornithologiae libri tres. London 1676. Fol. Description in a number of passages, eg. pp. 285-286, of trapping methods used in England.

German

1612

COLERUS, JOHANN: M. JOHANNIS COLERI OECONOMIAE oder Haussbuchs fünffte Theil: Darinnen allerley lustige, künstliche und nützliche Sachen gehandelt und beschrieben werden. I...II. Vom Vogelfang und derselbigen Zugehörung. III ... Wittemberg 1612. 4°.
Later editions: Wittemberg 1616, 1623, 1632 fol., Mainz
1645 fol., Frankfurt 1672, 1680, 1692 fol.
Deals with bird trapping in 74 chapters.

1626

AITINGER, JOHANN CONRAD: Kurtzer und Einfeltiger Bericht Von dem Vogelstellen Wie Raubvögel Habichte Velthüner Wachteln Crammet- und Andere Gross und Kleine Vögel mit geteckten und ungeteckten Netzen, im offenen Velte, geholtzen und Wassern, mit leim ruten hütten Kloben Schneissen fallen und Schleiffen gefangen werden. Allen Liebhabern denen das Vogelstellen zu gelassen und davon keinen sonderlichen bericht haben zu ehren Dienst und gefallen zu sammen getragen. Im Jahr nach Christi geburt 1626. kl. 4°.
(Staatsbibliothek München. T[1]. 570.)
2nd. edition: Rothenburg 1631 kl. 4°; enlarged editions: Cassel 1653. 4°, and (under the title: Vollständiges Jagd- und Weydbüchlein, von dem Vogelstellen) Cassel and Frankfurt 1681. obl. 8°. Facsimile Hamburg and Berlin 1972.

1682

HOCHBERG, WOLFGANG HELMHARD Frhr VON: Georgica curiosa, Das ist: Umständlicher Bericht und klarer Unterricht von dem Adelichen Land- und Feld-Leben (etc.). Nürnberg 1682. fol. 2 Vols. Later editions: Nürnberg 1688, 1701, 1715, 1749 fol.
The 12th. book ('Holtz- und Weidwerk') of the 2nd. volume deals with bird trapping over many chapters; this is based partly on the author's own experience and partly on the writings of OLINA, ANGELIO, AITINGER and others. With fine copperplates.

1703

(HOCHBERG, WOLFGANG HELMHARD Frhr VON): Waidmannschafft durchs gantze Jahr. 1703. 4°.
(Univ. Bibl. Halle Ta. 3793)
Reprint under the title "Herrn WOLFF HELMHARD FREIHERRN VON HOCHBERG etc. Waidmannschafft durchs ganze Jahr. 1707" appeared as a supplement to F. A. VON PERNAU: Unterricht (etc.). Coburg. 1707.

1716

(PERNAU, FERDINAND ADAM Frhr VON): Angenehmer Zeit-Vertreib, welchen das liebliche Geschöpf die Vögel auch außer dem Fang ... den Menschen schaffen können ... Nebst einem Anhang von der Waidmannschaft. Nürnberg 1716. 8°.

1717

SCHRÖDER, CASPAR: Neue lustige und vollständige Jagd-Kunst, so wol von den Vögeln als auch anderen Thieren. Franckfurt and Leipzig 1717.
Later editions: 1728, 1760, 1762.
A plagiarism. The text is almost word for word identical with Aitinger's book. The later editions are anonymous.

1719

FLEMING, HANNS FRIEDERICH VON: Der Vollkommene Teutsche Jäger (etc.). Leipzig 1719-1724 fol.
2nd. edition Leipzig 1749 fol.
Instruction in bird trapping in both volumes (compilation).

1720

(PERNAU, FERDINAND ADAM Frhr VON): Angenehme Land-Lust, deren man in Städten und auf dem Lande, ohne sonderbare Kosten, unschuldig geniessen kan, oder von Unterschied, Fang, Einstellung und Abrichtung der Vögel, Samt ... nöthigen Anmerkungen über Hervieux von Canarienvögeln und AITINGER vom Vogelstellen. Deme beygefüget JOSEPH MITELLI Jagd-Lust. Alles mit schönen Kupfern gezieret. Nürnberg 1720. 8°.
Further editions published 1754 and 1768 in Nürnberg under the title: Gründliche Anweisung alle Arten Vögel zu fangen, einzustellen, abzurichten, zahm zu machen (etc.). Nebst Anmerkungen über Hervieux von Canarienvögeln und Mitelli Jagdlust.
See also 1796: BECHSTEIN, J. M.
An important book, particularly because of the supplement concerning the writings of AITINGER and MITELLI.

SINCERUS, ALEXANDER: Der in allerley Ergötzlichkeiten vergnügte Land-Mann, vorgestellt in einen sonderbahren Kunst-Büchlein, darinnen Allerhand zum Vogelfang, Schiessen, Jagen und Fischerey dienliche Kunst-Stücke, nebst vielen andern Curiositäten enthalten, und denen Liebhabern des Land-Lebens treulich mitgetheilet. Nürnberg and Prag 1720. 12°.
(Bayr. Staatsbibliothek München. Oecon. 1052.)
Reprint: Nürnberg 1734. 12°.

1742

Zorn, Johann Heinrich: Petino-Theologie oder Versuch, die Menschen durch nähere Betrachtung der Vögel zur Bewunderung, Liebe und Verehrung ihres mächtigsten, weissesten und gütigsten Schöpffers aufzumuntern. I. Theil Pappenheim 1742. II. Theil Schwabach 1743. 8°. The work makes no new contribution to the study of bird trapping techniques but is of interest because of some passages in volume 2 (pp. 646-692) which discuss the ethics of trapping. The following sentence is a good example: "But it is another thing when a preacher, with his patron's permission; without neglect of his holy office; without annoyance to his parishioners and to the good of his health (even to an increase in his happiness) and, more importantly, recognising the wisdom, power and goodness of this Creator in the pretty creatures, at times enjoys a quiet hour of delight in catching birds; for this occupation, in its orderliness and propriety is pure and innocent, and what is sinful or harmful in it can only come from improper practice."

1746

Döbel, Heinrich Wilhelm: Eröffnete Jägerpractica, oder der wohlgeübte und erfahrene Jäger. Leipzig 1746. fol.
Later editions: Leipzig 1754, 1782 fol. New shortened edition (minus a large part of the sections dealing with the trapping of songbirds) published by Neumann, Neudamm (no date) 1910. 4°.
Contains detailed instructions on trapping.

1789

Naumann, Johann Andreas: Der Vogelsteller oder die Kunst allerley Arten von Vögeln sowohl ohne als auch auf dem Vogelheerd bequem und in Menge zu fangen. Nebst einer Naturgeschichte der bekannten und neuentdeckten Vögel. Leipzig 1789. kl. 8°.
Facsimile Leipzig 1972.

1791

Bechstein, Johann Matthaeus: Gemeinnützige Naturgeschichte Deutschlands nach allen drey Reichen. 2nd.-4th. vol. Leipzig 1791-1795. 8°.
The trapping methods described by the author are given in great detail for almost every species. The 26th. chapter of vol. 2 (pp. 102-135) is a summary, "On the Hunting and Trapping of Birds".

1795

Bechstein, Johann Matthaeus: Naturgeschichte der Stubenvögel oder Anleitung zur Kenntnis und Wartung derjenigen Vögel, welche man in der Stube halten kann. Gotha 1795. 8°.
A short passage on the trapping of almost every relevant species.

1796

Bechstein, Johann Matthaeus: Gründliche Anweisung alle Arten von Vögeln zu fangen, einzustellen, abzurichten, zahm zu machen (etc.). Nebst einem Anhange von Joseph Mitelli Jagdlust. Aufs neue ganz umgearbeitet von Johann Matthaeus Bechstein. Nürnberg and Altdorf 1796. 8°.
2nd. edition as above 1797.
A considerably altered and enlarged new edition of "Angenehme Land-Lust" by von Pernau (1720).

1797

Heppe, Johann Christoph: Der Vogelfang nach seinen verschiedenen Arten praktisch nach der Erfahrung beschrieben nebst Anleitung zur Jagd des Federwildprets. 2 vols. Nürnberg (1797) 1798-1805. 8°.
A 2nd. edition was published in Nürnberg in 1823, although the author's name was not given, under the title: "Der Vogelfänger und Vogeljäger in seinen verschiedensten Arten nach vieljähriger Erfahrung dargestellt von einem quiescirten Waidmann".

1803

Mayer, Franz Aloys: Vollständiger Unterricht, wie Nachtigallen, graue und gelbe Spottvögel, Rothkehlchen, Kanarienvögel, Finken, Lerchen, Gimpel, Zeisige, Stieglitze, Meisen und Tauben zu fangen, zu warten (etc.) sind. Wien 1803.
6th. edition, edited by D. J. Tscheiner, Pesth 1821.

1820

Naumann, Johann Friedrich: Johann Andreas Naumann's Naturgeschichte der Vögel Deutschlands nach eigenen Erfahrungen entworfen. Durchaus umgearbeitet ... aufs Neue herausgegeben von dessen Sohne Johann Friedrich Naumann. 13 Theile. Leipzig 1820-1854. 8°.
The "hunting" of each species is described in a seperate section.

Tscheiner, D. J.: Der Vogelfänger und Vogelwärter, oder Naturgeschichte, Fang, Zähmung, Pflege und Wartung unserer beliebtesten Sing- und Zimmervögel. Nebst einem Kalender für Liebhaber des Vogelfanges, und angehende Kabinetssammler. Pesth 1820. gr. 8°.
2nd. enlarged edition: Pesth 1828.

1836

Brehm, Christian Ludwig: Eine gründliche Anweisung zur Einrichtung des Drossel- und jeder andern Art des Vögelherdes, des Tränkherdes, des Lerchenstreichens, der Schneusse, der Schlingen, des Fanges mit dem Kauze, der Locke, der Heherhütte, des Wachtel- und Rebhühnerfanges, der verschiedenen Netz- und Raubvögelfallen u. s. w. nebst genauer Beschreibung aller zu fangenden Vögel; für Güter- und Waldbesitzer, Jäger, und Jagdliebhaber, Freunde des Vogelfanges und der Stubenvögel, mit Benutzung der in Europa gewöhnlichen Fangarten, nach eigenen Erfahrungen. In: Allgemeine Encyclopädie der gesammten Land- und Hauswirtschaft der Deutschen. Leipzig 1836.

1838

Schäffer, Otto: Der Finkler, oder: Deutliche und auf Erfahrung begründete Anweisung, die vorzüglichsten Singvögel zu fangen, zu pflegen, vor Krankheiten zu bewahren und von denselben zu heilen. Magdeburg 1838. 8°.

1855

Brehm, Christian Ludwig: Der vollständige Vogelfang. Eine gründliche Anleitung, alle europäischen Vögel auf dem Drossel-, Staaren-, Ortolan-, Regenpfeifer-, Strandläufer- und Entenheerde, mit Tag-, Nacht- und Zugnetzen, in Steck-, Klebe-, Hänge-, Glocken- und Deckgarnen, in Hühnersteigen, Nachtigall- und andern Gärnchen, auf dem Tränkheerde, der Krähen-, Heher- und Meisenhütte, in Raubvögelfallen und Habichtkörben, Tellereisen und Schwanenhälsen, auf den Milanscheiben und Salzlecken, in Erd- und Meisenkasten, Sprenkeln und Aufschlägen, Dohnen, Lauf- und Fußschlingen, mit Leimruthen und Leimhalmen, in Rohrfängen etc. zu fangen. Mit besonderer Berücksichtigung der Vogelstellerei der Franzosen und Afrikaner. Weimar 1855. 8°.
Facsimile: Heidelberg 1926.

1910

Otto, Hugo: Der Krammetsvogel. Seine Jagd unter besonderer Berücksichtigung des Vogelherdes. Neudamm 1910.

1919

Schäfer: Ägyptischer Vogelfang. In: Amtliche Berichte aus den Preuß. Kunstsammlungen 40, Nr. 8, pp. 163-184.
An attempted reconstruction of ancient Egyptian trapping equipment and comparison with that in use in contemporary Egypt. With 21 illustrations.

1926

Vallon, Gratiano: Die Fangarten der Vögel in Italien. In: Aquila XXXII-XXXIII, 1926, pp. 232-246.

1976

Lindner, Kurt: Bibliographie der deutschen und der niederländischen Jagdliteratur von 1480-1850. 840 pp. Berlin 1976.
This volume also contains a great many titles dealing with bird trapping.

English

1590

(Mascall, Leonard): A Booke of Fishing with Hooke and Line, and of all other instruments thereunto belonging. Another of Sundrie Engines and Trappes to take Polecats, Buzards (etc.). London 1590. kl. 8°.
Instruction on bird trapping in the 2nd. part.

1596

Gryndall, William: Hawking, Hunting, Fowling and Fishing; with the true measures of blowing. London 1596. kl. 4°.

1614

(Snodham, Thomas?): A Jewell for Gentrie. Being an exact Dictionary of true Method to make any man understand all the Arts, Secrets, and worthy Knowledges belonging to Hawking, Hunting, Fowling, and Fishing (etc.). London 1614. kl. 4°.

1621

Markham, Gervase: Hunger's Prevention; or the whole Arte of Fowling by Water and Land. Containing all the Secrets belonging to that Arte, etc. London 1621. 12°.
2nd. edition: London 1655.

1669

W(orlidge), J(ohn): Systema Agriculturae. The Mystery of Husbandry discovered (etc.). London 1669. fol.
Later editions: London 1677. 8°. London 1681 and 1697. fol. London 1716. 8°.
Chapter XII: "Of Fowling and Fishing".

1674

Anonymous: The Gentlemans Recreation. In four parts ... The second Part treats of Horse-manship, Hunting, Fowling, Fishing, and Agriculture. London 1674. 8°.
Further printing: London 1686, 1697, 1731.
The section dealing with bird trapping is mainly a compilation from the work of Fortin de Grandmont and Markham.

1714

(Ward, Thomas): The Bird-Fancier's Delight: or Choice Observations and Directions Concerning the Taking, Feeding (etc.) of all sorts of Singing Birds. London 1714. 12°.

1728

(Ward, Thomas): The Bird Fanciers Recreation. Being Curious Remarks of the Nature of Song-Birds, with choice Instructions Concerning the Taking, Feeding ... them (etc.). London 1728. 12°.

1768

Pennant, Thomas: British Zoology. Vol. II. London 1768.
Appendix V (On the small birds of flight; by Daines Barrington) deals with the trapping of song birds in England.

1859

Folkard, Henry Coleman: The Wild-Fowler; a Treatise of Ancient and Modern Wild-Fowling, historical and practical. London 1859. 8°. 2nd.-4th. editions: London 1864, 1875, 1899. 8°.
A monograph on bird trapping.

1867

Lloyd, L.: The game birds and wild fowl of Sweden and Norway. London 1867. 8°.
With much information on the trapping methods used in these two countries.

1875

Rowley, George Dawson: Ornithological Miscellany. Vol. I. London 1875.
An illustration of a larking-glass in shown on the plate opposite p. 89.

1876

Gurney, John Henry: Rambles of a Naturalist in Egypt and other Countries. London (1876).
pp. 280-282 ("Netting Sea-birds on the Wash") describe the netting of waders. Also trapping statistics.

1877

Rowley, George Dawson: On Bird-Nets. In: Rowley: Ornithological Miscellany, Vol. II, London 1877, pp. 353-373.
A short monograph on the once common technique in England of trapping birds in nets. The text contains 4 plates and many illustrations from old etchings as well as a large number of references.

Rowley, George Dawson: Saxicola oenanthe (Linn.) (The Wheatear). In: Rowley: Ornithological Miscellany, Vol. II, London 1877, pp. 397-404.
Description and illustration of the methods by which English shepherds catch the Wheatear.

1882

Payne-Gallwey, Sir Ralph: The Fowler in Ireland, or Notes on the Haunts and Habits of Wild Fowl and Sea Fowl, including Instructions in the Art of Shooting and Capturing Them. London 1882.

1886

Payne-Gallwey, Sir Ralph: The Book of Duck Decoys, their Construction, Managements and History. London 1886.
A comprehensive monograph on duck trapping.

1897

Macpherson, Hugh Alexander: A History of Fowling: being an account of the many curious devices by which wild birds are or have been captured in different parts of the world. Edinburgh 1897. 8°.
An excellent, richly illustrated monograph on bird trapping that leaves others far behind.

1903

Harper, E. W.: Bird Catching in India. In: Avicultural Magazine (2) I, pp. 262-268.
A description of bird trapping in India.

1911

Kershaw, J. C.: Chinese Bird-trap. In: The Zoologist (4) XV, 1911, pp. 392-393.

1919

Bonhote, J. L.: Bird-liming in Lower Egypt. Cairo (Ministery of Public Works, Egypt) 1919.
A description of the methods by which migrating birds are caught for food in Lower Egypt.

1924

Lincoln, Frederik, C.: Instructions for Banding Birds. United States Department of Agriculture. Miscell. Circular No. 18. Washington 1924.
Description and illustration of the trapping apparatus recommended for banding purposes; in addition a compilation of U. S. literature on the subject.

1926

Penard, Th. E. & A. P.: Bird Catching in Surinam. In: De West Indische Gids VII, pp. 545-566.

French

(Fortin de Grandmont, F. François): Les ruses innocentes, dans les quelles se voit, comment on prend les Oiseaux passagers et les non passagers, et de plusieurs sortes de bêtes a quatre pieds, avec les secrets de la pèche, et un traité pour la chasse et la maniere de faire touts les rets et les filets. Paris 1660. 4°.
Reprinted: Paris 1668 and 1700, 4°; Amsterdam 1695, 8°; Amsterdam 1700, 12°; enlarged reprint in 2 vols.: Amsterdam 1714, 12°.

1694

Anonymous: Delices de la Campagne, ou les Ruses de la Chasse et de la Pèche, ou l'on voit, comment on prend toute sorte d'oiseaux et des betes a quatre pieds (etc.). To. I, II. Amsterdam 1694. 8°.
Reprinted: Amsterdam 1699. 12°. and Amsterdam 1732. 8°; under the title: "Amusements de la Chasse et de la Pèche. Cinquième edition." Amsterdam 1742. Probably written using much of Fortin de Grandmont's work of 1660.

1709

Liger, Le Sieur Louis: Les Amusements de la campagne: ou Nouvelles Ruses Innocentes, qui enseignent la manière de prendre aux pièges toutes sortes d'oiseaux (etc.). 2nd. Vol. Paris 1709. kl. 8°.
Anonymous new edition: Paris 1764. 12°.

1774

(Buchoz, J. J.): Les amusements innocens, contenant

le traité des Oiseaux de volière, ou le parfait oiseleur, traduit en partie de l'ouvrage italien d'Olina, et mis en ordre d'après les avis des plus habiles oiseleurs. Paris 1774. 12°.

1778

(Bulliard, M.): Aviceptologie Françoise ou traité général de toutes les ruses dont on peut se servir pour prendre les oiseaux qui sont en France. Paris 1778. kl. 8°.
Further impressions: Paris yr. III (= 1795); 9th. edition (compiled by J. Cussac) Paris 1822. Has also appeared in a German translation.

1782

Anonymous: Description de la Chasse des Palombes, ou Pigeons Ramiers, dans les Pyrénées. In: Roziers Observations sur la Physique. Vol. XX pp. 306-312.
Deals with the trapping of Woodpigeons in Navarre in great detail.

1839

Blaze, E.: Le Chasseur aux filets. 1839. 8°.

c. 1850

G., J. J.: L'Oiseleur ou Secrets anciens et modernes de la Chasse aux Oiseaux.

1871

Quépat, Nerée: Le Chasseur d'Alouettes au miroir et au fusil. Paris 1871. 18°.
A monograph on the trapping of larks; illustrated.

1875

Quépat, Nerée: Monographie du Cini. Paris 1875. pp. 47-53.
Methods of trapping the Serin.

1876

De la Rue, A.: Procédé pour prendre les Oiseaux de proie et particulièrement l'Autour. In: L'Acclimatation 1876. pp. 330ff.

1896

Olphe-Galliard, Léon: Faune Ornithologique de l'Europe-occidentale. Vol. I-IV. Paris 1896.
A description of the contemporary trapping techniques used in the South of France to catch Turtle Dove, Woodpigeon, Ortolan, Skylark etc.

1921

Kehrig, Henri: La Protection des Oiseaux. Paris 1921.
On pp. 55-57 a description of the trapping of the White Wagtail on the Côte d'argent.

Italian

1601

Valli da Todi, Antonio: Il Canto de gl'Augelli ... Dove si dichiara la Natura de Sessante sorte di Uccelli, ... con il modo di pigliarli con facilità (etc.). Roma 1601. 4°.

1621

Raimondi, Eugenio: Delle Caccie et de gl'animali armate et disarmate libri quattro. Brescia 1621. 8°.
Further impressions: Brescia 1626, Napoli 1626, Paris 1630.

1622

Olina, Giovanni Pietro: Uccelliera, overo discorso della natura, e proprieta di diversi uccelli e in particolare di que'che cantano. Con il modo di prendergli, cnoscergli, allevargli, e mantenergli. Roma 1622. 4°.
With attractive copperplates. One of the classics of bird trapping literature. It is based on Valli's book of 1601. (Preuss. Staatsbibl. Lo. 4554).

1684

Mitelli, Gioseppe Maria: Caccia giocosa, invenzioni di Gioseppe Maria Mitelli Pittore Bolognese, da lui efettivamente sperimentate e dedicate a chi si diletta della Caccia. Bologna 1684. 4°.
Further impressions: Bologna 1704, 1745.
German translations: 1. Caccia Giocosa. Ruses Innocentes. Schertzendes Lufft-Waidwerck. Oder: Kurtzweil in der Waidmannschaft mit Vogelstellen. 1704. 4°.
2. Joseph, Mitelli Mahlers zu Bologna Jagd-Lust, welche er aus eigener Erfindung und würcklicher Erfahrung denenjenigen dedicirt, so die Jagd lieben. Als Anhang III. veröffentlicht in (F. A. von Pernau) Angenehme Landlust, Frankfurt & Leipzig 1720.

1724

Angelini, G. B.: La descrizione dell'uccellar col roccolo. 1724.

1735

Angelio, Pietro da Barga: L'Uccellatura a Vischio de P. Angelio Bargeo ... Poemetto dall'esametro Latino, all endecassilabo Italiano trasferito, ed interpretato ... di G. P. Bergantini. 4°. 1735.
A translation into Italian of the Latin poem of 1566.

1758

Pontini, Giovanni: La Cacciagione de Volatili osia l'arte di pigliare Uccelli in ogni maniera. Vicenza 1758. 8°.

1775

Tirabosco, Antonio: L'uccellagione. libri tre. Verona 1775.
Later editions: 1803, 1815, 1897, 1888.
A well-known poetical work.

1790

Popoleschi, G. A. (1551-1616): Del modo di piantare

e custodire una Ragnaia e di uccellare a ragna. Published by B. DAVANZATI 1790.

1869

RUGGERI, ANTONIO: L'Uccellatore, o Manuale di Ornitologia per la Sicilia. Messina 1869. 8°.
A manual for bird trappers in Sicily.

1886

TANARA, V.: La Caccia degli Uccelli, da ms. inedito della Bibl. Com. di Bologna, a cura di A. Bacchi della Lega. Bologna 1886. 8°.
A comprehensive text on trapping from the middle of the 17th. Century.

1894

ARRIGONI DEGLI ODDI, ETTORE: La Caccia di botte o di valle nelle lagune di Venezia. Milano 1894.
Description of the trapping of water birds on the lagoons around Venice.

1895

GIOLI, GIUSEPPE: Uccelli e caccie più comuni del Pisano e del Livornese. 1895.

1910

BACCHI DELLA LEGA, ALBERTO: Cacce e costumi degli uccelli silvani. 1910.

1923

FRANCESCHI, GUILIO: Manuale dell'uccellatore. 1923.

1925

GHIDINI, LUIGI: Il libro dell'uccellatore colle reti verticali ed orizzontali, ragnaia—quagliottara—bresciana—roccolo—paretaio—prodina—reti aperte, e col fucile al capanno, sulla scorta dei classici dell'uccellagione ... e con riferimenti alla pratica dell'uccellagione. Milano 1925. 8°.
A highly informative work which deals with the role of bird trapping in Italy as well as its history and methods, while at the close of the book detailed figures are given of the numbers of birds caught. Contains many instructive illustrations.

Spanish

1565

Anonymous: Libro de Cetreria, de Caça, de açor y arte quie se ha de tener en el conoscimiento y caça de las aves y sus curas, tan bien se trata de halcones, y de todas aves de Rapina. 4°.
Salamanca 1565.

1754

CALVO, PINTO y Velarde, AGUSTIN: Silva Venatoria. Modo de Cazar todo género de Aves y Animales; su naturaleza, virtudes, y noticias de los temporales (etc.). Madrid 1754. kl. 8°.

1879

AYALA, PERO LOPEZ DE: Libro de la Caza de las Aves. Madrid 1879.
Publication of a manuscript on bird trapping written in Portugal in 1386.

Flemish

1492

Anonymous: Dit Boecxken leert hoe men mach voghelen vanghen metten handen. Ende hoe men mach visschen vangen meten handen. Ende oeck andersins (etc.). Antwerpen (1492). 4°.
Some of the many later editions and translations should be mentioned here:
Diss buchleĩ sagt wie mã fisch uñ Vogel fahen soll. Mit den hendẽ und auch sũst mit vil bewertẽ recepte uñ pucktẽ (etc.). Erfurt 1498. 4°.
(Literal reprint by R. ZAUNICK in: Archiv für Fischereigeschichte, Supplement to Issue 7, 1916).
English translation of this earliest known book on fowling and fishing (etc.). Privately printed for ALFRED DENISON (London) 1872. 4°.

Dutch

1912

JURRIAANSE, J. H.: Iets over de Kvartelvangst in Noord-Egyptie. In: Ardea I, 1912, pp. 44-45.
Deals with the trapping of Quail in Egypt.

Danish

1896

RAMBUSCH, S. H. A.: Fuglekojerne paa Fanö. In: Naturen og Mennesket XV.
Description of the duck decoy on the island of Fanö. With catch records.

1913

MORTENSEN, H. Chr. C.: Maerkede Spidsaender. In: Dansk Ornithologisk Forenings Tidsskrift 8, pp. 113-159.
A further description of the Fanö decoy.

Swedish

1841

STENBECK, P.: Afhandling om bästa sättet att fånga skogsfågel med snaror. Stockholm 1841. 8°.
Instruction in how to catch woodland birds with snares.

Finnish

1934

SIRELIUS, U. T.: Jagd und Fischerei in Finnland. Berlin u. Leipzig.

1968

STORA, N.: Massfångst av Sjöfågel i Nordeurasien. Acta Acad. Aboensis, Ser. A. Vol. 34 No. 2. Åbo Akademi (Turku).

1973

SALONEN, A.: Vögel und Vogelfang im Alten Mesopotamien. Annales Academiae Scientiarum Fennicae. Vol. 180. Helsinki.

Russian

1813

LEVSHIN, V.: Des Sportsmannes Buch für den Fang von Säugetieren und Vögeln. In Russian; 4 vols. Moscow 1813-1814.
"The Sportman's Book on the Trapping of Mammals and Birds".

1900

MENZBIER, MICHAEL: Die jagdbaren Vögel des Europäischen Rußland und des Kaukasus. In Russian; 3 vols. Moscow 1900-1912.
"The Game Birds of European Russia and the Caucasus."
Contains much information on the various trapping techniques.

Japanese

1892

ODA, HIROUKI and MYZOGUCHI, DENZO: "Illustrated Hunting Techniques." In Japanese. 1892.

28. Bibliography

AICHELE, E. (1956): Meine Seidenschwänze. - Gef. Welt 80: 41-45; AITINGER, J. C. (1653): Vom Vogelstellen. - 2nd ed. Kassel; ALBARDA, H. (1885): Orn. Jahresbericht (1885) aus Holland (Friesland und Zuid-Holland). - Ornis I: 594-595; ANDERSON, R. (1966): Trapping Kookaburras. - Austral. Bird Band. 4: 61; ANELL, B. (1960): Hunting and Trapping Methods in Australia and Oceania. - Stud. ethn. Upsala 18; Anonym (1802): Die Kunst sich die zur Jagd und zum Vogelfang nötigen Netze selbst zu verfertigen. Leipzig; ANZINGER, F. (1902): Verschiedene Irisfärbungen bei ein und derselben Vogelart. - Gef. Welt 31: 370-371; APPENZELLER, E. (1957): Meine Erfahrungen beim Fang der Heidelerche *(Lullula arborea)*. - Vogelring 26: 17-19; ARMBRUST, W. (1967): Über Mehlwurmzucht. - Gef. Welt 91: 58; ARNOLD, E.-M. (1964): Beobachtungen bei Vogelhaltung und Vogelfang in Südwestafrika. - Gef. Welt 88: 76-77; ASCHENBRENNER, L., et al. (1957): Wasserralle tötet Vögel im Spiegelnetz! - Vogelk. Nachr. Österr. 7: 24-25; AUSOBSKY, A. (1964): Tonbandjagd auf Tierstimmen. Stuttgart; AUSTIN, L. jr. (1947): Mist netting for birds in Japan; Supreme commander for the Allied Powers, Natural Resources Section, Report No. 88: 21 pp

B., H. (1936): Die Ernährung unserer körnerfressenden Vögel. - Gef. Welt 65: 418-419, 429-430, 441-443, 452-453, 465-467; BACHMANN, A. (1902): Einiges über das Vogelleben auf Island. - Orn. Mschr. 27: 4-40; BÄHR, H. (1958): Der Vogel im Käfig. - Falke 5: 25-28; BAER, J. G. (1941/42): Pour baguer les oiseaux à la tour observatoire du Seeland. - Nos Oiseaux 16: 146-147; BALDWIN, S. P., & F. C. LINCOLN (1929): Manual for Bird Banders. Washington; BALSS, H. (1947): Albertus Magnus als Biologe. Stuttgart; BARBER, J. (1954): Wild Fowl Decoys. New York; BAUMANN, E. (1905): Ein seltener Fang. - Gef. Welt 34: 124-125; BEALS, W. V. (1939): An improved Drop Trap Mechanism. - Bird Banding Assu. 10: 157-159; BEAUX, O. de (1903): Vogelfang und Vogelschutzbestrebungen in Italien. - Orn. Mschr. 28: 122-135; BECHSTEIN, J. M. (1797): Gründliche Anweisung alle Arten von Vögeln zu fangen. Nuremberg and Altdorf; dgl. (1806): Handbuch der Jagdwissenschaft. Pt. 1, 3 vols., p. 599. ibid; dgl. (1821): Die Jagdwissenschaft nach allen ihren Teilen für Jäger und Jagdfreunde. III. Wildzucht und Wildjagd. Gotha; BECKER, P. (1979): Catching Rails by hand. - The Ring 98-99: 2-4; BECKER, W. J. (1931): Der Vogel in der Antike. - Gef. Welt 60: 248-249; BECKETT III, T. A. (1965): Furling device. - Ebba News 28: 88; dgl.: (1965): Parallel net sets. - ibid. 28: 150; BELOPOLSKY, L., & V. ERIK (1961): Mass trapping and ringing of birds on the Courland Spit, Ornitologine kogumik II. Tartu, pp. 189-201; BERG, B. (1925): Mein Freund der Regenpfeifer. Berlin; BERGER, A. (1928): Die Jagd aller Völker im Wandel der Zeit. ibid; BERGER, D. D., & H. C. MUELLER (1959): The Bal-chatri: A trap for the birds of prey. - Bird Band. Assu. 30: 18-26; same, & F. HAMERSTROM (1962): Protecting a trapping station from raptor predation. - J. Wildl. Man. 26: 203-206; BERGER, W. (1967): Die Mauser des Sprossers *(L. luscinia)*. - J. Orn. 108: 320-327; BERNDT, R., & W. MEISE (1958): Naturgeschichte der Vögel. Vol. I. Stuttgart; BERNHOFT-OSA, A. (1955): Trapping Waders at Revtangen. - Ring 4: 65-67; BEZZEL, E. (1961): Beobachtungen an farbig beringten Teichrohrsängern *(Acrocephalus scirpaceus)*. - Vogelwarte 21: 24-28; BICKEL, E. (1951): Die Norwegische Krähen-Massenfalle. - Anz. Schädlingsk. 24: 28-29; BIERMANN, W. H., & K. H. VOOUS (1950): Birds observed and collected in the Antarctic, 1946-47 and 1947-49. - Ardea 37, suppl., pp. 1-123; BIRKET-SMITH, S., & F. de LAGUNA (1938): The Eyak-Indians of the Copper River Delta, Alaska. Copenhagen; BIRKNER, W. (1639): Jagdbuch des Herzogs Johann Casimir von Sachsen-Coburg. Facsimile ed. 1968. Leipzig; BÖHME, B. B. (1952): Die Singvögel. Moscow; BÖSENBERG, K. (1954): Die Einrichtung von Vogeltränken gehört auch zu den Vogelschutzmaßnahmen. - Falke 1: 58-60; BÖTTGER, W.

(1960): Die ursprünglichen Jagdmethoden der Chinesen. Berlin; BOETTICHER, H. v. (1955a): Der Star in Gefangenschaft. - Gef. Welt 79: 26; dgl. (1955b): Allgemeines über die Haltung von Finkenvögeln. - ibid. 79: 215-216; BOHLKEN, H. (1934): Eine neue Prielfalle. - Vogelzug 5: 29-31; BOLAU, H. (1903): Einige Beobachtungen an Möwen und anderen Vögeln auf See. - Zool. Garten: 378-387; BOLEY, A. (1933a): Vogelschutz bei der Beringung. - Vogelring 5: 7-13; dgl. (1933b): Wacholderdrossel, Rotkopfwürger, Drosselrohrsänger usw. bei Fritzlar. - ibid. 5: 48; BOLLES (1890): Barred Owls in captivity. - Auk 7: 101-114; BONTEKOE, G. A. (1967): Ringproeven met houtsnippen in het Rysterboes. - Vanellus 20: 101-117; BOSSHARDT, A. (1940): Der Schrägscheiben-Fangkäfig. - Orn. Beob. Bern 37: 17; BRANDT, K. (1956): Eine rationelle Methode zum Sammeln von Puppen der Wiesenameise. - Gef. Welt 80: 208; BRAUN, F. (1916): Über den Wert der wichtigsten Futtersämereien. - ibid. 45: 321-323; BRAUN, W. (1942): Vielseitige Verwendbarkeit der Vogelfalle "Zwerg". - Vogelring 14: 10; BREHM, C. L. (1836): Der Vogelfang. Leipzig; same (1855): der vollständige Vogelfang. Weimar. Reprint Heidelberg, 1926; BREHM, R. (1858): Fang der Kalanderlerche in der Provinz Murcica (Spain). - Naumannia 8: 234-235; BRENNING, U., & W. MAHNKE (1971): Ornithologische Beobachtungen auf einer Reise in den Südatlantik von August bis Dezember 1966. - Beitr. Vogelk. 17: 89-103; BROWNLOW, H. G. (1952): The design construction and operation of Heligoland Traps. - Brit. Birds 45: 387-399; same: see HOLLOM (1955); BRUHN, J. F. W. (1965): Mist nets mounted on frames. - Ringers Bull. 27: 4-6; BRUNS, H. (1959): Das Problem der verwilderten Haustauben in den Städten. - Orn. Abh. No. 17: 26-27; BUB, H. (1969): Die Nahrungspflanzen des Berghänflings *(Carduelis fl. flavirostris)*. - Vogelwarte 25: 134-141; BÜTTIKER, W. (1959): Notizen über die Vogeljagd in Afghanistan. - Z. Jagdwiss. 5: 95-105; BULLIARD, M. (1778): Aviceptologie Françoise. Paris; same (1783): Aviceptologie. ibid; BURDETT-SCOUGALL, I. (1949): Pigeon netting sport of Basques. - Nat. Geogr. Mag. 96: 405-416; BURTT, H. E. (1965): Tame birds in a decoy trap. - Ebba News 28: 248-250; BUSE, M. (1915): Etwas über den Vogelfang und die Vogelwelt auf Helgoland. - Gef. Welt 44: 53-54, 62-63; BUSSIUS, E. (1956): Wildschafe wollte ich fangen und fing Flamingos. - ibid. 80: 175-176; BUYSSE, J. (1968): Twee nieuwe Vangwijzen voor "Rallen". - Ringers Bull., No. 5: 14-16

CAMPION, C. B. (1964): Little Tern banding progress. - Austral. Bird Band. 2: 23-25; CARRUTHERS, R. K. (1965): Mist Netting waders in daylight. - ibid. 3: 50-51; CHAIGNEAU, A. (1961): Les genres de Chasses. Paris; CHANNING, C. H. (1964): Two traps. - Ebba News 27: 75-76; CHOMEL, M. Noel (1743): Huishondelyk Woorbek. Leiden and Amsterdam; CLARK, W. S. (1967): Modification of the Bal-chatri trap for Shrikes. - Ebba News 30: 147-149; dgl. (1969): Migration trapping of Hawks at Cape May, N. J. - ibid. 32: 69-76; dgl. (1970): Migration trapping of Hawks (and Owls) at Cape May, N. J. Third year. - ibid. 33: 181-189; COLLENETTE, D. L. (1929/30): Bird-Trapping in a suburban garden. - Brit. Birds 23: 289-291; CONRAD, M. (1929): Nouveau Manuel Complet de L'Oiseleur. Revised edition. Paris; COOCH, G. (1957): Mass ringing of flightless Blue and Lesser Snow Geese in Canada's Eastern Arctic. - Wildf. Trust Ann. Rep. 8, 1954-1956, pp. 58-67; CREUTZ, G. (1942): Einige Erfahrungen mit dem "Zwerg". - Vogelring 14: 12; CROSBY, M. S. (1924): Bird Banding. - Nat. Hist. New York 24: 605-617; CROUSAZ, G. de: see GODEL; CURIO, E. (1959): Verhaltensstudien am Trauerschnäpper. - Z. Tierpsychol. Suppl. 3; same (1961): Rassenspezifisches Verhalten gegen einen Raubfeind. - Experientia 17: 188; dgl. (1963): Probleme des Feinderkennens bei Vögeln. - Proc. XIII. Int. Orn. Congr., pp. 206-239; dgl. (1969): Funktionsweise und Stammesgeschichte des Flugfeinderkennens einiger Darwinfinken (Geospizinae). - Z. Tierpsychol. 26: 394-487; CUSHMAN-MURPHY, R. (1955): Bird-Netting as a technique for banding shore-birds. - Bird Banding Assu. 26: 159-161

DÄHNICK, M. (1930): Die Zucht der Wachsmotte. - Gef. Welt 59: 157-159; dgl. (1930): Erfolgreiche Enchyträenzucht. - ibid. 59: 257-259; DAUMAS (1870): Straußenjagd bei den Arabern. In: F. HOBIRK, Neues Museum interessanter Scenen. Berlin; DELAY, J., & P. PICHOT (1968): Medizinische Psychologie. 2nd ed. Stuttgart; DENNIS, R. H. (1964): Capture of moulting Canada Geese in the Beauly Firth. - Wildf. Trust Ann. Rep. 15, 1962-1963, pp. 71-74; same (1966a): Fair Isle Bird Observatory. Report 1965; same (1966b): Catching wildfowl by artifical light. - ibid. 17, 1964-1965, pp. 98-100, DETMERS, E. (1905): Die Pflege, Zähmung, Abrichtung und Fortpflanzung der Raubvögel in der Gefangenschaft. Berlin; DIESSELHORST, G. (1968): Struktur einer Brutpopulation von *Sylvia communis*. - Bonn. zool. Beitr. 19: 307-321; DIETRICH, F. (1911): Die Vogelwelt der nordfriesischen Inseln und der Verein Jordsand. - Verh. V. Int. Orn. Kongr. Berlin 1910, pp. 867-868; dgl. (1925): Der Entenfang auf den nordfriesischen Inseln. - Naturschutz 6: 18-23; DIEWOCK, R. (1961/62): Elektromagnetische Fernauslösung für Schlagnetze. - Vogelring 30: 23-26; DOBBEN, W. H., & D. HOOS (1939): Vogeltrek en Vinkenbaan. Nederl. Jeugdbond voor Natuurst; DÖBEL, H. W. (1746): Jägerpractica. Reduced new edition, 1910. Neudamm; DORKA, V. (1966): Das jahres- und tageszeitliche Zugmuster von Kurz- und Langstreckenziehern nach Beobachtungen auf den Alpenpässen Cou/Bretolet (Wallis). - Orn. Beob. 63: 165-223; DOST, H. (1958a): Die einheimischen Stubenvögel. Leipzig; same (1958b): Stieglitze in Käfig und Voliere. - Falke 5: 135-137; DROST, R. (1925a): Die Käfigung von Vögeln zum Zwecke der Vogelzugsforschung. - Gef. Welt 54: 136-138; same (1925b): Eine gewaltige Zugnacht auf Helgoland als Folge ungünstiger Wetterverhältnisse im Frühjahr 1924. - Orn. Mber. 33: 11-13; same (1926): Über Arbeiten und Entwicklung der Vogelwarte Helgoland. - J. Orn. 74: 368-377; same (1928): Unermeßliche Vogelscharen über Helgoland. - Orn. Mber. 36: 3-6; same (1930a): Behälter für gefangene Vögel. - Vogelzug 1: 48-49; same (1930b): Unterbringung zu beringender Vögel beim Massenfang. - ibid. 1: 98-100; same (1932): Wurfscheiben als Hilfsmittel beim Netz- und Reusenfang. - ibid. 3: 143; same (1933a): Eine selbsttätige Kleinvogelreuse mit Wasser als Köder. - ibid. 37-38; same (1933b): Zur Akinese bei freilebenden Vögeln. - Orn. Mber. 41: 116-119; same (1940): Das "Zeisigfanghäuschen". - Vogelring 12: 38-39; same (1941): Gewaltiger Vogelzug und Massenberingung im Fanggarten der Vogelwarte auf Helgoland am 12. Oktober 1940. - Vogelzug 12: 24; dgl., I. GRITTNER (1942): Helgoland als "Brennpunkt" des Starenzuges. - Ber. Ver. Schles. Orn. 27: 10-19; same (1951): Kennzeichen für Alter und Geschlecht bei Sperlingsvögeln. Aachen; same (1952): Vogelleben und Vogelforschung auf Helgoland. In: Helgoland ruft. Hamburg; same, F. FOCKE & G. FREYTAG (1961): Entwicklung und Aufbau einer Population der Silbermöwe. - J. Orn. 102: 404-429; DROSTE, F. v. (1869): Enten- und Strandvogelfang in Stellnetzen. - ibid. 17: 279-283; DUBININ, W. B. (1953): Biotechnische Maßnahmen zum Schutz der Nutztierarten und Vögel im Astrachan-Reservat. In: Preobrasowanije fauny poswonotschnych naschei strany. Moscow (Russ.)

ECKE, H. (1924): Der Fang von Buchfinkenvorschlägern. - Gef. Welt 71: 27-30; EENSHUISTRA, O. (1956): De Goudplevier. - Vanellus 9; EGGELING, W. J. (1951): Ringing palaearctic waders and other birds on Lake Victoria. - Ibis 93: 312-313; ELIOT, S. A. jr. (1933): Banding Wilsons Petrels. - Bird Banding Assu. 4: 45-49; ENGELEN, G. D. (1959): Catching wintering gulls at the Hague, Netherlands. - Ring No. 21: 178-181; ENGELMANN, F. (1928): Die Raubvögel Europas. Neudamm; ERIK, V.: see BELOPOLSKY; EVANS, A. W. (1965): Further thoughts on boring holes. - Ringers Bull., No. 8: 10-11

FARMES, R. E. (1955): A new tip-top trap for taking Prairie Grouse. - Flicker 27: 123-125; FAUST, F. (1937): Befestigung von Mehlwürmern. - Vogelring 9: 42; FEHRINGER, O. (1950): Die Vögel Mitteleuropas. Vol. I. Heidelberg; FELDHAUS, F. M. (1970): Die Technik. 1th ed. 1914. Munich; FELTES, Ch. H. (1936): Trapping Cedar Waxwings in the San Joaquin Valley, California. - Condor 38: 18-23; FEUERSTEIN, W. (1939): Am Vogelherd. - Gef. Welt 68: 566-567; FIESS, E. (1969): Eichelhäherfallen schnell gebaut. - Wild u. Hund 72: 376-377; FILCHNER, W. (1922): Zum sechsten Erdteil. Berlin; FINSCH, O. (1879): Reise nach West-Sibirien im Jahre 1876. ibid; FIRDENHEIM, H. P. v. (1959): In: LINDNER, K., Deutsche Jagdtraktate des 15. und 16. Jahrhunderts. Part II. Berlin, pp. 135-216; FIRTH, C., & B. GUNN (1926): Excavations at Saqqara. Teti Pyramid Cemeteries I. Publ. Serv. Antiqu. Egypte. Cairo; FISKE, J. (1968): Woodpecker trap. - Ebba News 31: 154-155; FOG, Mette (1967): An investigation on the Brent Goose *(Branta bernicla)* in Denmark. - Danish Rev. Game Biol. 5: 1-40; FRANK, H. (1962): Das Fallenbuch. Hamburg and Berlin; FRAZIER, F. P. (1964): A portable net pole base. - Ebba News 27: 274-275; FREDGA, K., & I. FRYCKLUND (1965): Ringmärkningsverksamheten vid Ledskärs fågelstation 1957-1963. - Fågelvärld 24: 193-217; FREEMANN, B. (1956): Macht den Knechtsand zum Naturschutzgebiet. Bremerhaven; FRIELING, H. (1936): Exkursionsbuch zum Bestimmen der Vögel in freier Natur. 2nd ed. Berlin; FUCHS, W. (1951): Grünspechtfang am Ameisenhaufen. - Orn. Beob. Bern 48: 146-147; FUERTES, L. A. (1920): Falcony, the sport of kings. - Nat. Geogr. Mag. 38 (6): 429-460

GARDEN, E. A. (1964): Duck-Trapping methods. - The Wildfowl Trust 15. Report 1962/63, pp. 93-95; GAVRILOV, E. I. (1968): Die Ausnutzung von Salzlecken zum Vogelfang. - Ornitologija 9: 343-344 (Russ.); GAWRIN, W. F. (1964): Die Ökologie der Spießente in Kasachstan. - Proc. Zool. Ac. Sci. Kazakh. 24: 5-58 (Russ.); GEISLER, H. (1952): Zur Haltung des Blaukehlchens. - Gef. Welt 76: 92-93; same & K. RIEDEL (1956): Allgemeines über Haltung und Krankheiten der Vögel. - ibid. 80: 81-82 and following; GESSNER's Vogelbuch (1600): Edited by R. HEUSSLEIN. Frankfort on the Main; GHIGI, F. (1933): Methods of capturing birds at the Ornithological Station of Castel Fusano, Italy. - Bird Banding Assu. 4: 59-67; GIBSON, J. D., & SEFTON (1959): First Report of the New South Wales Albatross Study Group. - Emu 59: 73-82; same (1967): The wandering Albatross *(Diomedea exulans)*: Results of banding and observations in New South Wales coastral waters and the Tasman Sea. - Notornis 14: 47-57; GILL, D. E., W. J. L. SLADEN & C. E. HUNTINGTON (1970): A technique for capturing Petrels and Shearwaters at sea. - Bird Banding Ass. 41: 111-113; GLASEWALD, K. (1927/28): Ein Habichtsfang. - Naturschutz 9: 157; GLASGOW, L. L. (1958): The night-time capture of Woodcook and others birds for banding. - Ring 14: 10-11; GODEL, M., & G. DE CROUSAZ (1958): Studien über den Herbstzug auf dem Col de Cou-Bretolet. - Orn. Beob. Bern 55: 96-123; GOODHARDT, J., R. WEBBE & T. WRIGHT (1955): Goose-Ringing in West-Spitsbergen 1954. - Wildf. Trust Ann. Rep. 7: 53-54, 170-176; same, & T. WRIGHT (1958): North-East Greenland Expedition 1956. - ibid. 9, 1956-1957, pp. 180-192; GRÄFE, F.: see VAUK; GRISEZ, T. (1965): Soil sampling augers and tubes for setting net poles. - Ebba News 28: 236-237; GRANT, T. (1968): A method of catching swans. - Ringer's Bull. 3, No. 3: 6-7; GROFF, G. W., & T. C. LAU (1937): Landscaped Kwangsi. - Nat. Geogr. Mag. 52: 671-726; GROM, J. A. (1962): Banding thrushes at Maxada woodlands. - Ebba News 25: 39-41; GROSS (1924): Merkwürdiger

Fang eines Eisvogels. - Gef. Welt 53: 216; GROSS, A. O. (1947): Recoveries of banded Leach's Petrels. - Bird Banding Assu. 18: 117-126; GRUNER, D. (1972): Die Geschichte der Vogelforschung und der Vogelwarte auf Helgoland. - Der Helgoländer, no. 92; GÜTH, K. (1956): Der Fang des Stares. - Vogelring 25: 102-104; GUNDA, B. (1968/69): Die Jagd und Domestikation des Kranichs bei den Ungarn. - Anthropos 63-64: 473-496

HAARTMAN, L. v. (1949): Der Trauerschnäpper. - Acta zool. Fenn. 56: 19; HABLIZL, C. (1783): Bemerkungen in der persischen Landschaft Gilan und auf den Gilanischen Gebirgen. - N. Nord Beytr. 4: 3-104; HAGEN, Y. (1952): Birds of Tristan da Cunha. Results of the Norwegian Scientific Expedition to Tristan da Cunha. 1937-1938. Oslo. No. 20: 1-248; HAMERSTROM, Jr. F. N., & M. TRUAX (1938): Traps for pinnated and Sharp-Tailed Grouse. - Bird Banding Assu. 9: 177-183; same, F. (1957): The influence of a hawk's appetite on mobbing. - Condor 59: 192-194; same (1962): Winter visitors from the Far North. - Audubon Mag. 64: 12-15; dgl.: see BERGER (1962); same (1966): The use of Great Horned Owl in catching Marsh Hawks. - Proc. XIII. Int. Orn. Congr. Vol. II. Pp. 866-869; same (1970): In: Victor C. CAHALANE, ed. Alive in the wild. Prentice-Hall, New York; HARDENBERG, J. D. F. (1964): Vorläufige Mitteilung über den Einsatz von Fangkäfigen für Krähen. Festschr. 25jähr. Bestehen Vogelschutzwarte Essen-Altenhunden. Recklinghausen, pp. 91-95; HARDMAN, J. A. (1965): Decoys for mist netting. - Ebba News 28: 79; HARTLEY, P. H. T. (1950): An experimental analysis of interspecific recognition. - Symp. Soc. Exp. Biol. 4: 313-336; HARTMANN, C. (1933): Der Stieglitzfang am Lockfutterplatz. - Vogelring 5: 60-61; dgl. (1934): Häherfang am Futterkasten. - ibid. 6: 62-63; HARTWIG, G. (1870): Der Vogelfang auf St. Kilda. In: F. HOBIRK, Neues Museum interessanter Scenen. Berlin; HASSEL, H. J. (1969): Der Krähen- und Elsternfang in der Käfigfalle. Munich; HAUSBERGER, A. (1955): Heimchenzucht. - Gef. Welt 79: 156-157; same (1957): Heimchenzucht. - ibid. 81: 55-56; HAVESTADT, J. (1929): Über Fang und Eingewöhnung von Vögeln in Abessinien. - Vögel fern. Länd. 3: 150-161; HAVERSCHMIDT, F. (1943): De Goudplevierenvangst *(Charadrius apricarius L.)* in Nederland. - Ardea 32: 35-74; HEINROTH, O., & M. (1924-1933): Die Vögel Mitteleuropas. 4 vols. Berlin; HELD, B. (1959): Meine Grillenzucht. - Gef. Welt 83: 217-218; HENRY, T. R. (1955): Ice Age Man, the first American. - Nat. Geogr. Mag. 108: 781-806; HERDEMERTEN, K. (1939): Jakunguaq. Brunswick; HEWITT, A. D. S. (1966): Swallows at Rosherville and Vrischgewaard. - Bokmakierie 18: 90-91; HEYDER, R. (1954): Kreuzschnäbel als Salz- und Aschefresser. - Beitr. Vogelk. 4: 1-7; same (1960): Zur Aufnahme von Mineralsalzen durch Vögel. - ibid. 7: 1-6; same (1967): Der Vogelname "Wichtel". - ibid. 13: 198-204; HEYER, J. (1968): Rotkehlpieper und Sanderling beringt. - Thür. Orn. Rundbr., no. 13: 30-31; HILPRECHT, A. (1933): Erfahrungen beim Baumbesteigen. - Vogelfreund 2: 13-16; same (1937): Eine tragbare Garnreuse für den Nachtfang. - Vogelring 9: 10-11; same (1954): Nachtigall und Sprosser. - N. Brehm-Büch. 143; HINDE, R. A. (1954): Factors governing the changes in strength of a partially inborn response as shown by the mobbing behaviour of the Chaffinch *(Fringilla coelebs)*. I. - Proc. Roy. Soc. London 142: 306-331; HOCHEDER, L. (1956): Segler- und Schwalbenfang mit dem Japannetz. - Vogelwarte 18: 224; HÖGLUND, N. H. (1968): A method of trapping and marking Willow Grouse in winter. - Viltrevy 5: 95-101; HÖHN, E. O. (1969): Die Schneehühner. - N. Brehm-Büch. 408; HOESCH, W. (1956): Freuden und Sorgen beim Fang und Transport afrikanischer Vögel. - Gef. Welt 80: 3-5; same (1958): Vogelfang in Südwestafrika. - Vogelring 27: 149-150; HOFFMANN, A. (1960): Vogel und Mensch in China. - Nachr. Ges. Nat.-Völkerk. Ostas. Hamburg 88: 45-77; same, G. (1962): Eine praktische Methode zum Vorkeimen von Körnerfutter. - Gef. Welt 86: 64-65; HOFHANSL, F. (1939): Das Spiegelnetz. - Gef. Welt 68: 255-256; HOLGERSEN, H. (1956): *Anser arvensis brachyrhynchus* i den kalde Ettervinteren 1956. - Stavanger Mus. Arsb., pp. 151-158; HOLLOM, P. A. D. (1950): Trapping methods for Bird Ringers. British Trust for Orn., Field Guide Number One; dgl., & H. G. BROWNLOW (1955): Trapping methods for Bird Ringers. British Trust for Orn., Field Guide Number One. (Revised Edition); Hoos, D. (1937): De Vinkenbaan. - Ardea 26: 173-202; same: see W. H. DOBBEN; HOPPE, R. (1962): Über die Behandlung frischer Puppen. - Gef. Welt 86: 238; same, & M. (1963): Die Wachsmotte als ideales Lebendfutter. - ibid. 87: 77; dgl. (1969a): Die Nahrung des Stieglitz. - Vogelkosmos 6: 301-303; dgl. (1969b): Die Nahrung des Gimpels. - ibid. 6: 351-353; same (1969c): Die Nahrung des Birkenzeisigs. - ibid. 6: 384-385; same (1969d): Die Nahrung des Hänflings. - ibid. 6: 424-426; HOPPE-RAPP, M. (1969): Was fressen Zaun- und Dorngrasmücke? - Gef. Welt 93: 131-132; HUDSON, R. (1963): Bird-Ringing in British Antarctic Territory, 1957-1962. - Ring, no. 34: 171-173; dgl. (1967): British rings used abroad II. - Ring, no. 50/51: 35-42; HUMPHREY, P. S., D. BRIDGE & Th. E. LOVEJOY (1968): A technique for mist netting in the forest canopy. - Bird Banding Assu. 39: 43-50

IMLER, R. J. (1937): Methods of taking birds of prey for banding. - Bird Banding Assu. 8: 156-161; ISSEL, W. (1937): Beringung von Winterstaren in Bonn. - Vogelring 9: 62-64

JACOBS, K. F. (1958): A drop-net trapping technique for Greater Prairie Chickens. - Proc. Oklahoma Ac. Sci. 38: 154-157; JÄCKEL, H. (1941): Etwas über alte Vogelherde und Tränken in der Umgebung Rudolstadts. - Gef. Welt 70: 55-57, 66-68; JAHN, W. v. (1899): Zum Vogelzug und über Windverhältnisse. -

Orn. Mschr. 24: 321; JAHNKE, W. (1955): Fluchtreaktion und Vogelfang. - Orn. Mitt. 7: 70-71; JENA, A. (1919): Die Mehlwurmzucht. - Gef. Welt 48: 117-118; JENKINSON, M. A., & R. M. MENGEL (1970): A device for handling mist nets in the dark. - Bird Banding Assu. 41: 38-39; JOHNS, J. E. (1963): A new method of capture utilizing the mist net. - ibid. 34: 209-213; JOHOW, A. (1961): Fang und Haltung unserer Kolibris. - Gef. Welt 85: 141-142

KAEMPFER, E. (1924): Über das Vogelleben in Santa Domingo. - J. Orn. 72: 178-184; KALCHREUTER, H. (1971): Untersuchungen an der Krähenfalle. - Z. Jagdwiss. 17: 13-19; KALE II, H. W. (1966): Hand-Capture of roosting birds for banding. - Ebba News 29: 104-106; KALMBACH, E. R. (1930): English Sparrow Control. Leaflet no. 61. U. S. Dept. of the Interior Fish and Wildlife Service, Washington; KARTASCHEW, N. N. (1962): Mausernde Eiderenten fliegen ohne Schwungfedern. - J. Orn. 103: 297-298; KEIL, W. (1957): Vogeltränken. In: S. PFEIFER, Taschenbuch für Vogelschutz. 2nd ed. Frankfurt on the Main, pp. 53-55; same (1961): Benutzung von Nisthöhlen durch Vögel im Winter. - Angew. Orn. 1: 29-31; dgl. (1967a): Die Krähenmassenfalle, eine Möglichkeit zur Verminderung der Rabenkrähe *(Corvus c. corone)*. - Ges. Pflanzen 19: 56-62; same (1967b): Erfahrungen mit einer Falle zur Verminderung der Rabenkrähe. Dtsch. Bund f. Vogelschutz. Ann. Report 1967: 12-15; KELLER, E. (1936): Verbreitung der Fallenjagd in Afrika. - Z. Ethnol. 68: 1-118; KELM, H. (1970): Beitrag zur Methodik des Flügelmessens. - J. Orn. 111: 482-494; KIMMEL, H. (1957): Fang und Eingewöhnung des Rotkehlchens. - Gef. Welt 81: 33-34; KING, B. (1969): Swallow banding in Bangkok, Thailand. - Bird Banding Assu. 40: 95-104; KINLEN, L. (1961/62): Ringing Whooper Swans in Iceland. 1962. - Wildf. Trust Ann. Rep. 14: 107-114; KIRPITSCHEV, S. P. (1962): Der Fang, Ernährung und Transport von Auerhühnern. - Ornithologia 4: 385 (Russ.); KLEEN, V. M. (1967): A special bird-banding project in the Philippine Islands. - Ebba News 30: 161-164; KLEINSCHMIDT, O. (1897): Die palaearktischen Sumpfmeisen. - Orn. Jb. Hallein 8: 45-103; same (1922): A-B-C-Unterricht für ornithologische Sammler und solche, die es werden wollen. - Falco 18: 11; KLOER, B. (1967): Ein bewährter Krähenfang. - Dtsch. Jägerztg., pp. 276-277; KLOSE, W. (1956a): Meine Haubenmeise. - Gef. Welt 80: 74-76; same (1956b): Was füttern und sammeln wir im Herbst? - ibid. 80: 206-208; KNEUTGEN, J. (1969): "Musikalische" Formen im Gesang der Schamadrossel *(Kittacincla macroura Gm.)* und ihre Funktionen. - J. Orn. 110: 245-285; KNOBLOCH, H. (1957): Die Gartenammer als Käfigvogel. - Falke 4: 171-175; KNORR, B. (1963): Banding shorebirds. - Ebba News 26: 247-249; KÖNIG, C. (1969): Die Steinkäuze. In: Grzimeks Tierleben. Vol. 8, pp. 400-401; KOHN, A., & R. (1876): Sibirien und das Amurgebiet. Leipzig; KOLBMANN, H. J. (1961): Gefährlicher Vogelsand. - Gef. Welt 85: 215-216; KONING, F. J. (1970): Trapping ducks with a floating clapnet. - Ring, no. 64/65: 65-66; KOOY, A. (1967): Nogmaals de Lelystad-Kooi. - Op het Vinketouw, part 1, no. 9: 227-228; KORIDON, J. A. F. (1958): Het Zwarte Meer (Rayon-West). - Limosa 31: 1-17; KRACHT, W. (1924): Leicht zerlegbare Käfige für Frischfänge. - Gef. Welt 53: 182-183; same (1951): Anweisungen für jüngere Vogelliebhaber über die Fütterung von Insektenfressern. - ibid. 75: 1-4; same (1952): Einige Anweisungen zum Vogelfang. - ibid. 76: 91-92, 99-101; same (1958): Entnahme von Grillen aus Zuchtkästen. - ibid. 82: 178. KRÄMER, A. (1903): Die Samoa-Inseln. Stuttgart; same (1906): Ostmikronesien und Hawai. Stuttgart; KRÄTZIG, H. (1936): Auf Starenfang für die Vogelwarte Rossitten. - Gef. Welt 65: 437-440; same (1939): Untersuchungen zur Siedlungsbiologie waldbewohnender Höhlenbrüter. Berlin; KRAMBRICHT, A. (1953): Altes und Neues von Schwarzspecht und Hohltaube. - Vogelwelt 74: 136-139; KREMLITSCHKA, O. (1960): Über die Pflege der Heidelerche. - Gef. Welt 84: 28-29; KRIEGER, O. v. (1876): Über Lerchenjagd und Lerchenfang. - J. Orn. 24: 67-76; same (1878): Die hohe und niedere Jagd in ihrer vollen Blüte. Trier; KÜHLHORN, F. (1951): Eine praktische Fangeinrichtung für Kleinvögel. - Orn. Mitt. 3: 231; KUMERLOEVE, H. (1953): Vom "Hortulanenfang" bei Osnabrück im 17. und 18. Jahrhundert. - Veröff. Naturw. Ver. Osnabrück 26: 67-130; KUNZE, W. (1957): Von meinem Freund, dem Steinkauz (Wichtel). - Gef. Welt 81: 23-25

LABISKY, F. (1968): Nightlighting: Its use in capturing Pheasants, Prairie Chickens, Bobwhites and Cottontails. - Biol. Notes Illinois Nat. Hist. Surv., no. 62; LACHNER, R. (1963): Beiträge zur Biologie und Populationsdynamik der Türkentaube *(Streptopelia d. decaocto)*. - J. Orn. 104: 305-351; LAGERCRANTZ, S. (1933): Fallgropar hos den svenska Allmogen. Ymer, pp. 67-90; same (1937a): Skärgårdsbondens Sjöfagelsnät. In: Svenska Kulturbilder. Vol. 4, pt. 7, or 8, pp. 289-300; same (1937b): Contributions to the question of the origin of torsion traps. - Acta ethnol., pp. 105-130; same (1937c): Rezension von K. LINDNER, Die Jagd der Vorzeit (Berlin 1937). - ibid., pp. 179-191; same (1937d): Beiträge zur Jagdfallensystematik. - Ethnos 2: 361-366; same (1938): Beiträge zur Kulturgeschichte der afrikanischen Jagdfallen. Stockholm; same (1940): An East African accessory fishing implement. - Ethnos 5: 29-34; LAMBERT, K. (1971): Seevogelbeobachtungen auf zwei Reisen im östlichen Atlantik mit besonderer Berücksichtigung des Seegebietes vor Südwestafrika. - Beitr. Vogelk. 17: 1-32; LAMPE (1899): Angang zu Dr. LAMPE's Illustrierter Tierheilkunde: Fischzucht. Leipzig; LANE, S. G. (1963): A useful Holding Bag. - Austral. Bird Band. 1: 173-175, and 2: 14; same, & J. LIDDY (1965): Backyard Trapping. - ibid. 3: 9-13; same (1966): A proven method of trapping hawks. - ibid. 4: 56-57; LARSEN, T., & M. NORDERHAUG (1963): The ringing of Barnacle

Geese in Spitsbergen, 1962. - Wildf. Trust Ann. Rep. 1961-1962, no. 14: 98-104; LAWSON, W. J. (1966): Swallow-Ringing in Natal. - Bokmakierie 18: 92; LEHTONEN, L. (1947): Zur Winterbiologie der Kohlmeise. - Orn. Fenn. 24: 32-47; LEIJS, H. N. (1968): Het vangen van Oeverzwaluwen. - Op het Vinketouw, part 2, no. 2: 25-29; LENSKSI, E. (1932): Die Möwenklippe, ein historisches Fanggerät der Strandbewohner. - Mitt. Vogelwelt 31: 61-62; LIDDY, J. (1964): Evening mist-netting. - Austral. Bird Band. 1: 173-175, ibid. 2: 11-12; LIEBE, K. Th. (1893): Ornithologische Schriften (pp. 477-549). Ed. by C. R. HENNICKE. Leipzig; LINCOLN, F. C., & S. P. BALDWIN (1929): Manual for Bird Banders. Un. St. Dep. of Agriculture. Miscellaneous Publication no. 58. Washington; LINDAU, H. (1935): Wie kann man leicht junge Kiebitze fangen? - Mitt. orn. Ver. Magdeburg 9: 8; LINDBLOM, G. (1925/26): Jakt- och Fangstmetoder I and II. Stockholm, 26 pp; dgl. (1927): Die Schleuder in Afrika und anderwärts. - Riksmus. Ethnol. Avd. Medd., no. 2; same (1939): Der Lasso in Afrika. In: Kultur und Rasse, festive publication to O. RECHES. Berlin; LINDNER, H. (1972a): Fang und Beringung von Kreuzschnäbeln im Niestetal (Landkreis Kassel) 1959. - Vogelring 33: 37-40; same (1972b): Beringung von Nachtigall. - ibid. 33: 108; LINDNER, K. (1937): Die Jagd der Vorzeit. Berlin and Leipzig; same (1940): Die Jagd im frühen Mittelalter. Berlin; same (1959): Deutsche Jagdtraktate des 15. und 16. Jahrhunderts. Part I. ibid.; same (1964): Deutsche Jagdschriftsteller part I: HANS CASPAR RORDORF 1773-1843. ibid., pp. 259-310; dgl. (1967): Ein Ansbacher Beizbüchlein. ibid.; LIPS, J. (1927): Fallensysteme der Naturvölker. - Ethnologica 3: 123-283; LIPS, J. E. (1955): Vom Ursprung der Dinge. 3rd ed. Leipzig; LOCKLEY, R. M., & R. RUSSELL (1953): Bird-Ringing. London; LÖHNER, R. (1964): Zur Eingewöhnung von Frischfängen, besonders des Hänflings. - Gef. Welt 88: 193-195; Löhrl, H.: see SCHÜZ (1954); dgl. (1957): Der Kleiber. - N. Brehm-Büch. 196; same (1958): Das Verhalten des Kleibers. - Z. Tierpsych. 15: 191-252; same (1962a): Überwiegen Sperber-Weibchen in südwestdeutschen Fängen? - Vogelwelt 83: 49-52; same (1962b): Zusätzliche Futterquellen für Weichfresser. - Gef. Welt 86: 102-104; LÖTZSCH, E. (1937): Befestigung von Mehlwürmern. - Vogelring 9: 43; LORENZ, K. (1935): Der Kumpan in der Umwelt des Vogels. - J. Orn. 83: 137-213, 289-413; same (1966): Er redete mit dem Vieh, den Vögeln und den Fischen. Munich; LOVEJOY, T. E. (1967): Birds and viruses in lower Amazonia. - Ebba News 30: 116-122; LOW, S. H. (1935): Methods of Trapping Shore Birds. - Bird Banding Ass. 6: 16-22; LOYDS, L. (1867): The game birds and wild fowl of Sweden and Norway. London; LUDWIG, H. (1961): Der Seidenschwanz in Natur und Gefangenschaft. - Gef. Welt 85: 23-25; LUESHEN, W. (1962a): Holding the bird for banding. - Ebba News 25: 22-23; same (1962b): Cliff Swallow banding in Nebraska. - ibid. 25: 107-109; LUMHOLTZ, C. (1903): Unknown Mexico. Vol. II. London; LUTHER, M.: Klageschrift der Vögel gegen WOLF SIEBERGER; Weimar's edition, vol. 38, p. 290 and following. Erlangen's edition, vol. 64, p. 346 and following; LYON, W. I. (1926): Banding gulls and terns in Lake Michigan 1925. - Wilson Bull. 38: 240-248

MACK, J. (1966): More than one way to ruin a mist net: Find the Ass. Austral. Bird. Band. 4: 11; MACKEY, W. J. Jr. (1965): American Bird Decoys. New York; MACLEOD, J. G. R. (1966): Swallow-Ringing in the Western Cape. - Bokmakierie 18: 93-94; MACPHERSON, H. A. (1897): A History of Fowling, Edinburgh; MAGNUS, O. (1555): Historia de gentibus septentrionalibus. Rome; MAKATSCH, W. (1961): Die Vögel in Haus, Hof und Garten. Radebeul and Berlin; MAKOWSKI, H. (1961): Amsel, Drossel, Fink und Star. Stuttgart; MALEK, J. J. (1966): Walk-In ground trap. - Ebba News 29: 289; MANSFELD, K. (1950): Beiträge zur Erforschung der wissenschaftlichen Grundlagen der Sperlingsbekämpfung. - Nachr.bl. Dtsch. Pflanzenschutzdienst 4, no. 7/8/9; same (1955): Über Vogeltränken. - Falke 2: 135-137; MARINKELLE, C. J. (1957): Bird traps used in Indonesia and Western New Guinea. - Ring 10: 207-210; MARRIS, R., & M. A. OGILVIE (1962): The ringing of Barnacle Geese in Greenland in 1961. - Wildf. Trust Ann. Rep. 1960-1961, no. 13: 53-64; MATHIASSON, S. (1962): Ornitologiska observationer pa Myggenäs, Faröarna. - Fauna och Flora 50: 1-46; same (1973): Moulting Grounds of Mute Swans *(Cygnus olor)* in Sweden, their origin and relation to the population dynamics of Mute Swans in the Baltic Area. - Viltrevy 8: 399-452; MATSCHIE, P. (1897): Die von Herrn VAUGHAM STEVENS auf Malakka beobachteten Methoden des Vogelfangs. - Orn. Mber. 5: 137-142; MAUNDER (1852): Treasury of Natural History. 3rd ed. London; MCCLURE, E. (1956): Methods of Bird netting in Japan applicable to wildlife management problems. - Bird Banding Assu. 27: 67-73; same (1966): An Asian bird-bander's manual. Hong Kong; MEBS, T. (1964): Greifvögel Europas und die Grundzüge der Falknerei. Stuttgart; MENDALL, H. L. (1938): A technique for banding Woodcock. - Bird Banding Assu. 9: 153-155; MÉRITE, E. (1942): Les Pièges. Paris; MERKEL, K. (1927): Vogelfang und Vogelberingung auf Helgoland im Herbste 1926. - Gef. Welt 56: 199-202, 213-214, 224-227; MESTER, H. (1957): Ein winterlicher Schlafplatz des Wasserpiepers. - Vogelwelt 78: 185-189; MEWES, W., & E. F. v. HOMEYER (1886): Ornith. Beobachtungen im nordwestl. Rußland. - Ornis Jena 2: 207; MEYER-DEEPEN, H. (1951): Meine Feldlerchen. - Gef. Welt 75: 90-91; MEYLAN, A. (1966): The capture of Sand-Martins in colonies. - Ring, no. 49: 260-263; MINTON, C. D. T. (1965): Supports for mist nets over water. - Ebba News 28: 155; MODERSOHN, C. (1870): Der Fang von allerlei Vögeln auf dem Reisbaum. - J. Orn. 18: 394-397; MÖHRING, G. (1955): Vogeltränken ohne Zement und Blech. - Falke 2: 134-135; MÖRZER BRUIJNS, M. F. (1960): Over het vangen van Houtsnippen *(Scolopax rusticola)* met de flouw. - Levende Natuur 63: 73-77; MOGALL, K. (1951): Nestkontrolle und Beringung von Höhlenbrütern. - Vogelring 20: 33-37; same (1956): Ein zweiteiliger Vogelbehälter für den Beringer. - ibid. 25: 104-105; MOHR, R. (1966): Über den Fang der Kreuzschnäbel. - ibid. 32: 11-13; MOLLISON, B. C. (1957): A live-trap for birds. C. S. I. R. O. Wildl. Research. Vol. 2, no. 2, pp. 172-174; MOOIJMAN, J. G. J. (1955): Het weispel met kwartelen. - Wiek Sneb 3 (1): 4-5; MOORE, N. C. (1953): A balance for weighing Tits without capture. - Brit. Birds 46: 103-105; MORRIS, A. K. (1970): Quail banding at Mudgee, New South Wales. - Austral. Bird Band. 8: 35-36; MORTENSEN, H. Chr. (1950): Studies in Bird Migration. Copenhagen; MÜHLETHALER, F. (1952): Beobachtungen am Bergfinken-Schlafplatz bei Thun 1950/51. - Orn. Beob. Bern 49: 173-182; MUELLER, H. C., & D. D. BERGER (1961): Weather and fall migration of hawks at Cedar Grove, Wisconsin. - Wilson Bull. 73: 171-192; MÜLLER, W. (1903): Vertreibung der Milben aus der Mehlwurmhecke. - Gef. Welt 32: 311; MULZER, H. H. (1933): Der heilige Franz und die Vögel. - ibid. 62: 477-479, 489-490; MYRBERGET, S. (1960): Lundevangst pa Lovunden. Saertrykk av Jakt-Fiske-Friluftsliv, no. 8: 1-7; dgl. (1968): Metoder for Fangst av Sjofugle for Merking. - Sterna 8: 105-110

NAU, A. (1959): Das Keimenlassen von Hirse und anderen Sämereien. - Gef. Welt 83: 229; NAUMANN, J. A. (1789): Der Vogelsteller. Leipzig; NAUMANN, J. F. (1824): Naturgeschichte der Vögel Deutschlands. Vol. 4. Leipzig, pp. 177-182, 187; same (1900, 1902, 1903): Naturgeschichte der Vögel Mitteleuropas. Vols. 3, 10, 12. Gera-Untermhaus; same (1905a): Naturgeschichte der Vögel Mitteleuropas. Vol. I. ibid.; same (1905b): Naturgeschichte der Vögel Mitteleuropas. Vol. 3. ibid., pp. 27-28; NEEL, C. A. (1963): Portable net pole holes. - Ebba News 26: 32; NELSON, E. W. (1928): Bird Banding, the telltale of migratory flight. - Nat. Geogr. Mag. 75: 91-132; NEUNZIG, K. (1922): Die einheimischen Stubenvögel. 6th ed. Magdeburg; same (1927): Praxis der Vogelpflege und -züchtung. ibid.; NEWTON, J. (1967): The adaptive radation and feeding ecology of some british finches. - Ibis 109: 33-98; NICE, M. M., & J. TER PELKWYK (1941): Enemy recognition by the Song Sparrow. - Auk 58: 195-214; NICOLAI, J. (1956): Zur Biologie und Ethologie des Gimpels *(Pyrrhula pyrrhula)*. - Z. Tierpsychol. 13: 93-132; same (1965): Vogelhaltung - Vogelpflege. Das Vivarium. Stuttgart; same (1968): Käfig- und Volierenvögel. 2nd ed. Das Vivarium. ibid.; same, & H. E. WOLTERS (1960): Europäische Singvögel. Vol. I. In: Vögel in Käfig und Voliere. Aachen; NIESELT-LAUSA, E. (1924): Der Gelbspötter, sein Fang, seine Eingewöhnung und Überwinterung. - Gef. Welt 53: 108-109; NIETHAMMER, G. (1937): Handbuch der Deutschen Vogelkunde. Vol. I. Berlin; same (1955): Jagd auf Vogelstimmen. - J. Orn. 96: 115-118; dgl. (1972): Störche über Afghanistan. - Z. Kölner Zoo 15: 47-54; NIGMANN, E. (1908): Die Wahehe. Berlin; NORDERHAUG, M. (1963): Norsk Ornitologisk Ekspedisjon til Spitsbergen 1962. - Sterna 5: 184-189; same (1964): Ornitologiske feltarbeider pa Vestspitsbergen 1963-1964. - ibid. 6: 185-194; same (1966): Ornitologisk feltarbeid pa Vestspitsbergen 1965. - ibid. 7: 113-119; NØRREVANG, A., & T. J. MEYER (1960): Jeg ser pa Fugle. 2nd ed. Copenhagen

O. N. (1964): Three useful traps. - Ebba News 27: 42-44; ORFORD, N. (1967): Notes on netting Twite. - Ringers Bull. 3: 9-10; OSTAJOW, F. F. (1960): The Songbirds of our Country. Moskau (Russ.); OTTO, H. (1910): Der Krammetsvogel. Seine Jagd mit besonderer Berücksichtigung des Vogelherdes. Neudamm

PÄTZOLD, R. (1963): Die Feldlerche. - N. Brehm-Büch. 323; PALLAS, P. (ca. 1773): Reisen durch verschiedene Provinzen des russischen Reiches, part II; PAYNE-GALLWEY, R. (1886): The book of duck decoys. London; PEDERSEN, A. (1954): Die Vogelberge des Atlantik. Bern and Tübingen; PEITZMEIER, J. (1936): Die Akinese bei Vögeln ein Instinkt? - Orn. Mber. 44: 110-116; PELKWYK, J. ter (1941): Fowling in Holland. - Bird Banding 12: 1-8; PELTZER, J. (1967): Verhalten der Wasserpieper *(Anthus sp. spinoletta)* und Fangtechnik am Schlafplatz im Winter. - Regulus 47: 9-11; PEPPER, W. (1965): Problems of a gull bander. - Ebba News 28: 66-67; PERNAU, F. A. v. (1720): Angenehme Landlust ... Nuremberg; PERSSON, C. (1965): Sand Martin ringing in Scania, Sweden. - Ring, no. 43: 124-126; PETERMANN, K. (1875): Schilderungen des Fanges und Vogellebens im Brasilianischen Urwalde. - Gef. Welt 4: 127-129, 136-138, 145-146; PETERSEN, P. (1963): Mist Netting "Saw-Whet-Owls". - Ebba News 26: 184; PETTINGILL, S. (1962): Hawk migration around the Great Lakes. - Audubon Mag. 64: 44-45, 49; PFANNENSCHMID, E. (1882): Ornithologische Mitteilungen aus Ostfriesland. - Gef. Welt 11: 135 and 478; PFEIFER, S. (1957): Taschenbuch für Vogelschutz. 2nd ed. Frankfort on the Main; PFEIFFER, L. (1920): Die Werkzeuge der Steinzeitmenschen. Jena; PFITZENMAYER, E. W. (1926): Mammutleichen und Urwaldmenschen in Nordost-Sibirien. Leipzig; PIECHOCKI, R. (1961): Makroskopische Präparationstechnik. Leipzig; PLEYEL, J. v. (1901): Ein Beitrag zur Ornis vindobonensis. - Orn. Mschr. 26: 285-299 and following; POSTINIKOW, S. A. (1958): Arbeiten des Oksker Staatl. Naturschutzgebietes. 2nd ed. of the Ornithological Station, no. 1: 209-213; PRESCHER, H. (1933): Der Herbstvogelzug 1932 im Nordgebiet des Stettiner Haffs. - Dohrniana 12: 46-47; PREYWISCH, K. (1957): Wie schläft die Kohlmeise im Kasten? - Orn. Mitt. 9: 161-162; PUSCHNIGG, R. (1891): Die Blaumeise in der Gefangenschaft. - Gef. Welt 20: 423-424; PYLE, R. L. (1964): Predation by Cicada Killer Wasp. - Ebba News 27: 4

QUANTZ, B. (1941): Sperlingsreuse. - Vogelring 13:

9-11; QUINET, de (1904): Considerations sur les Oiseaux d'Egypte. - Ornis Paris 12: 1-74

RADDE, G. (1884): Ornis Caucasica. Cassel; same (1886): Reisen an der Persisch-Russischen Grenze. Leipzig; RAESFELD, F. v. (1921): Das deutsche Weidwerk. Berlin; same (1942): Das Deutsche Waidwerk. 5th ed. Berlin; RALPH, C. J., & F. C. SIBLEY (1970): A new method of capturing nocturnal Alcids. - Bird Banding Assu. 41; 124-127; REICHENOW, A. (1913): Die Vögel. Vol. I. Stuttgart; REINER, E. (1965): Eine besondere Art des Fledermausfanges in Neu-Guinea. - Natur Mus. 95: 463-464; RENDLE, M. (1902): Der Rotrückige Würger, dessen Schädlichkeit, Fang und Eingewöhnung. - Gef. Welt 31: 289-290; RIESENTHAL, O. v. (1876): Die Raubvögel Deutschlands und des angrenzenden Mitteleuropas. Cassel; RINGLEBEN, H. (1957): Schallplatte und Tonband im Dienste der Vogelkunde. - Falke 4: 57-60; RITTINGHAUS, H. (1961): Der Seeregenpfeifer. - N. Brehm-Büch. 282; ROBERTS, B. B. (1940): The life cycle of Wilson's Petrel. - Brit. Graham Exp. 1934-1937. - Sci. Rep. 1: 141-194; ROBERTS, B. (1967): Edward Wilson's Birds of the Antarctic. London; ROBERTS, T. S. (1955): A manual for the identification of the birds of Minnesota and neighbouring states. Minneapolis; ROBILLER, F. (1969): Kranke Stubenvögel. Berlin; ROER, H. (1957): Tagschmetterlinge als Vorzugsnahrung einiger Singvögel. - J. Orn. 98: 416-420; ROHWEDER, J. (1900): Der Vogelfang auf Helgoland. - Orn. Mschr. 25: 119-134; ROM, K. (1955): Unsere Meisen. - Gef. Welt 79: 21-24, 47-49, 71-72; same (1961/1962): Jahresrhythmus. - ibid. 85: 239-240, and 86: 7-9; RORDORF, H. C. (1836): Der Schweizer Jäger. Liestal (see LINDNER 1964); ROSER, A. (1960): Eine afrikanische Falle für Wachteln. - Vogelwarte 20: 233; ROTH, W. E. (1897): Ethnological studies among the North-West-Central Queensland Aborigines. Brisbane and London; same (1924): An introductory study of the arts, crafts and customs of the Guiana Indians. 38. Ann. Rep. Bur. Americ. Ethnol. Washington; ROTHENBÜCHER, M. (1918): Dr. Martin Luther als Vogelfreund. - Gef. Welt 47: 29-30; RÜPPELL, W. (1930): Über die Verbreitung des Krähenfangs am Kurischen Haff. - Vogelzug 1: 128-130; same (1936): Die Krähenfänger von Agilla. - Vogelring 8: 220-222; RUSCHKE, G. (1963): Freilandbeobachtungen an Zaunkönig und Rotkehlchen ... - Schr. Inst. Natursch. Darmstadt 7 (1): 12-17; RUSS, K. (1880): Bergmann's Futter- und Fangkasten. - Gef. Welt 9: 542-543; RUSSEL, E. S. (1943): Perceptual and sensory signs in instinctive behaviour. - Proc. Linn. Soc. London 154: 195-216; RUSSELL, R.: see LOCKLEY (1953); RUTHKE, P. (1935): Mit der Blendlaterne im Watt auf Vogelfang. - Vogelring 7: 23-25; RUTTER, R. J. (1962): Operation Rainfall. - Ebba News 25: 12-13

SABEL, K. (1967a): Wald- und Wiesenvögel in Käfig und Voliere. Das Vivarium. Stuttgart; same (1967b): Vogelfutterpflanzen. 3rd ed. Pfungstadt; same (1970): Zur Frage: Dompfaffenzucht animalisch oder vegetarisch? - Gef. Welt 94: 76-77; SAEMANN, D., & H. STÖTZER (1968): Eine billige und einfache Hochnetz-Fanganlage. - Falke 15: 422-424; SALVATOR, L. v. (1897): Die Balearen. Würzburg und Leipzig; SANDEN, W. v. (1936): Fischbunge als Vogelfangreuse. - Vogelzug 7: 87; SANTOS, J. R. (1965): A method of attaching mistnets to poles. - Ringers Bull. 2: 8-10; SAVAGE, Chr. (1961/62): Wildfowling in northern Iran. The Wildfowl Trust. 14 Annual Rep., pp. 30-46; SCHÄFER, H. (1918/19): Altägyptischer Vogelfang. - Amtl. Ber. preuß. Kunstsamml. 40; same (1939): Beobachtungen an den Schwalben meiner Heimat. - Vogelring 11: 58-73; SCHALOW, H. (1907): Im Protokoll einer D. O. G.-Sitzung. - J. Orn. 55: 160-161; SCHEITHE, K. (1956): Segler- und Schwalbenfang mit dem Schnellnetz. - Vogelwarte 18: 224-225; SCHELCHER, R. (1954): Zur Einrichtung von Vogeltränken. - Falke 1: 157; SCHENK, H. (1921): Der Fang kleiner Raubvögel mittels Schlingen. - Aquila 28: 221-222; same (1952): Merkwürdiger Vogelfang. - Columba 4: 39; SCHIFFERLI, A. (1932): Der Starenfang der Schweizer Vogelwarte Sempach. - Orn. Beob. Bern 29: 171-173; same (1934/35): Ein beweglicher Vogelherd mit Federkraft. - ibid. 32: 180-181; same (1936/37): Ergebnisse der Schweizer Bleßhuhnberingung. - ibid. 34: 93-99; same (1938): Schweizer Bade-Falle zum Zwecke der Vogelberingung. - Vogelring 10: 28-30; dgl. (1956): Vogelnetze aus Nylon. - Ring, no. 6: 108-109; SCHILDMACHER, H. (1965): Wir beobachten Vögel. Jena; SCHNEIDER, W. (1952): Beitrag zur Lebensgeschichte des Stars. - Beitr. Vogelk. 3: 27-52; SCHNURRE, O. (1942): Ein Beitrag zur Biologie des Sperlingskauzes. - Beitr. Fortpfl. Vögel 18: 45-51; SCHÜLER, L. (1960): Limikolenfang bei Marburg/Lahn. - Vogelring 29: 3-6; SCHÜNEMANN, K. E. (1957): Noch jagen die Adler. Berlin; SCHÜZ, E. (1930): Steigeisen. - Vogelzug 1: 136; same (1931): Die Beringungsstationen Italiens. - Orn. Beob. Bern 28: 133-140; dgl. (1932): Vogelwelt und Vogelwarte. In: Die Kurische Nehrung. Königsberg. Pp. 103-107; same (1942): Altersmerkmale und Zustand durchziehender Rauhfußbussarde, *Buteo l. lagopus*. - Vogelzug 13: 2-17; same, & H. LÖHRL (1954): Mehr Strenge gegenüber dem Stoff — gerade in der Ornithologie. - Vogelwarte 17: 1-6; same (1957): Bräuche von Vogelfang und Vogeljagd im südkaspischen Gebiet. - Z. Jagdwiss. 3: 107-114; same (1966): Über Stelzvögel (*Ciconiiformes* und *Gruidae*) im Alten Ägypten. - Vogelwarte 23: 263-284; SCHVINDT, T. (1905): Finsk Etnografisk Atlas, I. Jagt och Fiske. Helsingfors; SCHWEER, O. (1954): Ausgezeichnetes Naturfutter. - Gef. Welt 78: 233; SCHWENK, S. (1967): Zur Terminologie des Vogelfangs im Deutschen. Diss. Marburg/Lahn; SCHWERIN, F. v. (1934): Fasanen als Jagdwild. Neudamm; SCOTT, P. (1950): The Perry River Expedition, 1949. - Sev. Wildf. Trust Rep. 3, 1949-1950, pp. 56-64; same, J. FISHER & F. GUDMUNDSSON (1953a): The Severn Wildfowl Trust Expedition to Central Iceland, 1951. - ibid. 5: 79-115; same & same

(1953b): A Thousand Geese. London; same, H. Boyd & W. L. Sladen (1955): The Wildfowl Trust's second Expedition to Central Iceland, 1953. - Sev. Wildf. Trust Rep. 7, 1953-1954, pp. 63-97; same, & J. Fisher (1957): Geheimnis der Brutstätten. Eine Island-Expedition. Translated from English by R. Gerlach. Hamburg; Sellick, C. (1960): Duck catching in Iceland. The Wildfowl Trust. 11. Annual Report 1958-1959, pp. 114-149; Serventy, D. L., et al. (1962): Trapping and maintaining shore birds in captivity. - Bird Banding Assu. 33: 123-130; Seubert, J. L. (1963): Research on methods of trapping the Red-winged Blackbird *(Agelaius phoeniceus)*. - Angew. Orn. 1: 163-170; Shallenberger, R. (1971): A device for handling shearwaters. - Bird Banding Assu. 42: 125-127; Sharland, R. E. (1970): Fang von Weißstörchen mit Lockvögeln in Nigeria. - Vogelwarte 25: 359; Shaub, B. M. (1948): Combination window trap and feeding tray. - Bird Banding Assu. 19: 65-70; Sheldon, W. G. (1960): A method of mist netting woodcocks in summer. - ibid. 31: 130-135; Siegfried, R., & G. Broekhuysen (1971): Zum Verhalten des Falkenbussards *(Buteo b. vulpinus)* in der südwestlichen Kap-Provinz. - Vogelwarte 26: 78-86; Sirelius, U. T. (1934): Jagd und Fischerei in Finnland. Berlin and Leipzig; Skokowa, N. N. (1960): Über die jahreszeitliche Verbreitung und die Wanderungen des Kormorans am Kaspischen Meer. - Migrazii shiwotnych 2: 76-99 (Russ.); Sladen, W. J. L., & W. L. N. Tickell (1958): Antarctic birdbanding by the Falkland Island Depend. Survey 1945-1957. - Bird Banding Assu. 29: 1-26; dgl., R. C. Wood & E. P. Monaghan (1968): The USARP Bird Banding Program 1958-1965. In: Antarctic Birds. Edited by O. L. Austin Jr. Antarctic Res. Ser. 12, pp. 213-262. Amer. Geophys. Union Washington; Smythies, B. E. (1968): The birds of Borneo. 2nd ed. Edinburgh and London; Sohns, G., & H. W. Wawrzyniak (1973): Erfahrungen beim Fangen und Beringen von Seggenrohrsängern *(Acrocephalus paludicola)*. - Beitr. Vogelk. 19: 36-42; Spaepen, J. (1952): De Ortolaan *(Emberiza hortulana)* als Trek- en als Kooivogel. - Gerfaut 42: 164-214; dgl. (1953): De Trek van de Boompieper, *Anthus trivialis*, in Europa en Afrika. - ibid. 43: 178-230; Speek, B. J. (1971): How to catch more Swallows. - Ringers Bull. 3: 7-8; Spencer, A. W., & J. W. de Grazio (1962): Capturing Blackbirds and Starlings in Marsh Roosts with dip nets. - Bird Banding Ass. 33: 42-43; Spencer, R. (1959): Report on bird ringing for 1958. - Brit. Birds 52: 441-482; same (1963): "Tethered" mist nets. - Ebba News 26: 7-8; Spörer, J. (1867): Nowaja Semlä. Mitt. aus J. Perthes Geogr. Anstalt. Suppl. vol. V, 1867-1868, pp. 98-99; Springer, H. (1960): Studien an Rohrsängern. - Anz. orn. Ges. Bayern 5: 389-433; Stach, W. (1898): Raubzeugvertilgung im Interesse der Wildhege. Berlin; Steinbacher, G. (1953): Zur Biologie der Amsel. - Biol. Abh. no. 5: 4; Steinbacher, J. (1957): Vogelfang und Vogelberingung in Tunesien. - Vogelring 26: 98-105; Steinfatt, O. (1931): Die Vogelwarte Castel Fusano (Roma), ihre Ziele und bisherigen Resultate. - Mitt. Vogelwelt 30: 65-69; dgl. (1937): Aus dem Leben des Großen Buntspechtes. - Beitr. Fortpfl. Vögel 13: 101; Steiniger, F. (1935): Über Reaktionshemmung bei Vögeln. - Orn. Mber. 43: 66-73; same (1936): Über Reaktionshemmung bei jungen Möwen und Seeschwalben. - ibid. 44: 135-140; Stewart, F., & J. L. McKean (1962): Birds of prey in mist nets. - Austral. Bird Band. 1: 10; Stirling, M. W. (1937): America's first settlers, the Indians. - Nat. Geogr. Mag. 72: 535-596; Stoll, H. (1957): Natürliches Zusatzfutter für Körnerfresser. - Gef. Welt 81: 113-114; Stresemann, E. (1927/1934): Aves. In: Handbuch der Zoologie. Berlin and Leipzig; same (1949): Über den Raubvogelzug am Bosporus. - Vogelwarte 15: 109-110; Stromar, L. (1965): Interessante Vogelreaktionen beim Fangen zum Zwecke der Vogelberingung. - Larus 16-18: 155-158; Strusch, J. C. (1966): Meine Wachsmottenzucht. - Gef. Welt 90: 19; Sugden, L. G., & H. J. Poston (1970): A raft trap for Ducks. - Bird Banding Assu. 41: 128-129; Sunkel, W. (1920): Futterbeschaffung und Futterzuchten. - Gef. Welt 49: 161-162, 169-170; same (1927): Der Vogelfang für Wissenschaft und Vogelpflege. Hanover; same (1928): Vogelfangerlebnisse. 2. Dompfaffenfang. - Gef. Welt 57: 281-282; same (1934): Schlaggärnchen. - Vogelring 6: 60-62; same (1937): Befestigung von Mehlwürmern. - ibid. 9: 42; same (1938a): Eine Vogelfalle unter Benutzung einer Mause- oder Rattenfalle. - ibid. 10: 27; same (1938b): Vögel bei Frostwetter. - ibid. 10: 30-31; same (1940): Meine Erlebnisse mit Wasseramseln in Hessen-Nassau. - ibid. 12: 50-95; same (1942): Vogelfang-Allerlei. - ibid. 14: 10-12; same (1947/1950): Der kleine Vogelsteller, no. 1-6; same (1948a): ibid. 3: 22-24; same (1948b): ibid. 4: 26; same (1954a): "Mach die Wurzel nicht kaputt ...". - Vogelring 23: 21-24; same (1954b): Fangkäfig mit Lockabteil. - ibid. 23: 42; same (1954c): Der Vogelsteller. - ibid. 23: 68-69; same (1954d): Vogelfangtage am Neusiedler See. - ibid. 23: 86-90; same (1956a): Italienische Vogelherde. - ibid. 25: 19-20; same (1956b): Fangkäfige. - ibid. 25: 126-128; same (1958): Fangkäfig "Zwerg" in verbesserter Form. - ibid. 27: 87-89; same (1961/1962): Vogelfang für den Beringer im Sommer. - ibid. 30: 70-74; same (1966): Vogelfang für die Beringung. - ibid. 32: 25-29; Swinebroad, J. (1964): Net-Shyness and Wood Thrush populations. - Bird Banding Assu. 35: 196-202; Szederjei, A., M. Szederjei & L. Studinka (1959): Hasen, Rebhühner, Fasanen. Berlin

Taapken, J., & J. G. J. M. Mooijman (1960): Ervaringen opgedaan bij het vangen van vogels met kunstlicht. Mededelingenblad van de Contactgroep voor Vogelringstations 2 (6): 25-29; Tarshis, J. B. (1956): Traps and techniques for trapping California Quail. - Bird Banding Assu. 27: 1-9; Tenner, A. (1892): Vogelfang und Vogelliebhaberei in Thüringen. - Gef. Welt 21: 237; Thielke, G. (1969): Die Reaktion von Tannen- und Kohlmeise *(Parus ater, P. major)* auf den Gesang nahverwandter Formen. - J. Orn. 110: 148-157; THIENEMANN, J. (1908): VII. Jahresbericht der Vogelwarte Rossitten. - ibid. 56; same (1927): Rossitten. Neudamm; same (1928): Rossitten. 2nd ed. Neudamm; THIERSANT, P. D. de (1872): La pisciculture et la pêche en Chine. Paris; THOMPSON, M. C., & R. L. DELONG (1967): The use of cannon and rocket-projected nets for trapping shorebirds. - Bird Banding 38: 214-218; TIMMERMANN, G. (1938): Die Vögel Islands. Reykjavik; TOSCHI, A. (1938): Die Brauchbarkeit der Vogelherde mit Netzfang für ornithologische Studien. - Vogelzug 9: 174-175; TRUAX, M.: see HAMERSTROM. TSCHUDI, F. v. (1860): Das Thierleben der Alpenwelt. 5th ed. Leipzig; TSCHUDI, J. J. v. (1878): Winckell's Handbuch für Jäger und Jagdliebhaber. 2 vols. Leipzig

UNGER, W. (1971): Habicht und Sperber im Spiegel der Beringung. - Beitr. Vogelk. 17: 135-154; USINGER, A. (1960): Einheimische Säugetiere und Vögel in der Gefangenschaft. Hamburg and Berlin; dgl. (1963): Die Ruf-, Lock- und Reizjagd. 2nd ed. ibid.; USPENSKI, S. M. (1965): The geese of Wrangel Island. - Wildf. Trust Ann. Rep. 16, 1963-64, pp. 126-129

VALLON, G. (1882 and 1883): Über die in Italien zur Anwendung gebrachten Fangarten der Vögel. - Orn. Mschr. 7: 65-69, 288-290, and 8: 92-96; same (1926): Die Fangarten der Vögel in Italien. - Aquila 32/33: 232-246; VAUK, G., & F. GRÄFE (1962): Volierenfalle zum Türkentaubenfang. - Vogelwarte 21: 204-206; VIANDEN, J. (1955): Sperlinge entwichen der Schwingschen Spatzenfalle. - Falke 2: 215; VILKS, K. (1931): Zwei Apparate zum Einfangen von Altvögeln. - Vogelzug 2: 139-141; VÖMEL, F. W. (1938): Kranich in Weilburg beringt. - Vogelring 10: 31-32; same (1939): Beobachtungen aus einer sechsjährigen Beringungstätigkeit. - ibid. 11: 29-30; Vogelwarte Helgoland, Rossitten und Sempach (1935): Merkblatt über Fangverfahren für den Beringer. - Vogelzug 6: 138-152; Vogelwarte Helgoland, Radolfzell und Sempach (1955): Fangverfahren für den Beringer. Aachen; Vogelwarte Sempach (1932): Vogelfang zur Beringung. - Orn. Beob. Bern 29: 84; VOGT, W. (1951): Der Bergfink als Käfigvogel. - Gef. Welt 75: 185-186; VOLQUARDSEN, J. V. (1933): Die Vogelkojen der Insel Föhr. Wyk; Voss, A. (1967): Vogelfang in Südbrasilien. - Gef. Welt 91: 217-218; VUILLEUMIER, F. (1959): Activités de l'Observatoire ornithologique alpin du col de Bretolet en 1958. - Nos Oiseaux 25: 65-78

WAGNER, H. O. (1958): Naturschutz. Anpassung der Tierwelt an die Kulturlandschaft und Tierhandel in Australien. - Zool. Garten 25: 43-44; WALKE, J. E. S. (1965): A convenient swan hook. - Ringers Bull. 2(7): 7; WALLER, R. (1962): Der wilde Falk ist mein Gesell. 2nd. ed. Neudamm; WALTHER, G. (1935): Die Vogelberingung des Ornithologischen Vereins Dessau 1932-1934. - Beitr. Avifauna Anhalts. Pp. 1-2; WARGA, K. (1929/1930): Der Vogelfang im Dienste der Beringung. - Aquila 36/37: 150-159; dgl. (1939): Die *Bombycilla g. garrulus*-Invasion in den Jahren 1931/32 und 1932/33 und die Ergebnisse der Beringungsversuche. - ibid. 42-45: 410-528; WARNAT, H. (1936): Mein erster Möwen-Massenfang. - Vogelring 8: 67-70; WARNKE, G. (1933): Ein Beitrag zur "Hypnose" bei Vögeln. - Orn. Mber. 41: 71-74; dgl. (1934): Akineseversuche an Meisen. - J. Orn. 82: 247-256; WASCHINSKY, D. (1903): Die Mehlwurmzucht. - Gef. Welt 32: 125-126; WATSON, B. (1966): Use of local weather forecasts. - Ringers Bull. 2, no. 9: 10-11; WEBER, H. (1939): Die Beringung eines Steinkauzpaares. - Vogelring 11: 56-57; WEIGOLD, H. (1930): Der Vogelzug auf Helgoland. Graphisch dargestellt. Berlin; WENTWORTHDAY, J. (1949): The modern fowler. London; WEPLER, G. E. (1960): Neues Gerät zum Fang von Schwalben, Seglern usw. - Mitt. bl. orn. Arb. Gem. Oberrhein, no. 1: 6; WHITE, J. (1890): The ancient history of the Maori. Wellington; WHITEHEAD, L. C. (without year): Life Buzzard Trap. U. S. Fish a. Wildlife Service, N. C. State Coll., Raleigh, N. C.; WHITMAN, J. D. (1962): Sparrow Hawk banding with the Balchatri trap. - Ebba News 25: 5-11; WIEDERMANN (1914): Vogelfang mit dem Wichtel. - Gef. Welt 43: 250-252; WILCOX, L. (1963): Capturing sleeping Flickers. - Ebba News 26: 94-95; WILLIAMS, J. G. (1963): Freeing Flamingos from anklets of death. - Nat. Geogr. Mag. 124: 934-944; WILLIAMSON, K. (1951): Fair Isle Bird Observatory. Report 1950; WINCKELL, G. F. D. (1820/1822): Handbuch für Jäger, Jagdberechtigte und Jagdliebhaber. Revised by J. J. v. Tschudi. Vols. 1 and 2. Leipzig; WISSENBACH, W. (1956): Über Tierfotografie. Frankfort on the Main; WITHERBY, H. F. (1943): The Handbook of British Birds. London; WITTE, H. (1938): Wie befestigt man am besten Mehlwürmer? - Vogelring 10: 73-74; WITZIG, A. (1952): Die festen Vogelfanganlagen (Roccoli) im Tessin vor 1875. - Orn. Beob. Bern 49: 84-88; WÖHRMANN, C. (1959): Sperber-Attrappe, um Vögel ins Spannetz zu scheuchen. - Vogelring 28: 26; WÖLDECKE, R. (1943): Ein neues Gerät zum Besteigen von Waldbäumen. - Vogelzug 14: 116-117; WOLFF, Chr. S. (1727, 1731): Ausführliche Nachrichten von denen Ortolanen, deren Fang und gewöhnlicher Wartung. Büchners Misc. Phys.-medicomathem. 1th and 2nd quart. Erfurt

YOUNG, C. (1964): Shelduck trapping methods. The Wildfowl Trust, 15th Report 1962/63, pp. 95-96; YUNICK, R. P. (1965): Making and using shore birds silhouette decoys. - Ebba News 28: 7-15; same (1967): An apron for mist nets. - ibid. 30: 71-72; same (1970): On Bank Swallow banding. - ibid. 33: 85-96

29. Species Index